Basic Trigonometric Graphs

The graph of $y = c + a \sin b(x - d)$ or $y = c + a \cos b(x - d)$ has amplitude $|a|$, period $2\pi/b$, a vertical translation c units up if $c > 0$ or $|c|$ units down if $c < 0$, and a phase shift d units to the right if $d > 0$ or $|d|$ units to the left if $d < 0$. Throughout, we assume $b > 0$. (To find d, write the argument as $b(x - d)$. The graph of $y = a \tan bx$ or $y = a \cot bx$ has period π/b.

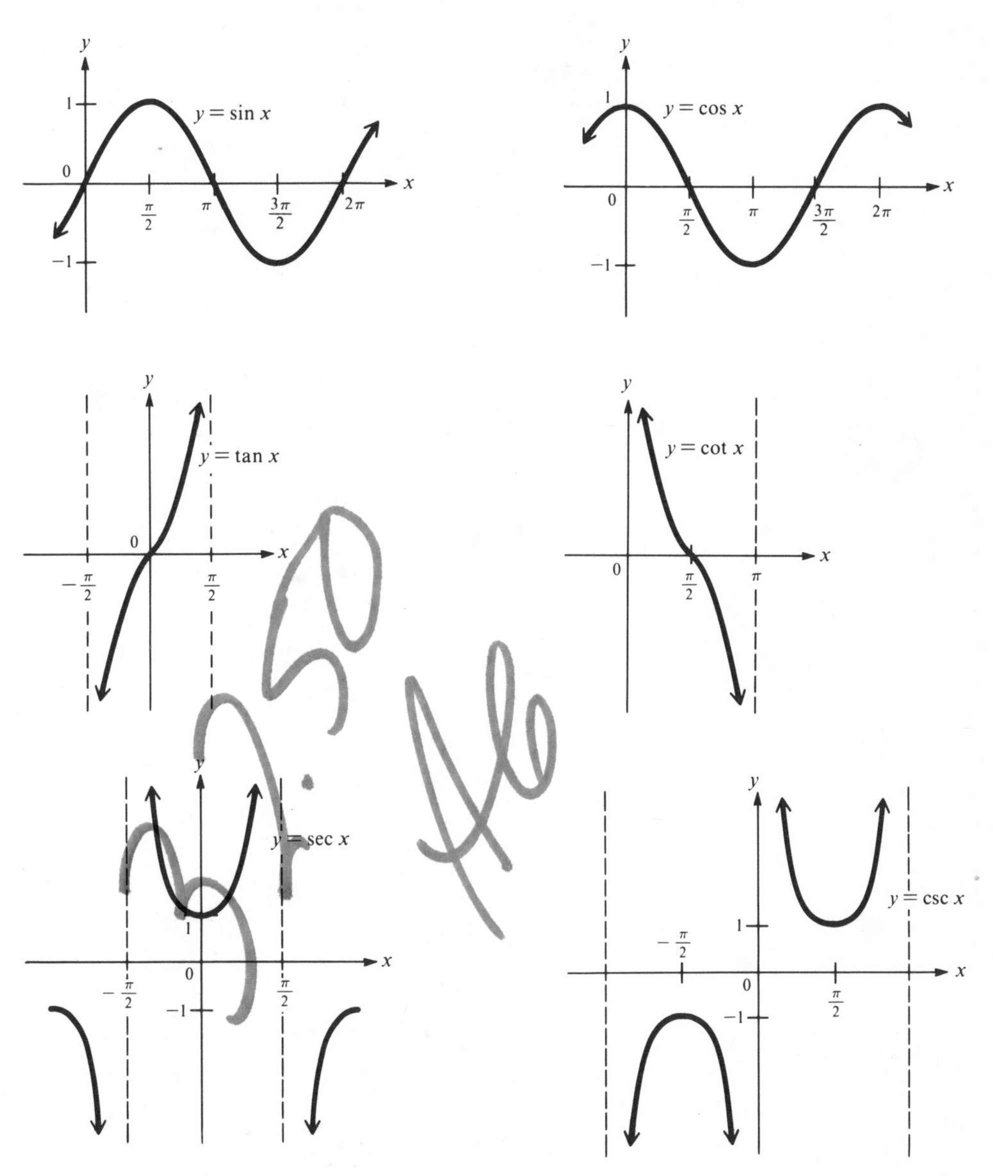

Trigonometry

4th Edition

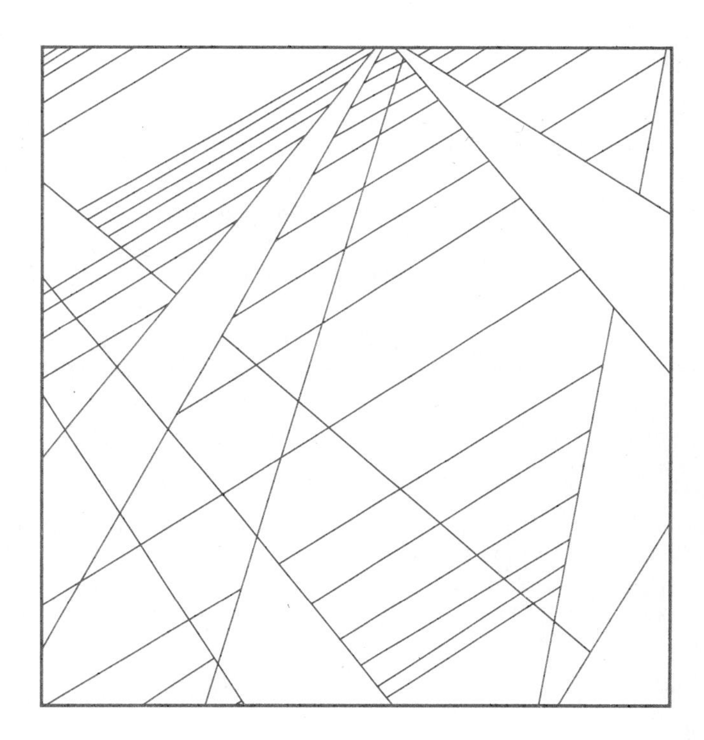

Margaret L. Lial
American River College

Charles D. Miller

SCOTT, FORESMAN AND COMPANY
Glenview, Illinois Boston London

To the Student

A *Student Solutions Manual* to accompany this book is available from your local college bookstore. This book can help you study and review the course material.

Photo Acknowledgments

Cover: Polarized light shining through folded pieces of clear acetate. Photo by Scott, Foresman staff.

ix, Scott, Foresman and Company; 21, 116 (2), Charles D. Miller; 28, National Park Service; 135, P. Dugan, M.D., and B. Gorham, R.N.

Library of Congress Cataloging-in-Publication Data

Lial, Margaret L.
Trigonometry.

Includes index.
1. Trigonometry. I. Miller, Charles David
II. Title.
QA531.L5 1989 516.2′4 88–18302
ISBN 0–673–38248–6

23456-RRC-9392919089

Preface

The Fourth Edition of *Trigonometry* is designed for a one-semester or one-quarter course that will prepare students either for calculus or for further work in electronics and other technical fields. Applications for both types of study are given throughout the text.

We have written the book assuming a background in algebra. A course in geometry is a desirable prerequisite, but many students reach trigonometry with little or no background in geometry. For this reason, we have included a section on geometry in chapter 1 and have explained the necessary ideas from geometry in the text as needed. In addition, the *Instructor's Guide* contains a geometry review unit, which can be reproduced and made available for students if desired.

Key Features

Examples More than 200 worked-out examples clearly illustrate concepts and techniques. Second color identifies pertinent steps within examples and highlights explanatory side comments.

Applications The number and variety of applications in examples and exercise sets have been increased in this edition. As before, these applications are spread throughout the text rather than being concentrated in one chapter. Applications from the fields of engineering, physics, biology, astronomy, navigation, and demography are included.

Calculators Without making the text a calculator instruction book, we do tell students how to use a calculator whenever appropriate throughout the text. Flexibility has not been sacrificed; table use and evaluation of problems is taught as well as calculator use, so that either approach can be used in the classroom.

Key Concepts New concise "Key Concept" summaries have been added at the end of each chapter to provide students with a quick look at the important concepts presented in the chapter.

Exercises

Graded Exercises The range of difficulty in the exercise sets not only affords ample practice with drill problems, but gradually eases students through exercises of increasing difficulty to problems that will challenge outstanding students. More than 3100 exercises, including approximately 2900 drill problems and 200 word problems, are provided in the text.

Calculator Exercises A large number of exercise sets feature calculator problems, and they are identified with the symbol ▦ . A scientific calculator is not essential for the course, however; most problems can be worked with a four-function calculator and the tables we supply.

Chapter Review Exercises A lengthy set of review exercises, about 525 problems in all, is given at the end of each chapter and reviews each section of the text thoroughly. These exercises provide further opportunity for mastery of the material before students take an examination.

Second Color

Second color is used pedagogically in the following ways.

- Screens set off key definitions, formulas, and procedures, helping students review easily.
- Color side comments within examples explain the structure of the problem.
- For clarity, the end of each example is indicated with a color symbol, ✣.

Content Features

The basic ideas of trigonometry are presented very early. **Angles** are discussed in section 1.2, and the **trigonometric functions** are defined in section 1.4. **Identities** are introduced briefly in chapter 1, so that students see an early discussion of this key idea.

Geometry The review of geometry has been expanded to include discussion of alternate interior and corresponding angles formed by parallel lines cut by a transversal.

Triangles are presented early. In chapter 2 we show how triangles are used to obtain the values of the trigonometric functions for acute angles and then show some applications of trigonometry. This is designed to give students a feel for the usefulness of trigonometry. All the right triangle applications are grouped together at the end of chapter 2.

Radian measure is introduced in chapter 3, and thereafter a balance between radian and degree measures is provided. Students going on to calculus will have received enough practice with radian measure to be comfortable using it in calculus.

Graphing In chapter 4, the presentation of graphing the trigonometric functions has been revised to cover amplitude, period, and translations of sine and cosine graphs before introducing graphs of the other trigonometric functions. The presentation also has been expanded to offer students more detailed explanations of the steps to follow in graphing the trigonometric functions.

Identities in chapter 5 are proven in a format that shows the reasons for key steps at the side. This gives students a clear understanding of the structure of each proof.

Inverse functions are treated in detail in chapter 6, which begins with a section on inverse functions in general for review before introducing the inverse trigonometric functions.

Although the **law of sines** and the **law of cosines** are presented in chapter 7, this chapter could be taught after chapter 2, except for some derivations.

The chapter on **logarithms** is written to emphasize the exponential and logarithmic functions and the solving of exponential and logarithmic equations.

Supplements

The ***Instructor's Guide*** gives a geometry review unit, four alternative forms of the chapter tests in a format ready for duplication, a test bank for each chapter, two final examinations, and answers to all of these. An ***Instructor's Solution Manual*** contains solutions to all even-numbered exercises, and an ***Answer Manual*** gives answers to all the exercises in the text.

A ***Student's Solution Manual*** has solutions to all the odd-numbered exercises in the text. Some students may want to use this book as an additional source of examples.

A semi-programmed ***Study Guide*** is also available.

Trigonometry, Fourth Edition, is part of a series of texts, including *Beginning Algebra,* Fifth Edition; *Intermediate Algebra,* Fifth Edition; *College Algebra,* Fifth Edition; *Algebra for College Students; Fundamentals of College Algebra,* Second Edition; *Fundamentals of Trigonometry* (a unit-circle approach) and *Algebra and Trigonometry,* Fourth Edition. The use of a series of texts increases student understanding by offering continuity of notation, definitions, and format.

We thank the many users of the previous editions of this book who were kind enough to share their experiences with us. This revision has benefited from their comments and suggestions.

We also thank those who reviewed all or part of the revised manuscript and gave us many helpful suggestions: Jule Connolly, Wake Forest University; Arthur Dull, Diablo Valley College; Mary Lou Hart, Brevard Community College-Melbourne; Marion Littlepage, Clark County Community College; William Walworth, C. S. Mott Community College.

Our appreciation also goes to the staff at Scott, Foresman and Company, who did an excellent job in working with us toward publication.

Margaret L. Lial
Charles D. Miller

Contents

Using a Calculator

1 The Trigonometric Functions 1

1.1 Basic Terms 1
1.2 Angles 10
1.3 Similar Triangles 20
1.4 Definitions of the Trigonometric Functions 29
1.5 Using the Definitions of the Trigonometric Functions 33
Chapter 1 Key Concepts 41
Chapter 1 Review Exercises 43

2 Acute Angles and Right Triangles 46

2.1 Trigonometric Functions of Acute Angles 46
2.2 Trigonometric Functions of Special Angles 51
2.3 Reference Angles and Trigonometric Tables 60
2.4 Significant Digits 71
2.5 Solving Right Triangles 76
2.6 Further Applications of Right Triangles 84
Appendix: Linear Interpolation 93
Chapter 2 Key Concepts 95
Chapter 2 Review Exercises 97

3 Radian Measure and Circular Functions 100

3.1 Radian Measure 100
3.2 Applications of Radian Measure 108
3.3 Circular Functions of Real Numbers 118
3.4 Linear and Angular Velocity 125
Chapter 3 Key Concepts 131
Chapter 3 Review Exercises 133

4 *Graphs of Trigonometric Functions* 135

4.1 *Graphs of Sine and Cosine* 136

4.2 *Horizontal Translations: Phase Shift* 148

4.3 *Graphs of Tangent and Cotangent* 156

4.4 *Graphing by Combining Functions (Optional)* 162

4.5 *Simple Harmonic Motion (Optional)* 168

Chapter 4 Key Concepts 174

Chapter 4 Review Exercises 175

5 *Trigonometric Identities* 177

5.1 *Fundamental Identities* 177

5.2 *Verifying Trigonometric Identities* 186

5.3 *Sum and Difference Identities for Cosine* 193

5.4 *Sum and Difference Identities for Sine and Tangent* 200

5.5 *Double-Angle Identities* 207

5.6 *Half-Angle Identities* 213

5.7 *Sum and Product Identities (Optional)* 219

5.8 *Reduction of* $a \sin \theta + b \cos \theta$ to $k \sin (\theta + \alpha)$ (Optional) 222

Chapter 5 Key Concepts 227

Chapter 5 Review Exercises 229

6 *Inverse Trigonometric Functions and Trigonometric Equations* 232

6.1 *Inverse Functions* 232

6.2 *Inverse Trigonometric Functions* 239

6.3 *Trigonometric Equations* 249

6.4 *Trigonometric Equations with Multiple Angles* 254

6.5 *Inverse Trigonometric Equations* 258

Chapter 6 Key Concepts 264

Chapter 6 Review Exercises 266

7 Triangles and Vectors 269

7.1 Oblique Triangles and the Law of Sines 269
7.2 The Ambiguous Case of the Law of Sines 277
7.3 The Law of Cosines 282
7.4 Vectors 291
7.5 Applications of Vectors 297
Chapter 7 Key Concepts 304
Chapter 7 Review Exercises 305

8 Complex Numbers 308

8.1 Operations on Complex Numbers 308
8.2 Trigonometric Form of Complex Numbers 314
8.3 Product and Quotient Theorems 319
8.4 Powers and Roots of Complex Numbers 323
8.5 Polar Equations 328
Chapter 8 Key Concepts 337
Chapter 8 Review Exercises 339

9 Logarithms 341

9.1 Exponential and Logarithmic Functions 341
9.2 Common and Natural Logarithms 348
9.3 Exponential and Logarithmic Equations 353
Chapter 9 Key Concepts 358
Chapter 9 Review Exercises 359

Appendix 361
Table 1 Squares and Square Roots 363
Table 2 Trigonometric Functions in Degrees and Radians 364
Table 3 Common Logarithms 368
Table 4 Natural Logarithms 370
Answers to Selected Exercises 371
Index 403

Using a Calculator

Buying a Calculator A scientific calculator is very helpful for trigonometry. These calculators once cost hundreds of dollars, but now can be purchased for less than the cost of a pizza. Scientific calculators can be recognized by the following keys (among others):

The first of these keys is used in trigonometry, the second to find the logarithm of a number, and the third to raise a number to a power. A scientific calculator has the advantage of doing away with the need for most tables; with a scientific calculator you seldom if ever need to use a table in a trigonometry course. (The general skill of table reading is still useful, however, so we include it in the book and many instructors discuss it in class.)

Some advanced calculators are *programmable:* instead of starting each new problem of a given type from scratch, only the necessary keystrokes are entered. Then the data for a new problem can be entered, with only one or two keys needed to get the result. A programmable calculator is not really necessary for this course. However, such calculators do offer two advantages: first, since you cannot program a calculator for a group of problems unless you completely understand the basic ideas of the problems, your understanding would be enhanced; and second, the programming skills taught in using such calculators are useful in further course work in science or mathematics.

There are two types of logic in common use in calculators today. Both algebraic and Reverse Polish Notation (RPN) have advantages and disadvantages. Algebraic logic is the easiest to learn. For example, the problem $8 + 17$ is entered into an algebraic machine by pressing

$$8 + 17 = .$$

On a machine with Reverse Polish Notation (named for the eminent Polish mathematicians who developed the system), this same problem would be entered as

$$8 \text{ ENTER } 17 + .$$

Some people claim that Reverse Polish machines work advanced problems more easily than algebraic machines. Others claim that algebraic machines are easier to use for the great bulk of ordinary, common problems. It is up to you to decide which to buy. You may wish to discuss your calculator purchase with your instructor for further guidance.

Calculator Errors A calculator can store only so many digits in its memory. Because of this, numbers that have more digits than can be stored must be rounded. For example, 1/3 is not stored as the exact fraction 1/3, but rather as a decimal, perhaps .3333333333333. Since this rounded form of 1/3 is used, errors can occur in calculations. To see how this happens, use a calculator to divide 1 by 3, and then multiply the result by 3. You should get 1 (exactly), but many machines produce

$$(1 \div 3) \times 3 = \left(\frac{1}{3}\right) \times 3 = .9999999999.$$

Some machines round this result to 1; however, the machine does not treat the number internally as 1. To see this, subtract 1 from the result above; you should get 0 but probably will not.

Another calculator error results when numbers of greatly different size are used in addition. For example,

$$10^9 + 10^{-5} - 10^9 = 10^{-5}.$$

However, most calculators would give

$$10^9 + 10^{-5} - 10^9 = 0.$$

These calculator errors seldom occur in realistic problems, but if they do occur you should know what is happening.

Using a Calculator While this introduction is not designed to replace your calculator instruction manual, we do list a few things to keep in mind as you use your calculator.

Parentheses Many calculators have parentheses keys, [( ] and [ )]. These are used as in algebra. For example, $(3 \cdot 5 + 8 \cdot 2) \cdot 4$ could be found as follows.

[( ] 3 [×] 5 [+] 8 [×] 2 [ )] [×] 4 [=] 124

Memory A memory key is like an electronic piece of scratch paper. Pressing [M] or [STO] will cause the number in the display to be stored, and pressing [MR] or [RCL] will cause it to be recalled.

Scientific Notation A key labeled [EE] permits numbers to be entered in scientific notation. For example, entering 9.68, pressing [EE], and then entering 5, followed by [+/−], results in the display

9.68 −05,

which represents 9.68×10^{-5}. With some calculators, pressing [INV] and then [EE] causes a number displayed in scientific notation to be written in regular notation.

1 The Trigonometric Functions

The foundations of trigonometry go back at least three thousand years. The ancient Egyptians, Babylonians, and Greeks developed trigonometry to find the lengths of the sides of triangles and the measures of their angles. In Egypt trigonometry was used to reestablish land boundaries after the annual flood of the Nile River. In Babylonia it was used in astronomy. The very word *trigonometry* comes from the Greek words for triangle *(trigon)* and measurement *(metry)*. Today trigonometry is used in electronics, surveying, and other engineering areas, and is necessary for further courses in mathematics, such as calculus.

Perhaps the most exciting thing to happen to trigonometry in the last few years is the widespread availability of inexpensive calculators. With these calculators, arithmetic calculations that formerly required many hours, or were not done at all, can now be done in a few moments. In using this book, while a calculator is not required, it would be a big help. Exercises where a calculator would be especially helpful are identified by the symbol ▦.

1.1 Basic Terms

Many ideas in trigonometry are best explained with a graph. Each point in the plane corresponds to an **ordered pair,** two numbers written inside parentheses, such as $(-2, 4)$. Graphs are set up with two axes, one for each number in an ordered pair. The horizontal axis is called the **x-axis,** and the vertical axis is the **y-axis.** The two axes cross at a point called the **origin.** To locate the point that corresponds to the ordered pair $(-2, 4)$, start at the origin, and go 2 units left and 4 units up. The point $(-2, 4)$ and other sample points are shown in Figure 1.1.

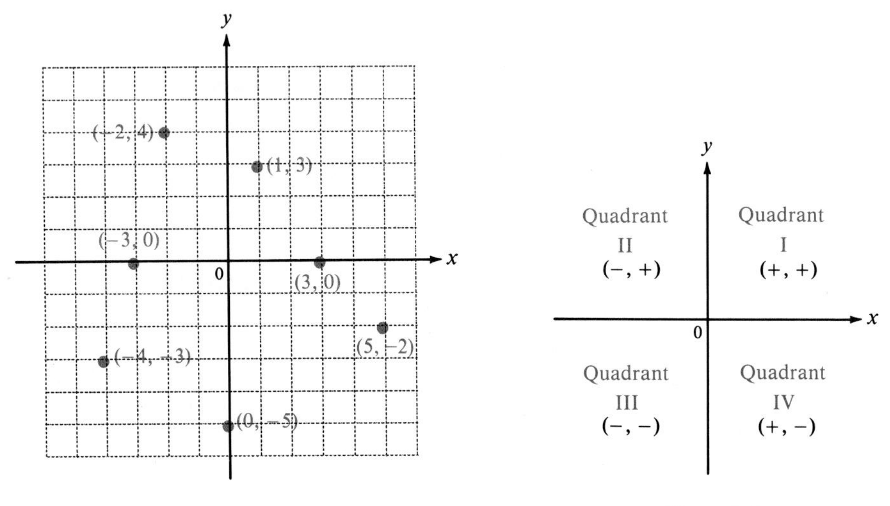

Figure 1.1 **Figure 1.2**

The axes divide the plane into four regions called **quadrants.** The quadrants are numbered in a counterclockwise direction, as shown in Figure 1.2. The points on the axes themselves belong to none of the quadrants. Figure 1.2 also shows that in quadrant I both the x-coordinate and the y-coordinate are positive; in quadrant II the value of x is negative while y is positive, and so on.

The distance between any two points on a plane can be found by using a formula derived from the **Pythagorean theorem.**

Pythagorean Theorem

If the two shorter sides (the **legs**) of a right triangle have lengths a and b, respectively, and if the length of the **hypotenuse** (the longest side, opposite the 90° angle) is c, then

$$a^2 + b^2 = c^2.$$

To find the distance between the two points (x_1, y_1) and (x_2, y_2), start by drawing the line segment connecting the points, as shown in Figure 1.3. Complete a right triangle by drawing a line through (x_1, y_1) parallel to the x-axis and a line through (x_2, y_2) parallel to the y-axis. The ordered pair at the right angle of this triangle is (x_2, y_1).

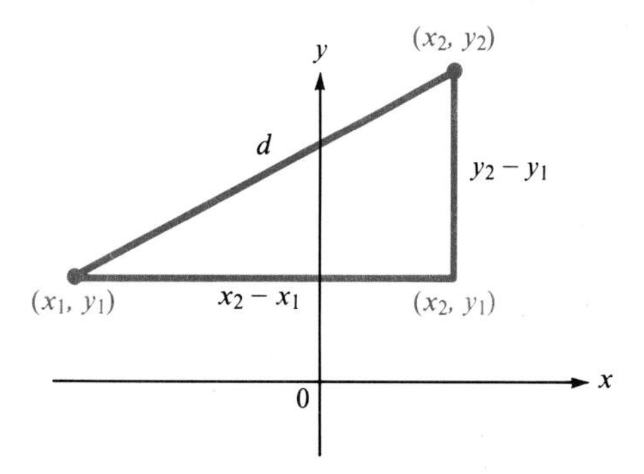

Figure 1.3

The horizontal side of the right triangle in Figure 1.3 has length $x_2 - x_1$, while the vertical side has length $y_2 - y_1$. If d represents the distance between the two original points, then by the Pythagorean theorem,

$$d^2 = (x_2 - x_1)^2 + (y_2 - y_1)^2,$$

or, upon solving for d,

Distance Formula

the distance between the points (x_1, y_1) and (x_2, y_2) is given by the **distance formula,**

$$d = \sqrt{(x_2 - x_1)^2 + (y_2 - y_1)^2}.$$

Example 1 Use the distance formula to find the distance, d, between each of the following pairs of points.

(a) (2, 6) and (5, 10)

Either point can be used as (x_1, y_1). If we choose (2, 6) as (x_1, y_1), then $x_1 = 2$, $y_1 = 6$, $x_2 = 5$, and $y_2 = 10$.

$$\begin{aligned} d &= \sqrt{(x_2 - x_1)^2 + (y_2 - y_1)^2} \\ &= \sqrt{(5 - 2)^2 + (10 - 6)^2} \\ &= \sqrt{3^2 + 4^2} \\ &= \sqrt{9 + 16} \\ &= \sqrt{25} \\ &= 5 \end{aligned}$$

(b) $(-7, 2)$ and $(3, -8)$

$$\begin{aligned} d &= \sqrt{[3-(-7)]^2 + (-8-2)^2} \\ &= \sqrt{10^2 + (-10)^2} \\ &= \sqrt{100 + 100} \\ &= \sqrt{200} \\ &= 10\sqrt{2} \end{aligned}$$

Here $\sqrt{200}$ was simplified as $\sqrt{200} = \sqrt{100} \cdot \sqrt{2} = 10\sqrt{2}$. ✣

Example 2 Use the distance formula to decide if the points $A(-1, 3)$, $B(3, 11)$, and $C(-4, -3)$ lie on a straight line.

If these points lie on a straight line, the sum of two of the distances AB, BC, and AC will equal the third. By the distance formula,

$$\begin{aligned} AB &= \sqrt{(3+1)^2 + (11-3)^2} \\ &= \sqrt{16+64} = \sqrt{80} = 4\sqrt{5} \\ BC &= \sqrt{(-4-3)^2 + (-3-11)^2} \\ &= \sqrt{49+196} = \sqrt{245} = 7\sqrt{5} \\ AC &= \sqrt{(-4+1)^2 + (-3-3)^2} \\ &= \sqrt{9+36} = \sqrt{45} = 3\sqrt{5}. \end{aligned}$$

Since $3\sqrt{5} + 4\sqrt{5} = 7\sqrt{5}$, the points do lie on a straight line. ✣

Functions A **relation** is defined as a set of ordered pairs. Many relations have a rule or formula showing the connection between the two components of the ordered pairs. For example, the formula

$$y = -5x + 6$$

shows that a value of y can be found from a given value x by multiplying the value of x by -5 and then adding 6. According to this formula, if $x = 2$, then $y = -5 \cdot 2 + 6 = -4$, so that $(2, -4)$ belongs to the relation. In the relation $y = -5x + 6$, the value of y depends on the value of x, so that y is the **dependent variable** and x is the **independent variable.**

Most of the relations in trigonometry are also *functions*.

Function

> A relation is a **function** if each value of the independent variable leads to exactly one value of the dependent value.

It is customary for x to be considered the independent variable and y the dependent variable, and we shall follow that convention.

For example, $y = -5x + 6$ is a function. For any one value of x that might be chosen, $y = -5x + 6$ gives exactly one value of y. In contrast, $y^2 = x$ is a relation that is not a function. If we choose the value $x = 16$, then $y^2 = x$ becomes $y^2 = 16$, from which $y = 4$ or $y = -4$. The one x-value, 16, leads to two y-values, 4 and -4, so that $y^2 = x$ is not a function.

Functions are often named with letters such as f, g, or h. For example, the function $y = -5x + 6$ can be written as

$$f(x) = -5x + 6,$$

where $f(x)$ is read "f of x." For the function $f(x) = -5x + 6$, if $x = 3$ then $f(x) = f(3) = -5 \cdot 3 + 6 = -15 + 6 = -9$, or

$$f(3) = -9.$$

Also,

$$f(-7) = -5(-7) + 6 = 41.$$

Recall that $|a|$ represents the absolute value of a. By definition, $|a| = a$ if $a \geq 0$ and $|a| = -a$ if $a < 0$. Thus $|4| = 4$ and $|-4| = 4$.

Example 3 Let $f(x) = -x^2 + |x - 5|$. Find each of the following.

(a) $f(0)$

Use $f(x)$ and replace x with 0.

$$f(0) = -0^2 + |0 - 5| = -0 + |-5| = 5$$

(b) $f(-4) = -(-4)^2 + |-4 - 5| = -16 + |-9| = -16 + 9 = -7$

(c) $f(a) = -a^2 + |a - 5|$ Each x was replaced with a.

(d) Is f a function?

For each value of x, there is exactly one value of $f(x)$; therefore f is a function. ✣

The set of all possible values that can be used as a replacement for the independent variable in a relation is called the **domain** of the relation. The set of all possible values for the dependent variable is the **range** of the relation.

Example 4 Find the domain and range for the following relations. Identify any functions.

(a) $y = x^2$

Here x, the independent variable, can take on any value, so the domain is the set of all real numbers. Since the dependent variable y equals the square of x, and since a square is never negative, the range is the set of all nonnegative numbers, $y \geq 0$. Each value of x leads to exactly one value of y, so $y = x^2$ is a function.

(b) $3x + 2y = 6$

In this relation x and y can take on any value at all. Both the domain and range are the set of all real numbers. For any value of x that might be chosen, the equation $3x + 2y = 6$ would lead to exactly one value of y. Therefore, $3x + 2y = 6$ is a function.

(c) $x = y^2 + 2$

For any value of y, the square of y is nonnegative; that is, $y^2 \geq 0$. Since $x = y^2 + 2$, this means that $x \geq 0 + 2 = 2$, making the domain of the relation $x \geq 2$. Any real number may be squared, so the range is the set of all real numbers. To decide whether the relation is a function, choose some sample values of x. Choosing 6 for x gives

$$6 = y^2 + 2$$
$$4 = y^2$$
$$y = 2 \quad \text{or} \quad y = -2.$$

Since one x-value, 6, leads to two y-values, 2 and -2, the relation $x = y^2 + 2$ is not a function.

(d) $y = \sqrt{1 - x}$

The domain of x is found from the requirement that the quantity under the radical, $1 - x$, must be greater than or equal to 0 for y to be a real number.

$$1 - x \geq 0$$
$$1 \geq x \quad \text{or} \quad x \leq 1$$

The domain is $x \leq 1$. To determine the range of y, note that the given radical is nonnegative, so $y \geq 0$. ✥

Example 5 Find the domain for the following.

(a) $y = \dfrac{1}{x - 2}$

Since division by 0 is undefined, x cannot equal 2. (This value of x would make the denominator become $2 - 2$, or 0.) Any other value of x is acceptable, so the domain is all values of x other than 2, written $x \neq 2$.

(b) $y = \dfrac{8 + x}{(2x - 3)(4x - 1)}$

This denominator is 0 if either

$$2x - 3 = 0 \quad \text{or} \quad 4x - 1 = 0 .$$

Solve each of these equations.

$$2x - 3 = 0 \qquad\qquad 4x - 1 = 0$$
$$2x = 3 \qquad\qquad 4x = 1$$
$$x = \frac{3}{2} \qquad\qquad x = \frac{1}{4}$$

The domain here includes all real numbers x such that $x \neq 3/2$ and $x \neq 1/4$. ✥

For a relation to be a function, each value of x in the domain of the function must lead to exactly one value of y. Figure 1.4 shows the graph of a relation. A point x_1 has been chosen on the x-axis. A vertical line drawn through x_1 cuts the graph in more than one point. Since the x-value x_1 leads to more than one value of y, this graph is not the graph of a function. This example suggests the *vertical line test* for a function.

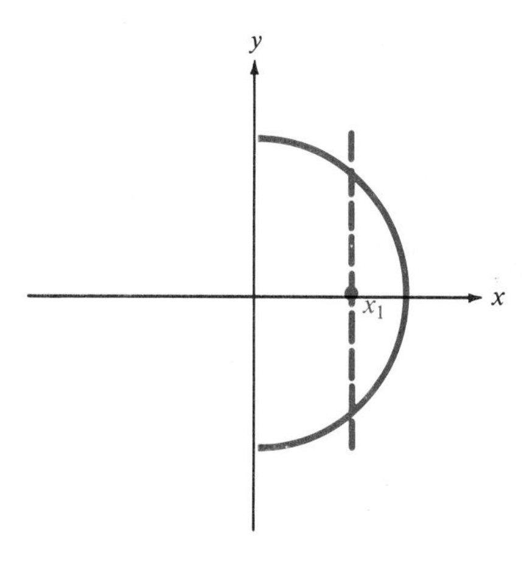

Figure 1.4

Vertical Line Test

If any vertical line cuts the graph of a relation in more than one point, then the graph is not the graph of a function.

1.1 Exercises

Find each of the following.

1. $|-6| + 2$

2. $-|-2| + 4$

3. $-|8| + |-3|$

4. $-|12| + |-9|$

5. $-|-8| - |7|$

6. $-|-3| - |-2|$

7. $-|-8 - 9| + 4$

8. $-|16 - (-2)| - 3$

9. $|-14 - (-3)| - |-2 + 1|$

10. $|-8 - (-9)| - |-4 - (-2)|$

11. $-9.4765 + |8.32179 - 12.4025|$

12. $-|-14.4720 - 3.8976| + |-4.1214|$

Find the quadrant in which each of the following points lie.

13. $(-4, 2)$

14. $(-3, -5)$

15. $(-9, -11)$

16. $(8, -5)$

17. $(9, 0)$

18. $(-2, 0)$

19. $(0, -2)$

20. $(0, 6)$

21. $(-5, \pi)$

22. $(\pi, -3)$

23. $(3\pi, -1)$

24. $(-2\sqrt{2}, 2\sqrt{2})$

Use the distance formula to find the distance between each of the following pairs of points. See Example 1.

25. $(-2, 7)$ and $(1, 4)$
26. $(8, -2)$ and $(4, -5)$
27. $(2, 1)$ and $(-3, -4)$
28. $(-5, 2)$ and $(3, -7)$
29. $(-1, 0)$ and $(-4, -5)$
30. $(-2, -3)$ and $(-6, 4)$
31. $(3, -7)$ and $(-2, -5)$
32. $(-5, 8)$ and $(-3, -7)$
33. $(-3, 6)$ and $(-3, 2)$
34. $(5, -2)$ and $(5, -4)$
35. $(\sqrt{2}, -\sqrt{5})$ and $(3\sqrt{2}, 4\sqrt{5})$
36. $(5\sqrt{7}, -\sqrt{3})$ and $(-\sqrt{7}, 8\sqrt{3})$
37. $(3.7492, 8.1125)$ and $(-2.9082, 3.4147)$
38. $(-1.2004, .8619)$ and $(.2438, -.2199)$

A triangle having sides of lengths a, b, and c is a right triangle if $a^2 + b^2 = c^2$. Use this result and the distance formula to decide if the following points are the vertices of right triangles.

39. $(-2, 5)$, $(1, 5)$, $(1, 9)$
40. $(-9, -2)$, $(-1, -2)$, $(-9, 11)$
41. $(-4, 0)$, $(1, 3)$, $(-6, -2)$
42. $(-8, 2)$, $(5, -7)$, $(3, -9)$
43. $(\sqrt{3}, 2\sqrt{3} + 3)$, $(\sqrt{3} + 4, -\sqrt{3} + 3)$, $(2\sqrt{3}, 2\sqrt{3} + 4)$
44. $(4 - \sqrt{3}, -2\sqrt{3})$, $(2 - \sqrt{3}, -\sqrt{3})$, $(3 - \sqrt{3}, -2\sqrt{3})$

Use the distance formula to decide if the following points lie on a straight line. See Example 2.

45. $(0, 7)$, $(3, -5)$, $(-2, 15)$
46. $(1, -4)$, $(2, 1)$, $(-1, -14)$
47. $(0, -9)$, $(3, 7)$, $(-2, -19)$
48. $(1, 3)$, $(5, -12)$, $(-1, 11)$

Find all values of x or y such that the distance between the given points is as indicated.

49. $(x, 7)$ and $(2, 3)$ is 5
50. $(5, y)$ and $(8, -1)$ is 5
51. $(3, y)$ and $(-2, 9)$ is 12
52. $(x, 11)$ and $(5, -4)$ is 17

53. Use the distance formula to write an equation for all points that are 5 units from $(0, 0)$. Sketch a graph showing these points.

54. Write an equation for all points 3 units from $(-5, 6)$. Sketch a graph showing these points.

*The following two exercises use the Pythagorean theorem.**

55. A 1000-ft section of railroad track expands 6 in because the day is very hot. This causes end C (see the figure) to break off and move to position B, forming right triangle ABC. Find BC. (The surprising answer to this simple problem explains why railroad tracks, bridges, and similar structures must be designed to allow for expansion.)

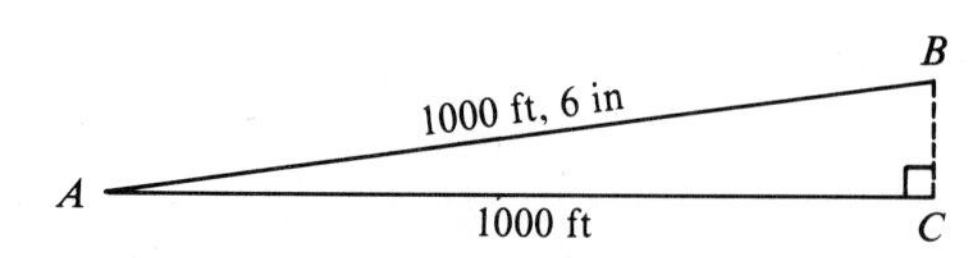

*Exercises are reprinted with permission of Macmillan Publishing Company from *Trigonometry with Calculators* by Lawrence S. Levy. Copyright © 1983 by Macmillan Publishing Company.

56. Clothing manufacturers sometimes cut their material "on the bias" (that is, at 45° to the direction the threads run) to give it more elasticity. A tie maker wants to cut twenty 8-in strips of silk on the bias from material that costs $10 per (linear) yd of material 42 in wide (see the figure). Find the total cost of the material. *Note:* This unappealing combination of units—inches, yards, and dollars—is typical of many practical problems, not just in the clothing industry. (*Hint:* First find length *AB*, using isosceles triangle *ABX*.)

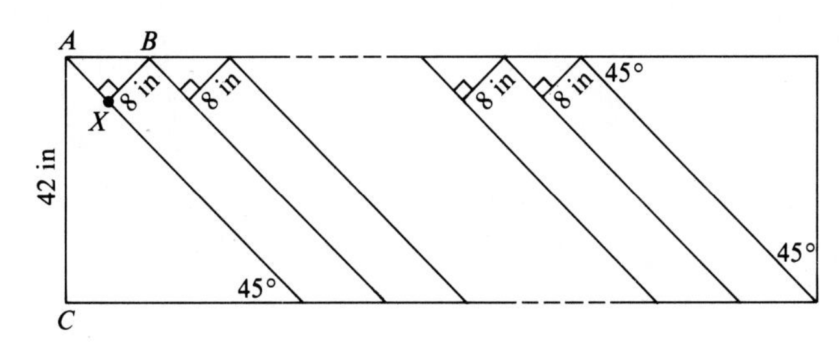

57. The height (h) of the Great Pyramid of Egypt is 144 m. The apothem (a in the figure) measures 184.7 m. Assuming the base is a square, find the length l of a side of the base.

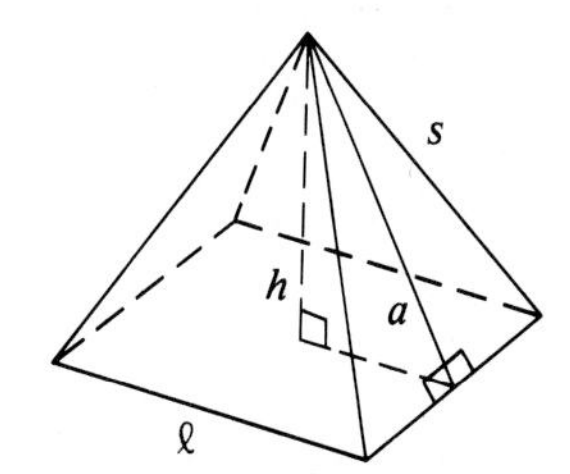

58. Use the result of Exercise 57 to find the length of the edge of the pyramid labeled s in the figure.

Let $f(x) = -2x^2 + 4x + 6$. *Find each of the following. See Example 3.*

59. $f(0)$ **60.** $f(-2)$ **61.** $f(-1)$ **62.** $f(3)$

63. $f(-3)$ **64.** $f(-5)$ **65.** $f(.9461)$ **66.** $f(-.2247)$

67. $f(a)$ **68.** $f(-m)$ **69.** $f(1 + a)$ **70.** $f(2 - p)$

For each of the following, replace x, in turn, by -2, -1, 0, 1, 2, *and* 3. *Then plot the resulting six points on a coordinate system.*

71. $y = -3x + 5$ **72.** $y = 2x - 4$

73. $y = -x^2 + 2x$ **74.** $y = x^2 - 4x + 1$

Find the domain and range of each of the following. Identify any that are functions. See Example 4.

75. $y = 4x - 3$ **76.** $2x + 5y = 10$ **77.** $y = x^2 + 4$

78. $y = 2x^2 - 5$ **79.** $y = -2(x - 3)^2 + 4$ **80.** $y = 3(x + 1)^2 - 5$

81. $x = y^2$ **82.** $-x = y^2$ **83.** $y = \sqrt{4 + x}$

84. $y = \sqrt{x - 2}$ **85.** $y = \sqrt{x^2 + 1}$ **86.** $y = \sqrt{1 - x^2}$

Find the domain and range of each of the following. Use the vertical line test to identify any functions.

87.

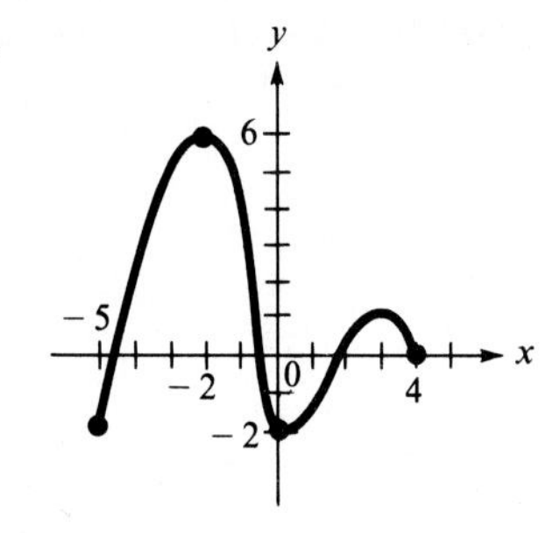

88.

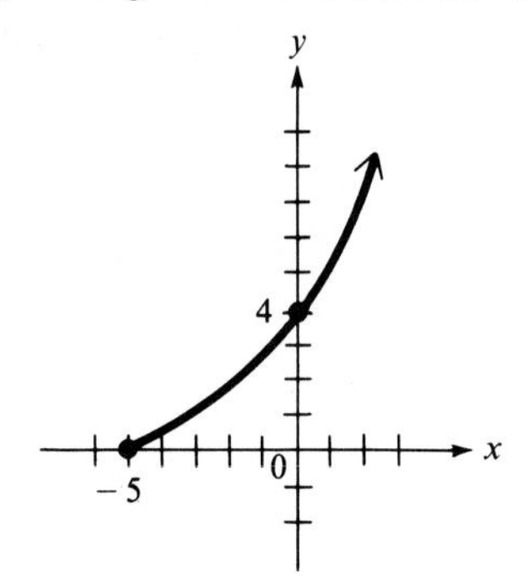

89.

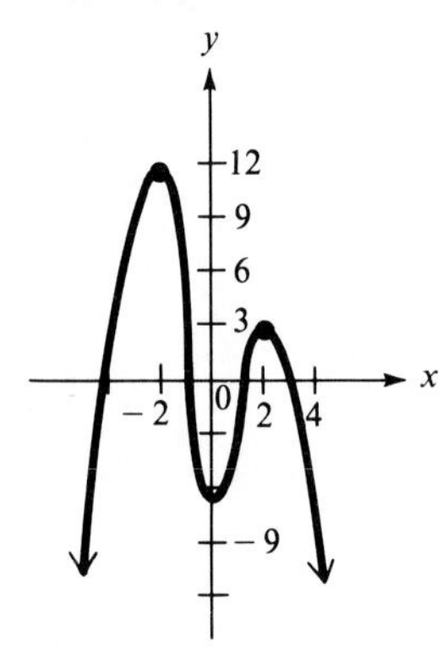

90.

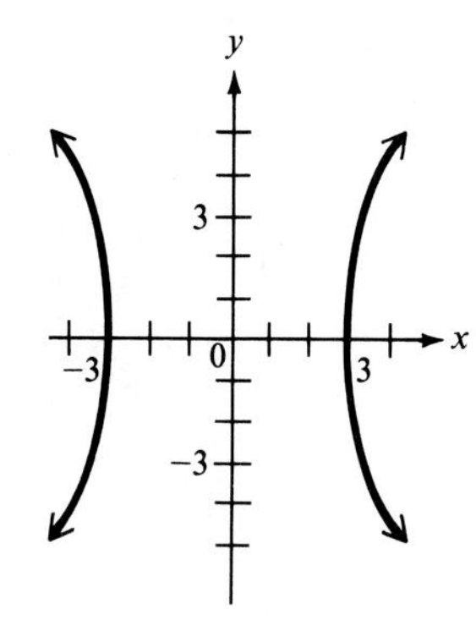

91.

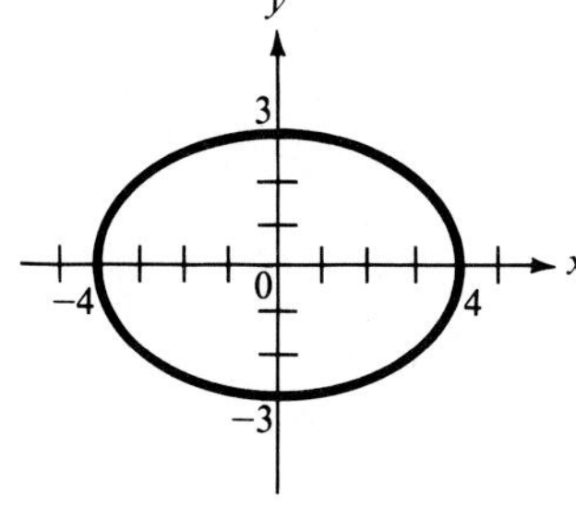

92.

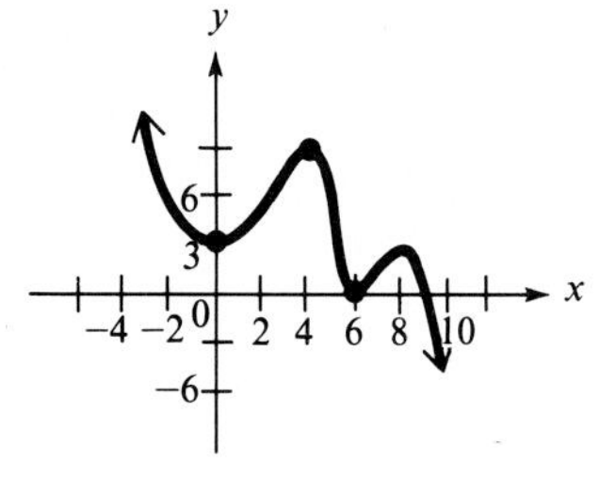

Find the domain for each of the following. See Example 5.

93. $y = \dfrac{1}{x}$

94. $y = \dfrac{-2}{x + 1}$

95. $y = \dfrac{3 + x}{(3x - 7)(2x + 1)}$

96. $y = \dfrac{4 + x^2}{(5x + 1)(3x + 8)}$

97. $y = \dfrac{7 + 5x}{x^2 + 4}$

98. $y = \dfrac{10 + 3x}{-9 - x^2}$

1.2 Angles

Figure 1.5 shows the line through the two distinct points A and B. This line is called **line *AB***. The portion of the line between A and B, including points A and B themselves, is **segment *AB***. The portion of line AB that starts at A and continues through B, and on past B, is called **ray *AB***. Point A is the endpoint of the ray. (See Figure 1.5.)

An **angle** is formed by rotating a ray around its endpoint. The ray in its initial position is called the **initial side** of the angle, while the ray in its location after the rotation is the **terminal side** of the angle. The endpoint of the ray is the **vertex** of the angle. Figure 1.6 shows the initial and terminal sides of an angle with vertex A.

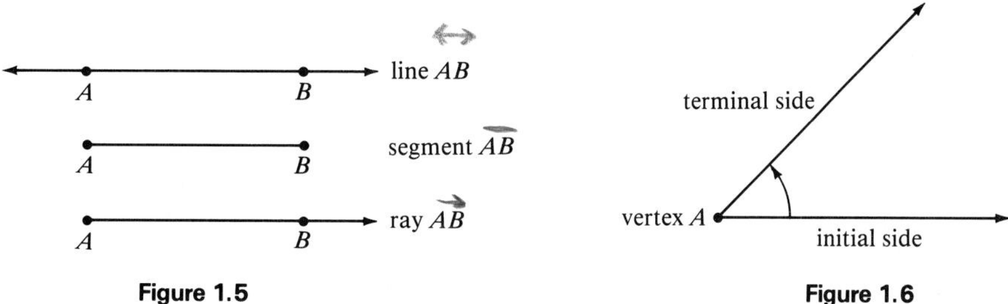

Figure 1.5

Figure 1.6

If the rotation of the terminal side is counterclockwise, the angle is **positive.** If the rotation is clockwise, the angle is **negative.** Figure 1.7 shows two angles, one positive and one negative.

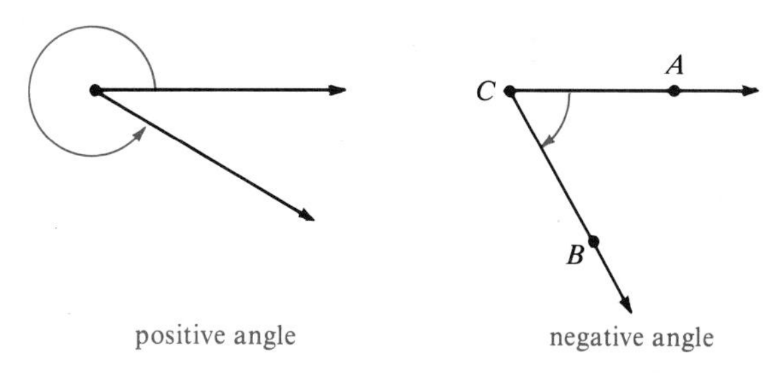

Figure 1.7

An angle can be named by using the name of its vertex. For example, the angle on the right in Figure 1.7 can be called angle C. Alternatively, an angle can be named using three letters, with the vertex letter in the middle. For example, the angle on the right also could be named angle ACB or angle BCA.

There are two systems in common use for measuring the size of angles. The most common unit of measure is the **degree.** (The other common unit of measure is called the *radian,* which is discussed in chapter 3.) Degree measure was developed by the Babylonians, four thousand years ago. To use degree measure, assign 360 degrees to a complete rotation of a ray. In Figure 1.8, notice that the terminal side of the angle corresponds to its initial side when it makes a complete rotation.

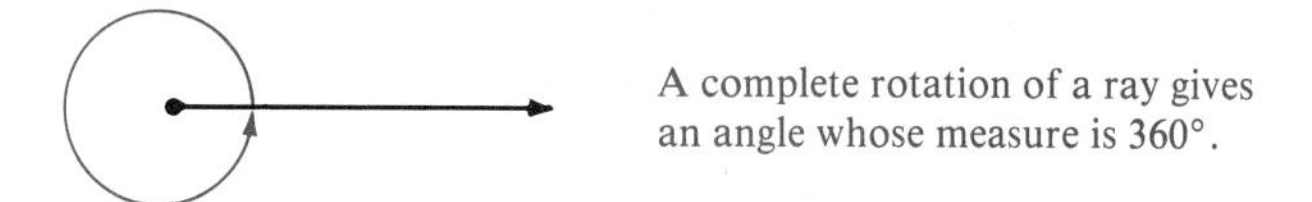

Figure 1.8

One degree, written 1°, represents 1/360 of a rotation. For example, 90° represents 90/360 = 1/4 of a complete rotation, and 180° represents 180/360 = 1/2 of a complete rotation. Angles of measure 1°, 90°, and 180° are shown in Figure 1.9.

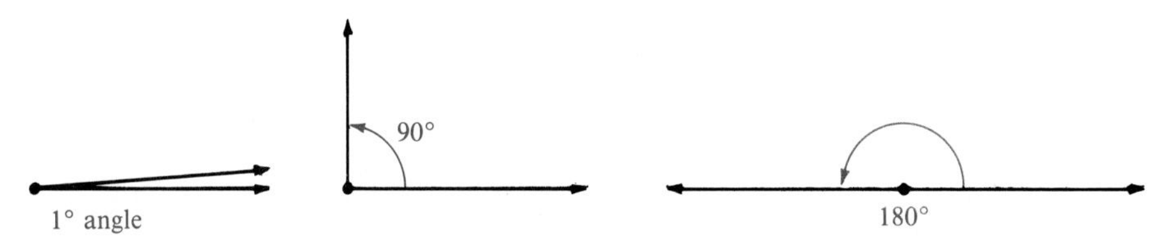

Figure 1.9

Angles are named as shown in the following chart.

Types of Angles

Name	*Angle measure*	*Example*
Acute angle	Between 0° and 90°	60° 82°
Right angle	Exactly 90°	90°
Obtuse angle	Between 90° and 180°	97° 138°
Straight angle	Exactly 180°	180°

If the sum of the measures of two angles is 90°, the angles are called **complementary.** Two angles with measures that add up to 180° are **supplementary.**

Example 1 Find the measure of each angle in Figures 1.10 and 1.11.

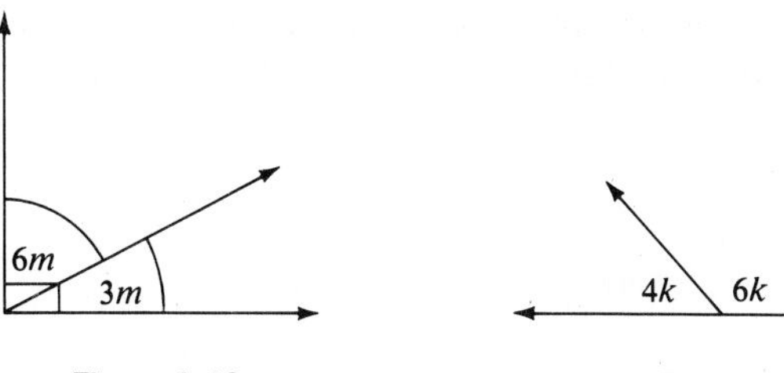

Figure 1.10

Figure 1.11

(a) In Figure 1.10, since the two angles form a right angle,

$$\begin{aligned} 6m + 3m &= 90^\circ \\ 9m &= 90^\circ \\ m &= 10^\circ. \end{aligned}$$

The two angles have measures of $6 \cdot 10^\circ = 60^\circ$ and $3 \cdot 10^\circ = 30^\circ$.
(b) The angles in Figure 1.11 are supplementary, so

$$\begin{aligned} 4k + 6k &= 180^\circ \\ 10k &= 180^\circ \\ k &= 18^\circ. \end{aligned}$$

These angle measures are $4(18^\circ) = 72^\circ$ and $6(18^\circ) = 108^\circ$. ✣

Angles are measured with a tool called a **protractor.** Figure 1.12 shows a protractor measuring an angle of 35°.

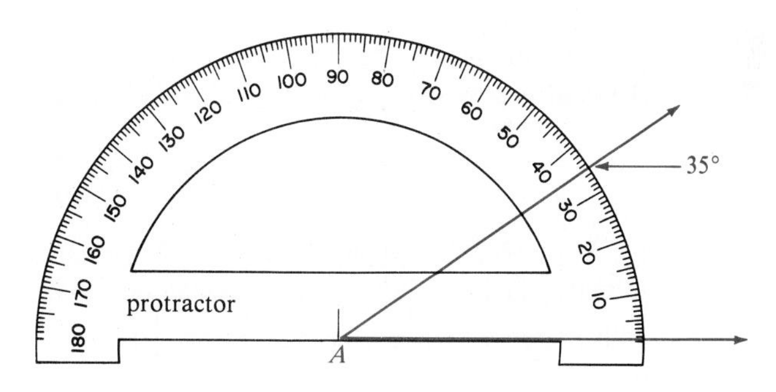

Figure 1.12

Do not confuse an angle with its measure. Angle A of Figure 1.12 is a rotation; the measure of the rotation is 35°. This measure is often expressed by saying that m(angle A) is 35°, where m(angle A) is read "the measure of angle A." It saves a lot of work, however, to abbreviate $m(\text{angle } A) = 35^\circ$ as simply angle $A = 35^\circ$.

Traditionally, portions of a degree have been measured with minutes and seconds. One **minute,** written $1'$, is 1/60 of a degree.

$$1' = \frac{1}{60}^{\circ} \qquad \text{or} \qquad 60' = 1^{\circ}$$

One **second,** $1''$, is 1/60 of a minute.

$$1'' = \frac{1'}{60} = \frac{1}{3600}^{\circ} \qquad \text{or} \qquad 60'' = 1'$$

The measure $12^{\circ}\ 42'\ 38''$ represents 12 degrees, 42 minutes, 38 seconds.

The next example shows how to perform calculations with degrees, minutes, and seconds.

Example 2 Perform each calculation.

(a) $51^{\circ}\ 29' + 32^{\circ}\ 46'$

Add the degrees and the minutes separately.

$$51^{\circ}\ 29' + 32^{\circ}\ 46' = (51^{\circ} + 32^{\circ}) + (29' + 46') = 83^{\circ}\ 75'$$

Since $75' = 60' + 15' = 1^{\circ}\ 15'$, the sum is written

$$83^{\circ}\ 75' = 83^{\circ} + (1^{\circ}\ 15') = 84^{\circ}\ 15'.$$

(b) $90^{\circ} - 73^{\circ}\ 12'$

Write 90° as $89^{\circ}\ 60'$. Then

$$90^{\circ} - 73^{\circ}\ 12' = 89^{\circ}\ 60' - 73^{\circ}\ 12' = 16^{\circ}\ 48'.$$ ✣

With the increasing use of calculators, it is now common to measure angles in **decimal degrees.** For example, 12.4238° represents

$$12.4238^{\circ} = 12\frac{4238}{10{,}000}^{\circ}.$$

The next example shows how to change between decimal degrees and degrees, minutes, and seconds. Some calculators will make these conversions automatically, often with a key labeled DMS.

Example 3 **(a)** Convert $74^{\circ}\ 8'\ 14''$ to decimal degrees. Round to the nearest thousandth of a degree.

Since $1' = \frac{1}{60}^{\circ}$ and $1'' = \frac{1}{3600}^{\circ}$,

$$\begin{aligned} 74^{\circ}\ 8'\ 14'' &= 74^{\circ} + \frac{8}{60}^{\circ} + \frac{14}{3600}^{\circ} \\ &= 74^{\circ} + .1333^{\circ} + .0039^{\circ} \\ &= 74.137^{\circ} \quad \text{(rounded).} \end{aligned}$$

A calculator does the preceding example as follows.

Enter	Press	Display
74.0814	DMS	74.13722222

(b) Convert 34.817° to degrees, minutes, and seconds.

$$
\begin{aligned}
34.817^\circ &= 34^\circ + .817^\circ \\
&= 34^\circ + (.817)(60') \\
&= 34^\circ + 49.02' \\
&= 34^\circ + 49' + .02' \\
&= 34^\circ + 49' + (.02)(60'') \\
&= 34^\circ + 49' + 1'' \quad \text{(rounded)} \\
&= 34^\circ\ 49'\ 1''
\end{aligned}
$$

To convert 34.817° to degrees, minutes, and seconds, use a calculator as follows. (The column headed "Result" shows how to interpret the display.)

Enter	Press	Display	Result
34.817	INV DMS	34.4901	34° 49′ 1″

An angle is in **standard position** if its vertex is at the origin and its initial side is along the positive x-axis. The two angles in Figure 1.13 are in standard position. An angle in standard position is said to lie in the quadrant in which its terminal side lies. For example, an acute angle is in quadrant I and an obtuse angle is in quadrant II. Angles in standard position having their terminal side along the x-axis or y-axis, such as angles of 90°, 180°, 270°, and so on, are called **quadrantal angles.**

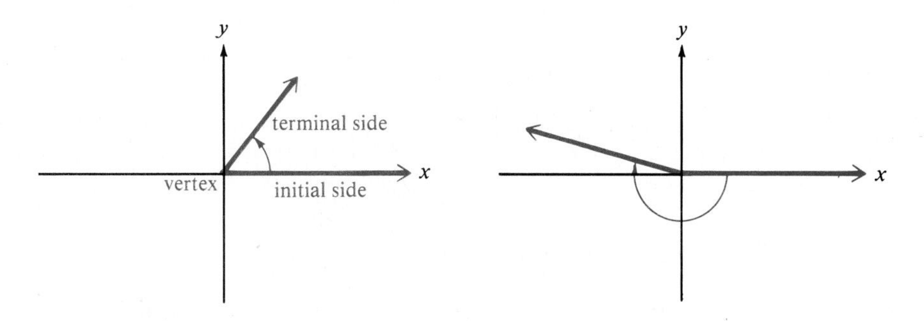

Figure 1.13

A complete rotation of a ray results in an angle of measure 360°. But there is no reason why the rotation need stop at 360°. By continuing the rotation, angles of measure larger than 360° can be produced. The angles in Figure 1.14(a) have measures 60° and 420°. These two angles have the same initial side and the same terminal side, but different amounts of rotation. Angles that have the same initial side and the same terminal side are called **coterminal angles.** As shown in Figure 1.14(b), angles with measures 110° and 830° are coterminal.

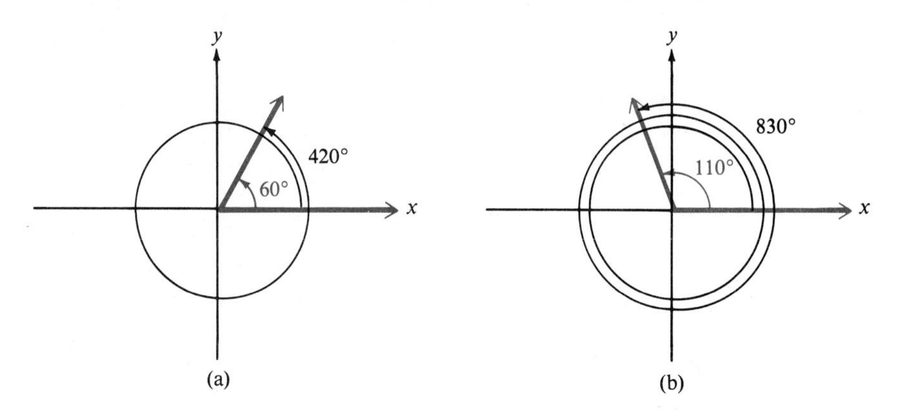

Figure 1.14

Example 4 Find the angles of smallest possible positive measure coterminal with the following angles.

(a) 908°

Add or subtract 360° as many times as needed to get an angle with measure at least 0° but less than 360°. Since $908° - 2 \cdot 360° = 908° - 720° = 188°$, an angle of 188° is coterminal with an angle of 908°. See Figure 1.15.

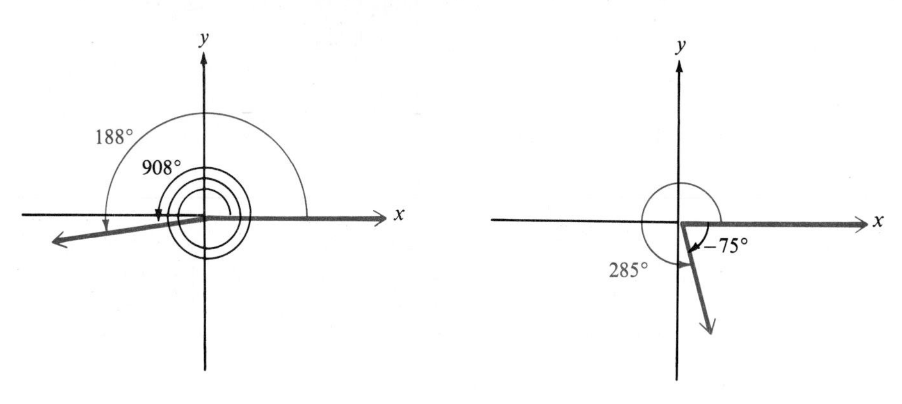

Figure 1.15 **Figure 1.16**

(b) −75°

Use a rotation of $360° + (-75°) = 285°$. See Figure 1.16.

Example 5 A phonograph record makes 45 revolutions per minute. Through how many degrees will a point on the edge of the record move in 2 seconds?

The record revolves 45 times per minute or $45/60 = 3/4$ times per second (since there are 60 seconds in a minute). In 2 seconds, the record will revolve

$2 \cdot (3/4) = 3/2$ times. Each revolution is 360°, so a point on the edge will revolve $(3/2) \cdot 360° = 540°$ in 2 seconds. ✥

The following sketches are three examples of naturally occurring angles.

A plant will grow fastest if none of its leaves shade other leaves on the plant from sunlight. It turns out that to ensure the maximum amount of sunlight to its leaves, a plant should have its leaves in a spiral, with the angle between successive leaves given by

$$360°\left(\frac{3 - \sqrt{5}}{2}\right) \approx 137.5°.$$

For further information on this, see The Curves of Life, *by T. A. Cook, a Dover reprint, 1979.*

Cinder Cone, in Lassen Volcanic National Park, is one of the most symmetric such cones in the world. The sides slope at an angle of 35°, the angle of repose. The sides cannot be any steeper; if they were, the cinders would roll down. This cone is on an old immigrant trail. It was not on the trail in 1850; when the immigrants of 1851 came through the area, they found it.

A set of parallel lines with equidistant spacing intersects an identical set, but at a small angle. The result is a moiré pattern, named after the fabric moiré ("watered") silk. You often see similar effects looking through window screens with bulges. Moiré patterns are related to periodic functions, which describe regular, recurring phenomena (wave patterns such as biorhythms or business cycles). Moirés thus apply to the study of electromagnetic, sound, and water waves, and to crystal structure.

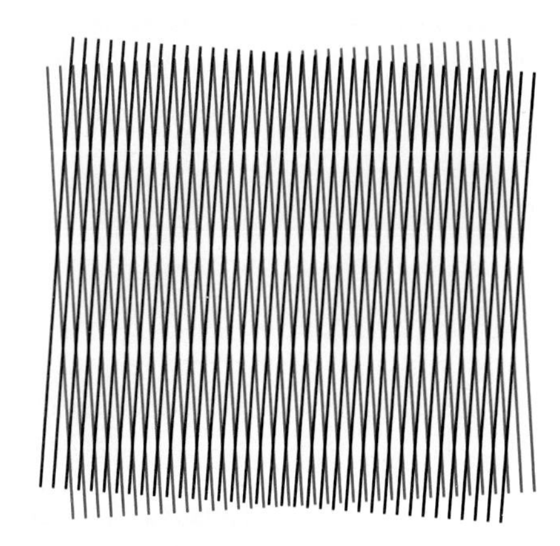

1.2 Exercises

Find the angles of smallest possible positive measure coterminal with the following angles. See Example 4.

1. $-40°$ **2.** $-98°$ **3.** $-125°$ **4.** $-203°$

5. $450°$ **6.** $489°$ **7.** $539°$ **8.** $699°$

9. $850°$ **10.** $1000°$ **11.** $-985.4063°$ **12.** $-1762.3974°$

Place the following angles in standard position. Draw an arrow representing the correct amount of rotation. Find the measure of two other angles, one positive and one negative, that are coterminal with the given angle. Give the quadrant of each angle.

13. $75°$ **14.** $89°$ **15.** $122°$ **16.** $174°$

17. $234°$ **18.** $250°$ **19.** $300°$ **20.** $324°$

21. $438°$ **22.** $593°$ **23.** $512°$ **24.** $624°$

25. $-52°$ **26.** $-61°$ **27.** $-159°$ **28.** $-214°$

Locate the following points in a coordinate system. Draw a ray through the given point, starting at the origin. Use the ray you draw, along with the positive x-axis, to determine an angle. Use a protractor to measure the angle to the nearest degree. (Hint: $\sqrt{3}$ is approximately 1.7.)

29. $(-3, -3)$ **30.** $(-5, 2)$ **31.** $(-3, -5)$

32. $(\sqrt{3}, 1)$ **33.** $(-2, 2\sqrt{3})$ **34.** $(4\sqrt{3}, -4)$

Find the measure of each angle in Exercises 35–40. See Example 1.

35.

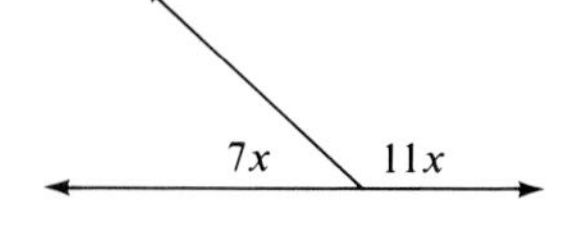

36.

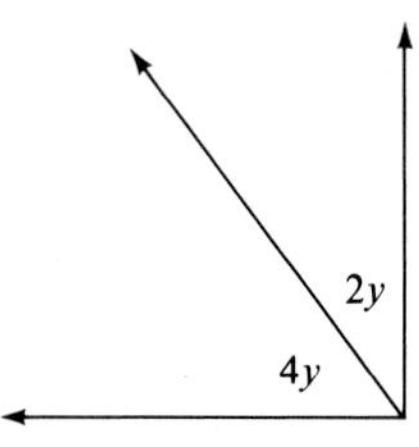

37.

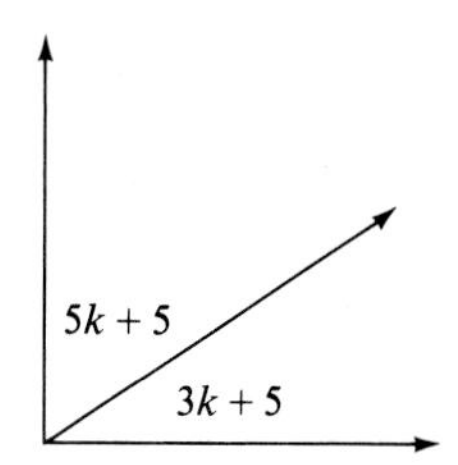

38. Supplementary angles with measures $10m + 7$ and $7m + 3$

39. Supplementary angles with measures $6x - 4$ and $8x - 12$

40. Complementary angles with measures $9z + 6$ and $3z$

Perform each calculation. See Example 2.

41. $75° 15' + 83° 32'$ **42.** $62° 18' + 21° 41'$ **43.** $89° + 23° 42'$

44. $71° 58' + 47° 29'$ **45.** $90° - 51° 28'$ **46.** $90° - 73° 48'$

47. $180° - 152° 43'$ **48.** $180° - 124° 51'$ **49.** $90° - 36° 18' 47''$

50. $90° - 72° 58' 11''$ **51.** $180° - 120° 42' 37''$ **52.** $180° - 86° 39' 54''$

The sum of the angles in any triangle is 180°. In Exercises 53–58, two angles of a triangle are given. Find the third angle.

53. $74° 15'$, $83° 57'$ **54.** $29° 42'$, $83° 47'$ **55.** $147° 12'$, $30° 19'$

56. $136° 50'$, $41° 38'$ **57.** $74° 12' 59''$, $80° 58' 05''$ **58.** $29° 51' 37''$, $49° 28' 50''$

Convert each angle measure to decimal degrees. Round to the nearest thousandth of a degree. See Example 3.

59. $20° 54'$ **60.** $38° 42'$ **61.** $91° 35' 54''$

62. $34° 51' 35''$ **63.** $274° 18' 59''$ **64.** $165° 51' 09''$

Convert each angle measure to degrees, minutes, and seconds. See Example 3.

65. 31.4296° **66.** 59.0854° **67.** 89.9004°

68. 102.3771° **69.** 178.5994° **70.** 122.6853°

Solve each of the following. See Example 5.

71. A tire is rotating 600 times per minute. Through how many degrees does a point on the edge of the tire move in 1/2 second?

72. An airplane propeller rotates 1000 times per minute. Find the number of degrees that a point on the edge of the propeller will rotate in 1 second.

73. A pulley rotates through 75° in one minute. How many rotations does the pulley make in an hour?

74. One student in a surveying class measures an angle as 74.25°, while another student measures the same angle as 74° 20′. Find the difference in these measurements, both to the nearest minute and to the nearest hundredth of a degree.

Let n represent any integer. Give an expression using n for all angles coterminal with angles of the following measures.

75. 95° **76.** −112°

1.3 Similar Triangles

Triangles are classified as shown in the following chart.

Types of Triangles

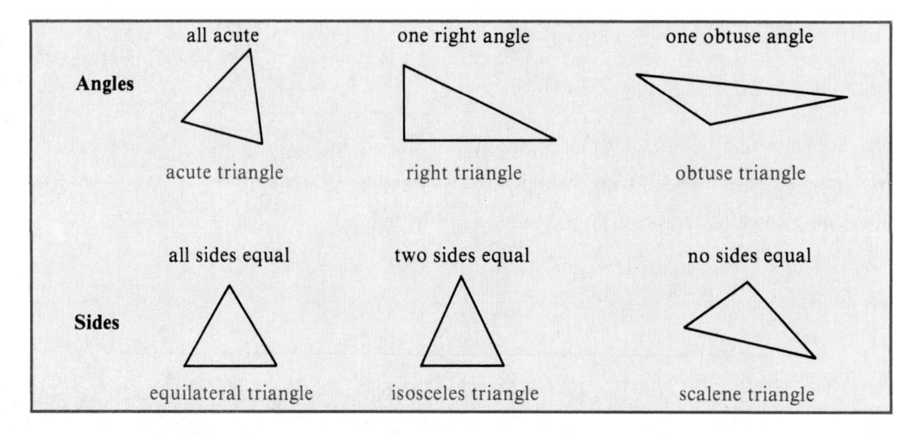

Many of the key ideas of trigonometry depend on **similar triangles,** which are triangles of exactly the same shape but not necessarily the same size. Figure 1.17 shows three pairs of similar triangles.

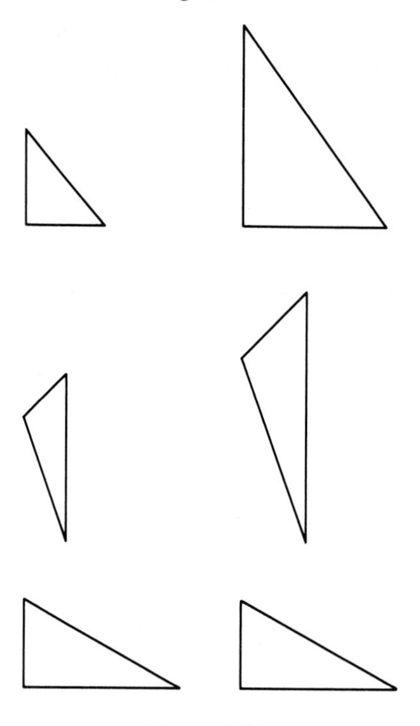

Figure 1.17

The two triangles in the third pair have not only the same shape but also the same size. Triangles that are both the same size and the same shape are called **congruent triangles.** If two triangles are congruent, then it is possible to pick one of them up and place it on top of the other so that they coincide. If two triangles are congruent, then they must be similar. However, two similar triangles need not be congruent.

The triangle supports for a child's swing are congruent triangles, machine-produced with exactly the same dimensions each time. These supports are just one example of similar triangles. The supports of a long bridge, all the same shape but decreasing in size toward the center of the bridge, are an example of similar triangles. Another example of similar figures is shown in Figure 1.18.

Figure 1.18 *These rock carvings on the island of Hawaii were carved by Polynesians who settled the islands beginning about the year 500. These carvings suggest the ideas of similarity and congruence.*

The usefulness of similar triangles depends on the result given below.

Similar Triangles

> If two triangles are similar, then corresponding sides are in proportion and corresponding angles are equal.

Example 1 Find the missing angles in the larger triangle of Figure 1.19. Assume that the triangles are similar.

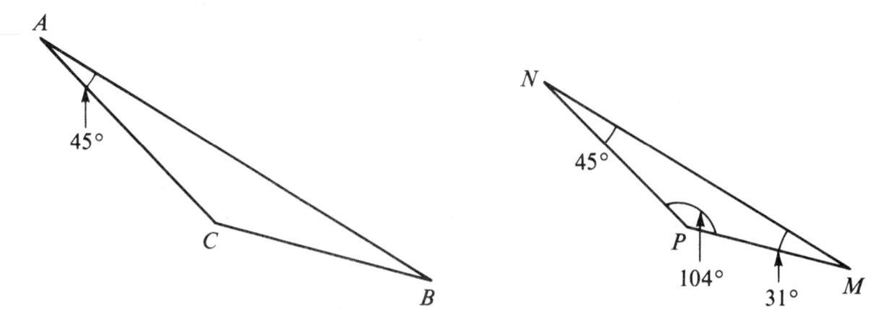

Figure 1.19

Since the triangles are similar, corresponding angles are equal. The figure suggests that angles C and P correspond, making angle $C = 104°$. Also, angles B and M correspond so that angle $B = 31°$. ✣

Triangles are named by their vertices. For example, the triangles in Figure 1.19 are triangle ABC and triangle NMP. It is customary to abbreviate "triangle ABC" as "$\triangle ABC$."

Example 2 Given that $\triangle ABC$ and $\triangle DFE$ in Figure 1.20 are similar, find the lengths of the missing sides of $\triangle DFE$.

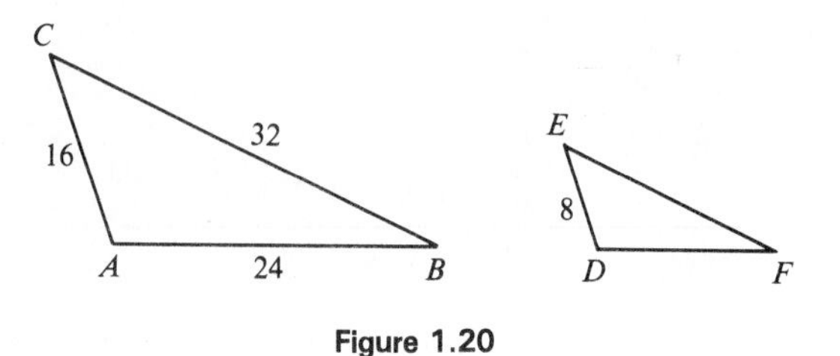

Figure 1.20

As mentioned before, similar triangles have corresponding sides in proportion. Use this fact to find the missing sides in $\triangle DFE$. Side DF of $\triangle DFE$ corresponds to side AB of $\triangle ABC$, and sides DE and AC correspond. This leads to the proportion

$$\frac{8}{16} = \frac{DF}{24}.$$

Recall, in the proportion

$$\frac{a}{b} = \frac{c}{d}, \quad \text{that} \quad ad = bc.$$

This step is called *cross-multiplication*. Use cross-multiplication to solve the equation for DF.

$$\frac{8}{16} = \frac{DF}{24}$$

$$8 \cdot 24 = 16 \cdot DF \qquad \text{Cross-multiply}$$

$$192 = 16 \cdot DF$$

$$12 = DF$$

Side DF has a length of 12.

Side EF corresponds to side BC. This leads to another proportion:

$$\frac{8}{16} = \frac{EF}{32}.$$

Cross-multiplication gives

$$8 \cdot 32 = 16 \cdot EF$$

$$16 = EF.$$

Side EF has a length of 16. ✣

Example 3 Find the missing parts of the similar triangles in Figure 1.21.

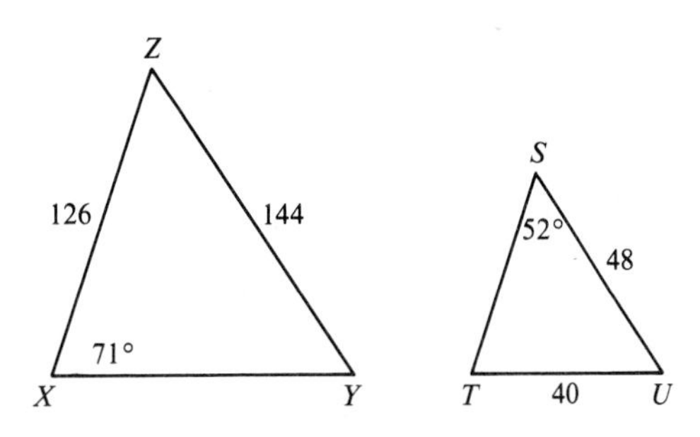

Figure 1.21

Here angles X and T correspond, as do angles Y and U and angles Z and S. Since angles Z and S correspond and since angle S is 52°, angle Z also must be 52°. The sum of the angles of any triangle is 180°. In $\triangle XYZ$, $X = 71°$ and $Z = 52°$. To find Y, set up an equation and solve for Y.

$$X + Y + Z = 180$$

$$71 + Y + 52 = 180$$

$$123 + Y = 180$$

$$Y = 57$$

Angle Y is 57°. Since angles Y and U correspond, $U = 57°$ also.

Now we can find the missing sides. Sides SU and ZY correspond, as do XZ and TS, leading to the proportion

$$\frac{48}{144} = \frac{ST}{126}.$$

Simplify the work by writing 48/144 in lowest terms as 1/3. (A similar process could have been used in the earlier examples.)

$$\frac{1}{3} = \frac{ST}{126}$$

Now cross-multiply.

$$1 \cdot 126 = 3 \cdot ST$$
$$42 = ST$$

Also,

$$\frac{XY}{40} = \frac{144}{48},$$

which can be solved most readily by writing 144/48 as 3/1.

$$\frac{XY}{40} = \frac{3}{1}$$

Cross-multiply to get

$$XY = 120.$$

Side ST has a length of 42, and side XY has a length of 120. ✣

Example 4 The people at the Arcade Fire Station need to measure the height of the station flagpole. They notice that at the instant when the shadow of the station is 18 m long, the shadow of the flagpole is 99 m long. The station is 10 m high. Find the height of the flagpole.

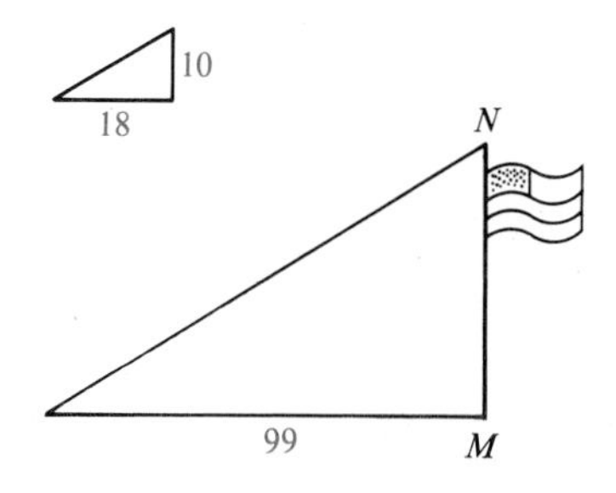

Figure 1.22

Figure 1.22 shows the information given in the problem. The two triangles shown there are similar, so that corresponding sides are in proportion, with

$$\frac{MN}{10} = \frac{99}{18},$$

or

$$\frac{MN}{10} = \frac{11}{2}.$$

Cross-multiply.

$$2 \cdot MN = 110$$
$$MN = 55$$

The flagpole is 55 m high.

When two parallel lines are crossed by a third line, sometimes called a **transversal,** several pairs of equal angles are formed. For example, in Figure 1.23 AE and DB are parallel lines crossed by AB and DE. Angles 3 and 4, called **vertical** angles, are equal. Angles 5 and 1, called **alternate interior** angles, are equal, as are angles 6 and 2. Thus, triangles ACE and BCD are similar triangles.

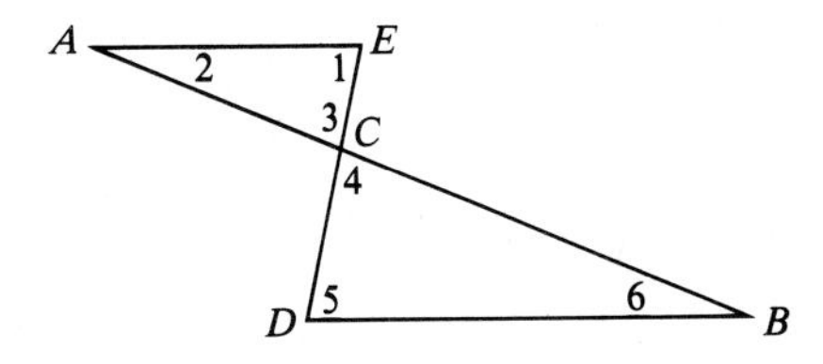

Figure 1.23

1.3 Exercises

Classify each triangle in Exercises 1–12 as either acute, right, *or* obtuse. *Also classify each as either* equilateral, isosceles, *or* scalene.

1.

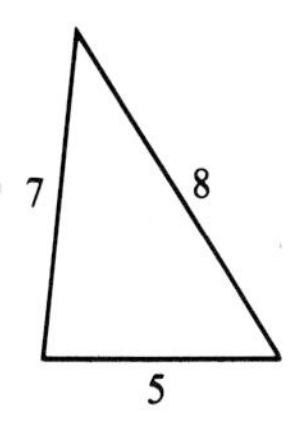

2.

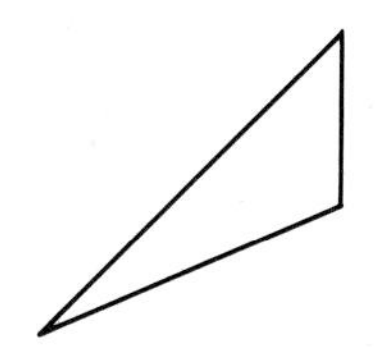

3.

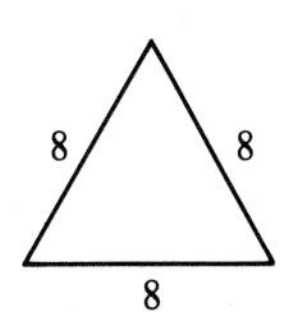

4.

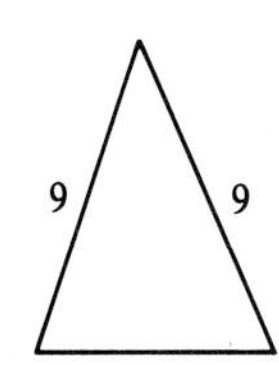

5.

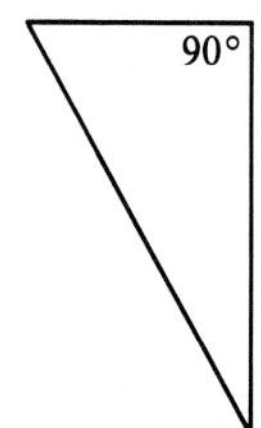

6.

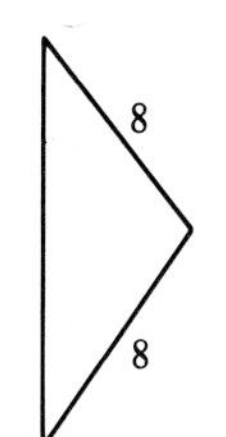

7.

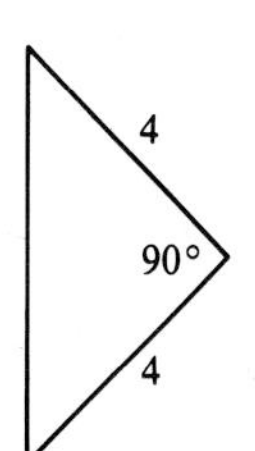

8.

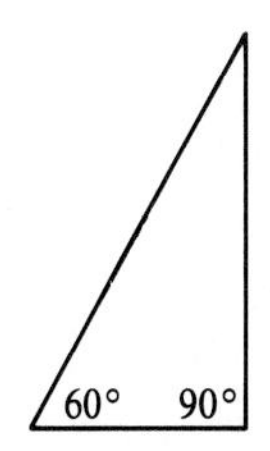

9.

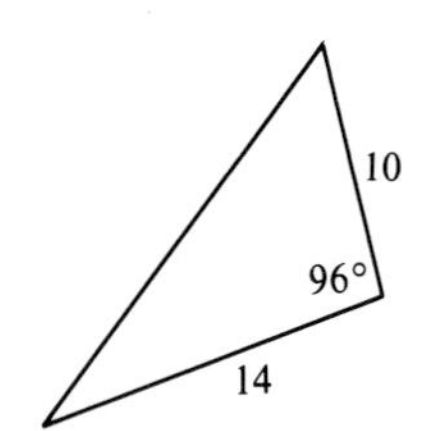

10.

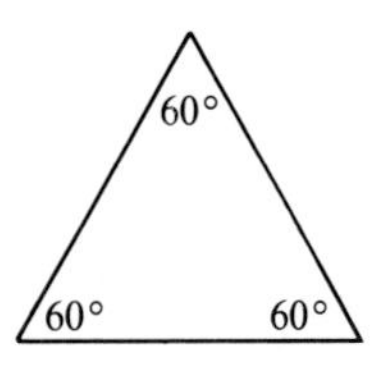

11.

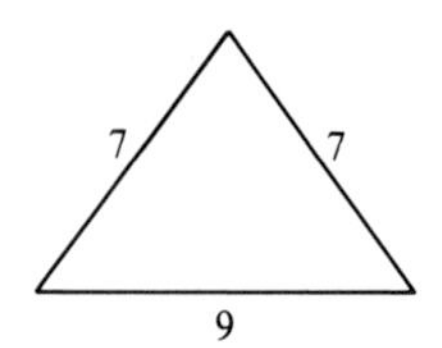

12.

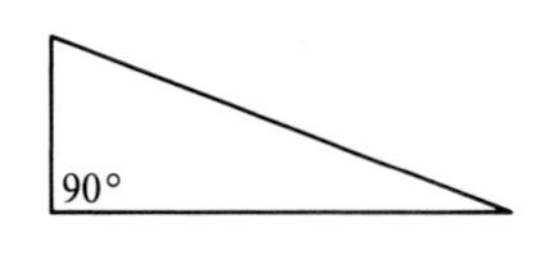

Name the corresponding angles and the corresponding sides for each of the following pairs of similar triangles.

13.

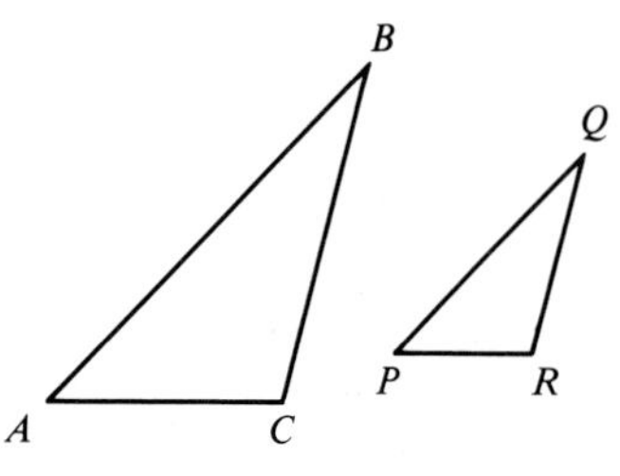

14.

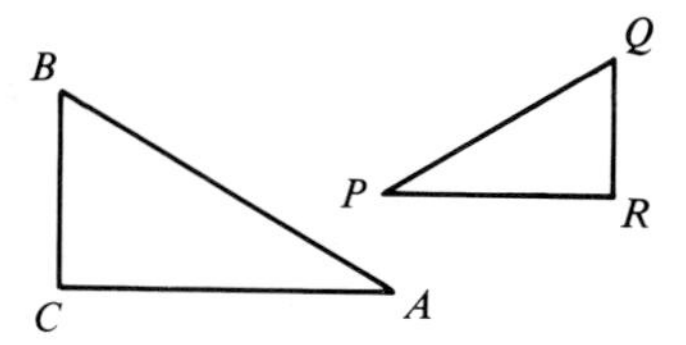

15.

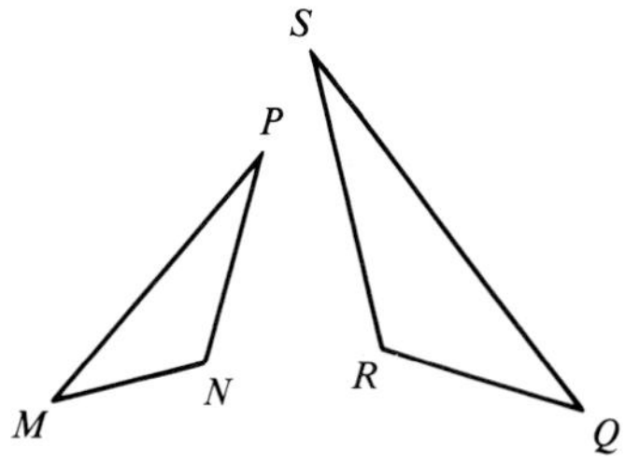

16.

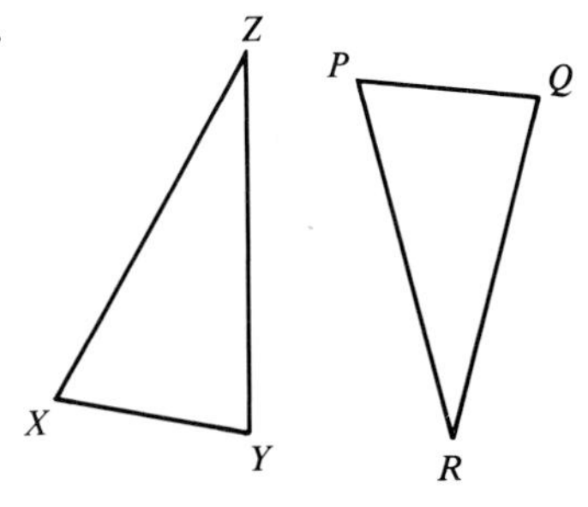

17.

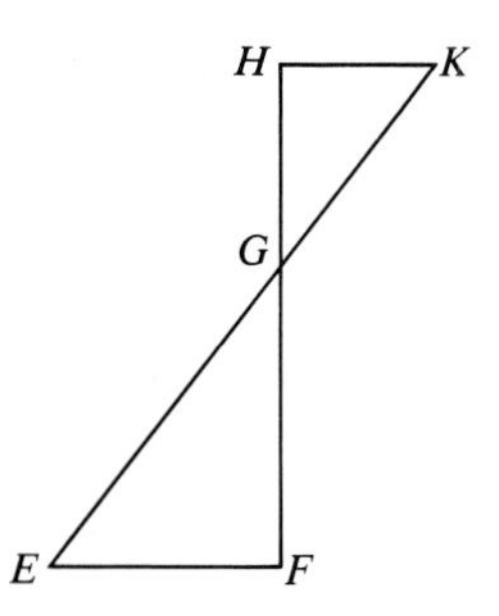

18.

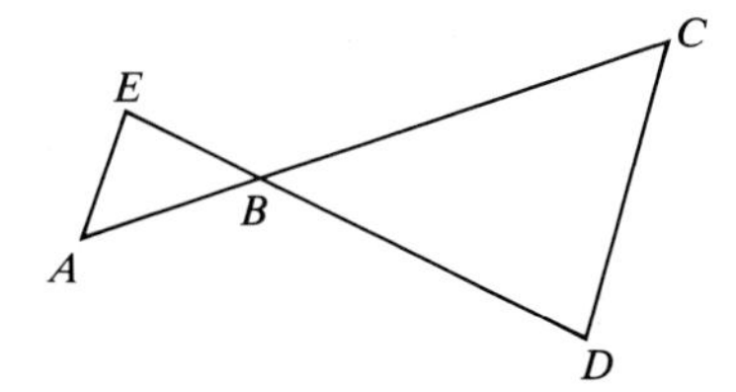

Tell whether each pair of triangles is similar *or* not *similar.*

19.

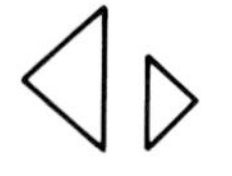

20.

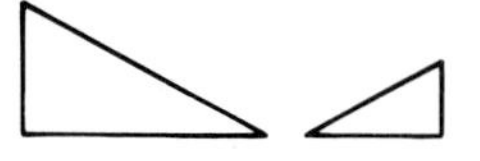

21.

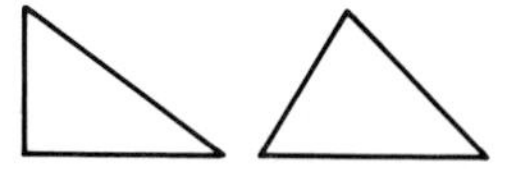

22.

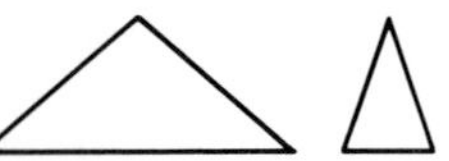

23.

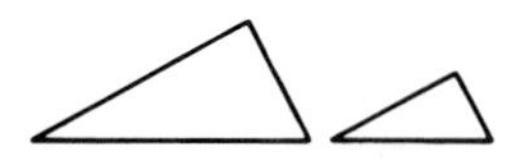

24.

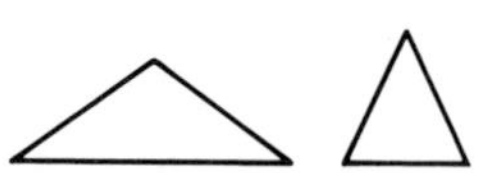

Find all missing angles in each pair of similar triangles. See Example 1.

25.

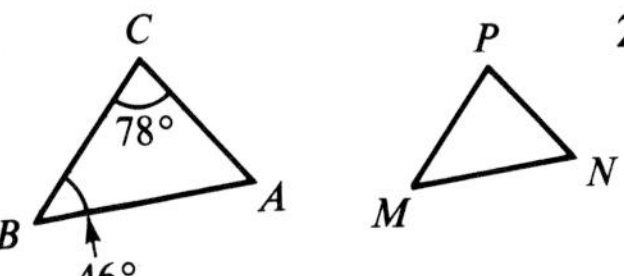

26.

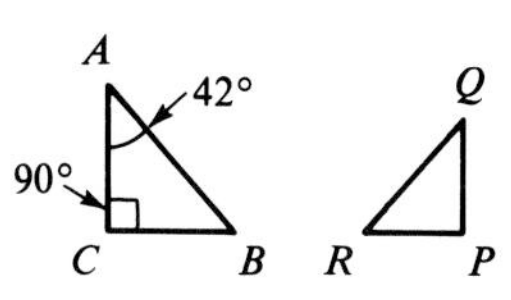

27.

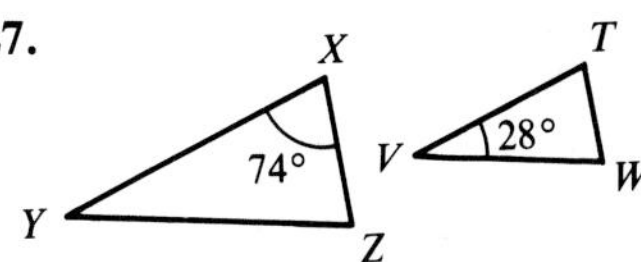

28.

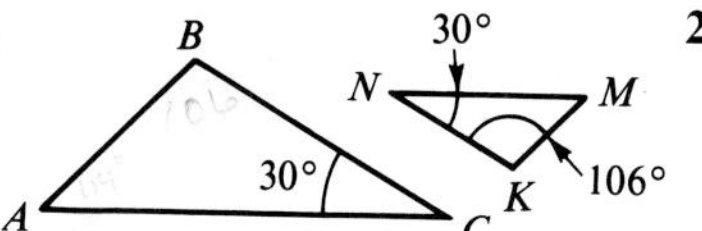

29.

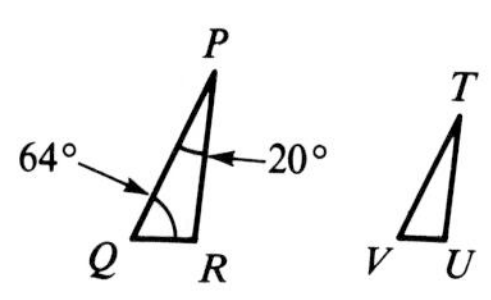

30.

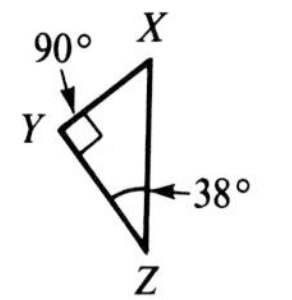

Find the lengths of the missing sides in each pair of similar triangles. See Examples 2 and 3.

31.

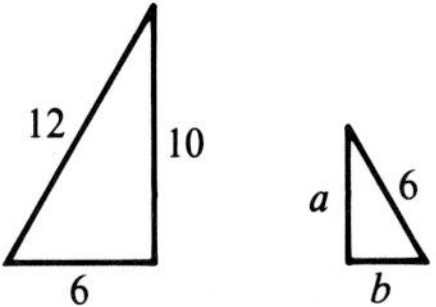

32.

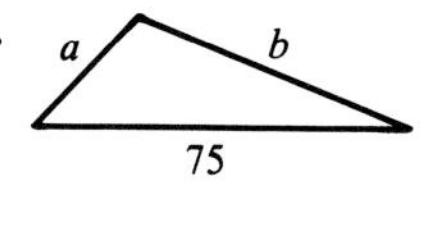

33.

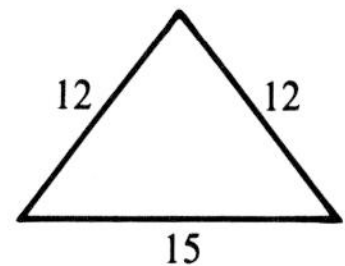

For the similar triangles in Exercises 34–36, find the measure of each side labeled with a letter. See Examples 2 and 3.

34.

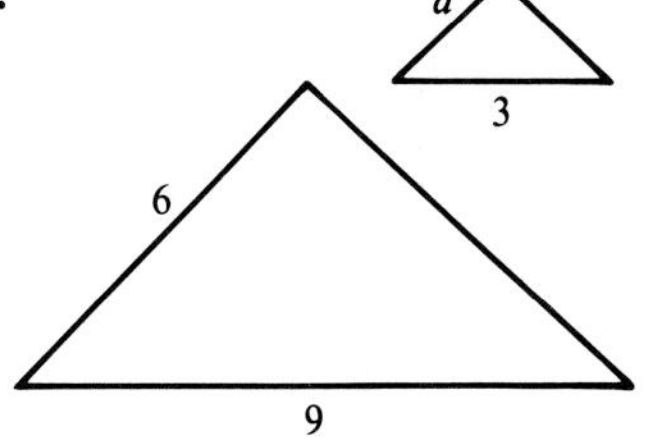

35.

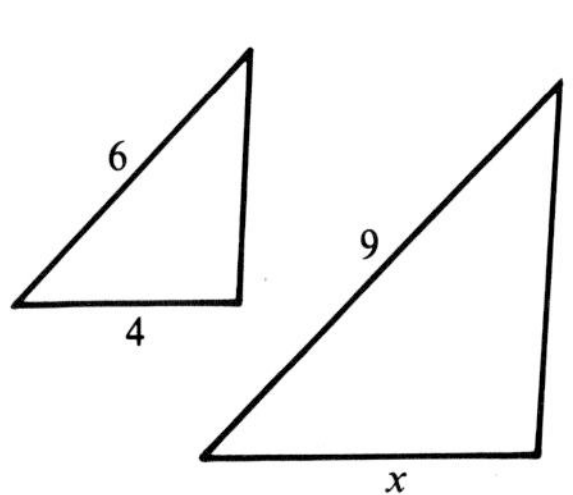

36.

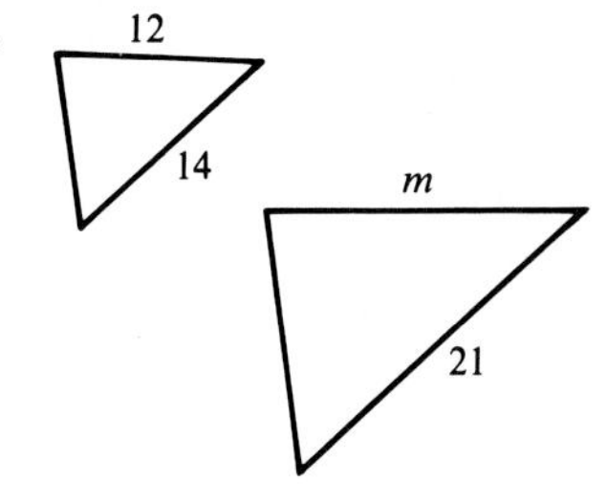

Solve the word problems in Exercises 37–40. See Example 4.

37. On a photograph of a triangular piece of land, the lengths of the three sides measure 4 cm, 5 cm, and 7 cm, respectively. The shortest side of the actual piece of land is 400 m in length. Find the lengths of the other two sides.

38. By drawing lines on a map, a triangle can be formed by the cities of Phoenix, Tucson, and Yuma. On the map, the distance between Phoenix and Tucson is 8 cm, the distance between Phoenix and Yuma is 12 cm, and the distance between Tucson and Yuma is 17 cm. The actual straight-line distance from Phoenix to Yuma is 230 km. Find the distances between the other pairs of cities.

39. A tree casts a shadow 45 m long. At the same time, the shadow cast by a vertical 2-meter stick is 3 m long. Find the height of the tree.

40. The Santa Cruz lighthouse is 14 m tall and casts a shadow 28 m long at 7 P.M. At the same time, the shadow of the lighthouse keeper is 3.5 m long. How tall is she?

Find the missing measurement in each of the following. (Hint: Sketch each of the triangles separately. In the sketch for Exercise 41, the side of length 100 *in the small triangle corresponds to a side of length* 100 + 120 = 220 *in the larger triangle.)*

41.

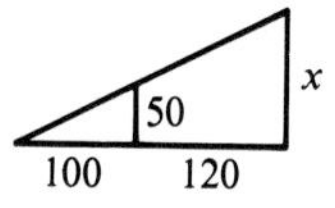

42.

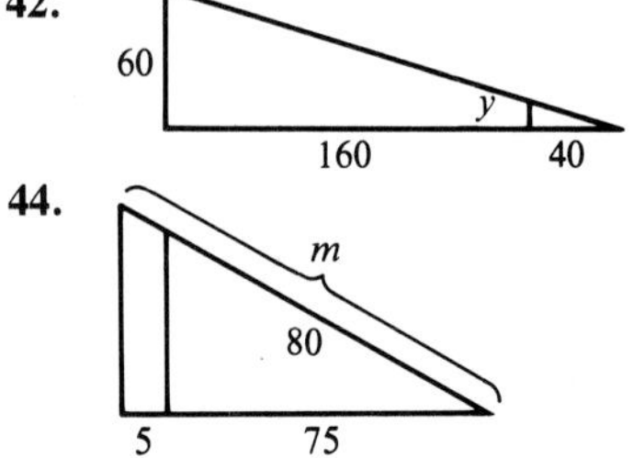

43.

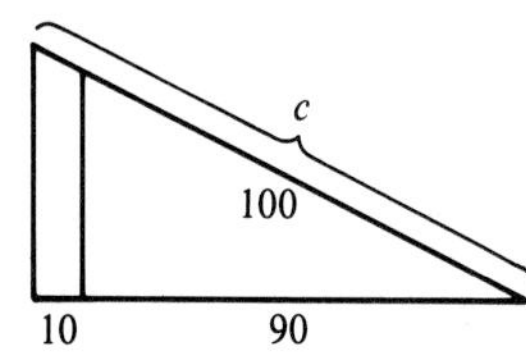

44.

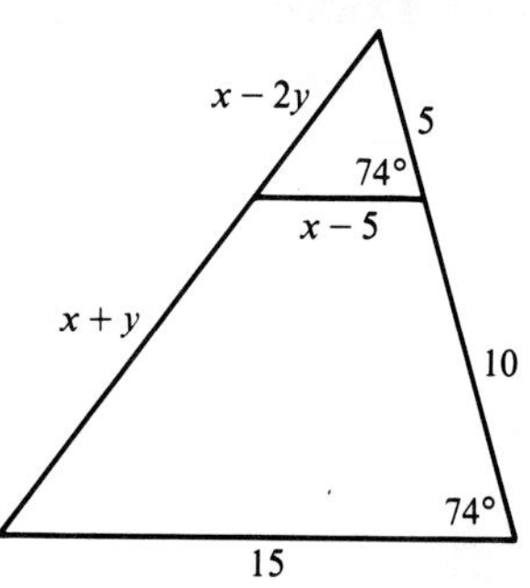

Work each of the following exercises.

45. Two quadrilaterals (four-sided figures) are similar. The lengths of the three shortest sides of the first quadrilateral are 18 cm, 24 cm, and 32 cm. The lengths of the two longest sides of the second quadrilateral are 48 and 60 cm. Find the missing lengths of the sides of these two figures.

46. The photograph shows the construction of the Mount Rushmore head of Lincoln. The man standing beside Lincoln's nose is 6 ft tall. Assume that Lincoln was roughly 6 ft tall and his head 3/4 ft long. Knowing that the carved head of Lincoln is 60 ft tall, find out how tall his entire body would be if it were carved into the mountain.

Find the value of each variable in the following figures.

47.

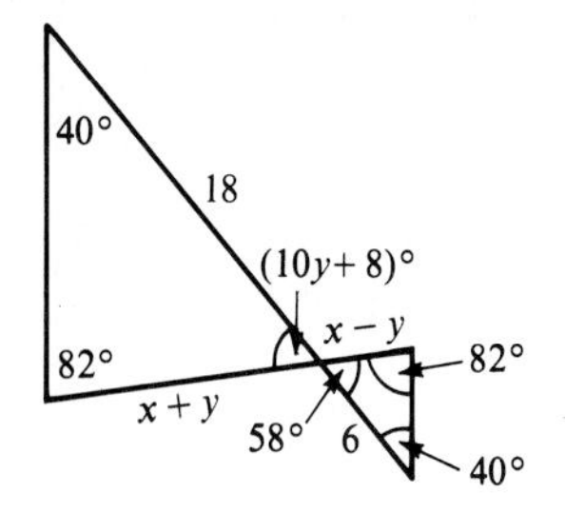

48.

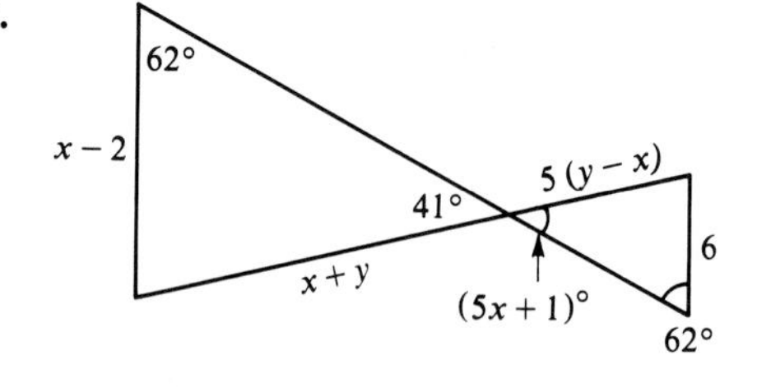

49.

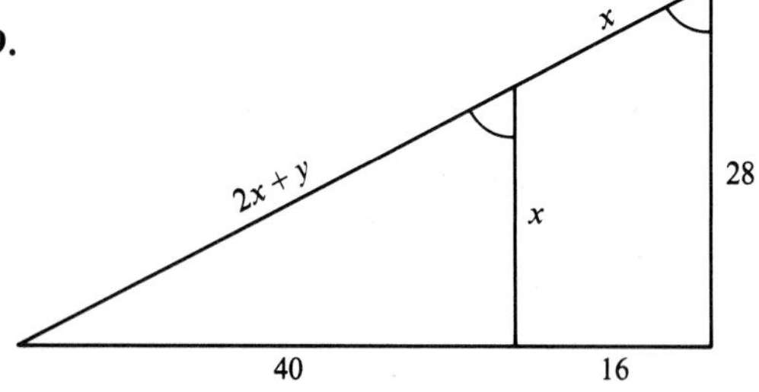

50.

1.4 Definitions of the Trigonometric Functions

The study of trigonometry covers the six trigonometric functions defined in this section. Most sections in the remainder of this book involve at least one of these functions. To define these six basic functions, start with an angle θ (the Greek letter *theta**) in standard position. Choose any point P having coordinates (x, y) on the terminal side of angle θ. (The point P must not be the vertex of the angle.) See Figure 1.24.

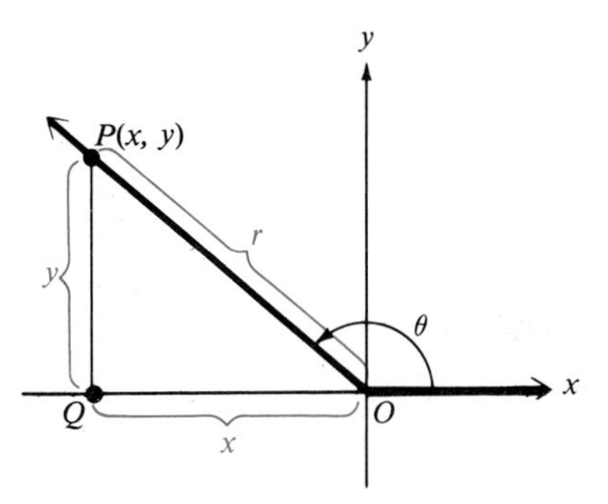

Figure 1.24

A perpendicular from P to the x-axis at point Q determines a triangle having vertices at O, P, and Q. Let the distance from P to O be r. Since distance is never negative, $r > 0$. The six trigonometric functions of angle θ are called **sine, cosine, tangent, cotangent, secant,** and **cosecant.** These functions are defined and abbreviated below.

Trigonometric Functions

Let (x, y) be a point other than the origin on the terminal side of an angle θ in standard position. Let r be the distance from the origin to (x, y). Then

$$\sin\theta = \frac{y}{r} \qquad \csc\theta = \frac{r}{y}$$

$$\cos\theta = \frac{x}{r} \qquad \sec\theta = \frac{r}{x}$$

$$\tan\theta = \frac{y}{x} \qquad \cot\theta = \frac{x}{y}.$$

*Greek letters are often used to name angles. A list of Greek letters appears inside the back cover of the book.

Example 1 The terminal side of an angle α in standard position goes through the point (8, 15). Find the values of the six trigonometric functions of angle α.

Figure 1.25 shows angle α and the triangle formed by dropping a perpendicular from the point (8, 15). The point (8, 15) is 8 units to the right of the y-axis and 15 units above the x-axis, so that $x = 8$ and $y = 15$. Find r with the Pythagorean theorem.

$$r^2 = x^2 + y^2$$
$$r = \sqrt{x^2 + y^2}$$

(Recall: $\sqrt{a}$ represents the nonnegative square root of a.) Substitute the known values, $x = 8$ and $y = 15$.

$$\begin{aligned} r &= \sqrt{8^2 + 15^2} \\ &= \sqrt{64 + 225} \\ &= \sqrt{289} \end{aligned}$$

From a calculator or a square root table, $r = 17$. The values of the six trigonometric functions of angle α can now be found with the definitions given above.

$$\sin\alpha = \frac{y}{r} = \frac{15}{17} \qquad \csc\alpha = \frac{r}{y} = \frac{17}{15}$$
$$\cos\alpha = \frac{x}{r} = \frac{8}{17} \qquad \sec\alpha = \frac{r}{x} = \frac{17}{8}$$
$$\tan\alpha = \frac{y}{x} = \frac{15}{8} \qquad \cot\alpha = \frac{x}{y} = \frac{8}{15}$$

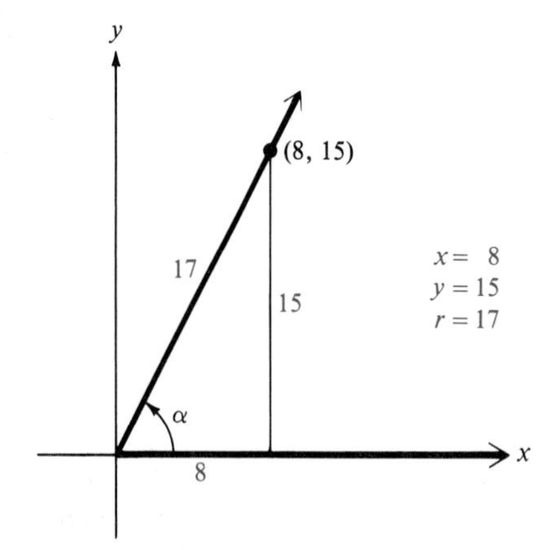

Figure 1.25

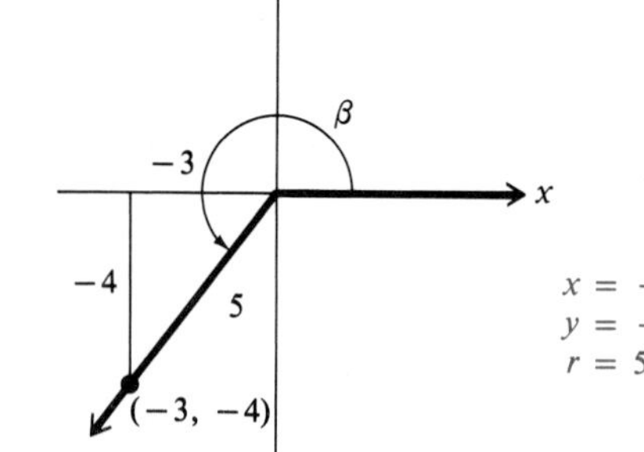

Figure 1.26

Example 2 The terminal side of angle β in standard position goes through $(-3, -4)$. Find the values of the six trigonometric functions of β.

As shown in Figure 1.26, $x = -3$ and $y = -4$. Use the Pythagorean theorem to find that $r = 5$. (Remember: $r > 0$.) Then, by the definitions given above,

$$\sin \beta = \frac{-4}{5} = -\frac{4}{5} \qquad \csc\beta = \frac{5}{-4} = -\frac{5}{4}$$

$$\cos \beta = \frac{-3}{5} = -\frac{3}{5} \qquad \sec \beta = \frac{5}{-3} = -\frac{5}{3}$$

$$\tan \beta = \frac{-4}{-3} = \frac{4}{3} \qquad \cot \beta = \frac{-3}{-4} = \frac{3}{4}.$$

The six trigonometric functions can be found from *any* point on the terminal side of the angle other than the origin. To see why any point may be used, refer to Figure 1.27, which shows an angle θ and two distinct points on its terminal side. Point P has coordinates (x, y) and point P' (read ''P-prime'') has coordinates (x', y'). Let r be the length of the hypotenuse of triangle OPQ, and let r' be the length of the hypotenuse of triangle $OP'Q'$. Since corresponding sides of similar triangles are in proportion,

$$\frac{y}{r} = \frac{y'}{r'},$$

so that $\sin \theta = y/r$ is the same no matter which point is used to find it. A similar result holds for the other five functions.

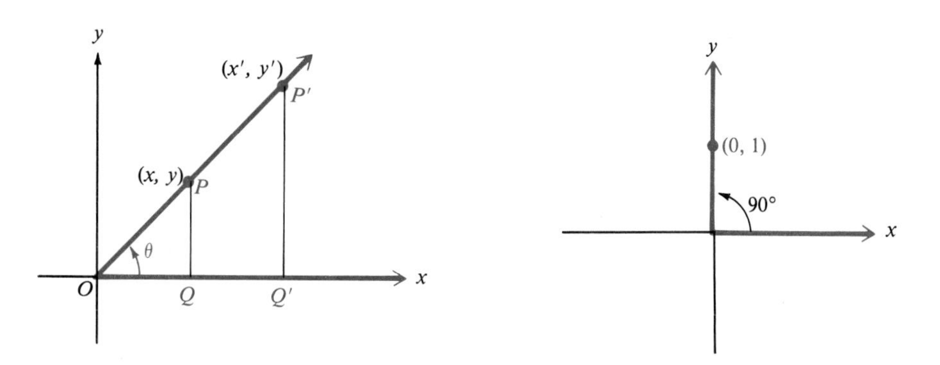

Figure 1.27

Figure 1.28

Example 3 Find the values of the six trigonometric functions for an angle of 90°.

First, select any point on the terminal side of a 90° angle. Let us select the point (0, 1), as shown in Figure 1.28. Here $x = 0$ and $y = 1$. Verify that $r = 1$.

Then, by the definition of the trigonometric functions,

$$\sin 90^\circ = \frac{1}{1} = 1 \qquad \csc 90^\circ = \frac{1}{1} = 1$$

$$\cos 90^\circ = \frac{0}{1} = 0 \qquad \sec 90^\circ = \frac{1}{0} \text{ (undefined)}$$

$$\tan 90^\circ = \frac{1}{0} \text{ (undefined)} \qquad \cot 90^\circ = \frac{0}{1} = 0.$$

(Recall: Division by zero is not defined.) ✦

A similar procedure could be used to find the values of the six trigonometric functions for the quadrantal angles 0°, 180°, 270°, and 360°. These results are summarized in the following table. This table is for reference only; you should either memorize it or be able to reproduce it quickly.

Quadrantal Angles

θ	$\sin \theta$	$\cos \theta$	$\tan \theta$	$\cot \theta$	$\sec \theta$	$\csc \theta$
0°	0	1	0	Undefined	1	Undefined
90°	1	0	Undefined	0	Undefined	1
180°	0	−1	0	Undefined	−1	Undefined
270°	−1	0	Undefined	0	Undefined	−1
360°	0	1	0	Undefined	1	Undefined

The values given in this table can also be found with a calculator that has trigonometric function keys. First, make sure the calculator is set for *degree measure*. Then, for example, cos 90° can be found by entering 90 and pressing the [COS] key.

Enter: 90 [COS] **Display:** 0

Trying to find tan 90° would produce the following.

Enter: 90 [TAN] **Display:** Error message

This error message, which sometimes is a flashing display, shows that tan 90° is undefined. There are no calculator keys for finding the function values of cotangent, secant, or cosecant. The next section shows how to find these function values with a calculator.

1.4 Exercises

Evaluate each of the following. An expression such as $\cot^2 90°$ *means* $(\cot 90°)^2$.

1. $\cos 90° + 3 \sin 270°$

2. $\tan 0° - 6 \sin 90°$

3. $3 \sec 180° - 5 \tan 360°$

4. $4 \csc 270° + 3 \cos 180°$

5. $\tan 360° + 4 \sin 180° + 5 \cos^2 180°$

6. $2 \sec 0° + 4 \cot^2 90° + \cos 360°$

7. $\sin^2 180° + \cos^2 180°$

8. $\sin^2 360° + \cos^2 360°$

9. $\sec^2 180° - 3 \sin^2 360° + 2 \cos 180°$

10. $5 \sin^2 90° + 2 \cos^2 270° - 7 \tan^2 360°$

11. $2 \sec^2 360° - 4 \sin^2 90° + |5 \cos 180°|$

12. $3 \csc^2 270° + 2 \sin^2 270° - |3 \sin 270°|$

13. $-4|\sin 90°| + 3|\cos 180°| + 2|\csc 270°|$

14. $-|\cos 270°| - 2|\sin 90°| + 5|\cos 180°|$

Find the values of the six trigonometric functions for the angles in standard position having the following points on their terminal sides. See Examples 1 and 2.

15. $(-3, 4)$

16. $(-4, -3)$

17. $(5, -12)$

18. $(-12, -5)$

19. $(6, 8)$

20. $(-9, -12)$

21. $(-7, 24)$

22. $(24, 7)$

23. $(0, 2)$

24. $(-4, 0)$

25. $(8, 0)$

26. $(0, -9)$

27. $(1, \sqrt{3})$

28. $(-2\sqrt{3}, -2)$

29. $(5\sqrt{3}, -5)$

30. $(8, -8\sqrt{3})$

31. $(2\sqrt{2}, -2\sqrt{2})$

32. $(-2\sqrt{2}, 2\sqrt{2})$

33. $(\sqrt{5}, -2)$

34. $(-\sqrt{7}, \sqrt{2})$

35. $(-\sqrt{13}, \sqrt{3})$

36. $(-\sqrt{11}, -\sqrt{5})$

37. $(\sqrt{15}, -\sqrt{10})$

38. $(-\sqrt{12}, \sqrt{13})$

39. $(8.7691, -3.2473)$

40. $(-5.1021, 7.6132)$

41. $(-.04716, -.03219)$

42. $(126.89, 104.21)$

43. $(9.713\sqrt{12.4}, -8.765\sqrt{10.2})$

44. $(-5.114\sqrt{286}, 2.1094\sqrt{395})$

Suppose that the point (x, y) *is, in turn, in each of the following quadrants. Decide whether* x *is positive or negative and whether* y *is positive or negative in each case.*

45. I

46. II

47. III

48. IV

Suppose that r *is a positive number and that the point* (x, y) *is in the indicated quadrant. Decide whether the given ratio is positive or negative. (Hint: It may be helpful to draw a sketch.)*

49. II, y/r

50. II, x/r

51. III, y/r

52. III, x/r

53. III, y/x

54. III, x/y

55. IV, x/r

56. IV, y/r

57. IV, y/x

58. IV, x/y

59. III, r/x

60. II, r/y

1.5 Using the Definitions of the Trigonometric Functions

In this section several useful results are derived from the definitions of the trigonometric functions given in the previous section. First, recall the definition of a reciprocal: the **reciprocal** of the nonzero number x is $1/x$. For example, the reciprocal of 2 is 1/2, and the reciprocal of 8/11 is 11/8. There is no reciprocal for 0.

Many calculators have a reciprocal key, usually labeled [1/x]. Using this key gives the reciprocal of any nonzero number entered in the display, as shown in the following examples.

Enter:		**Display:**
5.4	[1/x]	.18518519
.25	[1/x]	4
0	[1/x]	Error message

The definitions of the trigonometric functions in the previous section were written so that functions on the same line are reciprocals of each other. Since sin $\theta = y/r$ and csc $\theta = r/y$,

$$\sin \theta = \frac{1}{\csc \theta} \quad \text{and} \quad \csc \theta = \frac{1}{\sin \theta}.$$

Also, cos θ and sec θ are reciprocals, as are tan θ and cot θ. In summary,

Reciprocal Identities

$$\sin \theta = \frac{1}{\csc \theta} \qquad \csc \theta = \frac{1}{\sin \theta}$$

$$\cos \theta = \frac{1}{\sec \theta} \qquad \sec \theta = \frac{1}{\cos \theta}$$

$$\tan \theta = \frac{1}{\cot \theta} \qquad \cot \theta = \frac{1}{\tan \theta}$$

These formulas, called the **reciprocal identities,** hold for any angle θ that does not lead to a 0 denominator. **Identities** are equations that are true for all meaningful values of the variable. For example, both $(x + y)^2 = x^2 + 2xy + y^2$ and $2(x + 3) = 2x + 6$ are identities. Identities are studied in more detail in chapter 5.

Example 1 Find each function value

(a) cos θ, if sec $\theta = 5/3$

Since cos $\theta = 1/\sec \theta$,

$$\cos \theta = \frac{1}{5/3} = \frac{3}{5}.$$

(b) sin θ, if csc $\theta = -\sqrt{12}/2$

$$\sin \theta = \frac{1}{-\sqrt{12}/2} = \frac{-2}{\sqrt{12}}$$

Remove $\sqrt{12}$ from the denominator (called *rationalizing the denominator*) as shown below.

$$\sin\theta = \frac{-2}{\sqrt{12}} = \frac{-2\sqrt{3}}{\sqrt{12}\cdot\sqrt{3}} = \frac{-2\sqrt{3}}{6} = \frac{-\sqrt{3}}{3}$$ ✤

In the definition of the trigonometric functions, r is the distance from the origin to the point (x, y). Distance is never negative, so $r > 0$. If we choose a point (x, y) in quadrant I, then both x and y will be positive. Since $r > 0$, all six of the fractions used in the definitions of the trigonometric functions will be positive, so that the values of all six functions will be positive in quadrant I.

A point (x, y) in quadrant II has $x < 0$ and $y > 0$. This makes the values of sine and cosecant positive for quadrant II angles, while the other four functions take on negative values. Similar results can be obtained for the other quadrants, as summarized below. Again, these results should be memorized.

Signs of Function Values

θ in quadrant	$\sin\theta$	$\cos\theta$	$\tan\theta$	$\cot\theta$	$\sec\theta$	$\csc\theta$
I	+	+	+	+	+	+
II	+	−	−	−	−	+
III	−	−	+	+	−	−
IV	−	+	−	−	+	−

II
sine and cosecant positive

I
all functions positive

III
tangent and cotangent positive

IV
cosine and secant positive

Example 2 Identify the quadrant (or quadrants) for any angle θ that satisfies $\sin\theta > 0$, $\tan\theta < 0$.

Since $\sin\theta > 0$ in quadrants I and II, while tangent $\theta < 0$ in quadrants II and IV, both conditions are met only in quadrant II. ✤

Figure 1.29 shows an angle θ as it increases in size from near 0° toward 90°. In each case, the value of r is the same. As the size of the angle increases, y increases but never exceeds r, so that $y \leq r$. Dividing both sides by the positive number r gives

$$y \leq r$$
$$\frac{y}{r} \leq 1.$$

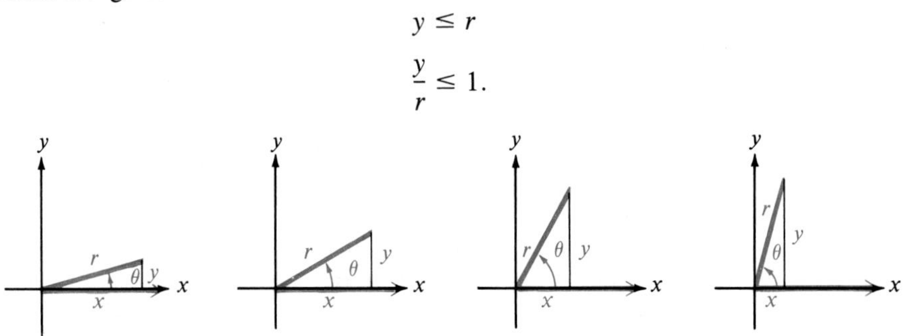

Figure 1.29

In a similar way, angles in the fourth quadrant suggest that

$$-1 \leq \frac{y}{r},$$

so

$$-1 \leq \frac{y}{r} \leq 1.$$

Since $y/r = \sin \theta$,

$$-1 \leq \sin \theta \leq 1$$

for any angle θ. In the same way,

$$-1 \leq \cos \theta \leq 1.$$

The tangent of an angle is defined as y/x. It is possible that $x < y$, that $x = y$, or that $x > y$. For this reason y/x can take on any value at all, so $\tan \theta$ can be any real number, as can $\cot \theta$.

The functions $\sec \theta$ and $\csc \theta$ are reciprocals of the functions $\cos \theta$ and $\sin \theta$, respectively, making

$$\sec \theta \leq -1 \quad \text{or} \quad \sec \theta \geq 1,$$
$$\csc \theta \leq -1 \quad \text{or} \quad \csc \theta \geq 1.$$

In summary, the possible values of the trigonometric functions are as follows.

Values of Trigonometric Functions

For any angle θ for which the indicated functions exist:
1. $-1 \leq \sin \theta \leq 1$ and $-1 \leq \cos \theta \leq 1$;
2. $\tan \theta$ and $\cot \theta$ may be equal to any real number;
3. $\sec \theta \leq -1$ or $\sec \theta \geq 1$ and $\csc \theta \leq -1$ or $\csc \theta \geq 1$.
(Notice that $\sec \theta$ and $\csc \theta$ are *never* between -1 and 1.)

Example 3 Decide whether the following statements are *possible* or *impossible*.

(a) $\sin \theta = \sqrt{8}$

For any value of θ, $-1 \le \sin \theta \le 1$. Since $\sqrt{8} > 1$, there is no value of θ with $\sin \theta = \sqrt{8}$.

(b) $\tan \theta = 110.47$

Tangent can take on any value. Thus, $\tan \theta = 110.47$ is possible.

(c) $\sec \theta = .6$

Since $\sec \theta \le -1$ or $\sec \theta \ge 1$, the statement $\sec \theta = .6$ is impossible. ✣

The six trigonometric functions are defined in terms of x, y, and r, where the Pythagorean theorem shows that $r^2 = x^2 + y^2$ and $r > 0$. With these relationships, knowing the value of only one function and the quadrant in which the angle lies makes it possible to find the values of all six of the trigonometric functions. This process is shown in the next example.

Example 4 Suppose the angle α is in quadrant II and $\sin \alpha = 2/3$. Find the values of the other five functions.

We can choose any point on the terminal side of angle α. For simplicity, choose the point with $r = 3$. Since $\sin \alpha = y/r$,

$$\frac{y}{r} = \frac{2}{3}.$$

If $r = 3$, then y will be 2. To find x, use the result $x^2 + y^2 = r^2$.

$$x^2 + y^2 = r^2$$
$$x^2 + 2^2 = 3^2$$
$$x^2 + 4 = 9$$
$$x^2 = 5$$
$$x = \sqrt{5} \quad \text{or} \quad x = -\sqrt{5}$$

Since α is in quadrant II, x must be negative, as shown in Figure 1.30. Reject $\sqrt{5}$ for x, so $x = -\sqrt{5}$. This puts the point $(-\sqrt{5}, 2)$ on the terminal side of α.

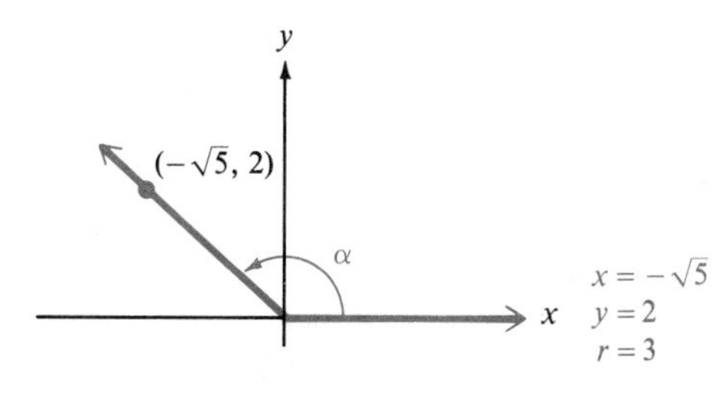

Figure 1.30

Now that the values of x, y, and r are known, the values of the remaining trigonometric functions can be found.

$$\cos\alpha = \frac{x}{r} = \frac{-\sqrt{5}}{3}$$

$$\tan\alpha = \frac{y}{x} = \frac{2}{-\sqrt{5}} = \frac{2\sqrt{5}}{-\sqrt{5}\cdot\sqrt{5}} = \frac{-2\sqrt{5}}{5}$$

$$\cot\alpha = \frac{x}{y} = \frac{-\sqrt{5}}{2}$$

$$\sec\alpha = \frac{r}{x} = \frac{3}{-\sqrt{5}} = \frac{-3\sqrt{5}}{5}$$

$$\csc\alpha = \frac{r}{y} = \frac{3}{2}$$ ✣

As previously mentioned, $x^2 + y^2 = r^2$. Dividing both sides by r^2 gives

$$\frac{x^2}{r^2} + \frac{y^2}{r^2} = \frac{r^2}{r^2},$$

or

$$\left(\frac{x}{r}\right)^2 + \left(\frac{y}{r}\right)^2 = 1.$$

Since $\sin\theta = y/r$ and $\cos\theta = x/r$, this result becomes

$$(\sin\theta)^2 + (\cos\theta)^2 = 1,$$

or, as it is usually abbreviated,

$$\sin^2\theta + \cos^2\theta = 1.$$

Starting with $x^2 + y^2 = r^2$ and dividing through by x^2 gives

$$\frac{x^2}{x^2} + \frac{y^2}{x^2} = \frac{r^2}{x^2}$$

$$1 + \left(\frac{y}{x}\right)^2 = \left(\frac{r}{x}\right)^2$$

$$1 + (\tan\theta)^2 = (\sec\theta)^2$$

or

$$\tan^2\theta + 1 = \sec^2\theta.$$

On the other hand, dividing through by y^2 leads to

$$1 + \cot^2\theta = \csc^2\theta.$$

These three identities are called the **Pythagorean identities.** (The name is used since the Pythagorean theorem is used to get $x^2 + y^2 = r^2$.)

Pythagorean Identities

$$\sin^2\theta + \cos^2\theta = 1 \qquad \tan^2\theta + 1 = \sec^2\theta$$
$$1 + \cot^2\theta = \csc^2\theta$$

Example 5 Find $\sin\alpha$ if $\cos\alpha = -\sqrt{3}/4$ and α is in quadrant II.

Start with $\sin^2\alpha + \cos^2\alpha = 1$, and replace $\cos\alpha$ with $-\sqrt{3}/4$.

$$\sin^2\alpha + \left(-\frac{\sqrt{3}}{4}\right)^2 = 1$$
$$\sin^2\alpha + \frac{3}{16} = 1$$
$$\sin^2\alpha = \frac{13}{16}$$
$$\sin\alpha = \pm\frac{\sqrt{13}}{4}$$

Since α is in quadrant II, $\sin\alpha > 0$, and

$$\sin\alpha = \frac{\sqrt{13}}{4}.$$

1.5 Exercises

Find each function value. See Example 1.

1. $\sin\theta$, if $\csc\theta = 3$
2. $\cos\alpha$, if $\sec\alpha = -2.5$
3. $\cot\beta$, if $\tan\beta = -1/5$
4. $\sin\alpha$, if $\csc\alpha = \sqrt{15}$
5. $\csc\alpha$, if $\sin\alpha = \sqrt{2}/4$
6. $\sec\beta$, if $\cos\beta = -1/\sqrt{7}$
7. $\tan\theta$, if $\cot\theta = -\sqrt{5}/3$
8. $\cot\theta$, if $\tan\theta = \sqrt{11}/5$
9. $\sin\theta$, if $\csc\theta = 1.42716$
10. $\cos\alpha$, if $\sec\alpha = 9.80425$
11. $\tan\alpha$, if $\cot\alpha = .439002$
12. $\csc\theta$, if $\sin\theta = -.376908$

Find the tangent of each of the following angles. See Example 1.

13. $\cot\gamma = 2$
14. $\cot\phi = -3$
15. $\cot\omega = \sqrt{3}/3$
16. $\cot\theta = \sqrt{6}/12$
17. $\cot\alpha = -.01$
18. $\cot\beta = .4$

Find a value of the variable in each of the following.

19. $\cos(6A + 5°) = \dfrac{1}{\sec(4A + 15°)}$
20. $\tan(3B - 4°) = \dfrac{1}{\cot(5B - 8°)}$
21. $\sin(4\theta + 2°)\csc(3\theta + 5°) = 1$
22. $\sec(2\alpha + 6°)\cos(5\alpha + 3°) = 1$
23. $\dfrac{1}{\sin(3\theta - 1°)} = \csc(2\theta + 3°)$
24. $\dfrac{1}{\tan(2k + 1°)} = \cot(4k - 3°)$

Identify the quadrant or quadrants for the angles satisfying the following conditions. See Example 2.

25. $\sin\alpha > 0$, $\cos\alpha < 0$
26. $\cos\beta > 0$, $\tan\beta > 0$
27. $\sec\theta < 0$, $\csc\theta < 0$
28. $\tan\gamma > 0$, $\cot\gamma > 0$
29. $\sin\beta < 0$, $\cos\beta > 0$
30. $\cos\beta > 0$, $\sin\beta > 0$

31. $\tan \omega < 0$, $\cot \omega < 0$ **32.** $\csc \theta < 0$, $\cos \theta < 0$ **33.** $\sin \alpha > 0$

34. $\cos \beta < 0$ **35.** $\tan \theta > 0$ **36.** $\csc \alpha < 0$

Give the signs of the six trigonometric functions for each of the following angles.

37. 74° **38.** 129° **39.** 183° **40.** 298°

41. 302° **42.** 372° **43.** 406° **44.** 412°

45. −82° **46.** −14° **47.** −121° **48.** −208°

Decide whether each of the following statements is possible *or* impossible. *See Example 3.*

49. $\sin \theta = 2$ **50.** $\cos \alpha = -1.001$ **51.** $\tan \beta = .92$ **52.** $\cot \omega = -12.1$

53. $\csc \alpha = 1/2$ **54.** $\sec \alpha = 1$ **55.** $\tan \theta = 1$ **56.** $\sin \alpha = -.82$

57. $\sin \beta + 1 = .6$ **58.** $\sec \omega + 1 = 1.3$ **59.** $\csc \theta - 1 = -.2$ **60.** $\tan \alpha - 4 = 7.3$

61. $\sin \alpha = 1/2$ and $\csc \alpha = 2$ **62.** $\cos \theta = 3/4$ and $\sec \theta = 4/3$

63. $\tan \beta = 2$ and $\cot \beta = -2$ **64.** $\sec \gamma = .4$ and $\cos \gamma = 2.5$

65. $\sin \alpha = 3.251924$ and $\csc \alpha = .3075103$ **66.** $\tan \alpha = 4.67129$ and $\cot \alpha = .214074$

Use the proper Pythagorean identity to find the indicated function value. See Example 5.

67. $\cos \theta$, if $\sin \theta = 2/3$, with θ in quadrant II

68. $\tan \alpha$, if $\sec \alpha = 3$, with α in quadrant IV

69. $\csc \beta$, if $\cot \beta = -1/2$, with β in quadrant IV

70. $\sin \alpha$, if $\cos \alpha = -1/4$, with α in quadrant II

71. $\sec \theta$, if $\tan \theta = \sqrt{7}/3$, with θ in quadrant III

72. $\cos \alpha$, if $\sin \alpha = -\sqrt{5}/7$, with α in quadrant IV

73. $\sin \theta$, if $\cos \theta = .914276$, with θ in quadrant IV

74. $\cot \alpha$, if $\csc \alpha = -3.589142$, with α in quadrant III

Find all the other trigonometric functions for each of the following angles. See Example 4.

75. $\cos \alpha = -3/5$, with α in quadrant III **76.** $\tan \alpha = -15/8$, with α in quadrant II

77. $\sin \beta = 7/25$, with β in quadrant II **78.** $\cot \gamma = 3/4$, with γ in quadrant III

79. $\csc \theta = 2$, with θ in quadrant II **80.** $\tan \beta = \sqrt{3}$, with β in quadrant III

81. $\cot \alpha = \sqrt{3}/8$, with $\sin \alpha > 0$ **82.** $\sin \beta = \sqrt{5}/7$, with $\tan \beta > 0$

83. $\tan \theta = 3/2$, with $\csc \theta = \sqrt{13}/3$ **84.** $\csc \alpha = -\sqrt{17}/3$, with $\cot \alpha = 2\sqrt{2}/3$

85. $\sin \alpha = .164215$, with α in quadrant II **86.** $\cot \theta = -1.49586$, with θ in quadrant IV

87. $\sin \gamma = a$, with γ in quadrant I **88.** $\tan \omega = m$, with ω in quadrant III

Use the definitions of the trigonometric functions to derive each identity.

89. $\tan \theta = \dfrac{\sin \theta}{\cos \theta}$ **90.** $\cot \theta = \dfrac{\cos \theta}{\sin \theta}$

91. Derive the identity $1 + \cot^2\theta = \csc^2\theta$ by dividing $x^2 + y^2 = r^2$ by y^2.

Chapter 1 *Key Concepts*

Types of Angles

Name	*Angle measure*	*Example*
Acute angle	Between 0° and 90°	60°, 82°
Right angle	Exactly 90°	90°
Obtuse angle	Between 90° and 180°	97°, 138°
Straight angle	Exactly 180°	180°

Definitions of the Trigonometric Functions

Let (x, y) be a point other than the origin on the terminal side of an angle θ in standard position. Let r be the distance from the origin to (x, y). Then

$$\sin \theta = \frac{y}{r} \qquad \csc \theta = \frac{r}{y}$$

$$\cos \theta = \frac{x}{r} \qquad \sec \theta = \frac{r}{x}$$

$$\tan \theta = \frac{y}{x} \qquad \cot \theta = \frac{x}{y}.$$

Trigonometric Values for Quadrantal Angles

θ	0°	90°	180°	270°	360°
$\sin \theta$	0	1	0	-1	0
$\cos \theta$	1	0	-1	0	1
$\tan \theta$	0	undefined	0	undefined	0
$\cot \theta$	undefined	0	undefined	0	undefined
$\sec \theta$	1	undefined	-1	undefined	1
$\csc \theta$	undefined	1	undefined	-1	undefined

Chapter 1 *Key Concepts (continued)*

Reciprocal Identities

$$\sin \theta = \frac{1}{\csc \theta} \qquad \csc \theta = \frac{1}{\sin \theta}$$

$$\cos \theta = \frac{1}{\sec \theta} \qquad \sec \theta = \frac{1}{\cos \theta}$$

$$\tan \theta = \frac{1}{\cot \theta} \qquad \cot \theta = \frac{1}{\tan \theta}.$$

Pythagorean Identities

$$\sin^2 \theta + \cos^2 \theta = 1 \qquad \tan^2 \theta + 1 = \sec^2 \theta$$

$$1 + \cot^2 \theta = \csc^2 \theta$$

Signs of Trigonometric Functions

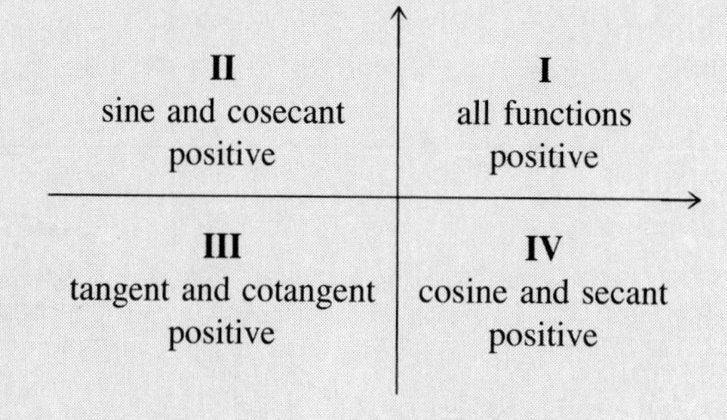

Chapter 1 *Review Exercises*

Find each of the following.

1. $-|-7 - 11| - (-4 + 6)$

2. $|-3| + |-2| - |-5 + 14|$

Find the distance between each of the following pairs of points.

3. $(4, -2)$ and $(1, -6)$

4. $(-6, 3)$ and $(-2, -5)$

5. $(6\sqrt{2}, -\sqrt{3})$ and $(2\sqrt{2}, \sqrt{3})$

6. $(2\sqrt{7}, -\sqrt{11})$ and $(3\sqrt{7}, 2\sqrt{11})$

Let $f(x) = -x^2 + 3x + 2$. *Find each of the following.*

7. $f(0)$

8. $f(-2)$

9. $f(.479)$

10. $f(a)$

Find the domain and range of each of the following. Identify any functions.

11. $y = 9x + 2$

12. $4x - 7y = 1$

13. $y = -(x - 4)^2 + 3$

14. $y = 6 - (2x - 1)^2$

15. $x + 1 = y^2$

16. $x = \sqrt{y + 3}$

17. $y = \sqrt{1 + x^2}$

18. $y = -\sqrt{4 - x^2}$

Find the angles of smallest possible positive measure coterminal with the following angles.

19. $-51°$

20. $-174°$

21. $792°$

22. $1479°$

Work each of the following exercises.

23. A pulley is rotating 320 times per minute. Through how many degrees does a point on the edge of the pulley move in 2/3 second?

24. The propeller of a speedboat rotates 650 times per minute. Through how many degrees will a point on the edge of the propeller rotate in 2.4 second?

Convert decimal degrees to degrees, minutes, seconds, and convert degrees, minutes, seconds to decimal degrees. Round to the nearest second or the nearest thousandth of a degree, as appropriate.

25. $47° 25' 11''$

26. $119° 08' 03''$

27. $74.2983°$

28. $-61.5034°$

29. $183.0972°$

30. $275.1005°$

Find all missing angles in each pair of similar triangles.

31.

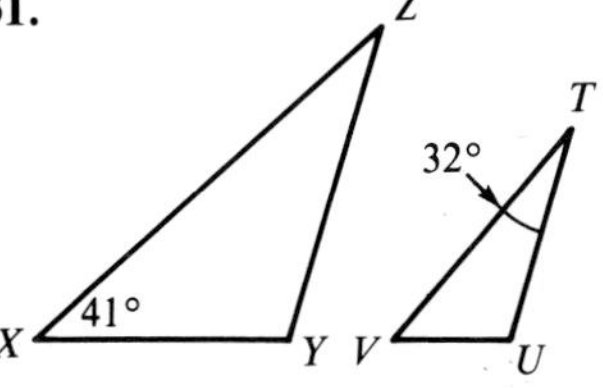

32.

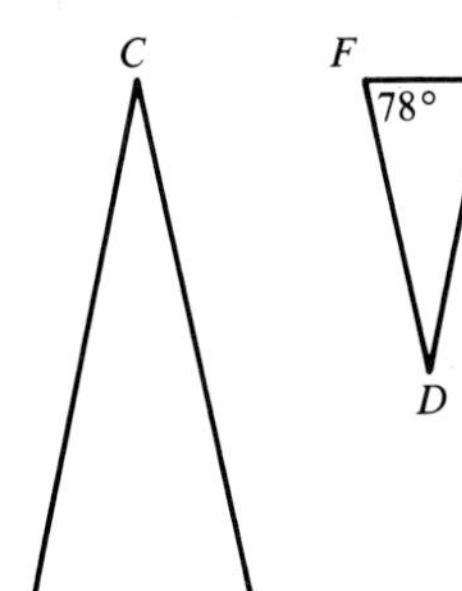

33.

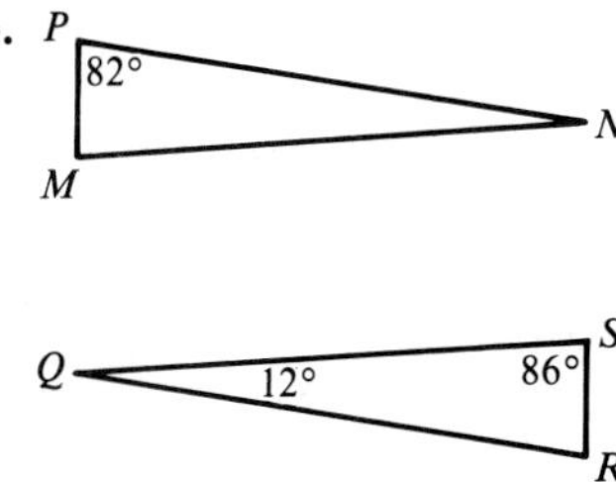

Find the lengths of the missing sides in each pair of similar triangles.

34.

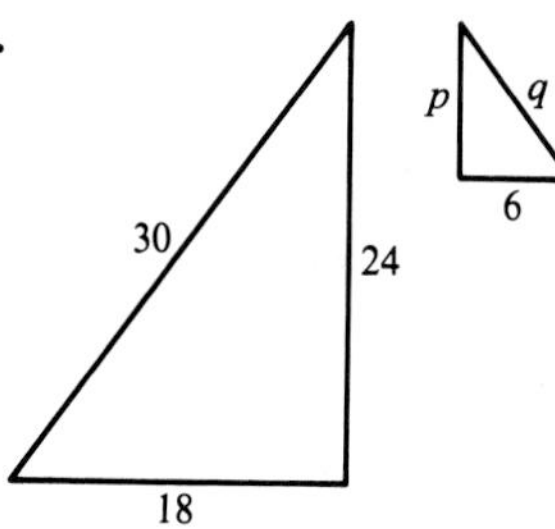

35.

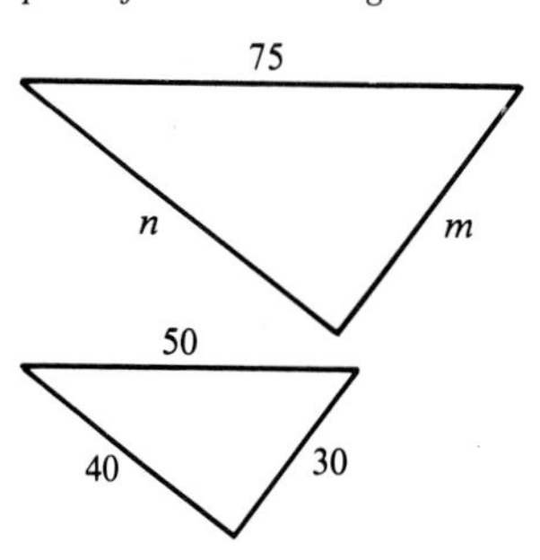

36.

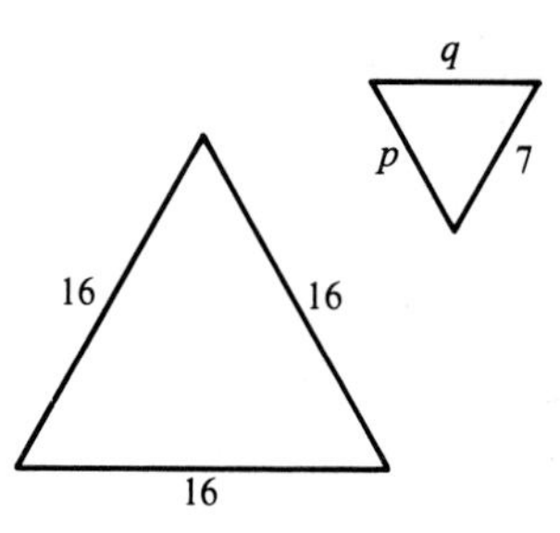

Find the missing measurement in each of the following.

37.

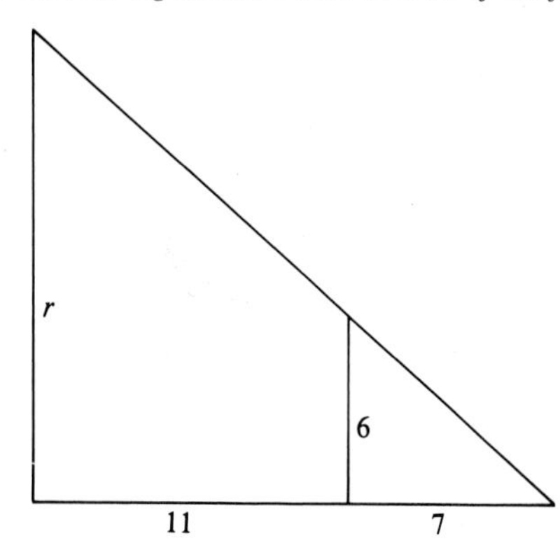

38.

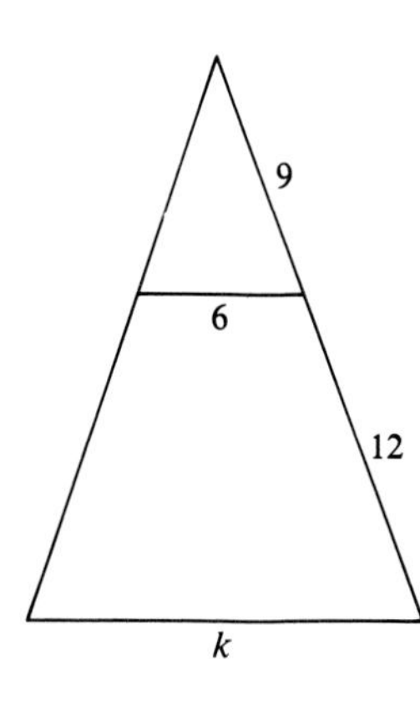

Find x and y.

39.

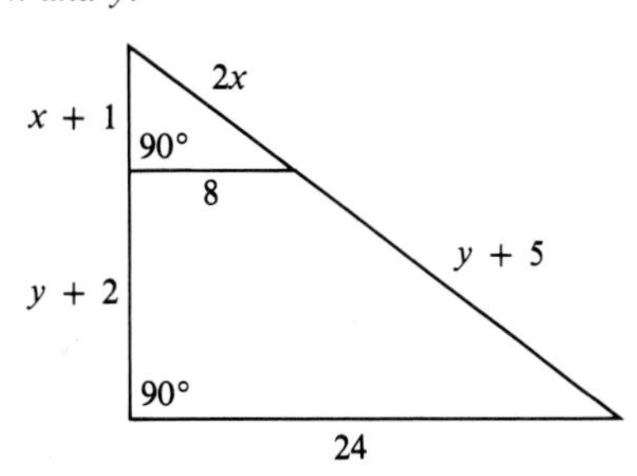

40.

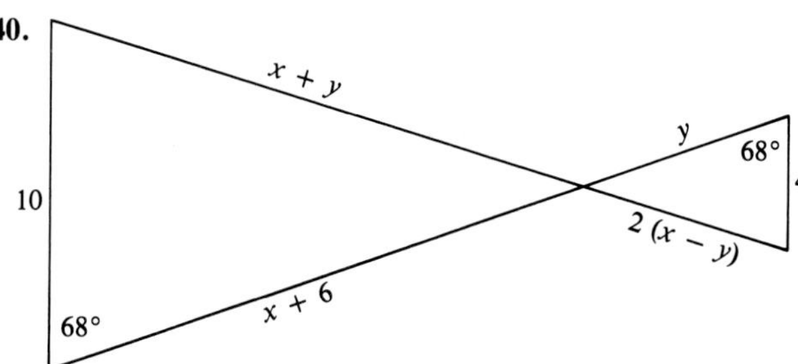

Find the sine, cosine, and tangent of each of the following angles.

41.

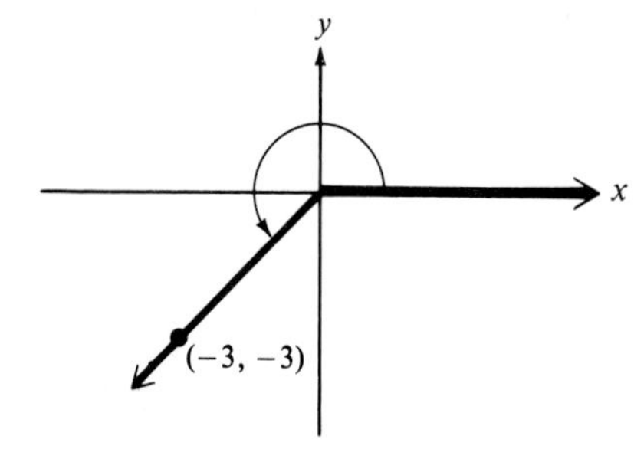

42.

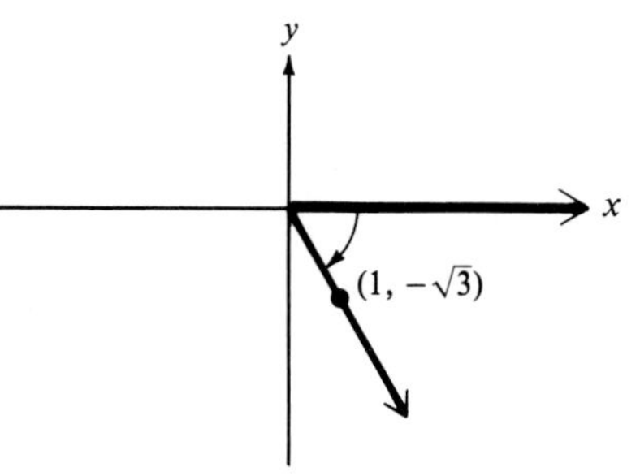

43. $180°$

44. $360°$

Find the values of all the trigonometric functions for angles in standard position having the following points on their terminal sides.

45. $(-8, 15)$

46. $(3, -4)$

47. $(1, -5)$

48. $(9, -2)$

49. $(6\sqrt{3}, -6)$

50. $(-2\sqrt{2}, 2\sqrt{2})$

Evaluate each of the following.

51. $4 \sec 180° - 2 \sin^2 270°$

52. $-\cot^2 90° + 4|\cos 180°|$

53. $5 \sin^2 360° + 5 \cos^3 180° + |-2 \sin 270°|$

54. $-3|\cot 270°| + 6 \sec 180° - 3 \csc^2 270°$

2 *Decide whether each of the following statements is* possible *or* impossible.

55. $\sin \theta = 3/4$ and $\csc \theta = 4/3$

56. $\sec \theta = -2/3$

57. $\tan \theta = 1.4$

58. $\cos \theta = .25$ and $\sec \theta = -4$

25 *Find the values of the other trigonometric functions, given the following.*

59. $\sin \theta = \sqrt{3}/5$ and $\cos \theta < 0$

60. $\cos \gamma = -5/8$, with γ in quadrant III

61. $\tan \alpha = 2$, with α in quadrant III

62. $\sec \beta = -\sqrt{5}$, with β in quadrant II

Use the Pythagorean identities to find the following.

63. $\cos \theta$, if $\sin \theta = -2/5$, with θ in quadrant III

64. $\tan \alpha$, if $\sec \alpha = 5/4$, with α in quadrant IV

Acute Angles and Right Triangles

So far, the definitions of the trigonometric functions have been used only for angles such as 0°, 90°, 180°, or 270°. This chapter extends this work to include finding the values of trigonometric functions of other useful angles, such as 30°, 45°, and 60°. We also discuss the use of tables and calculators for angles whose function values cannot be found directly. The chapter ends with some applications of trigonometry.

2.1 Trigonometric Functions of Acute Angles

Figure 2.1 shows an acute angle A in standard position. The definitions of the trigonometric function values of angle A require x, y, and r. As drawn in Figure 2.1, x and y are the lengths of the two legs of right triangle ABC, and r is the length of the hypotenuse.

The side of length y is called the **side opposite** angle A, and the side of length x is called the **side adjacent** to angle A. The lengths of these sides can be used to replace x and y in the definition of the trigonometric functions, with r replaced with the length of the hypotenuse, to get the following alternative definitions.

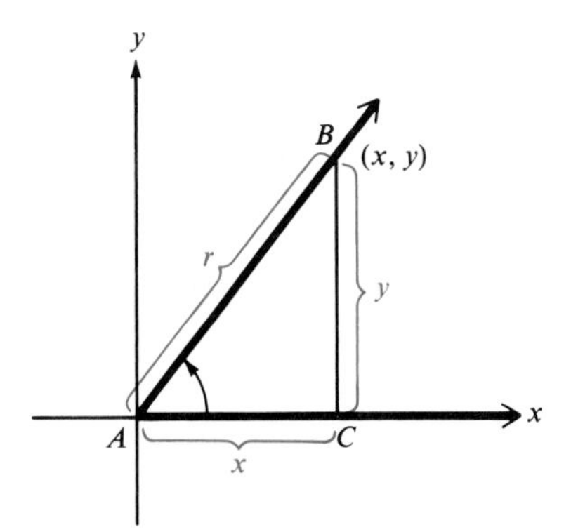

Figure 2.1

Alternative Definitions of Trigonometric Functions

For any acute angle A in standard position.

$$\sin A = \frac{y}{r} = \frac{\text{side opposite}}{\text{hypotenuse}} \qquad \csc A = \frac{r}{y} = \frac{\text{hypotenuse}}{\text{side opposite}}$$

$$\cos A = \frac{x}{r} = \frac{\text{side adjacent}}{\text{hypotenuse}} \qquad \sec A = \frac{r}{x} = \frac{\text{hypotenuse}}{\text{side adjacent}}$$

$$\tan A = \frac{y}{x} = \frac{\text{side opposite}}{\text{side adjacent}} \qquad \cot A = \frac{x}{y} = \frac{\text{side adjacent}}{\text{side opposite}}.$$

Example 1 Find the values of the trigonometric functions for angles A and B in the right triangle in Figure 2.2.

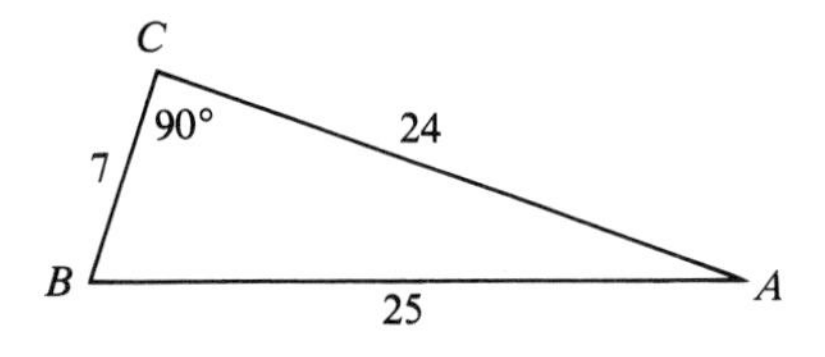

Figure 2.2

The length of the side opposite angle A is 7. The length of the side adjacent to angle A is 24, and the length of the hypotenuse is 25. Using the relationships given above,

$$\sin A = \frac{\text{side opposite}}{\text{hypotenuse}} = \frac{7}{25} \qquad \csc A = \frac{\text{hypotenuse}}{\text{side opposite}} = \frac{25}{7}$$

$$\cos A = \frac{\text{side adjacent}}{\text{hypotenuse}} = \frac{24}{25} \qquad \sec A = \frac{\text{hypotenuse}}{\text{side adjacent}} = \frac{25}{24}$$

$$\tan A = \frac{\text{side opposite}}{\text{side adjacent}} = \frac{7}{24} \qquad \cot A = \frac{\text{side adjacent}}{\text{side opposite}} = \frac{24}{7}$$

The length of the side opposite angle B is 24, while the length of the side adjacent to B is 7, making

$$\sin B = \frac{24}{25} \qquad \tan B = \frac{24}{7} \qquad \sec B = \frac{25}{7}$$

$$\cos B = \frac{7}{25} \qquad \cot B = \frac{7}{24} \qquad \csc B = \frac{25}{24}.$$

In Example 1, you may have noticed that $\sin A = \cos B$, $\cos A = \sin B$, and so on. Such relationships are always true for the two acute angles of a right triangle. Figure 2.3 shows a right triangle with acute angles A and B and a right angle at C. (Whenever we use A, B, and C to name the angles in a right triangle, C will be the right angle.) The length of the side opposite angle A is a, and the length of the side opposite angle B is b. The length of the hypotenuse is c.

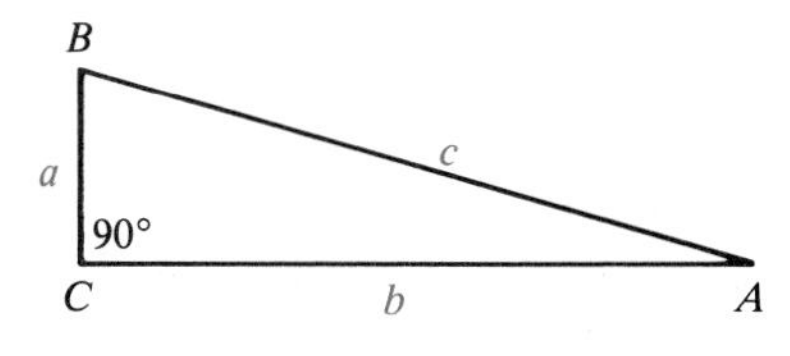

Figure 2.3

By the definitions given above, $\sin A = a/c$. Since $\cos B$ is also equal to a/c,

$$\sin A = a/c = \cos B.$$

In a similar manner,

$$\tan A = a/b = \cot B \qquad \sec A = c/b = \csc B.$$

The sum of the three angles in any triangle is 180°. Since angle C equals 90°, angles A and B must have a sum of $180° - 90° = 90°$. As mentioned in chapter 1, angles with a sum of 90° are called complementary angles. Since angles A and B are complementary and $\sin A = \cos B$, the functions sine and cosine are called **cofunctions.** Also, tangent and cotangent are cofunctions, as are secant and cosecant. Since angles A and B are complementary, $A + B = 90°$, or

$$B = 90° - A,$$

giving

$$\sin A = \cos B = \cos(90° - A).$$

Similar results, called the **cofunction identities,** are true for the other trigonometric functions.

Cofunction Identities

For any acute angle A,

$$\sin A = \cos(90° - A) \qquad \csc A = \sec(90° - A)$$

$$\cos A = \sin(90° - A) \qquad \sec A = \csc(90° - A)$$

$$\tan A = \cot(90° - A) \qquad \cot A = \tan(90° - A).$$

(These identities will be extended to *any* angle A, and not just acute angles, in chapter 5.) It would be wise to *memorize* all the identities presented in this book.

Example 2 Write each of the following in terms of cofunctions.

(a) $\cos 52^\circ$

Since $\cos A = \sin (90^\circ - A)$,

$$\cos 52^\circ = \sin (90^\circ - 52^\circ) = \sin 38^\circ.$$

(b) $\tan 71^\circ = \cot 19^\circ$

(c) $\sec 24^\circ = \csc 66^\circ$ ✥

Example 3 Find a value of θ satisfying each of the following. Assume that all angles involved are acute angles.

(a) $\cos (\theta + 4^\circ) = \sin (3\theta + 2^\circ)$

Since sine and cosine are cofunctions, this equation is true if the sum of the angles is 90°, or

$$\begin{aligned} (\theta + 4^\circ) + (3\theta + 2^\circ) &= 90^\circ \\ 4\theta + 6^\circ &= 90^\circ \\ 4\theta &= 84^\circ \\ \theta &= 21^\circ. \end{aligned}$$

(b) $\tan (2\theta - 18^\circ) = \cot (\theta + 18^\circ)$

$$\begin{aligned} (2\theta - 18^\circ) + (\theta + 18^\circ) &= 90^\circ \\ 3\theta &= 90^\circ \\ \theta &= 30^\circ \end{aligned}$$ ✥

Figure 2.4 shows three right triangles. From left to right, the length of each hypotenuse is the same, but angle A increases in size. As angle A increases in size from 0° to 90°, the length of the side opposite angle A also increases. Since

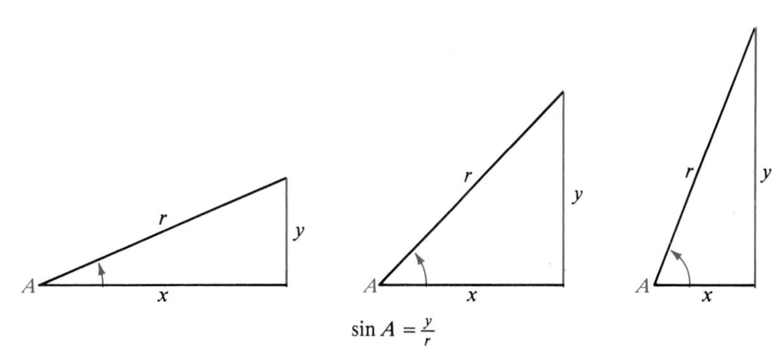

Figure 2.4

$$\sin A = \frac{\text{side opposite}}{\text{hypotenuse}},$$

as angle A increases, the numerator of this fraction also increases, while the denominator is fixed. This means that $\sin A$ *increases* as A increases from 0° to 90°.

As angle A increases from 0° to 90°, the length of the side adjacent to A decreases. Since r is fixed, the ratio x/r will decrease. This ratio gives $\cos A$, showing that the values of cosine *decrease* as the angle measure changes from 0° to 90°. Finally, increasing A from 0° to 90° causes y to increase and x to decrease, making the values of $y/x = \tan A$ increase.

A similar discussion shows that as A increases from 0° to 90°, the values of $\sec A$ increase, while the values of $\cot A$ and $\csc A$ decrease.

Example 4 Tell whether each of the following is *true* or *false*.

(a) $\sin 21° > \sin 18°$

In the interval from 0° to 90°, as the angle increases, so does the sine of the angle, which makes $\sin 21° > \sin 18°$ a true statement.

(b) $\cos 49° \leq \cos 56°$

As the angle increases, the cosine decreases. The given statement $\cos 49° \leq \cos 56°$ is false. ✤

2.1 Exercises

In each of the following exercises, find the values of the six trigonometric functions for angle A. Leave answers as fractions. See Example 1.

1.

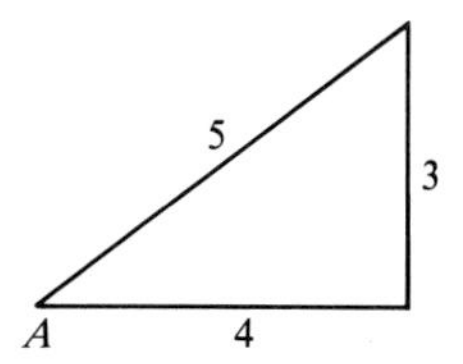

2.

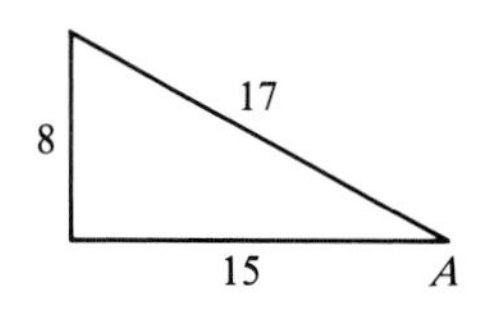

3.

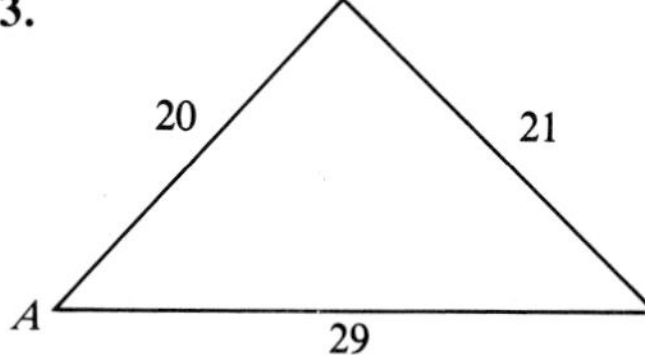

4.

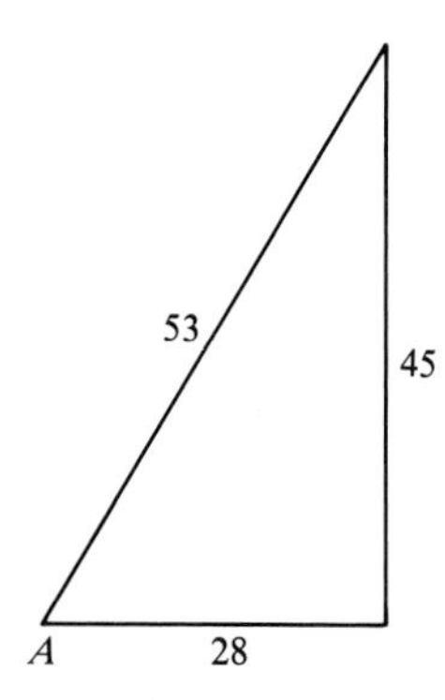

5.

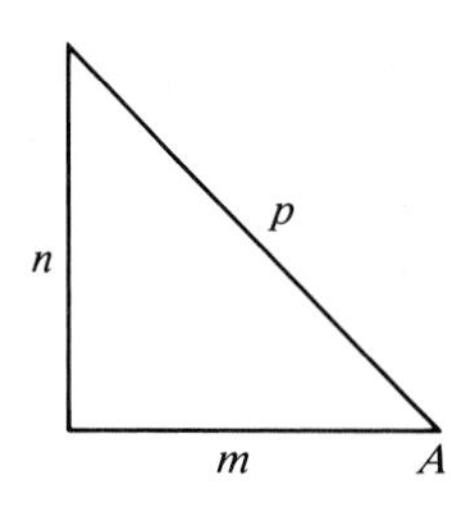

6.

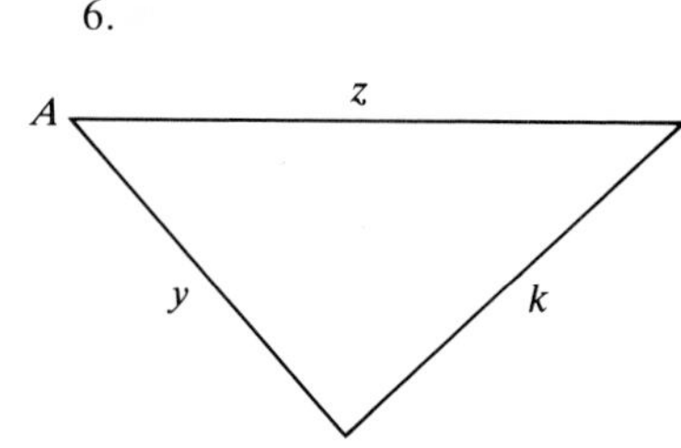

Use a calculator to find the values of the six trigonometric functions for each angle A. Round answers to four decimal places. See Example 1.

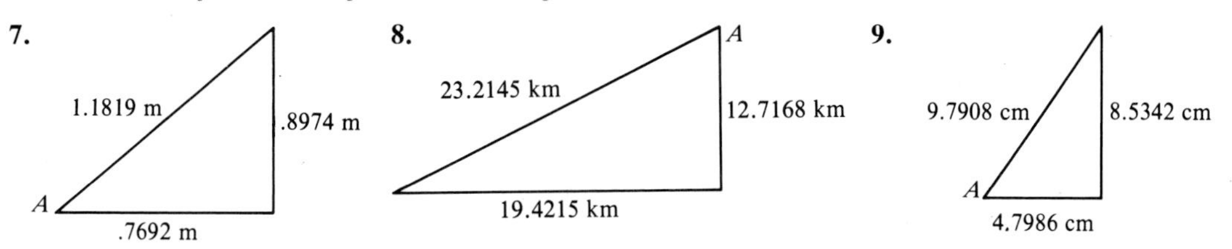

Write each of the following in terms of the cofunction. Assume that all angles are acute angles. See Example 2.

10. cot 73° **11.** tan 50° **12.** sec 39° **13.** csc 47°

14. cos 43° **15.** cos 52° 49′ **16.** sin 38° 29′ **17.** tan 25° 43′

18. sec 75° 58′ **19.** $\sin \gamma$ **20.** $\cot \alpha$ **21.** sec 2*A*

22. tan 3*B* **23.** $\cos (\alpha + 20°)$ **24.** $\cot (\beta - 10°)$

Find a solution for each of the following equations. Assume that all angles are acute angles. See Example 3.

25. $\cos \theta = \sin 2\theta$

26. $\tan \alpha = \cot (\alpha + 10°)$

27. $\sec (\beta + 10°) = \csc (2\beta + 20°)$

28. $\sin (2\gamma + 10°) = \cos (3\gamma - 20°)$

29. $\cot (5\theta + 2°) = \tan (2\theta + 4°)$

30. $\csc (\beta + 6°) = \sec (2\beta + 21°)$

31. $\sin (6A + 2°) = \cos (4A + 8°)$

32. $\tan (3B + 4°) = \cot (5B - 10°)$

33. $\sec \left(\frac{\beta}{2} + 5°\right) = \csc \left(\frac{\beta}{2} + 15°\right)$

34. $\sin \left(\frac{3A}{2} - 5°\right) = \cos \left(\frac{2}{3}A + 30°\right)$

Tell whether each of the following is true *or* false. *See Example 4.*

35. $\tan 28° \leq \tan 40°$

36. $\sin 50° > \sin 40°$

37. $\sec 80° < \sec 82°$

38. $\cot 52° > \cot 58°$

39. $\sin 46° < \cos 46°$
(*Hint:* $\cos 46° = \sin 44°$)

40. $\cos 28° < \sin 28°$

41. $\tan 41° < \cot 41°$

42. $\cot 30° < \tan 40°$

43. $\sin 60° \leq \cos 30°$

44. $\sec 80° \geq \csc 10°$

2.2 Trigonometric Functions of Special Angles

Certain special angles, such as 30°, 45°, and 60°, occur so often in trigonometry and in more advanced mathematics that they deserve careful study. The trigonometric function values for 30° and 60° are found by using a 30°–60° right triangle. Figure 2.5(a) on the next page shows an **equilateral triangle,** a triangle with all sides of equal length. Each angle of such a triangle has a measure of 60°.

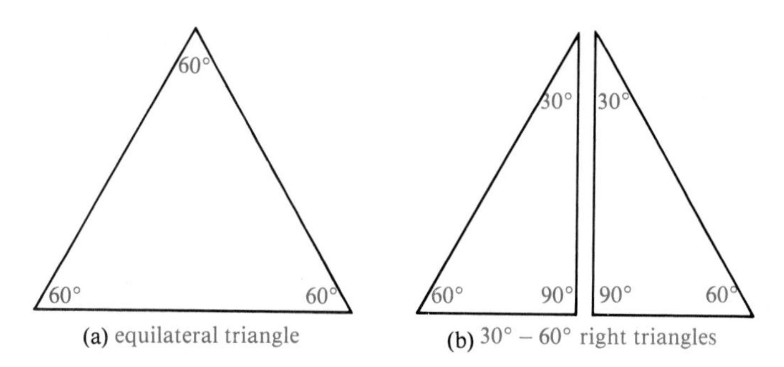

(a) equilateral triangle (b) 30° – 60° right triangles

Figure 2.5

Bisecting one angle of an equilateral triangle leads to two right triangles, each of which has angles of 30°, 60°, and 90°, as shown in Figure 2.5(b). If the hypotenuse of one of these right triangles has a length of 2, then the shortest side will have a length of 1. (Why?) If x represents the length of the medium side, then, by the Pythagorean theorem,

$$\begin{aligned} 2^2 &= 1^2 + x^2 \\ 4 &= 1 + x^2 \\ 3 &= x^2 \\ \sqrt{3} &= x. \end{aligned}$$

The length of the medium side is $\sqrt{3}$. Thus the ratio of the sides opposite the 30°, 90°, and 60° angles respectively is $1:2:\sqrt{3}$.

30°–60° Right Triangle

In a 30°–60° right triangle, the hypotenuse is always twice as long as the shorter leg, and the longer leg has a length that is $\sqrt{3}$ times as long as that of the shorter leg. Also, the shorter leg is opposite the 30° angle, and the longer leg is opposite the 60° angle.

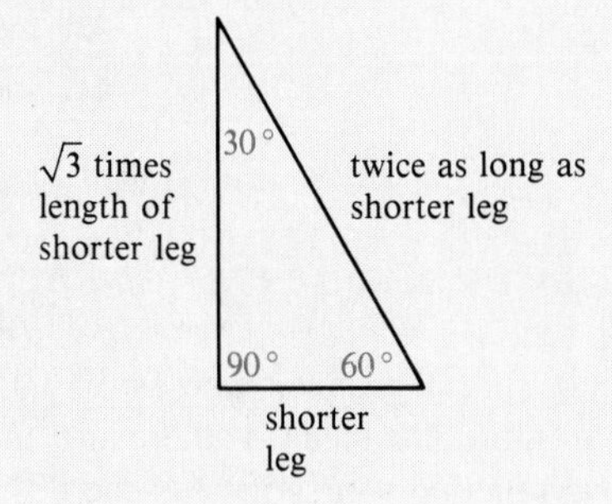

Example 1 The hypotenuse of a 30°–60° right triangle has a length of 7. Find the lengths of the two legs.

As mentioned above, the shorter leg is half the length of the hypotenuse, so the shorter leg has a length of 7/2. Since the longer leg has a length $\sqrt{3}$ times that of the shorter leg, the length of the longer leg is $7\sqrt{3}/2$. ✥

Example 2 Find the trigonometric function values for 30° and 60°.

Figure 2.6 shows a 30°–60° right triangle having a hypotenuse of length 2.

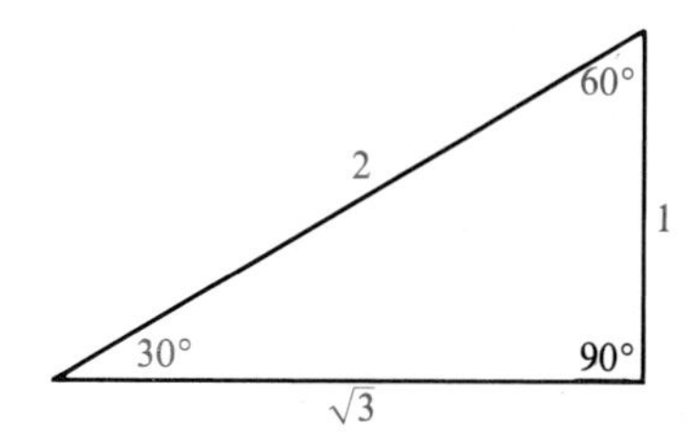

Figure 2.6

As shown above, the side opposite the 30° angle has length 1; that is, for the 30° angle,

$$\text{hypotenuse} = 2, \qquad \text{side opposite} = 1, \qquad \text{side adjacent} = \sqrt{3}.$$

Using the definitions of the trigonometric functions,

$$\sin 30^\circ = \frac{\text{side opposite}}{\text{hypotenuse}} = \frac{1}{2} \qquad \csc 30^\circ = \frac{2}{1} = 2$$

$$\cos 30^\circ = \frac{\text{side adjacent}}{\text{hypotenuse}} = \frac{\sqrt{3}}{2} \qquad \sec 30^\circ = \frac{2}{\sqrt{3}} = \frac{2\sqrt{3}}{3}$$

$$\tan 30^\circ = \frac{\text{side opposite}}{\text{side adjacent}} = \frac{1}{\sqrt{3}} = \frac{\sqrt{3}}{3} \qquad \cot 30^\circ = \frac{\sqrt{3}}{1} = \sqrt{3}.$$

The denominator was rationalized for tan 30° and sec 30°.

In a similar manner,

$$\sin 60^\circ = \frac{\sqrt{3}}{2} \qquad \tan 60^\circ = \sqrt{3} \qquad \sec 60^\circ = 2$$

$$\cos 60^\circ = \frac{1}{2} \qquad \cot 60^\circ = \frac{\sqrt{3}}{3} \qquad \csc 60^\circ = \frac{2\sqrt{3}}{3}. \quad ✥$$

It is important to memorize the twelve *exact* function values found in Example 2. If you have a calculator that finds trigonometric function values at the touch of a key, you may wonder why we spend so much time in finding values for special angles. We do this because a calculator gives only *approximate* values in most cases, while we need *exact* values. For example, tan 30° can be found on a calculator by first setting the machine for *degree measure,* then entering 30 and pressing the TAN key.

Enter: 30 [TAN] **Display:** .57735027,

so that

$$\tan 30° \approx .57735027$$

(the symbol $\approx$ means "is approximately equal to"). Earlier, however, we found the exact value:

$$\tan 30° = \frac{\sqrt{3}}{3}.$$

Since an exact value is often more useful than an approximation, you should know the exact values of all the trigonometric functions for the special angles.

Example 3 Find the values of the trigonometric functions for 210°.

Even though a 210° angle is not an angle of a right triangle, the ideas mentioned above can still be used to find the trigonometric function values for this angle. To do so, draw an angle of 210° in standard position, as shown in Figure 2.7. Choose point P on the terminal side of the angle so that the distance from the origin O to P is 2. By the results from 30°–60° right triangles, the coordinates of point P become $(-\sqrt{3}, -1)$, with $x = -\sqrt{3}$, $y = -1$, and $r = 2$. Then, by the definitions of the trigonometric functions,

$$\sin 210° = -\frac{1}{2} \qquad \tan 210° = \frac{\sqrt{3}}{3} \qquad \sec 210° = -\frac{2\sqrt{3}}{3}$$

$$\cos 210° = -\frac{\sqrt{3}}{2} \qquad \cot 210° = \sqrt{3} \qquad \csc 210° = -2.$$

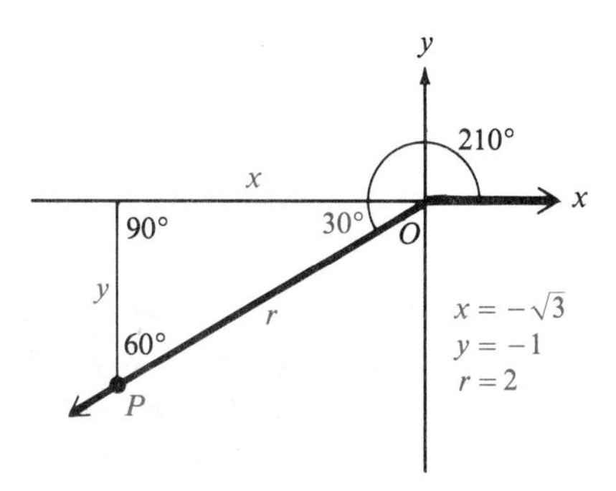

Figure 2.7

The values of the trigonometric functions for 45° can be found by starting with a 45°–45° right triangle, as shown in Figure 2.8. This triangle is isosceles, with two sides of equal length.

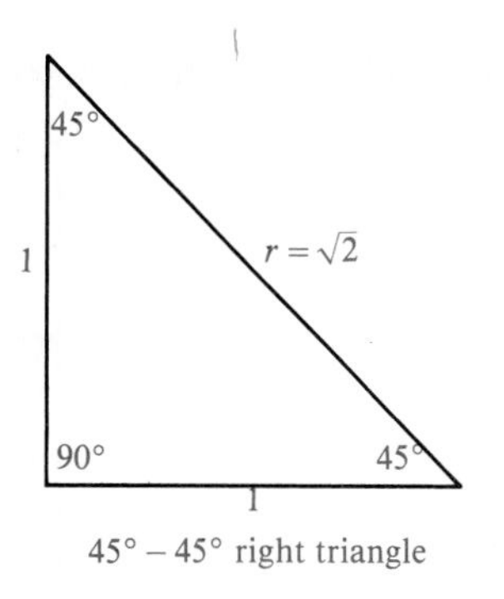

45° – 45° right triangle

Figure 2.8

If the shorter sides each have length 1 and if r represents the length of the hypotenuse, then

$$\begin{aligned} 1^2 + 1^2 &= r^2 \\ 2 &= r^2 \\ \sqrt{2} &= r, \end{aligned}$$

so the ratio of the sides opposite the 45°, 45°, and 90° angles is $1:1:\sqrt{2}$.

45°–45° Right Triangle

In a 45°–45° right triangle, the hypotenuse has a length that is $\sqrt{2}$ times as long as the length of either leg.

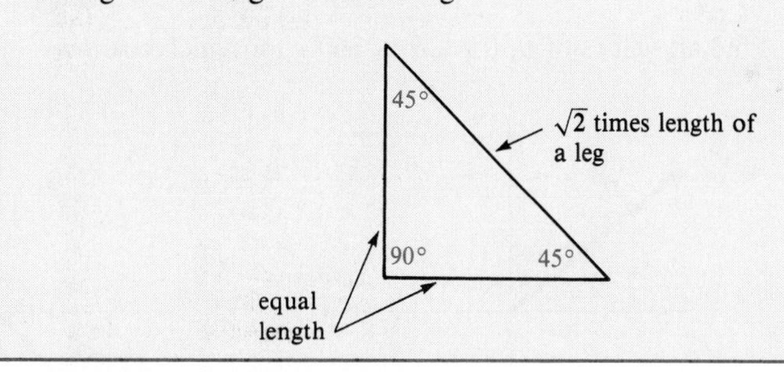

Example 4 Find the values of the trigonometric functions for 45°.

Using the measures indicated on the 45°–45° right triangle in Figure 2.8,

$$\sin 45° = \frac{1}{\sqrt{2}} = \frac{\sqrt{2}}{2} \qquad \tan 45° = \frac{1}{1} = 1 \qquad \sec 45° = \frac{\sqrt{2}}{1} = \sqrt{2}$$

$$\cos 45° = \frac{1}{\sqrt{2}} = \frac{\sqrt{2}}{2} \qquad \cot 45° = \frac{1}{1} = 1 \qquad \csc 45° = \frac{\sqrt{2}}{1} = \sqrt{2}.$$

These six exact function values should also be memorized. ✥

Example 5 Find the values of the trigonometric functions for an angle of 315°.

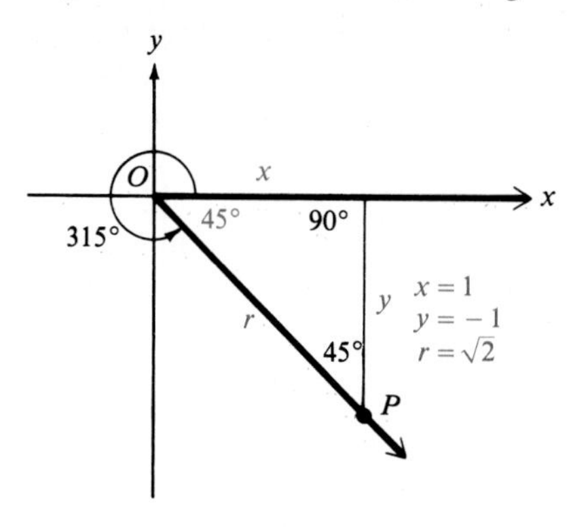

Figure 2.9

Figure 2.9 shows an angle of measure 315°. Choose point P on the terminal side of the angle so that the distance from O to P is $r = \sqrt{2}$. The 45°–45° right triangle in Figure 2.9 shows that $x = 1$ and $y = -1$, with

$$\sin 315^\circ = -\frac{\sqrt{2}}{2} \qquad \tan 315^\circ = -1 \qquad \sec 315^\circ = \sqrt{2}$$

$$\cos 315^\circ = \frac{\sqrt{2}}{2} \qquad \cot 315^\circ = -1 \qquad \csc 315^\circ = -\sqrt{2}.$$

Example 6 Find all values of θ, $0^\circ \le \theta \le 360^\circ$, for which $\cos\theta = -1/\sqrt{2}$.

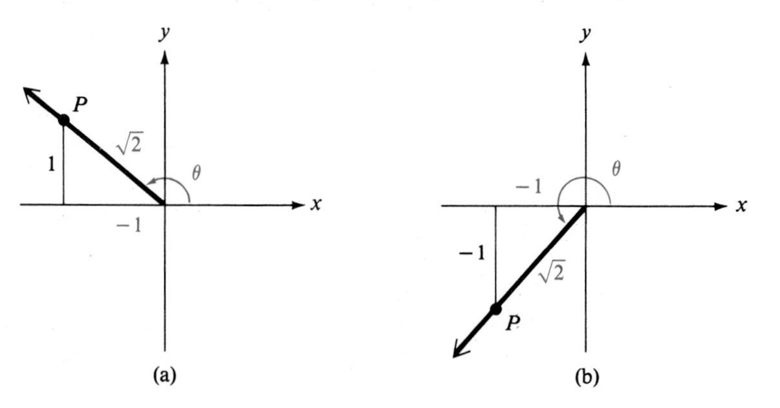

Figure 2.10

Recall from section 1.5 that cosine is negative for angles in quadrants II and III. As shown in Figure 2.10, draw a second quadrant angle θ and a third quadrant angle θ. Since

$$\cos\theta = \frac{x}{r} = \frac{-1}{\sqrt{2}} = -\frac{1}{\sqrt{2}},$$

for each angle, choose a point P on the terminal side so that $OP = r = \sqrt{2}$ and $x = -1$. By the Pythagorean theorem, the length of the third side of each triangle is 1, so both are 45°–45° right triangles. Then, in Figure 2.10(a), $\theta = 180° - 45° = 135°$ and in Figure 2.10(b), $\theta = 180° + 45° = 225°$. ✥

Example 7 Evaluate $\cos 60° + 2\sin^2 60° - \tan^2 30°$.

Since $\cos 60° = 1/2$, $\sin 60° = \sqrt{3}/2$, and $\tan 30° = \sqrt{3}/3$,

$$\cos 60° + 2\sin^2 60° - \tan^2 30° = \frac{1}{2} + 2\left(\frac{\sqrt{3}}{2}\right)^2 - \left(\frac{\sqrt{3}}{3}\right)^2$$
$$= \frac{1}{2} + 2\left(\frac{3}{4}\right) - \frac{3}{9}$$
$$= \frac{5}{3}.$$ ✥

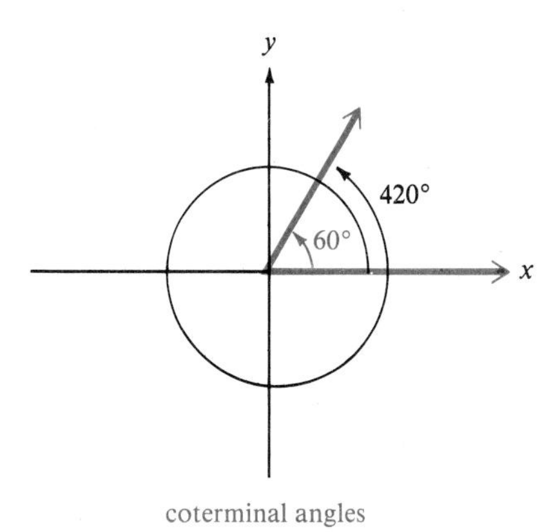

coterminal angles

Figure 2.11

Coterminal Angles Recall from chapter 1 that coterminal angles have the same initial side and the same terminal side. For example, Figure 2.11 shows an angle of 60°. Angles of $360° + 60° = 420°$, or $720° + 60° = 780°$, or $-360° + 60° = -300°$, and so on, are coterminal with this angle. Because the trigonometric functions are defined by a point on the terminal side of the angle (and not by the amount of rotation),

$$\sin 60° = \sin 420° = \sin 780° = \sin(-300°),$$

and so on. Similar statements are true for the other five functions. Rotating the terminal side of an angle through 360° results in a new angle having its terminal side coterminal with that of the original angle. The values of the trigonometric functions of both angles are the same, leading to the following **identities for coterminal angles.**

Identities for Coterminal Angles

For any angle θ,

$$\sin \theta = \sin (\theta + 360^\circ) \qquad \cot \theta = \cot (\theta + 360^\circ)$$
$$\cos \theta = \cos (\theta + 360^\circ) \qquad \sec \theta = \sec (\theta + 360^\circ)$$
$$\tan \theta = \tan (\theta + 360^\circ) \qquad \csc \theta = \csc (\theta + 360^\circ).$$

All these identities could be extended; for example, $\theta + 360^\circ$ could be replaced with $\theta + 720^\circ = \theta + 2 \cdot 360^\circ$, or $\theta + 1080^\circ = \theta + 3 \cdot 360^\circ$. In fact, for each identity in the box, $\theta + 360^\circ$ could be replaced with $\theta + n \cdot 360^\circ$, for any integer n.

Example 8 Evaluate each of the following by first expressing the function in terms of an angle between 0° and 360°.

(a) cos 780°

Add or subtract 360° as many times as necessary so that the final angle is between 0° and 360°. Subtracting 720°, which is $2 \cdot 360^\circ$, gives

$$\begin{aligned} \cos 780^\circ &= \cos (780^\circ - 720^\circ) \\ &= \cos 60^\circ \\ &= 1/2. \end{aligned}$$

(b) tan (−210°)

Adding 360° gives an angle between 0° and 360°.

$$\begin{aligned} \tan (-210^\circ) &= \tan (-210^\circ + 360^\circ) \\ &= \tan 150^\circ \\ &= -\sqrt{3}/3 \end{aligned}$$

2.2 Exercises

Find the exact values of the six trigonometric functions for each of the following angles. Do not use tables or a calculator. See Examples 2–5 and 8.

1. 120°	**2.** 135°	**3.** 150°	**4.** 225°
5. 240°	**6.** 300°	**7.** 330°	**8.** 390°
9. 420°	**10.** 495°	**11.** 510°	**12.** 570°
13. 750°	**14.** 1320°	**15.** 1500°	**16.** 2670°
17. −390°	**18.** −510°	**19.** −1020°	**20.** −1290°

Complete the following table. Do not use tables or a calculator. See Examples 2–5.

	θ	$\sin\theta$	$\cos\theta$	$\tan\theta$	$\cot\theta$	$\sec\theta$	$\csc\theta$
21.	30°	$1/2$	$\sqrt{3}/2$	____	____	$2\sqrt{3}/3$	2
22.	45°	____	____	1	1	____	____
23.	60°	____	$1/2$	$\sqrt{3}$	____	2	____
24.	120°	$\sqrt{3}/2$	____	$-\sqrt{3}$	____	____	$2\sqrt{3}/3$
25.	135°	$\sqrt{2}/2$	$-\sqrt{2}/2$	____	____	$-\sqrt{2}$	$\sqrt{2}$
26.	150°	____	$-\sqrt{3}/2$	$-\sqrt{3}/3$	____	____	2
27.	210°	$-1/2$	____	$\sqrt{3}/3$	$\sqrt{3}$	____	-2
28.	240°	$-\sqrt{3}/2$	$-1/2$	____	____	-2	$-2\sqrt{3}/3$

Evaluate each of the following. See Example 7.

29. $\sin^2 120° + \cos^2 120°$

30. $\sin^2 225° + \cos^2 225°$

31. $2\tan^2 120° + 3\sin^2 150° - \cos^2 180°$

32. $\cot^2 135° - \sin 30° + 4\tan 45°$

33. $\sin^2 225° - \cos^2 270° + \tan 60°$

34. $\cot^2 90° - \sec^2 180° + \csc^2 135°$

35. $\cos^2 60° + \sec^2 150° - \csc^2 210°$

36. $\cot^2 135° + \tan^4 60° - \sin^4 180°$

37. $\sec 30° - \sin 60° + \cos 210°$

38. $\cot 30° + \tan 60° - \sin 240°$

Tell whether each of the following is true *or* false.

39. $\sin 30° + \sin 60° = \sin(30° + 60°)$

40. $\sin(30° + 60°) = \sin 30° \cdot \cos 60° + \sin 60° \cdot \cos 30°$

41. $\cos 60° = 2\cos^2 30° - 1$

42. $\cos 60° = 2\cos 30°$

43. $\sin 120° = \sin 150° - \sin 30°$

44. $\sin 210° = \sin 180° + \sin 30°$

45. $\sin 120° = \sin 180° \cdot \cos 60° - \sin 60° \cdot \cos 180°$

46. $\cos 300° = \cos 240° \cdot \cos 60° - \sin 240° \cdot \sin 60°$

47. $\cos 150° = \cos 120° \cdot \cos 30° - \sin 120° \cdot \sin 30°$

48. $\sin 120° = 2\sin 60° \cdot \cos 60°$

Find all values of the angle θ, *when* $0° \le \theta < 360°$, *for which the following are true. See Example 6.*

49. $\sin\theta = \frac{1}{2}$

50. $\cos\theta = \frac{\sqrt{3}}{2}$

51. $\tan\theta = \sqrt{3}$

52. $\sec\theta = \sqrt{2}$

53. $\cos\theta = -\frac{1}{2}$

54. $\cot\theta = -\frac{\sqrt{3}}{3}$

55. $\sin\theta = -\frac{\sqrt{3}}{2}$

56. $\cos\theta = -\frac{\sqrt{2}}{2}$

57. $\tan\theta = -1$

58. $\cot\theta = -\sqrt{3}$

59. $\cos\theta = 0$

60. $\sin\theta = 1$

61. $\cot\theta$ is undefined

62. $\csc\theta$ is undefined

Use a calculator with sine and tangent keys to find each of the following. (Be sure to set the machine for degree measure.) Then explain why these answers are not really "correct" if the exact value has been requested.

63. sin 45°

64. tan 60°

Find the exact value of each labeled part in each of the following figures.

65.

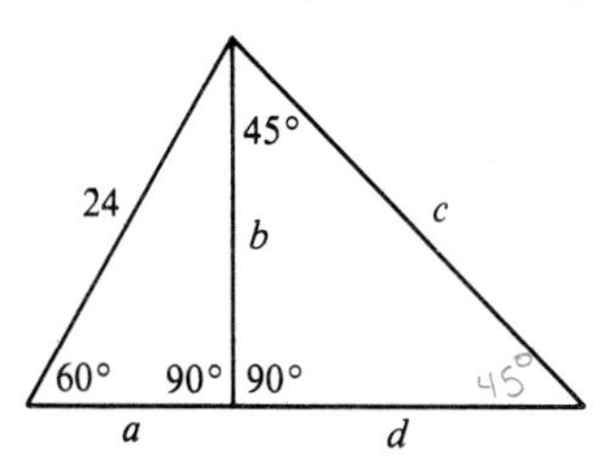

66.

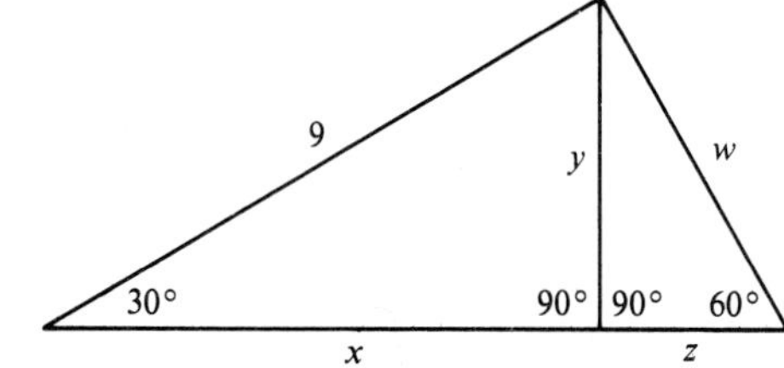

67.

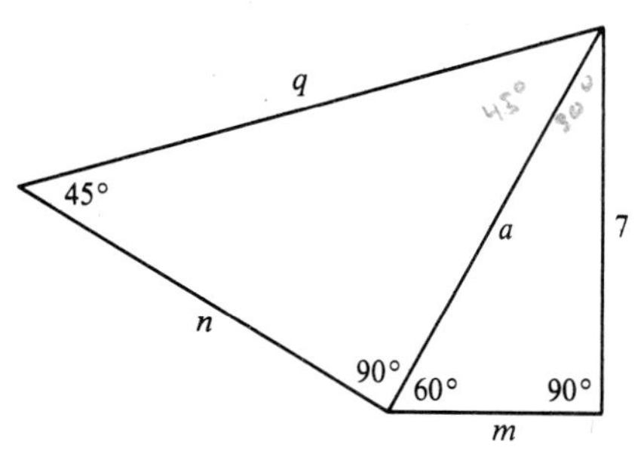

68.

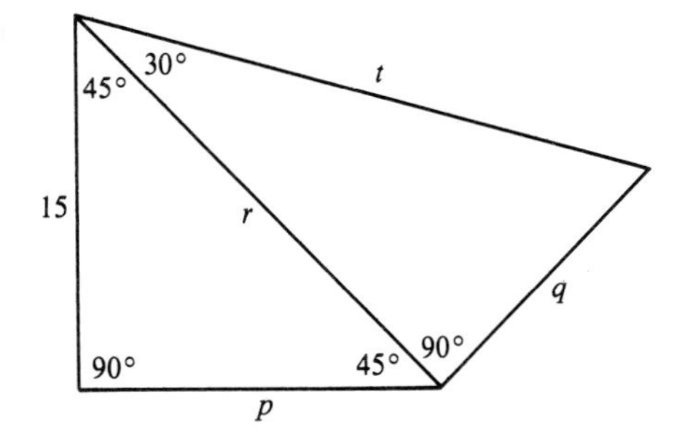

Find a formula for the area of each figure in terms of s.

69.

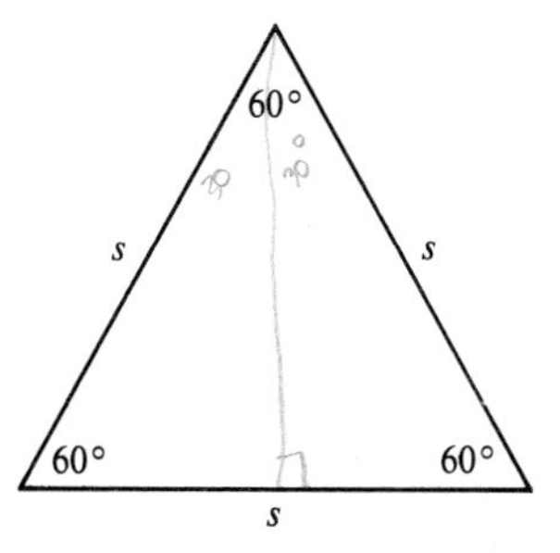

70.

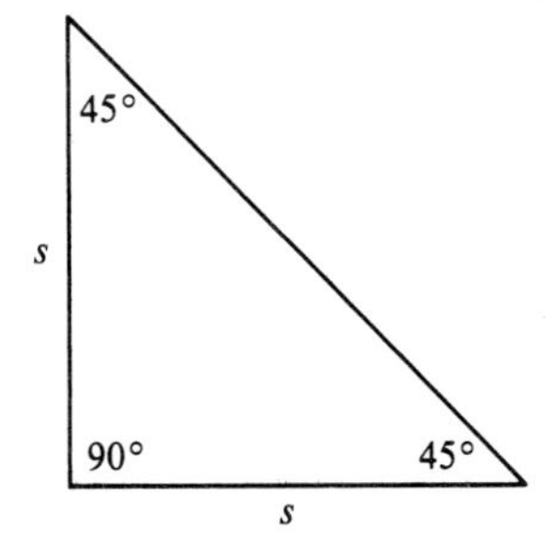

2.3 Reference Angles and Trigonometric Tables

Trigonometric function values for angles other than the quadrantal angles, 0°, 90°, 180°, 270°, and so on, and the special angles, 30°, 45°, 60°, and so on, must be found by using a calculator or a table.

Example 1 Use a calculator to find each value.

(a) sin 49° 12′

Convert 49° 12′ to decimal degrees, as explained in chapter 1.

$$49^\circ\ 12' = 49\frac{12^\circ}{60} = 49.2^\circ$$

(As mentioned earlier, some calculators will do this conversion automatically.) Then press the SIN key.

Enter: 49.2 SIN **Display:** .75699506

To eight decimal places,

$$\sin 49^\circ\ 12' = .75699506.$$

(b) $\sec 97^\circ\ 58'\ 37'' = \sec\left(97 + \frac{58}{60} + \frac{37}{3600}\right)^\circ$

$$= \sec 97.976944^\circ$$

Calculators do not have secant keys. However,

$$\sec\theta = \frac{1}{\cos\theta}$$

for all angles θ when $\cos\theta$ is not 0. So, find sec 97.976944° by pressing the cosine key and then taking the reciprocal, by pressing the 1/x key.

Enter: 97.976944 COS 1/x **Display:** −7.2059291

(c) cot 51.4283°

This angle is already in decimal degrees. Use the identity $\cot\theta = 1/\tan\theta$.

Enter: 51.4283 TAN 1/x **Display:** .79748114

(d) sin (−246°)

Enter 246, and then use the key that changes the sign of the displayed number. (This key is often labeled +/− .)

Enter: 246 +/− SIN **Display:** .91354546. ✥

It is a common error to evaluate the values of trigonometric functions with a calculator not set for degree measure. One way to avoid this problem is to get in the habit of always starting work as follows.

Enter: 90 SIN

If the displayed answer is 1, the calculator is set for degree measure; otherwise it is not.

Tables If your calculator does not have sine, cosine, and tangent keys, the values of trigonometric functions can be found by using the tables in the back of this book. Table 2 is used to find four-digit approximations of the trigonometric function values for angles measured to the nearest 10 minutes. A portion of Table 2 is shown below. (Note: The column in Table 2 headed "radians" will be used in chapter 3.)

Table 2 (Portion)

θ (degrees)	θ (radians)	$\sin\theta$	$\cos\theta$	$\tan\theta$	$\cot\theta$	$\sec\theta$	$\csc\theta$		
36°00′	.6283	.5878	.8090	.7265	1.376	1.236	1.701	.9425	**54°00′**
10	.6312	.5901	.8073	.7310	1.368	1.239	1.695	.9396	50
20	.6341	.5925	.8056	.7355	1.360	**1.241**	1.688	.9367	**40**
36°30′	.6370	.5948	.8039	.7400	1.351	1.244	1.681	.9338	**53°30′**
40	.6400	**.5972**	.8021	.7445	1.343	1.247	1.675	.9308	20
50	.6429	.5995	.8004	.7490	1.335	1.249	1.668	.9279	10
37°00′	.6458	.6018	.7986	.7536	1.327	1.252	1.662	.9250	**53°00′**
		$\cos\theta$	$\sin\theta$	$\cot\theta$	$\tan\theta$	$\csc\theta$	$\sec\theta$	θ (radians)	θ (degrees)

Example 2 Use the given portion of Table 2, when appropriate, to find the following.

(a) sin 36° 30′

For angles between 0° and 45°, read down the left column of the table and use the function names at the top of the table. Doing this here gives

$$\sin 36^\circ\, 30' = .5948.$$

(b) csc 53° 40′

For angles between 45° and 90°, find the angle in the column at the right. Use the function names at the bottom. Notice that 53° 40′ is above the entry for 53°.

$$\csc 53^\circ\, 40' = 1.241$$

(c) $\tan 82^\circ\, 00' = 7.115$

(d) $\csc 14^\circ\, 10' = 4.086$ ✥

It is not practical to give tables that can be used for all of the infinite number of angle measures that exist. For this reason, most tables give values only for angles between 0° and 90°. Function values for an angle having a larger or smaller measure must be calculated using an angle in this range. As we saw in the previous section, by adding or subtracting multiples of 360°, the values of the trigonometric functions can be found by considering only angles from 0° to 360°.

Angles from 0° to 90° can be read directly from Table 2. Angles from 90° to 360° require a *reference angle*. A **reference angle,** for an angle θ, written θ', is the positive acute angle made by the terminal side of angle θ and the x-axis.

Figure 2.12 shows angles θ in quadrants II, III, and IV, respectively, with the reference angle θ' also shown. (In quadrant I, θ and θ' are the same.)

A very common error is to find the reference angle with reference to the y-axis. The reference angle is always found with reference to the x-axis.

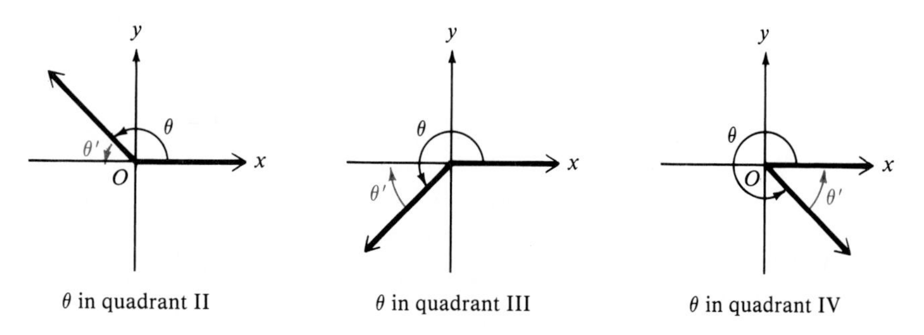

Figure 2.12

Example 3 Find the reference angles for the following three angles.

(a) 218°

As shown in Figure 2.13, the positive acute angle made by the terminal side of this angle and the x-axis is $218° - 180° = 38°$. For $\theta = 218°$, the reference angle $\theta' = 38°$.

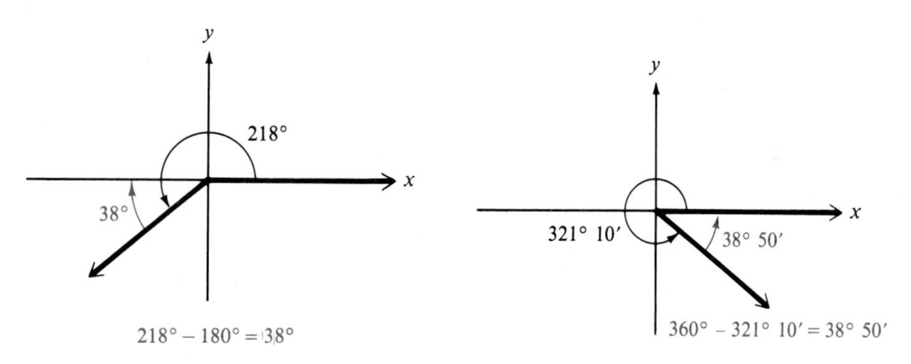

Figure 2.13

Figure 2.14

(b) 321° 10′

As shown in Figure 2.14, the positive acute angle made by the terminal side of this angle and the x-axis is $360° - 321°\,10'$. Write 360° as 359° 60′ so that the reference angle is

$$359°\,60' - 321°\,10' = 38°\,50'.$$

Note that an angle of $-38°\,50'$, which has the same terminal ray as $321°\,10'$, also has a reference angle of $38°\,50'$.

(c) 1387°

First find a coterminal angle between 0° and 360°. Divide 1387° by 360° to get a quotient of about 3.9. Begin by subtracting 360° three times (because of the 3 in 3.9):

$$1387° - 3 \cdot 360° = 307°.$$

The reference angle for 307° is $360° - 307° = 53°$. ✣

The preceding examples suggest the following table for finding the reference angle θ' for any angle θ between 0° and 360°.

Reference Angles

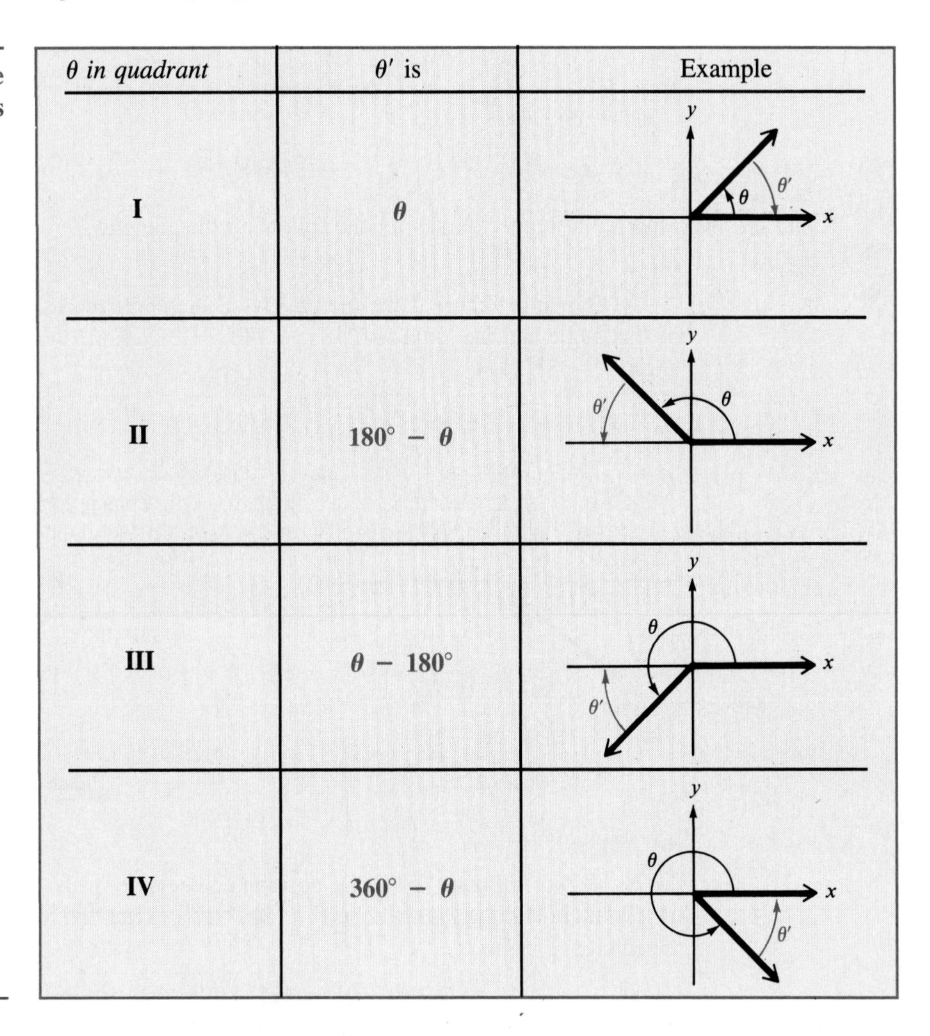

θ *in quadrant*	θ' is	Example
I	θ	
II	$180° - \theta$	
III	$\theta - 180°$	
IV	$360° - \theta$	

Figure 2.15 shows an angle θ in quadrant II and related angle θ' in quadrant I, drawn so that θ' is in standard position. Point P, with coordinates (x_1, y_1), has been located on the terminal side of angle θ. Let r be the distance from O to P.

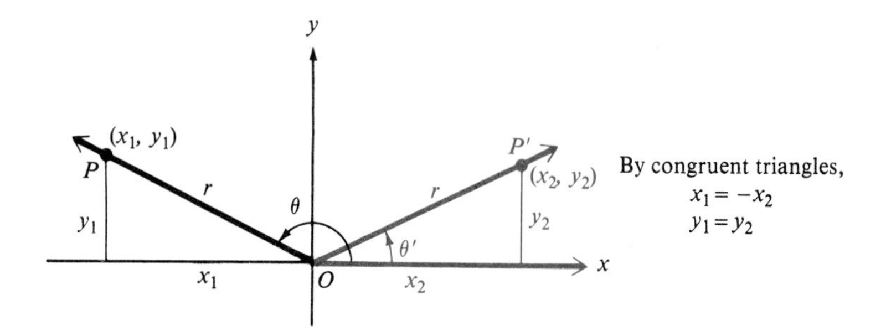

Figure 2.15

Choose point P' on the terminal side of angle θ' so that the distance from O to P' is also r. Let P' have coordinates (x_2, y_2). By congruent triangles,

$$x_1 = -x_2 \qquad \text{and} \qquad y_1 = y_2,$$

making

$$\sin \theta = \frac{y_1}{r} = \frac{y_2}{r} = \sin \theta',$$

$$\cos \theta = \frac{x_1}{r} = \frac{-x_2}{r} = -\cos \theta',$$

$$\tan \theta = \frac{y_1}{x_1} = \frac{y_2}{-x_2} = -\tan \theta',$$

and so on. These results show that the values of the trigonometric functions of the reference angle θ' are the same as those of angle θ, except perhaps for signs. We have shown the truth of this statement for an angle θ in quadrant II; similar results can be established for angles in the other quadrants.

For example, if $\theta = 120°$, as shown in Figure 2.16, then $\theta' = 60°$, and

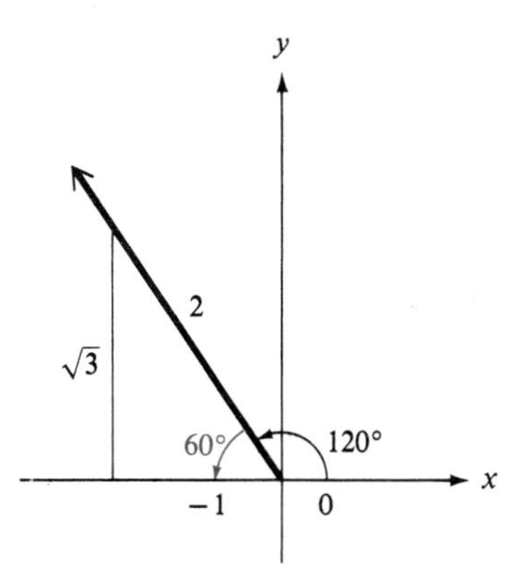

Figure 2.16

$$\sin 120° = \sin 60° = \frac{\sqrt{3}}{2},$$

because sine is positive in quadrant II;

$$\cos 120° = -\cos 60° = -\frac{1}{2},$$

because cosine is negative in quadrant II;

$$\tan 120° = -\tan 60° = -\sqrt{3},$$

because tangent is negative in quadrant II, and so on.

Based on this work, the values of the trigonometric functions for any angle θ can be found by finding the function value for an angle between 0° and 90°. To do this, go through the following steps.

Finding Trigonometric Function Values for Any Angle

1. If $\theta \geq 360°$, or if $\theta < 0°$, add or subtract 360° as many times as needed to get an angle of at least 0° but less than 360°.
2. Find the reference angle θ'. (Use the table given above.)
3. Find the necessary values of the trigonometric functions for the reference angle θ'.
4. Find the correct signs for the values found in Step 3. (Use the table of signs in the box on page 35). This result gives the value of the trigonometric function for angle θ.

Example 4 Find the following values.

(a) cos 210°

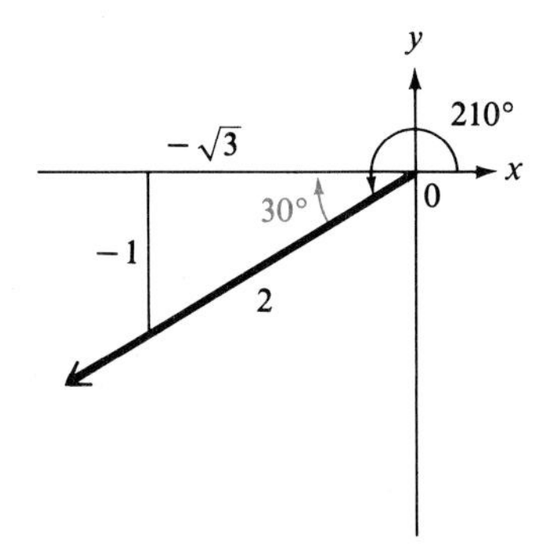

Figure 2.17

The reference angle is $210° - 180° = 30°$, as shown in Figure 2.17. Since cosine is negative in quadrant III,

$$\cos 210° = -\cos 30° = -\frac{\sqrt{3}}{2}.$$

(b) tan 675°

Begin by subtracting 360° to get an angle between 0° and 360°.

$$675° - 360° = 315°$$

As shown in Figure 2.18, the reference angle is $360° - 315° = 45°$. An angle of 315° is in quadrant IV, so the tangent will be negative, and

$$\tan 675° = \tan 315° = -\tan 45° = -1.$$

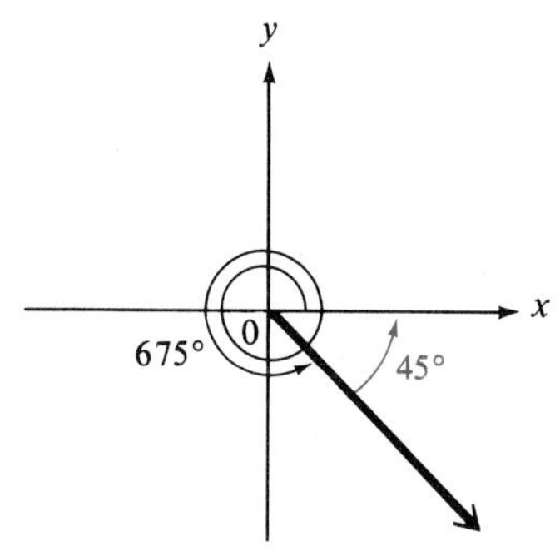

Figure 2.18

Example 5 Use a calculator or Table 2 to find each of the following.

(a) sin 190° 10′

Figure 2.19 shows the reference angle 10° 10′. An angle of 190° 10′ is in quadrant III, where sine is negative. Thus,

$$\sin 190°\ 10' = -\sin 10°\ 10' = -.1765.$$

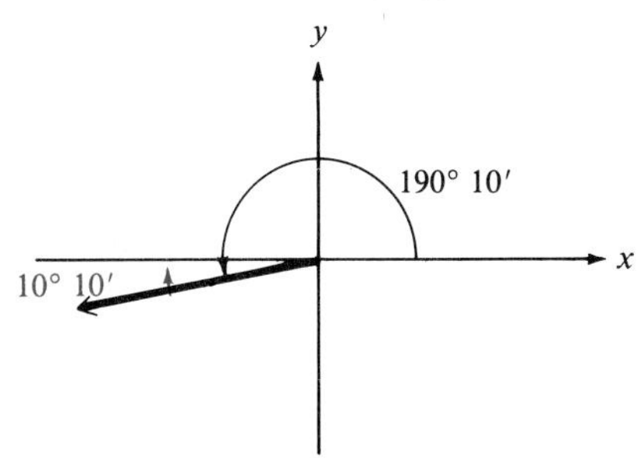

Figure 2.19

(b) tan 305° 20′

The reference angle is 54° 40′. Since 305° 20′ is in quadrant IV, tangent is negative and

$$\tan 305° 20' = -\tan 54° 40' = -1.411.$$

Reference angles are not needed with most calculators. With a calculator, tan 305° 20′ from the previous example could be found simply by converting to decimal degrees and pressing the TAN button.

Finding Angles with a Calculator So far we have used a calculator or table to find the trigonometric function values for a given angle. This process can be reversed: An angle can be found from the value of a trigonometric function, as shown in the next examples.

Example 6 Use a calculator to find a value of θ in the interval $0° \leq \theta \leq 90°$ satisfying each of the following. Leave answers in decimal degrees.

(a) $\sin \theta = .818150$

Find θ using a key labeled ARC or INV together with the SIN key. Some calculators require a different sequence of keystrokes, or may use a key labeled SIN^{-1}. As always, make sure the calculator is set for degree measure.

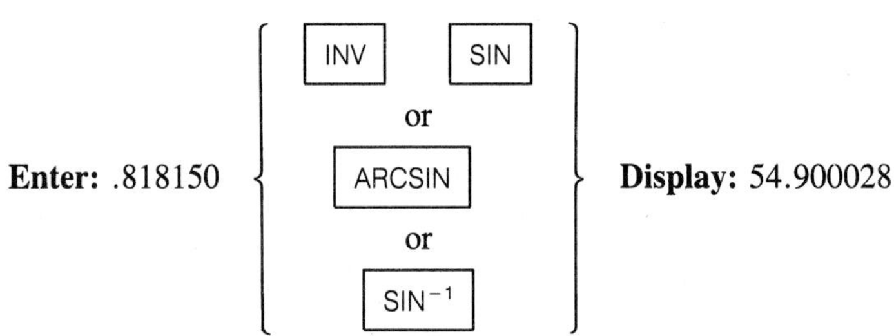

(The *arc* and sin^{-1} notations are explained in chapter 6.) Finally,

$$\theta = 54.900028°.$$

The second function key may be required in addition to one of the combinations shown above.

(b) $\sec \theta = 1.0545829$

Use the identity $\cos \theta = 1/\sec \theta$.

Enter: 1.0545829 1/x Use appropriate keys involving cos **Display:** 18.514704

The result: $\theta = 18.514704°.$

Note that the [1/x] key is used *before* the [cos] key when finding the angle, but *after* the [cos] key when finding the trigonometric function given the angle.

Finding Angles with a Table The next example shows how to use Table 2 to find an angle from a given trigonometric function value.

Example 7 Use Table 2 to find a value of θ in the interval $0° \leq \theta \leq 90°$ satisfying each of the following.

(a) $\sin \theta = .5807$

Use Table 2 and read columns having *sine* at either the top or the bottom. Since .5807 is in a column having sine at the top, use the angles listed at the left.

$$\theta = 35°\ 30'$$

(b) $\tan \theta = 2.699$

Since 2.699 is in a column having tangent at the bottom, use the angles listed at the right.

$$\theta = 69°\ 49'$$

2.3 Exercises

Find an angle θ, where $0° \leq \theta < 360°$, *that is coterminal with each of the following. See Example 3(c).*

1. 615° **2.** 708° **3.** 483° **4.** 592°

5. 458° 20′ **6.** 506° 40′ **7.** 738° 30′ **8.** 948° 50′

9. −250° 30′ **10.** −321° 50′ **11.** −198° 10′ **12.** −243° 30′

Find a reference angle for each of the following. See Example 3.

13. 98° **14.** 143° **15.** 212° **16.** 239°

17. 285° 30′ **18.** 314° 50′ **19.** 389° 10′ **20.** 427° 20′

21. 538° 39′ **22.** 619° 25′ **23.** 114° 47′ **24.** 138° 58′

25. 215° 46′ **26.** 312° 25′ **27.** −110° 10′ **28.** −183° 20′

29. −214° 30′ **30.** −429° 10′ **31.** −579° 12′ **32.** −682° 41′

Use Table 2 or a calculator to find the value of each of the following. See Examples 1 and 5.

33. sin 38° 40′ **34.** tan 29° 30′ **35.** cot 41° 20′ **36.** cos 27° 10′

37. sin 58° 30′ **38.** cos 46° 10′ **39.** tan 17° 20′ **40.** sin 39° 40′

41. cot 128° 30′ **42.** tan 153° 20′ **43.** sin 179° 20′ **44.** cos 124° 10′

45. sin 204° 20′ **46.** sec 218° 50′ **47.** cos 251° 10′ **48.** cot 298° 30′

49. sin 274° 30′ **50.** cos 304° 50′ **51.** csc 421° 10′ **52.** cot 512° 20′

53. sec (−108° 20′) **54.** csc (−29° 30′) **55.** tan (−197° 50′) **56.** cos (−299° 40′)

Use Table 2 or a calculator to find a value of θ in the interval $0° \leq \theta \leq 90°$ *satisfying each of the following. See Examples 6 and 7.*

57. $\sin \theta = .8480$ **58.** $\tan \theta = 1.473$ **59.** $\cos \theta = .8616$ **60.** $\cot \theta = 1.257$

61. $\sin \theta = .7214$ **62.** $\sec \theta = 2.749$ **63.** $\tan \theta = 6.435$ **64.** $\sin \theta = .2784$

Use a calculator to find each of the following to five decimal places.

65. $\sin 59.642°$ **66.** $\cos 38.1219°$ **67.** $\tan(-80.612°)$ **68.** $\sec(-19.702°)$

69. $\cos 258.409°$ **70.** $\sin 315.768°$ **71.** $\cot 109.713°$ **72.** $\csc 278.143°$

73. $\cos 74° 11'$ **74.** $\tan 58° 46'$ **75.** $\cot 125° 52' 10''$ **76.** $\csc 211° 40' 38''$

77. $\sin(-15° 28' 17'')$ **78.** $\cos(-29° 03' 05'')$ **79.** $\sec(-121° 53' 42'')$ **80.** $\cot(-276° 18' 34'')$

Use a calculator to find each of the following. As shown in chapter 5, all these answers should be integers.

81. $\cos 100° \cos 80° - \sin 100° \sin 80°$ **82.** $\sin 35° \cos 55° + \cos 35° \sin 55°$

83. $\sin 28° 14' \cos 61° 46' + \cos 28° 14' \sin 61° 46'$ **84.** $\cos 75° 29' \cos 14° 31' - \sin 75° 29' \sin 14° 31'$

When a light ray travels from one medium, such as air, to another medium, such as water or glass, the speed of the light changes, and the direction that the ray is traveling changes. (This is why a fish under water is in a different position than it appears to be.) These changes are given by Snell's law.

$$\frac{c_1}{c_2} = \frac{\sin \theta_1}{\sin \theta_2},$$

where c_1 *is the speed of light in the first medium,* c_2 *is the speed of light in the second medium, and* θ_1 *and* θ_2 *are the angles shown in the figure.*

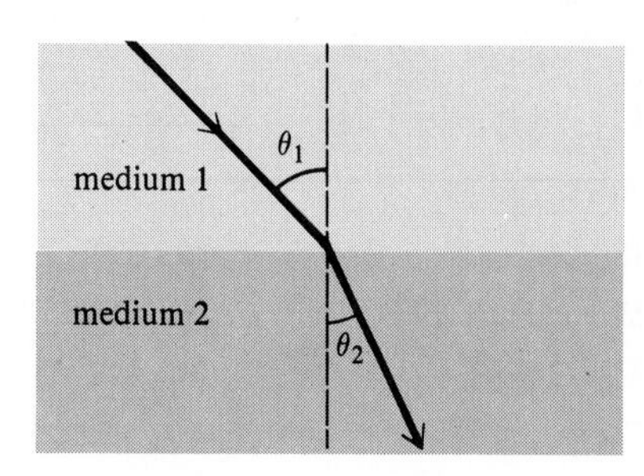

If this medium is less dense, light travels at a faster speed, c_1.

If this medium is more dense, light travels at a slower speed, c_2.

In the following exercises, assume that $c_1 = 3 \times 10^8$ *m per sec. Find the speed of light in the second medium.*

85. $\theta_1 = 46°$, $\theta_2 = 31°$ **86.** $\theta_1 = 39°$, $\theta_2 = 28°$

Find θ_2 for the following values of θ_1 and c_2. Round to the nearest degree.

87. $\theta_1 = 40°$, $c_2 = 1.5 \times 10^8$ m per sec **88.** $\theta_1 = 62°$, $c_2 = 2.6 \times 10^8$ m per sec

The figure below shows a fish's view of the world above the surface of the water. Suppose that a light ray comes from the horizon, enters the water, and strikes the fish's eye.*

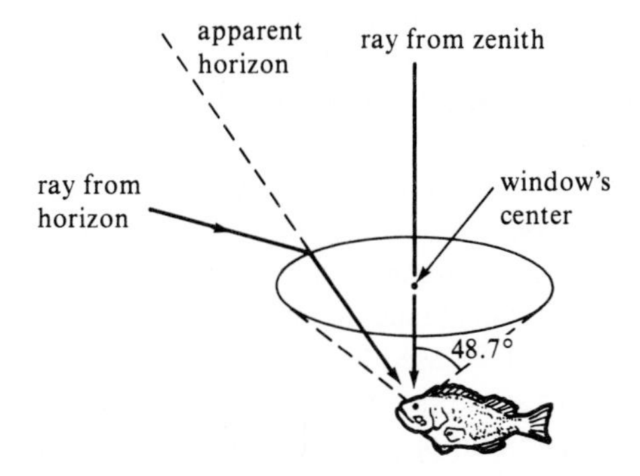

89. Let us assume that this ray gives a value of 90° for angle θ_1 in the formula for Snell's law. (In a practical situation this angle would probably be a little less than 90°.) The speed of light in water is about 2.254×10^8 m per sec. Find angle θ_2.

(Your result should have been about 48.7°. This means that a fish sees the world above the water as a cone, making an angle of 48.7° with the vertical.)

90. Suppose that an object is located at a true angle of 29.6° above the horizon. Find the apparent angle above the horizon to a fish.

2.4 Significant Digits

Suppose that we glance quickly at a room and guess that it is 15 ft by 18 ft. To calculate the length of a diagonal of the room, the Pythagorean theorem can be used.

$$d^2 = 15^2 + 18^2$$
$$d^2 = 549$$
$$d = \sqrt{549}$$

On a calculator,

$$\sqrt{549} = 23.430749.$$

Should this answer be given as the length of the diagonal of the room? Of course not. The number 23.430749 contains 6 decimal places, while the original data of 15 ft and 18 ft are only accurate to the nearest foot. Since the results of a problem can be no more accurate than the least accurate number in any calculation, we really should say that the diagonal of the 15-by-18-ft room is 23 ft.

*From "The Amateur Scientist" by Jearl Walker in *Scientific American,* March, 1984.

If a wall is measured to the nearest foot and is found to be 18 ft long, actually this means that the wall has a length between 17.5 ft and 18.5 ft. If the wall is measured more accurately and found to be 18.3 ft long, then its length is really between 18.25 ft and 18.35 ft. A measurement of 18.00 ft would indicate that the length of the wall is between 17.995 ft and 18.005 ft. The measurement 18 ft is said to have two **significant digits** of accuracy; 18.0 has three significant digits, and 18.00 has four.

A significant digit is a digit obtained by actual measurement. A number that represents the result of counting, or a number that results from theoretical work and is not the result of a measurement, is an **exact number.** There are fifty states in the United States, so 50 is an exact number. The number of states is not 49 3/4 or 50 1/4; nor is the number 50 used here to represent "some number between 45 and 55." In the formula for the perimeter of a rectangle, $P = 2L + 2W$, the 2's are obtained from the definition of perimeter, and are exact numbers.

Example 1 The following chart shows some numbers, the number of significant digits in each number, and the range represented by each number.

Number	*Number of significant digits*	*Range represented by number*
29.6	3	29.55 to 29.65
1.39	3	1.385 to 1.395
.000096	2	.0000955 to .0000965
.03	1	.025 to .035
100.2	4	100.15 to 100.25

Not all digits in a measurement are significant; some of the digits serve only as placeholders. For example, the measurement 19.00 m is a measurement to the nearest hundredth meter. What about the measurement 93,000 m? Does it represent a measurement to the nearest meter, to the nearest ten meters, hundred meters, or thousand meters? We cannot tell by the way the number is written. To avoid this problem, write the number in **scientific notation,** the product of a number between 1 and 10 and a power of 10. Depending on the accuracy of the measurement, write 93,000 using scientific notation as follows.

Measurement to nearest	*Scientific notation*	*Numbers of significant digits*
Meter	9.3000×10^4	5
Ten meters	9.300×10^4	4
Hundred meters	9.30×10^4	3
Thousand meters	9.3×10^4	2

Virtually all values of trigonometric functions are approximations, and virtually all measurements are approximations. To perform calculations on such approximate numbers, follow the rules given below.

Calculation with Significant Digits

For *adding* and *subtracting,* round the answer so that the last digit you keep is in the right-most column in which all the numbers have significant digits.

For *multiplying* or *dividing,* round the answer to the least number of significant digits found in any of the given numbers.

For *powers* and *roots,* round the answer so that it has the same number of significant digits as the number whose power or root you are finding.

Example 2 Perform each calculation.

(a) $23.4 + 17.92 + 19.004$

Adding the numbers gives 60.324. The last place in which all three of the numbers have significant digits is the tenths column, so the answer should be written 60.3

(b) $(2.9328)(30.531)(12.82) = 1148$

The answer was written with four significant digits, the *least* number of significant digits found in the three numbers.

(c) $\sqrt{284.8} = 16.88$

The answer has the same number of significant digits as the number whose root is being found. ✣

Example 3 **(a)** Find the perimeter of a field 125.5 m by 38.50 m.

The perimeter is found with the formula $P = 2L + 2W$. Substituting 125.5 for L and 38.50 for W gives

$$P = 2(125.5) + 2(38.50) = 328.0 \text{ m}.$$

The answer is written with four significant digits, as are the given measurements. As mentioned above, the 2's in the formula are exact numbers, which are assumed to have as many significant digits as desired.

(b) Find the area of the same field, by using the formula $A = LW$.

$$A = 125.5 \times 38.50 = 4831.75 \text{ sq m}$$

Rounding to four significant digits gives the area as

$$4832 \text{ sq m}.$$ ✣

The next example shows significant digits used with a calculator.

Example 4 Perform each calculation. Write each answer with the proper number of significant digits.

(a) $\dfrac{16.9^2 + 14.8^2 - 11.7^2}{2.00(16.9)(14.8)}$

Use the $\boxed{x^2}$ key if your calculator has one. Keep all digits in the intermediate steps, and round only at the end (to three significant digits).

$$\frac{16.9^2 + 14.8^2 - 11.7^2}{2.00(16.9)(14.8)} = .735.$$

(b) $\dfrac{79.469 \sin 122.84°}{58.72}$

Make sure the calculator is set for degree measure. The smallest number of significant digits in the given numbers is 4, so

$$\frac{79.469 \sin 122.84°}{58.72} = 1.137.$$

(c) $\dfrac{(7.894 \times 10^5) \times (3.9168 \times 10^{-2})}{1.0035 \times 10^6}$

Most scientific calculators can handle the scientific notation of this example (often with a key labeled $\boxed{\text{EE}}$ or $\boxed{\text{EEX}}$). If your calculator does not have provision for scientific notation, enter the numbers without scientific notation. Using either method gives the result

$$3.081 \times 10^{-2}, \quad \text{or} \quad .03081.$$

2.4 Exercises

The following numbers represent approximate measurements. State the range represented by each of the measures. See Example 1.

1. 5 lb **2.** 8 ft **3.** 9.6 ton **4.** 7.8 qt

5. 8.95 m **6.** 2.37 km **7.** 19.7 L **8.** 32 cm

9. 253.741 m **10.** 74.358 oz **11.** .02 ft **12.** .0302 mi

Work each of the following exercises.

13. When Mt. Everest was first surveyed, the surveyors obtained a height of 29,000 ft to the nearest foot. State the range represented by this number. (The surveyors thought that no one would believe a measurement of 29,000 ft, so they reported it as 29,002.)

14. At Denny's, a chain of restaurants, the Low-Cal Special is said to have "approximately 472 calories." What is the range of calories represented by this number? By claiming "approximately 472 calories," they are probably claiming more accuracy than is possible. In your opinion, what might be a better claim?

Give the number of significant digits in each of the following. See Example 1.

15. 21.8 **16.** 37 **17.** 42.08 **18.** 600.9

19. 31.00 **20.** 20,000 **21.** 3.0×10^7 **22.** 5.43×10^3

23. 2.7100×10^4 **24.** 3.7000×10^3 **25.** -9.12×10^7 **26.** 3.005×10^{-8}

Round each of the following numbers to three significant digits. Then refer to the original numbers and round each to two significant digits.

27. 768.7 **28.** 921.3 **29.** 12.53 **30.** 28.17

31. 149.8732 **32.** 915.7514 **33.** 9.003 **34.** 1.700

35. 7.125 **36.** 9.375 **37.** 11.55 **38.** 9.155

Find the error in each statement.

39. I have 2 bushel baskets, each containing 65 apples. I know that $2 \times 65 = 130$, but 2 has only one significant figure, so I must write the answer as 1×10^2, or 100. I therefore have 100 apples.

40. The formula for the circumference of a circle is $C = 2\pi r$. My circle has a radius of 54.98 cm, and my calculator has a [π] key, giving fifteen digits of accuracy. Pressing the right buttons gives 345.44953. Because 2 has only one significant digit, however, the answer must be given as 3×10^2, or 300 sq cm. (What is the correct answer?)

Use a calculator to decide whether the following statements are true *or* false. *It may be that a true statement will lead to results that differ in the last decimal place, due to rounding error.*

41. $\sin 10° + \sin 10° = \sin 20°$

42. $\cos 40° = 2 \cos 20°$

43. $\sin 50° = 2 \sin 25° \cdot \cos 25°$

44. $\cos 70° = 2 \cos^2 35° - 1$

45. $\cos 40° = 1 - 2 \sin^2 80°$

46. $\cos^2 28° 42' + \sin^2 28° 42' = 1$

47. $2 \cos 38° 22' = \cos 76° 44'$

48. $\tan 29° 32' = \dfrac{\sin 29° 32'}{\cos 29° 32'}$

49. $\sin 39° 48' + \cos 39° 48' = 1$

50. $\cot 76° 43' = \dfrac{1}{\tan 76° 43'}$

51. $\sin 78.4944° \cos 11.5056° + \sin 11.5056° \cos 78.4944° = 1$

52. $\dfrac{\tan 52.3081° + \tan 37.6919°}{1 - \tan 52.3081° \tan 37.6919°}$ does not exist

Use a calculator to work the following problems to the correct number of significant digits. Never round intermediate results. Keep all digits and then round to the proper number of digits at the end. See Examples 2 and 4.

53. $(8.742)^2$

54. $(.98352)^2$

55. $\dfrac{.746}{.391}$

56. $\dfrac{.375}{.005792}$

57. $(.425)(89.3)(746.91)$

58. $\dfrac{2.000}{(74.83)(.0251)}$

59. $\dfrac{1.000}{(.0900)^2 + (3.21)^2}$

60. $\dfrac{(6.93)^2 + (21.74)^2}{(38.76)^2 - 29.4}$

61. $\sqrt{74.689}$

62. $\sqrt{253{,}000}$

63. $\dfrac{4.00}{\sqrt{59.7} - \sqrt{74.6}}$

64. $\dfrac{-5.000(2.143)}{\sqrt{.009826}}$

65. $\dfrac{54.92 \sin 48.6514°}{71.853}$

66. $\dfrac{1.409 \sin 71.0972°}{1.583}$

67. $128.9(3.02) + 97.6(.0589) - \sqrt{700.9}$

68. $\left[\dfrac{(.00900)(74)}{1.0 - (.0382)\sqrt{741.6} + \sqrt{98.32}}\right]^2$

Use a calculator to evaluate each of the following. Write each answer in scientific notation with the proper number of significant digits. See Example 4(c).

69. $(1.66 \times 10^4)(2.93 \times 10^3)$

70. $(6.92 \times 10^7)(8.14 \times 10^5)$

71. $(4.08 \times 10^{-9})(3.172 \times 10^4)$

72. $(9.113 \times 10^{12})(8.72 \times 10^{-11})$

73. $\dfrac{(-4.389 \times 10^4)(2.421 \times 10^{-2})}{1.76 \times 10^{-9}}$

74. $\dfrac{-5.421 \times 10^{-7}}{(4.2803 \times 10^{-4})(9.982 \times 10^{-8})}$

2.5 Solving Right Triangles

One of the main applications of trigonometry is solving triangles. **To solve a triangle** means to find the measures of all the angles and sides of the triangle. This section and the next discuss methods of solving right triangles. (Methods for solving other triangles are presented in chapter 7.)

When solving triangles, use the following table for deciding on significant digits in angle measure.

ACUTE ∠S

Significant Digits for Angles

Number of significant digits	*Angle measure to nearest:*
2	Degree
3	Ten minutes, or nearest tenth of a degree 14° 30′
4	Minute, or nearest hundredth of a degree
5	Tenth of a minute, or nearest thousandth of a degree

For example, an angle measuring 52° 30′ has three significant digits (assuming that 30′ is measured to the nearest ten minutes).

In using trigonometry to solve triangles, it is convenient to use a to represent the length of the side opposite angle A, b for the length of the side opposite angle B, and so on. As mentioned earlier, in a right triangle the letter c is reserved for the hypotenuse.

Example 1 Solve right triangle ABC, with $A = 34° 30'$ and $c = 12.7$. See Figure 2.20.

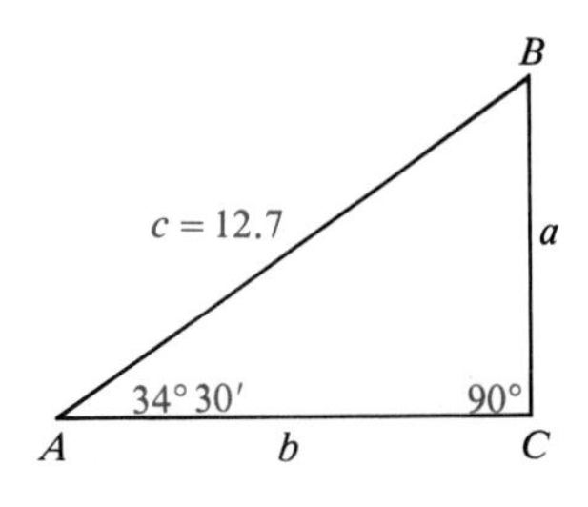

Figure 2.20

To solve the triangle, find the measures of the remaining sides and angles. The value of a can be found with a trigonometric function involving the known values of angle A and side c. Since the sine of angle A is given by the quotient of the side opposite A and the hypotenuse, use $\sin A$.

$$\sin A = \frac{a}{c}$$

Substituting known values gives

$$\sin 34° 30' = \frac{a}{12.7},$$

or, upon multiplying both sides by 12.7,

$$a = 12.7 \sin 34° 30'$$

$$a = 12.7(.5664) \qquad \text{Use a calculator or table}$$

$$a = 7.19.$$

The value of b could be found with the Pythagorean theorem. It is better, however, to use the information given in the problem rather than a result just calculated. If a mistake were to be made in finding a, then b also would be incorrect. Also, rounding more than once may cause the result to be less accurate. Using $\cos A$ gives

$$\cos A = \frac{\text{side adjacent}}{\text{hypotenuse}} = \frac{b}{c}$$

$$\cos 34° 30' = \frac{b}{12.7}$$

$$b = 12.7 \cos 34° 30'$$

$$b = 10.5.$$

Once b has been found, the Pythagorean theorem could be used as a check. All that remains to solve triangle ABC is to find B. Since $A + B = 90°$ and $A = 34° 30'$.

$$A + B = 90°$$
$$B = 90° - A$$
$$B = 89° 60' - 34° 30'$$
$$B = 55° 30'.$$

Example 2 Solve right triangle ABC if $a = 29.43$ cm and $c = 53.58$ cm.

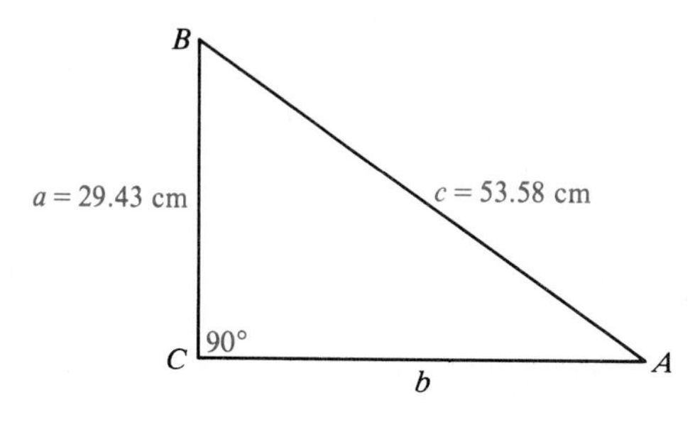

Figure 2.21

Draw a sketch showing the given information, as in Figure 2.21. Find angle A by using the sine.

$$\sin A = \frac{\text{side opposite}}{\text{hypotenuse}}$$

$$\sin A = \frac{29.43}{53.58} = .549272$$

A calculator would give $A = 33.32°$. Tables, and the process of interpolation explained in the appendix to this chapter, would give $A = 33° 19'$. Since $A + B = 90°$, angle $B = 56.68°$ or $56° 41'$.

Find b from the Pythagorean theorem, $a^2 + b^2 = c^2$, or $b^2 = c^2 - a^2$. Since $c = 53.58$ and $a = 29.43$.

$$b^2 = 53.58^2 - 29.43^2 = 2004.6915,$$

giving

$$b = 44.77 \text{ cm}.$$

Many problems with right triangles involve the angle of elevation or the angle of depression. The **angle of elevation** from point X to point Y (above X) is the angle made by line XY and a horizontal line through X. The angle of elevation is always measured from the horizontal. See Figure 2.22. The **angle of depression** from point X to point Y (below X) is the angle made by line XY and a horizontal line through X.

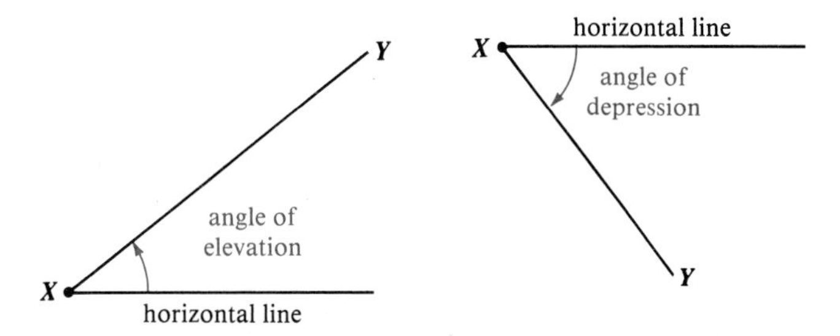

Figure 2.22

Example 3 Donna Spence knows that when she stands 123 ft from the base of a flagpole, the angle of elevation to the top is 26° 40′. If her eyes are 5.30 ft above the ground, find the height of the flagpole.

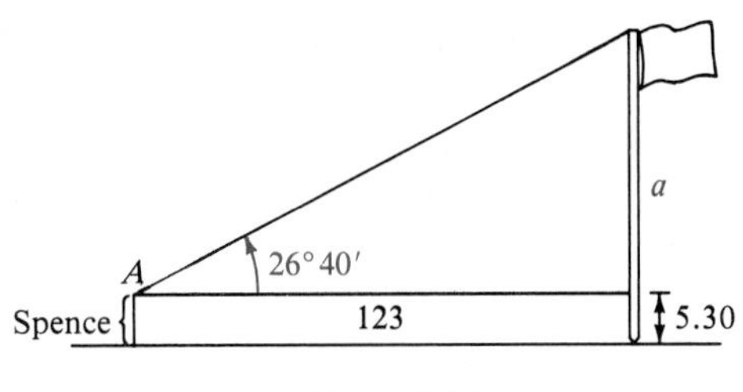

Figure 2.23

The length of the side adjacent to Spence is known and the length of the side opposite her is to be found. See Figure 2.23. The ratio that involves these two values is the tangent.

$$\tan A = \frac{\text{side opposite}}{\text{side adjacent}}$$

$$\tan 26^\circ\, 40' = \frac{a}{123}$$

$$a = 123 \tan 26^\circ\, 40'$$

$$a = 61.8 \text{ ft}$$

Since Spence's eyes are 5.30 feet above the ground, the height of the flagpole is

$$61.8 + 5.30 = 67.1 \text{ ft.}$$

Example 4 The length of the shadow of a building 34.09 m tall is 37.62 m. Find the angle of elevation of the sun. See Figure 2.24 on the following page.

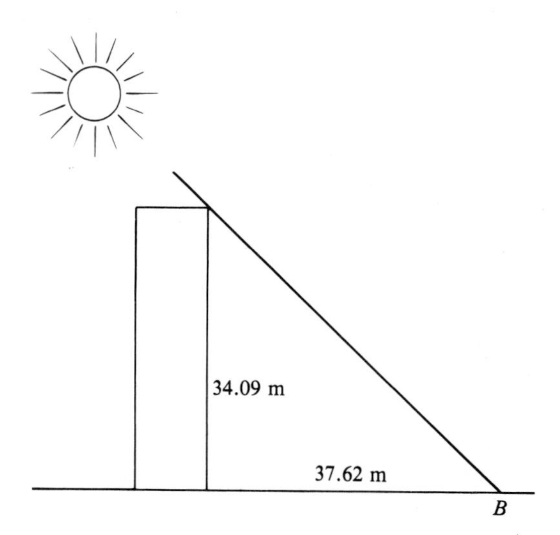

Figure 2.24

As shown in Figure 2.24, the angle of elevation of the sun is angle B. Since the side opposite B and the side adjacent to B are known, use the tangent ratio to find B.

$$\tan B = \frac{34.09}{37.62}$$

$$B = 42.18^\circ$$

The angle of elevation of the sun is 42.18°.

2.5 Exercises

Solve each right triangle. See Examples 1 and 2.

1.

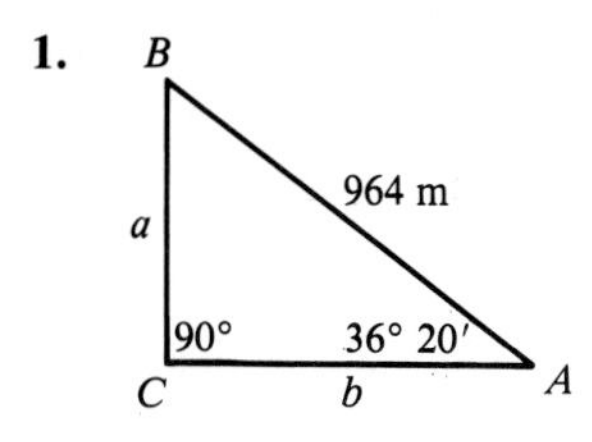

2.

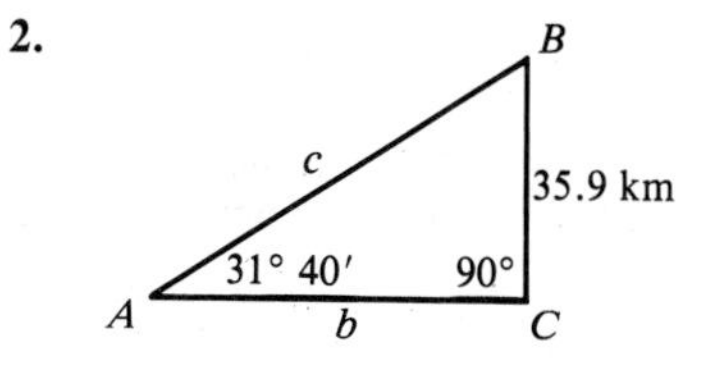

3.

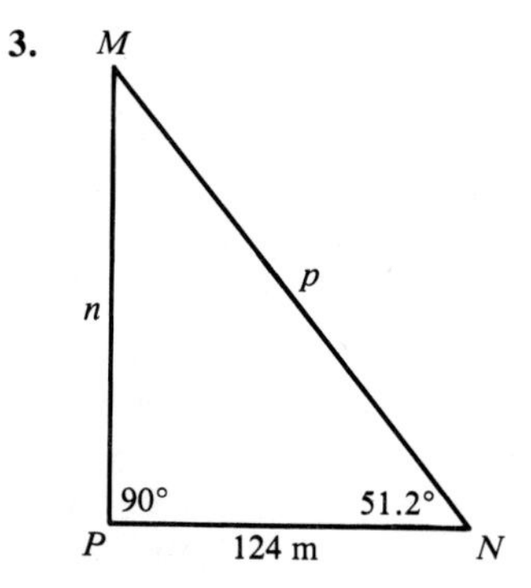

4.

Y

89.6 cm

x

47.8°

90°

X y Z

5.

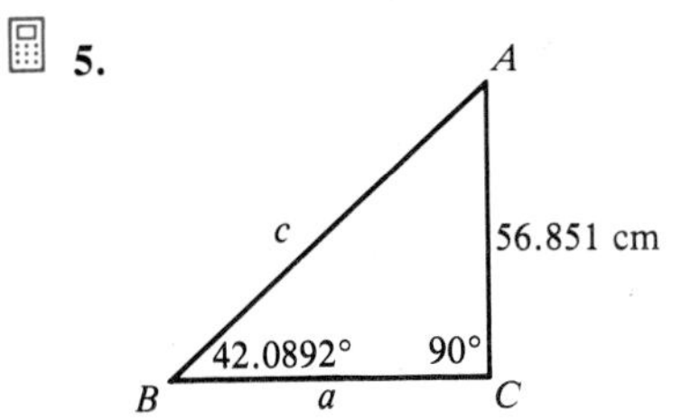

6.

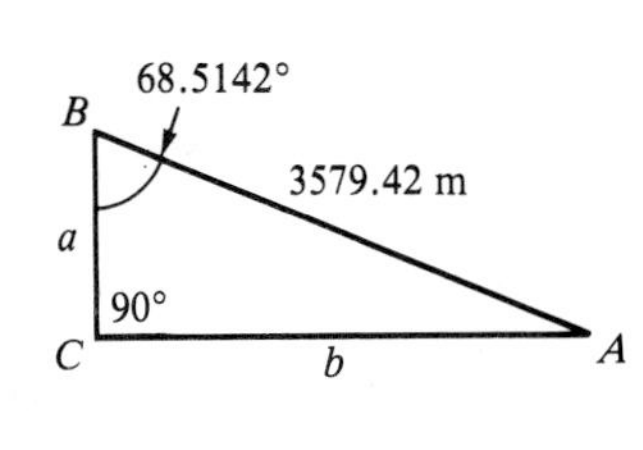

Solve each of the following right triangles. See Examples 1 and 2.

7. $A = 28° 00'$, $c = 17.4$ ft

8. $B = 46° 00'$, $c = 29.7$ m

9. $B = 73° 00'$, $b = 128$ in

10. $A = 61° 00'$, $b = 39.2$ cm

11. $a = 76.4$ yd, $b = 39.3$ yd

12. $a = 958$ m, $b = 489$ m

13. $a = 18.9$ cm, $c = 46.3$ cm

14. $b = 219$ m, $c = 647$ m

15. $A = 53° 24'$, $c = 387.1$ ft

16. $A = 13° 47'$, $c = 1285$ m

17. $B = 39° 09'$, $c = .6231$ m

18. $B = 82° 51'$, $c = 4.825$ cm

19. $c = 7.813$ m, $b = 2.467$ m

20. $c = 44.91$ mm, $a = 32.71$ mm

21. $B = 42.432°$, $a = 157.49$ m

22. $A = 36.704°$, $c = 1461.3$ cm

23. $A = 57.209°$, $c = 186.49$ cm

24. $B = 12.099°$, $b = 7.0463$ m

25. $b = 173.921$ m, $c = 208.543$ m

26. $a = 864.003$ cm, $c = 1092.84$ cm

Solve each of the following. See Example 3.

27. A 39.4-m fire-truck ladder is leaning against a wall. Find the distance the ladder goes up the wall if it makes an angle of 42° 30′ with the ground.

28. A swimming pool is 50.0 ft long and 4.00 ft deep at one end. If it is 12.0 ft deep at the other end, find the total distance along the bottom.

29. A guy wire 87.4 m long is attached to the top of a tower that is 69.4 m high. Find the angle that the wire makes with the ground.

30. Find the length of a guy wire that makes an angle of 42° 10′ with the ground if the wire is attached to the top of a tower 79.6 m high.

31. To measure the height of a flagpole, José finds that the angle of elevation from a point 24.73 ft from the base to the top is 38° 12′. Find the height of the flagpole.

32. A rectangular piece of land is 629.3 ft by 872.6 ft. Find an acute angle made by the diagonal of the rectangle.

33. To find the distance RS across a lake, a surveyor lays off $RT = 72.6$ m, with angle $T = 32° 10'$, and angle $S = 57° 50'$. Find length RS.

34. A surveyor must find the distance QM across a depressed freeway. She lays off $QN = 769$ ft along one side of the freeway, with angle $N = 21° 50'$, and with angle $M = 68° 10'$. Find QM.

35. The length of the base of an isosceles triangle is 37.41 in. Each base angle is 49.74°. Find the length of each of the two equal sides of the triangle. (*Hint:* Divide the triangle into two right triangles.)

36. Find the altitude of an isosceles triangle having a base of 125.6 cm if the angle opposite the base is 72° 42′.

Work the following problems involving angles of elevation or depression. See Examples 3 and 4.

37. Suppose that the angle of elevation of the sun is 28.4°. Find the length of the shadow cast by a man 6.00 ft tall.

38. The shadow of a vertical tower is 58.2 m long when the angle of elevation of the sun is 36.51°. Find the height of the tower.

39. Find the angle of elevation of the sun if a 53.9-ft flagpole casts a shadow 74.6 ft long.

40. The angle of depression from the top of a building to a point on the ground is 34° 50′. How far is the point on the ground from the top of the building if the building is 368 m high?

41. An airplane is flying 10,000 ft above the level ground. The angle of depression from the plane to a tree is 13° 50′. How far horizontally must the plane fly to be directly over the tree?

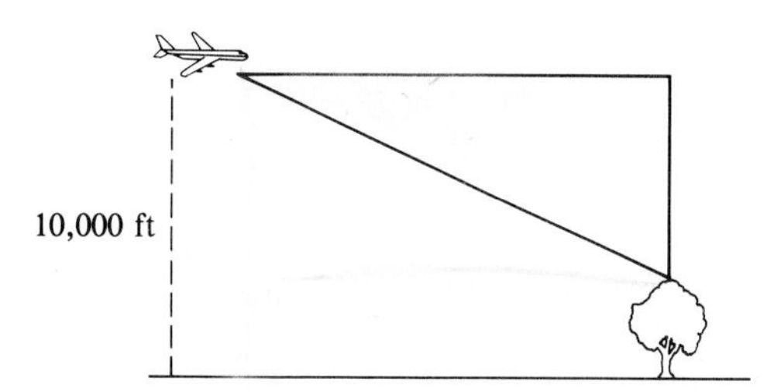

42. The angle of elevation from the top of a small building to the top of a nearby taller building is 46° 40′, while the angle of depression to the bottom is 14° 10′. If the smaller building is 28.0 m high, find the height of the taller building.

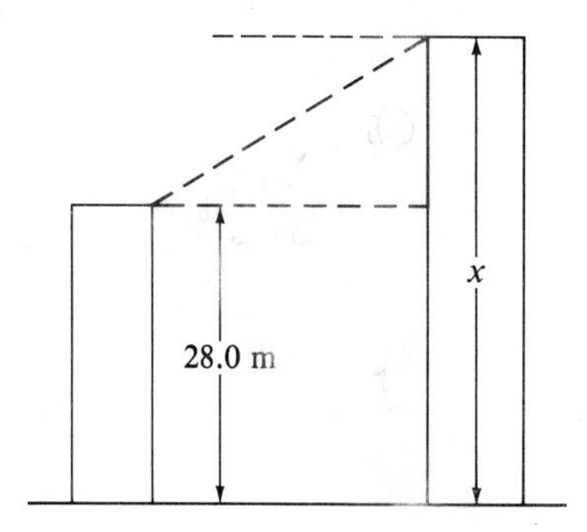

43. A television camera is to be mounted on a bank wall so as to have a good view of the head teller (see the figure). Find the angle of depression that the lens should make with the horizontal.

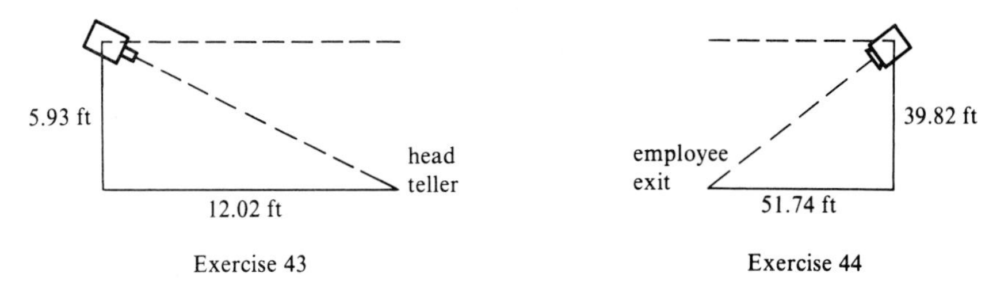

Exercise 43 Exercise 44

44. A company safety committee has recommended that a floodlight be mounted in a parking lot so as to illuminate the employee exit (see the figure). Find the angle of depression of the light.

45. Atoms in metals can be arranged in patterns called *unit cells*. One such unit cell, called a *primitive cell,* is a cube with an atom at each corner. A right triangle can be formed from one edge of the cell, a face diagonal, and a cube diagonal, as shown in the figure. If each cell edge is 3×10^{-8} cm and the face diagonal is 4.24×10^{-8} cm, what is the angle between the cell edge and a cube diagonal?

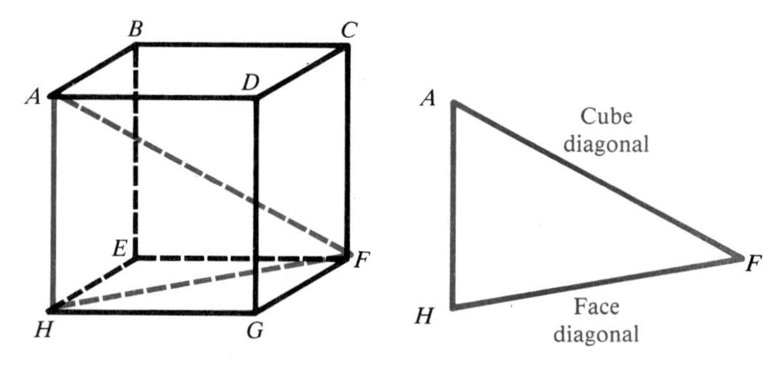

46. To determine the diameter of the sun, an astronomer might sight with a transit (a device used by surveyors for measuring angles) first to one edge of the sun and then to the other, finding that the included angle equals 1° 4′. Assuming that the distance from the earth to the sun is 92,919,800 mi, calculate the diameter of the sun. (See the figure.)

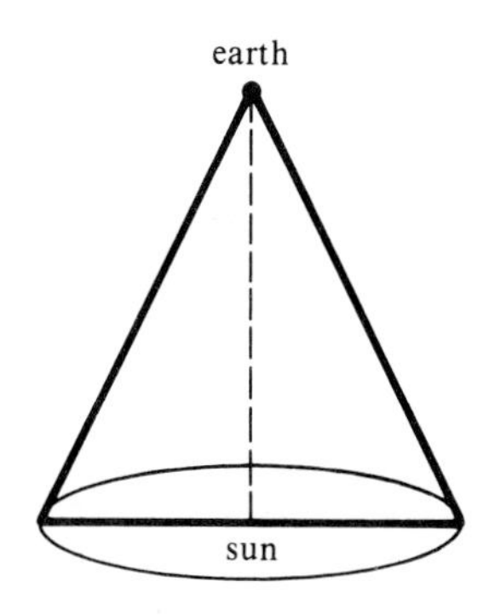

47. Very accurate measurements have shown that the distance between California's Owens Valley Radio Observatory and the Haystack Observatory in Massachusetts is 2441.2938 mi. Suppose that the two observatories focus on a distant star and find that angles E and E' in the figure are both 89.99999°. Find the distance to the star from Haystack. (Assume that the earth is flat.)

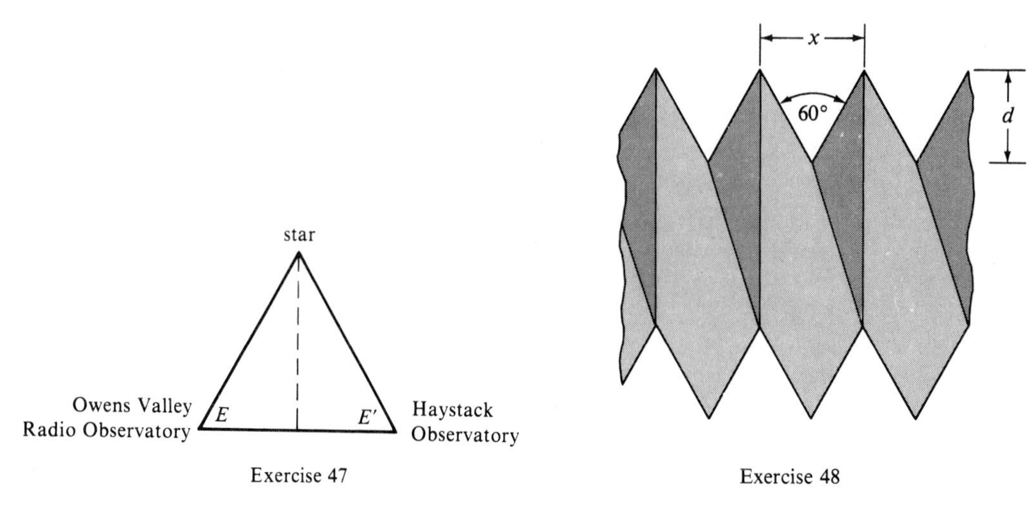

Exercise 47

Exercise 48

48. The figure shows a magnified view of the threads of a bolt. Find x if d is 2.894 mm.

2.6 Further Applications of Right Triangles

Other applications of right triangles involve **bearing,** an important idea in navigation. There are two common ways to express bearing. *When a single angle is given, such as* 164°, *it is understood that the bearing is measured in a clockwise direction from due north.* Several sample bearings using this system are shown in Figure 2.25.

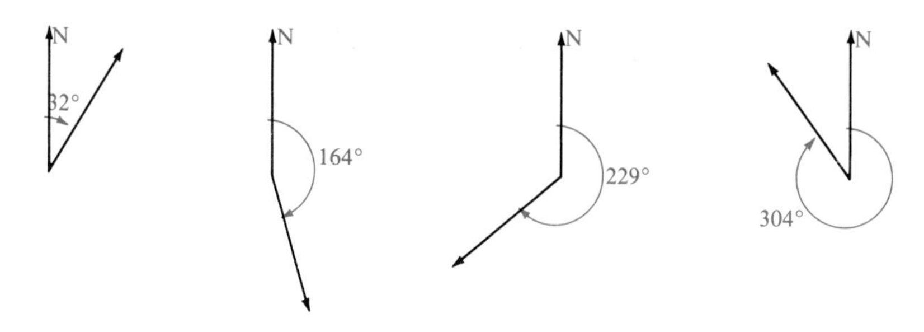

Figure 2.25

In the following examples and exercises, the problems all result in right triangles, so the methods of the previous section apply. Chapter 7 will include problems involving bearing that result in triangles that are *not* right triangles and require other methods to solve.

Example 1 Radar stations A and B are on an east-west line, 3.7 km apart. Station A detects a plane at C, on a bearing of 61°. Station B simultaneously detects the same plane, on a bearing of 331°. Find the distance from A to C.

Draw a sketch showing the given information, as in Figure 2.26. Since a line drawn due north is perpendicular to an east-west line, right angles are formed at A and B, so that angles CAB and CBA can be found. Angle C is a right angle because angles CAB and CBA are complementary. (If C were not a right angle, the methods of chapter 7 would be needed.) Find distance b by using the cosine function.

$$\cos 29° = \frac{b}{3.7}$$

$$3.7 \cos 29° = b$$

$$3.7(.8746) = b$$

$$b = 3.2 \text{ km}$$

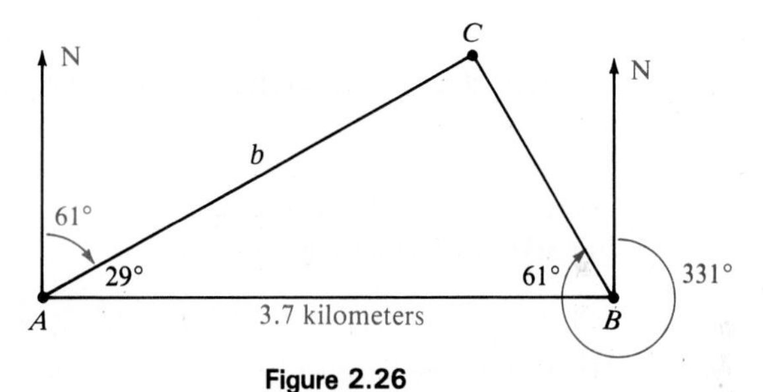

Figure 2.26

The other common system for expressing bearing starts with a north-south line and uses an acute angle to show the direction, either east or west, from this line. Figure 2.27 shows several sample bearings using this system. Either N or S always comes first, followed by an acute angle, and then E and W.

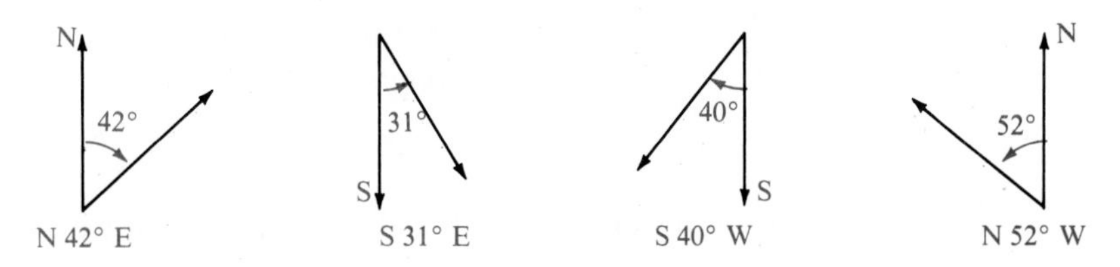

Figure 2.27

Example 2 The bearing from A to C is S 52° E. The bearing from A to B is N 84° E. The bearing from B to C is S 38° W. A plane flying at 250 mph takes 2.4 hr to go from A to B. Find the distance from A to C.

Make a sketch of the situation. First draw the two bearings from point A. Choose a point B on the bearing N84° E from A and draw the bearing to C. Point C will be located where the bearings from A to B intersect as shown in Figure 2.28.

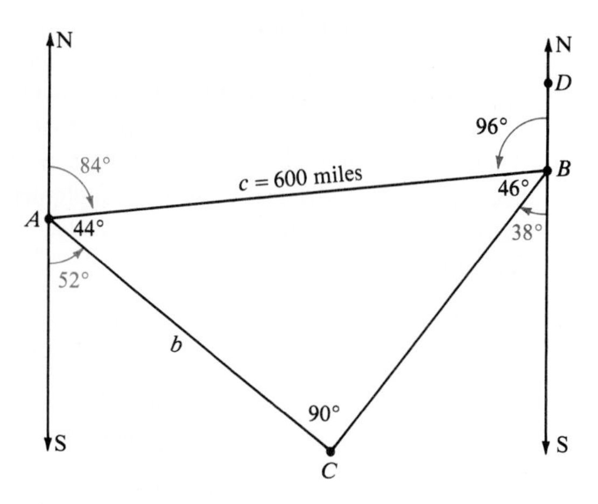

Figure 2.28

Since the bearing from A to B is N 84° E, angle ABD is $180° - 84° = 96°$. Thus, angle ABC is 46°. Also, angle BAC is $180° - (84° + 52°) = 44°$. Angle C is $180° - (44° + 46°) = 90°$. From the statement of the problem, a plane flying at 250 mph takes 2.4 hr to go from A to B. The distance from A to B is the product of rate and time, or

$$c = \text{rate} \times \text{time} = 250(2.4) = 600 \text{ mi.}$$

To find b, the distance from A to C, use the sine. (The cosine could also have been used.)

$$\sin 46° = \frac{b}{c}$$

$$\sin 46° = \frac{b}{600}$$

$$600 \sin 46° = b$$

$$600(.7193) = b$$

$$b = 430 \text{ mi}$$

The next example uses the idea of the angle of elevation first discussed in the previous section.

Example 3 Francisco needs to know the height of a tree. From a given point on the ground he finds that the angle of elevation to the top of the tree is 36° 40′. He then moves back 50 ft. From the second point, the angle of elevation to the top of the tree is 22° 10′. See Figure 2.29. Find the height of the tree.

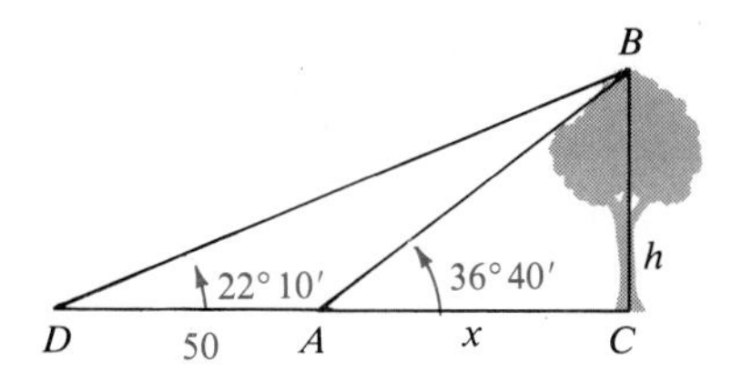

Figure 2.29

The figure shows two unknowns, x, the distance from the center of the trunk of the tree to the point where the first observation was made, and h, the height of the tree. Since nothing is given about the length of the hypotenuse of either triangle ABC or triangle BCD, use a ratio that does not involve the hypotenuse—the tangent.

In triangle ABC, $\tan 36° \, 40' = \dfrac{h}{x}$ or $h = x \tan 36° \, 40'$.

In triangle BCD, $\tan 22° \, 10' = \dfrac{h}{50 + x}$ or $h = (50 + x) \tan 22° \, 10'$.

Since each of these expressions equals h, these expressions must be equal. Thus,

$$x \tan 36° \, 40' = (50 + x) \tan 22° \, 10'.$$

Now use algebra to solve for x.

$$\begin{aligned} x \tan 36° \, 40' &= 50 \tan 22° \, 10' + x \tan 22° \, 10' \\ x \tan 36° \, 40' - x \tan 22° \, 10' &= 50 \tan 22° \, 10' \\ x(\tan 36° \, 40' - \tan 22° \, 10') &= 50 \tan 22° \, 10' \\ x &= \frac{50 \tan 22° \, 10'}{\tan 36° \, 40' - \tan 22° \, 10'} \end{aligned}$$

We saw above that $h = x \tan 36° \, 40'$. Substituting for x,

$$h = \left(\frac{50 \tan 22° \, 10'}{\tan 36° \, 40' - \tan 22° \, 10'}\right)(\tan 36° \, 40').$$

From a calculator, or using Table 2

$$\begin{aligned} \tan 36° \, 40' &= .7445 \\ \tan 22° \, 10' &= .4074, \end{aligned}$$

so,

$$\tan 36° \, 40' - \tan 22° \, 10' = .7445 - .4074 = .3371,$$

and

$$h = \left(\frac{50(.4074)}{.3371}\right)(.7445) = 45 \text{ ft.}$$ ✣

2.6 Exercises

Solve these problems involving right triangles.

1. Find the angle of the sun above the horizon when a person 5.94 ft tall casts a shadow 9.74 ft long. (Find θ in the figure.)

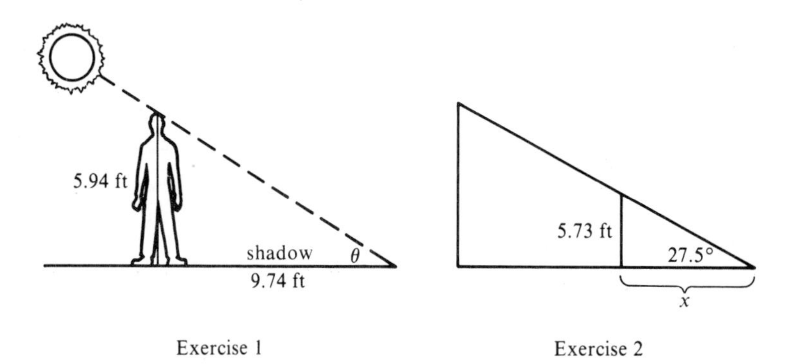

Exercise 1 Exercise 2

2. The angle of the sun above the horizon is 27.5°. Find the length of the shadow of a person 5.73 ft tall. (Find x in the figure.)

3. The solar panel shown in the figure must be tilted so that angle θ is 94° when the angle of elevation of the sun is 38°. (The sun's rays are assumed to be parallel, and they make an angle of 38° with the horizontal.) The panel is 4.5 ft long. Find h.

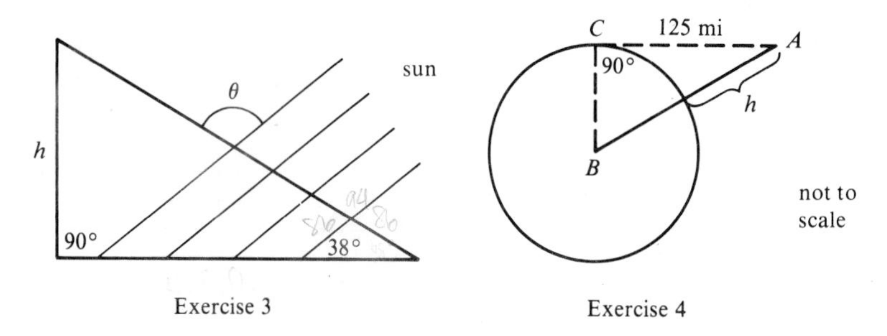

Exercise 3 Exercise 4

4. Find the minimum height h above the surface of the earth so that a pilot at point A in the figure can see an object on the horizon at C, 125 mi away. Assume that the radius of the earth is 4.00×10^3 mi.

5. In one area, the lowest angle of elevation of the sun in winter is 22° 40′. Find the minimum distance x that a plant needing full sun can be placed from a fence 5.58 ft high (see the figure).

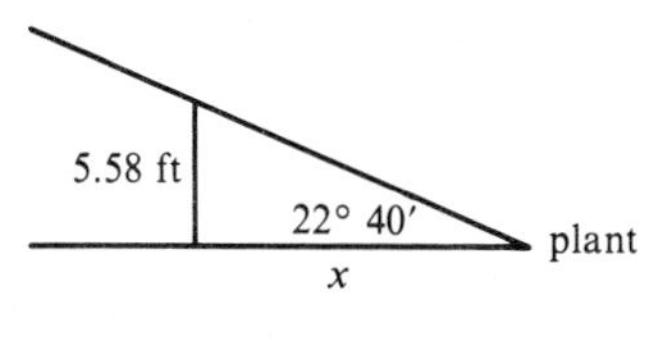

Exercise 5

6. A tunnel is to be dug from A to B (see the figure on the left). Both A and B are visible from C. If AC is 1.4923 mi and BC is 1.0837 mi, and if C is 90°, find the necessary angles A and B.

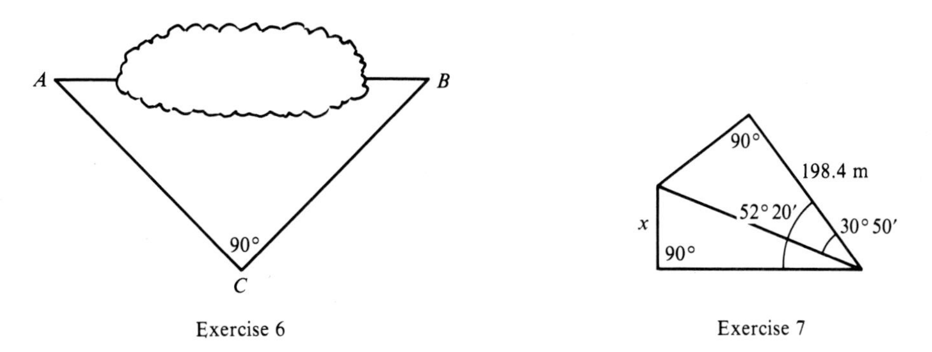

Exercise 6 Exercise 7

7. A piece of land has the shape shown on the right. Find x.

8. Find the value of x in the figure shown below.

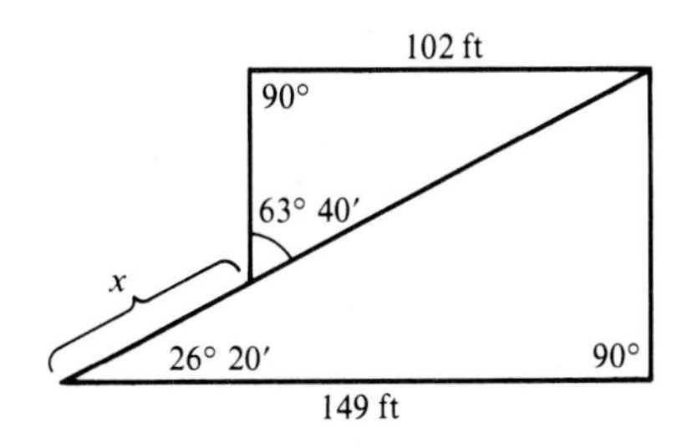

Exercise 8

9. A representation of an aerial photograph of a complex of buildings is shown in the figure.* If the sun was at an angle of 26.5° when the photograph was taken, how high is the rectangular-shaped building? (*Hint:* In the photograph the length of the shadow is .48 cm.)

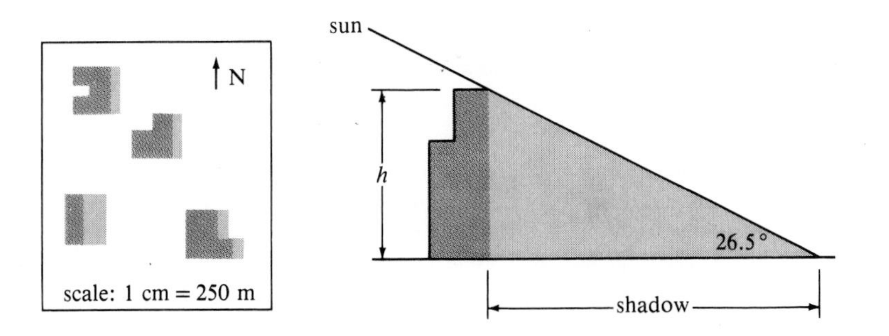

*Exercises 9 and 10 are from *Plane Trigonometry* by Bernard J. Rice and Jerry D. Strange. Copyright © 1981 by Prindle, Weber & Schmidt. Reprinted by permission.

10. The figure represents an aerial photograph of a cliff in a remote region of Antarctica. Compute the height of the cliff if the angle of elevation of the sun was 19.0° when the photograph was taken.

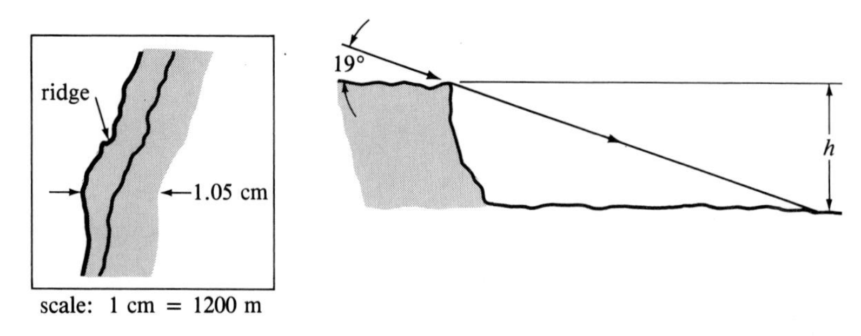

11. Find h.

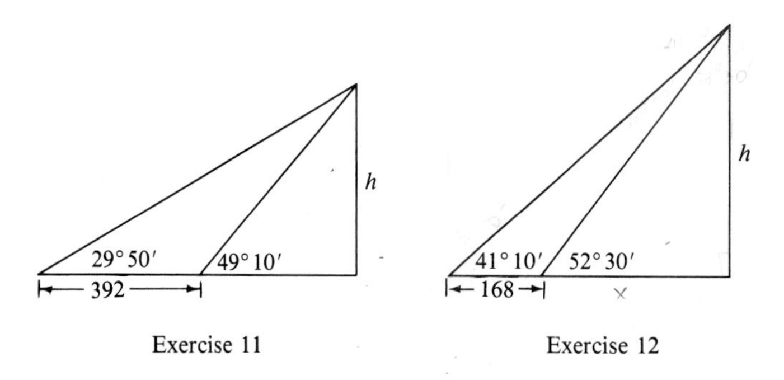

Exercise 11

Exercise 12

12. Find h.

Solve each of the following. See Examples 1–3. Drawing a sketch for these problems where one is not given may be helpful.

13. The angle of elevation from a point on the ground to the top of a pyramid is 35° 30′. The angle of elevation from a point 135 ft farther back to the top of the pyramid is 21° 10′. Find the height of the pyramid.

14. A lighthouse keeper is watching a boat approach the lighthouse directly. When she first begins watching the boat, the angle of depression of the boat is 15° 50′. Just as the boat turns away from the lighthouse, the angle of depression is 35° 40′. If the height of the lighthouse is 68.7 m, find the distance traveled by the boat as it approaches the lighthouse.

15. A television antenna is on top of the center of a house. The angle of elevation from a point 28.0 m from the center of the house to the top of the antenna is 27° 10′, and the angle of elevation to the bottom of the antenna is 18° 10′. Find the height of the antenna.

16. The angle of elevation from Lone Pine to the top of Mt. Whitney is 10° 50′. A driver traveling 7.00 km from Lone Pine along a straight, level road toward Mt. Whitney finds the angle of elevation to be 22° 40′. Find the height of the top of Mt. Whitney above the level of the road.

17. A plane is found by radar to be flying 6000 m above the ground. The angle of elevation from the radar to the plane is 80°. Fifteen seconds later, the plane is directly over the station. (See the figure.) Find the speed of the plane, assuming that it is flying level.

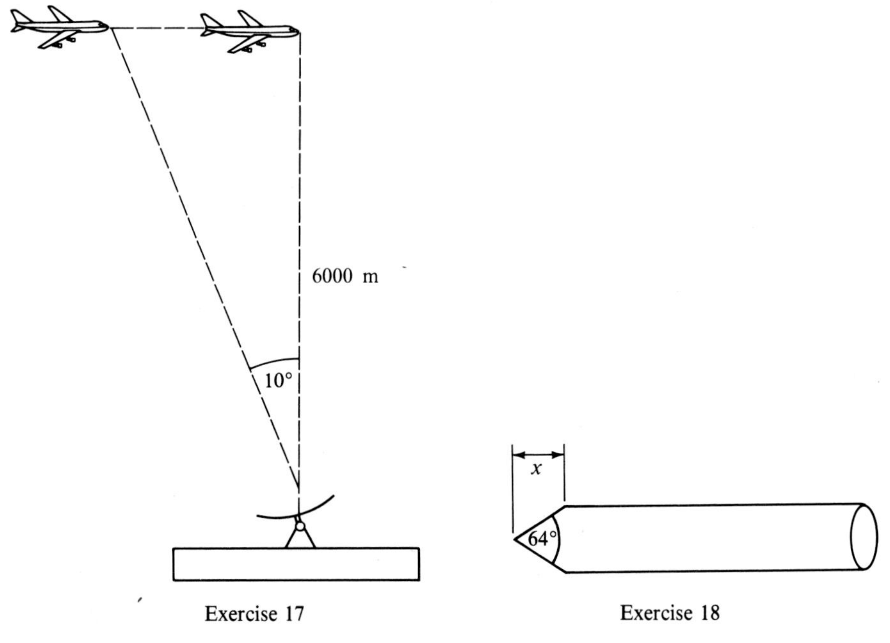

Exercise 17 Exercise 18

18. A chisel is to be made from a steel rod with a diameter of 4.6 cm. If the angle at the tip is 64°, how long will the tip be? (See the figure).

19. A plane flies 2.4 hr at 110 mph on a bearing of 40°. It then turns and flies 1.7 hr at the same speed on a bearing of 130°. How far is the plane from its starting point?

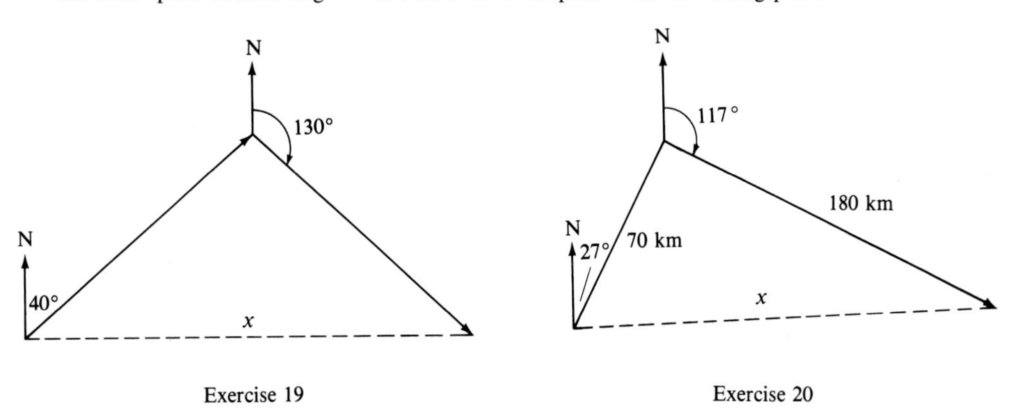

Exercise 19 Exercise 20

20. A ship travels 70 km on a bearing of 27°, and then travels on a bearing of 117° for 180 km. Find the distance of the end of the trip from the starting point.

21. Forest fire lookouts are located at points A and B, which are 12.4 mi apart on an east-west line. From A, the bearing of a fire is 127° 10′. From B, the bearing of the same fire is 217° 10′. Find the distance from B to the fire.

22. Two lighthouses are located on a north-south line. From lighthouse A the bearing of a ship 3742 m away is 129° 43′. From lighthouse B the bearing of the ship is 39° 43′. Find the distance between the lighthouses.

23. A ship leaves port and sails on a bearing of N 28° 10′ E. Another ship leaves the same port at the same time and sails on a bearing of S 61° 50′ E. If the first ship sails at 20.0 mph and the second sails at 24.0 mph, find the distance between the two ships after 5 hrs.

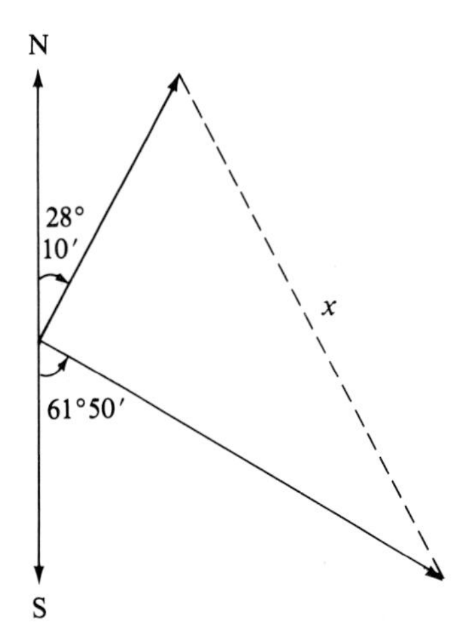

24. Radio direction finders are set up at points A and B, which are 2.00 mi apart on an east-west line. From A it is found that the bearing of the signal from a radio transmitter is N 36° 20′ E, while from B the bearing of the same signal is N 53° 40′ W. Find the distance of the transmitter from B.

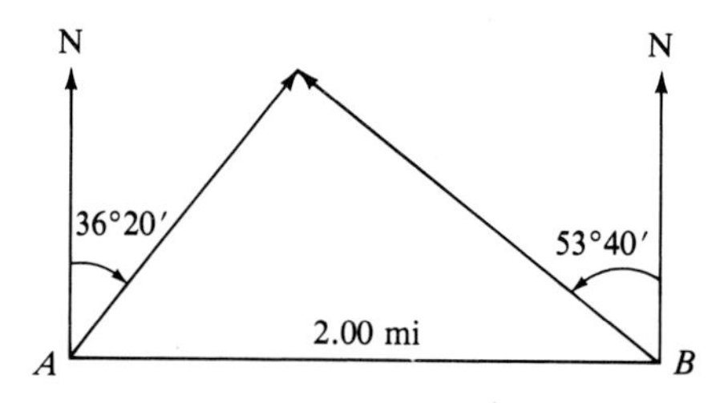

25. The bearing from Winston-Salem, North Carolina, to Danville, Virginia, is N 42° E. The bearing from Danville to Goldsboro, North Carolina, is S 48° E. A small plane piloted by Byron Hopkins, traveling at 60 mph, takes 1 hr to go from Winston-Salem to Danville and 1.8 hr to go from Danville to Goldsboro. Find the distance from Winston-Salem to Goldsboro.

26. The bearing from Atlanta to Macon is S 27° E, and the bearing from Macon to Augusta is N 63° E. A plane traveling at 60 mph needs 1 1/4 hr to go from Atlanta to Macon and 1 3/4 hr to go from Macon to Augusta. Find the distance from Atlanta to Augusta.

Appendix: *Linear Interpolation*

Table 2 gives values of the trigonometric functions for angles measured to the nearest ten minutes, but the table also can be used to find values for angles measured to the nearest minute. This is done with a processs called **linear interpolation,** as explained in the next few examples.

Example 1 Find $\sin 42° 37'$.

First, locate $\sin 42° 30'$ and $\sin 42° 40'$ in Table 2. Recall that in the first quadrant the values of sine increase as the angle measures increase, so

$$\sin 42° 30' < \sin 42° 37' < 42° 40'.$$

Using the values from Table 2 gives the inequality

$$.6756 < \sin 42° 37' < .6777.$$

The difference between the two sine values, .6777 and .6756, is

$$.6777 - .6756 = .0021.$$

Notice that the angle $42° 37'$ is 7/10 of the way between $42° 30'$ and $42° 40'$, so $\sin 42° 37'$ is

$$(.0021)\frac{7}{10} = .00147 \approx .0015 \quad \text{(rounded to four places)}$$

greater than $\sin 42° 30'$.

$$\begin{aligned} \sin 42° 37' &= \sin 42° 30' + \mathbf{.0015} \\ &= .6756 + .0015 \\ &= .6771 \end{aligned}$$

Using a more accurate table or a pocket calculator would show that to five decimal places, $\sin 42° 37' = .67709$. ✣

The work in Example 1 can be arranged in the following compact form.

$$10' \left[\begin{array}{l} 7' \left[\begin{array}{l} \sin 42° 30' = .6756 \\ \sin 42° 37' = \ ? \end{array} \right] d \\ \sin 42° 40' = .6777 \end{array} \right] .0021$$

$$\begin{aligned} \frac{d}{.0021} &= \frac{7}{10} \\ d &= \frac{7(.0021)}{10} \\ d &\approx .0015 \\ \sin 42° 37' &= .6756 + .0015 \\ &= .6771 \end{aligned}$$

Example 2 Find $\cos 58° 54'$.

In the first quadrant, values of cosine *decrease* as the angle measures *increase,* so

$$\cos 58° 50' > \cos 58° 54' > \cos 59° 00'.$$

Because of this, the procedure used above must be modified to *subtract* at the last step. Work as shown in the following compact form.

$$10'\left[\begin{array}{l} 4'\left[\begin{array}{l}\cos 58° 50' = .5175 \\ \cos 58° 54' = \;\;?\end{array}\right] d \\ \cos 59° 00' = .5150\end{array}\right] .0025$$

$$\frac{d}{.0025} = \frac{4}{10}$$

$$d = \frac{4(.0025)}{10} = .0010$$

$$\cos 58° 54' = .5175 - \mathbf{.0010}$$
$$= .5165$$

Therefore $\cos 58° 54' = .5165$ ✣

Example 3 Find a value of θ where $0° \leq \theta \leq 90°$ so that $\tan \theta = .6720$.

To begin, find the two numbers closest to .6720 in Table 2. Here

$$.6703 < .6720 < .6745$$
$$\tan 33° 50' < \tan \theta < \tan 34° 00'.$$

Then proceed as follows.

$$10'\left[\begin{array}{l} d\left[\begin{array}{l}\tan 33° 50' = .6703 \\ \tan \;\;? \;\; = .6720\end{array}\right] .0017 \\ \tan 34° 00' = .6745\end{array}\right] .0042$$

$$\frac{d}{10} = \frac{.0017}{.0042} = \frac{17}{42}$$

$$d = \frac{10(17)}{42} \approx 4$$

$$33° 50' + 4' = 33° 54'$$

Therefore $\theta = 33° 54'$ ✣

Appendix Exercises

Use interpolation to find each of the following values. Use related angles in Exercises 17–20. See Examples 1 and 2.

1. $\tan 29° 42'$ **2.** $\sin 56° 38'$ **3.** $\tan 49° 17'$ **4.** $\sin 78° 32'$

5. $\sec 62° 34'$ **6.** $\sin 74° 08'$ **7.** $\tan 42° 09'$ **8.** $\sec 12° 14'$

9. $\cos 14° 24'$ **10.** $\cos 29° 37'$ **11.** $\cot 71° 12'$ **12.** $\cot 38° 29'$

13. $\cos 78° 45'$ **14.** $\cos 82° 24'$ **15.** $\csc 42° 36'$ **16.** $\csc 71° 08'$

17. $\cot 212° 38'$ **18.** $\cot 257° 44'$ **19.** $\cos 324° 18'$ **20.** $\cos 342° 38'$

Find θ, $0° \leq \theta \leq 90°$, in each of the following. See Example 3.

21. $\sin \theta = .5840$ **22.** $\tan \theta = 1.420$ **23.** $\tan \theta = .4850$ **24.** $\sin \theta = .5160$

25. $\cos \theta = .9285$ **26.** $\cos \theta = .4670$ **27.** $\cot \theta = 2.340$ **28.** $\cot \theta = .7050$

29. $\sec \theta = 3.510$ **30.** $\csc \theta = 2.708$

Chapter 2 *Key Concepts*

Alternative Definitions of the Trigonometric Functions

For any acute angle A in standard position,

$$\sin A = \frac{y}{r} = \frac{\text{side opposite}}{\text{hypotenuse}}$$

$$\cos A = \frac{x}{r} = \frac{\text{side adjacent}}{\text{hypotenuse}}$$

$$\tan A = \frac{y}{x} = \frac{\text{side opposite}}{\text{side adjacent}}$$

$$\csc A = \frac{r}{y} = \frac{\text{hypotenuse}}{\text{side opposite}}$$

$$\sec A = \frac{r}{x} = \frac{\text{hypotenuse}}{\text{side adjacent}}$$

$$\cot A = \frac{x}{y} = \frac{\text{side adjacent}}{\text{side opposite}}.$$

Cofunction Identities

For any acute angle A,

$$\sin A = \cos (90° - A)$$
$$\cos A = \sin (90° - A)$$
$$\tan A = \cot (90° - A)$$
$$\cot A = \tan (90° - A)$$
$$\csc A = \sec (90° - A)$$
$$\sec A = \csc (90° - A).$$

30°–60° Right Triangle and 45°–45° Right Triangle

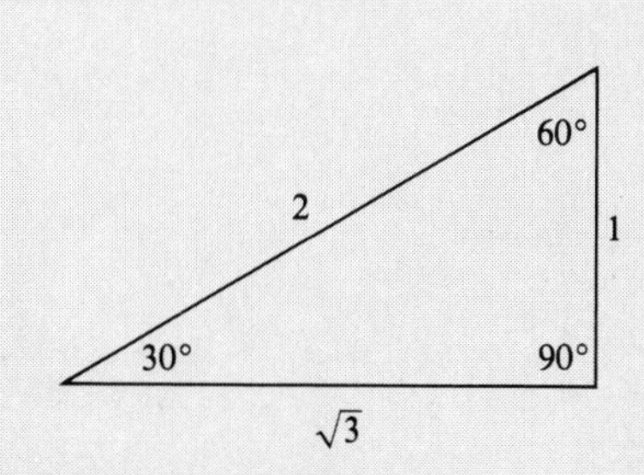

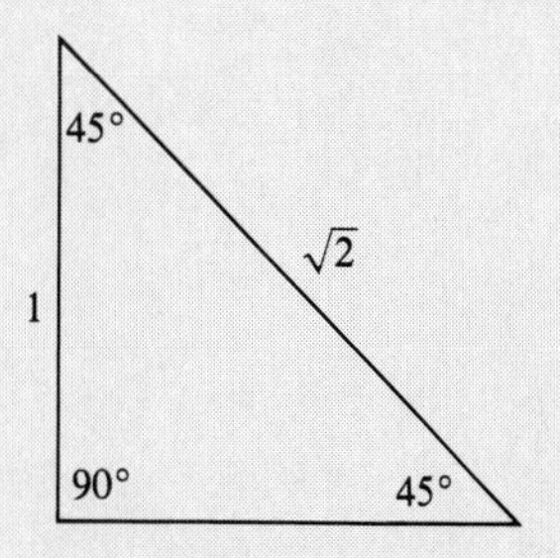

Chapter 2 *Key Concepts (continued)*

Reference Angles

θ *in quadrant*	θ' *is*
I	θ
II	$180° - \theta$
III	$\theta - 180°$
IV	$360° - \theta$

Finding Trigonometric Function Values For Any Angle

1. Add or subtract 360° as many times as needed to get an angle at least 0° but less than 360°.
2. Find the reference angle θ'.
3. Find the trigonometric function value for θ'.
4. Find the correct sign.

Bearing Type 1

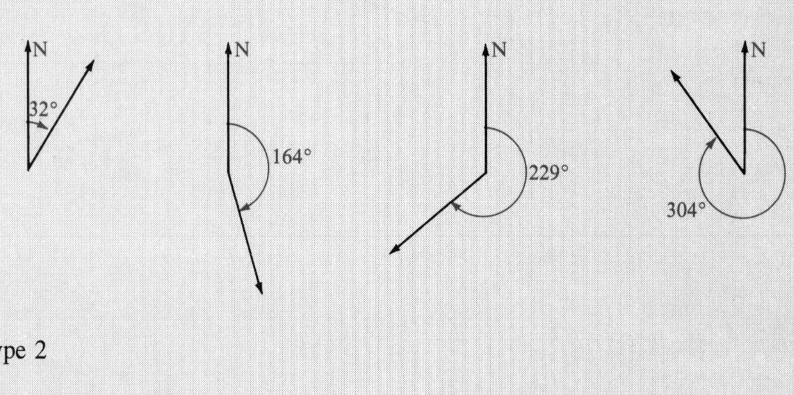

Type 2

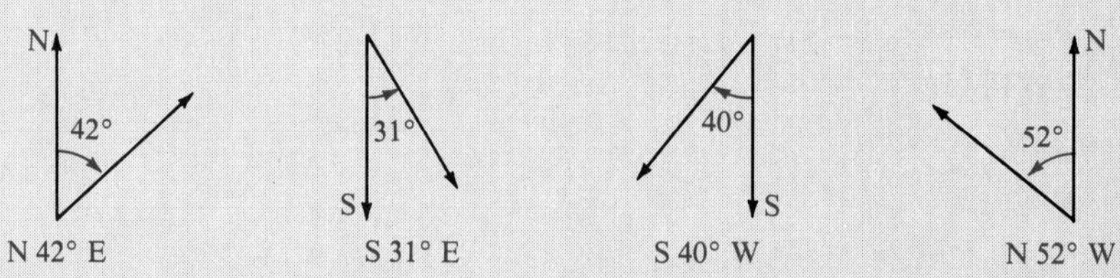

Chapter 2 *Review Exercises*

Find the values of the trigonometric functions for each of the following angles. Do not use tables or a calculator. Give the exact values.

1. 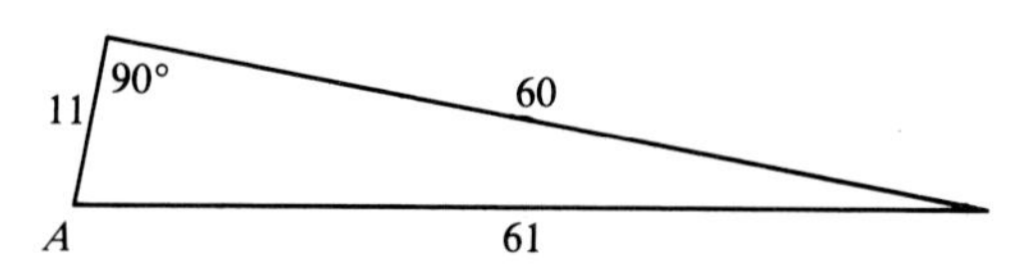

2. 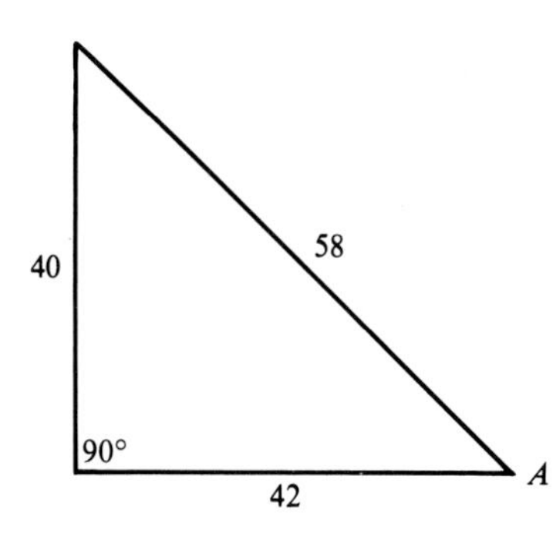

3. $120°$ **4.** $225°$ **5.** $300°$

6. $750°$ **7.** $1230°$ **8.** $2130°$

Solve each of the following. Assume that all angles are positive acute angles.

9. $\sin 4\beta = \cos 5\beta$

10. $\sec(2\gamma + 10°) = \csc(4\gamma + 20°)$

11. $\tan(5x + 11°) = \cot(6x + 2°)$

12. $\cos\left(\frac{3\alpha}{5} + 11°\right) = \sin\left(\frac{7\beta}{10} + 40°\right)$

Tell whether each of the following is true *or* false.

13. $\sin 46° < \sin 58°$ **14.** $\cos 47° < \cos 58°$ **15.** $\sec 48° \geq \cos 42°$ **16.** $\sin 22° \geq \csc 68°$

Evaluate each of the following. Do not use tables or a calculator. Give the exact values.

17. $\cos 60° + 2\sin^2 30°$

18. $\tan^2 120° - 2\cot 240°$

19. $\cot^2 300° + \cos^2 120° - 3\sin^2 240°$

20. $\sec^2 300° - 2\cos^2 150° + \tan 45°$

Tell whether each of the following is true *or* false. *Use a calculator as necessary.*

21. $\sin 50° + \sin 40° = \sin 90°$

22. $\cos 210° = \cos 180° \cdot \cos 30° - \sin 180° \cdot \sin 30°$

23. $\sin 240° = 2\sin 120° \cdot \cos 120°$

24. $\sin 42° + \sin 42° = \sin 84°$

25. $\sin 72.581° \cos 17.419° + \cos 72.581° \sin 17.419° = 1$

26. $\frac{\tan 37.804° + \tan 52.196°}{1 - \tan 37.804° \tan 52.196°}$ does not exist

Use Table 2 or a calculator to find the value of each of the following.

27. $\sin 72° 30'$ **28.** $\sec 222° 30'$ **29.** $\cot 305.6°$ **30.** $\sin 47° 24'$

31. $\cot 32° 42'$ **32.** $\csc 78° 21'$ **33.** $\sec 58.9041°$ **34.** $\tan 11.7689°$

35. $\sin 89.0043°$ **36.** $\cot 1.49783°$

Find the reference angle θ′ for each of the following.

37. $142° 20'$ **38.** $251.9°$ **39.** $578.94°$ **40.** $680° 30'$

Use a calculator or Table 2 to find a value of θ, where $0° \leq \theta \leq 90°$, *for each of the following.*

41. $\sin \theta = .8258$ **42.** $\cot \theta = 1.124$ **43.** $\cos \theta = .9754$ **44.** $\sec \theta = 1.263$

45. $\tan \theta = 1.963$ **46.** $\csc \theta = 9.567$

Find the number of significant digits in each of the following.

47. 28.000 **48.** 1007 **49.** 9.70×10^{-7} **50.** -8.3725×10^{-6}

Round each of the following to three significant digits.

51. 975.5 **52.** .000143259 **53.** 43.980 **54.** 357.658

Solve each of the following right triangles. The right angle is at C.

55.

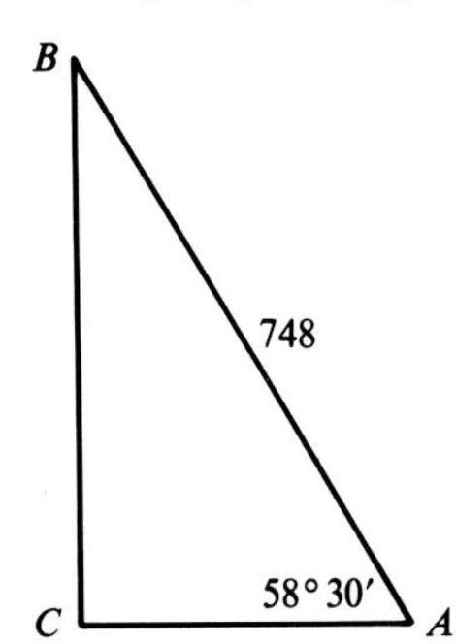

56.

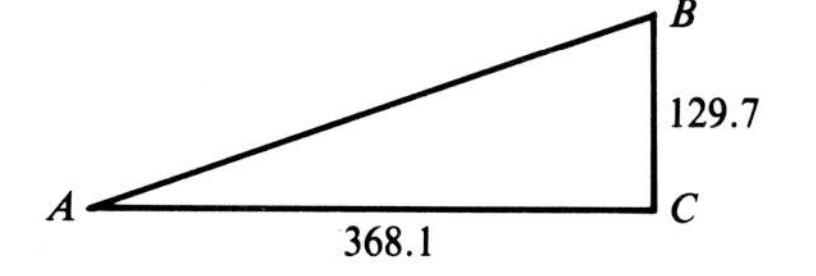

57. $A = 39.72°$, $b = 38.97$ m **58.** $B = 47° 53'$, $b = 298.6$ m

Work each of the following exercises.

59. The angle of elevation from a point 89.6 ft from the base of a tower to the top of the tower is 42° 40′. Find the height of the tower.

60. The angle of depression of a television tower to a point on the ground 32.5 m from the bottom of the tower is 29.5°. Find the height of the tower.

61. A rectangle has adjacent sides measuring 10.93 cm and 15.24 cm. The angle between the diagonal and the longer side is 35.65°. Find the length of the diagonal.

62. As isosceles triangle has a base of length 49.28 m. The angle opposite the base is 58.746°. Find the length of each of the two equal sides.

63. The bearing of B from C is 254°. The bearing of A from C is 344°. The bearing of A from B is 32°. The distance from A to C is 780 m. Find the distance from A to B.

64. A ship leaves a pier on a bearing of S 55° E and travels for 120 km. It then turns and continues on a bearing of N 35° E for 62 km. How far is the ship from the pier?

65. Two cars leave an intersection at the same time. One heads due south at 55 mph. The other travels due west. After two hours, the bearing of the car headed west from the car headed south is 324°. How far apart are they at that time?

66. The figure is an illustration of one of the first engines.* Steam pressure forced the beam to pivot up and down at point X. The beam, in turn, moved the shaft. Gear A then revolved in a path around a gear B of equal diameter, causing gear B to rotate and turn the attached wheel. Suppose that the beam moved from Y to Z, sweeping out an angle of 40°. If YZ equals the diameter of the path of gear A around gear B, and if $XY = 203$ cm, find the diameter of gear B.

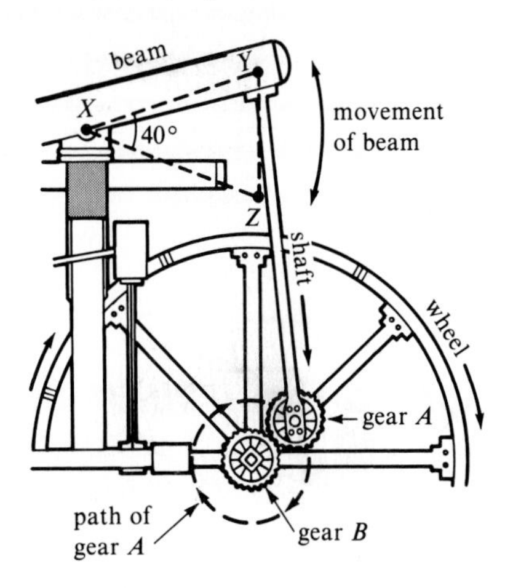

*John A. Graham and Robert H. Sorgenfrey, *Trigonometry with Applications*. Copyright © 1983 by Houghton Mifflin Company. Used with permission.

Radian Measure and Circular Functions

In most work in applied trigonometry, angles are measured in degrees, minutes, and seconds, or in decimal degrees. For several thousand years, the number 360 has been used to represent the number of degrees around a point. Originally it may have been chosen because it is close to the number of days in a year.

In more advanced work in mathematics, especially in calculus, the use of degree measure for angles makes many formulas very complicated. These formulas can be simplified by measuring angles with *radian measure* instead of degrees.

3.1 Radian Measure

Figure 3.1 shows an angle θ in standard position along with a circle of radius r. The vertex of θ is at the center of the circle. Angle θ cuts an arc on the circle equal in length to the radius of the circle. Because of this, angle θ is said to have a measure of one radian.

Radian

An angle that has its vertex at the center of a circle and that cuts an arc on the circle equal in length to the radius of the circle has a measure of **one radian.**

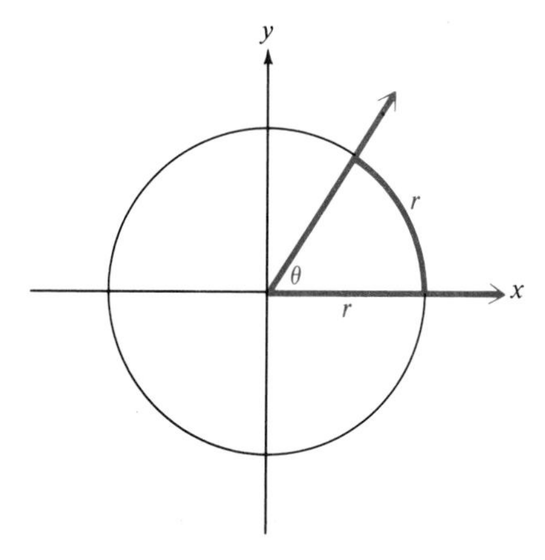

Figure 3.1 *The length of the arc cut by θ is equal to the radius of the circle. Therefore, θ has a measure of 1 radian.*

It follows that an angle of measure 2 radians cuts an arc equal in length to twice the radius of the circle, an angle of measure 1/2 radian cuts an arc equal in length to half the radius of the circle, and so on.

The circumference of a circle, the distance around the circle, is given by $C = 2\pi r$, where r is the radius of the circle. The formula $C = 2\pi r$ shows that the radius can be laid off 2π times around a circle. Therefore, an angle of 360°, which corresponds to a complete circle, cuts an arc equal in length to 2π times the radius of the circle. Because of this, an angle of 360° has a measure of 2π radians:

$$360° = 2\pi \text{ radians.}$$

The radian measure of an angle is the ratio of the arc length cut by the angle to the radius of the circle. In this ratio, the units of measure "cancel," leaving only a number. For this reason, radian measure is just a real number; there are no units associated with a radian measure. Because of this, $360° = 2\pi$ radians is written as just

$$360° = 2\pi.$$

An angle of 180° is half the size of an angle of 360°, so an angle of 180° has half the radian measure of an angle of 360°.

$$180° = \frac{1}{2}(2\pi) \text{ radians}$$

$$\mathbf{180° = \pi \text{ radians}}$$

Dividing both sides of $180° = \pi$ radians by π leads to

$$\textbf{1 radian} = \frac{\mathbf{180°}}{\pi},$$

or, approximately,

$$1 \text{ radian} \approx \frac{180°}{3.14159} \approx 57.296° \approx 57°\ 17'\ 45''.$$

Since $180° = \pi$ radians, dividing both sides by 180 gives

$$\mathbf{1°} = \frac{\pi}{\mathbf{180}} \textbf{ radians},$$

or, approximately,

$$1° \approx \frac{3.14159}{180} \text{ radians} \approx .0174533 \text{ radians}.$$

Angle measures can be converted back and forth between degrees and radians by either of two methods.

Converting Between Degrees and Radians

1. Proportion: $\dfrac{\text{radian measure}}{\pi} = \dfrac{\text{degree measure}}{180}$

2. Formulas:

From	*To*	*Multiply by*
Radians	Degrees	$\dfrac{180°}{\pi}$
Degrees	Radians	$\dfrac{\pi}{180°}$

Example 1 Convert each degree measure to radians.

(a) 45°

By the proportion method,

$$\frac{\text{radian measure}}{\pi} = \frac{45}{180}.$$

Multiply both sides by π.

$$\text{radian measure} = \frac{45\pi}{180} = \frac{\pi}{4}$$

To use the formula method, multiply by $\pi/180°$.

$$45° = 45°\left(\frac{\pi}{\mathbf{180°}}\right) = \frac{45\pi}{180} = \frac{\pi}{4} \text{ radians}$$

(b) Using the formula, $240° = 240°\left(\frac{\pi}{180°}\right) = \frac{4\pi}{3}$ radians ✥

Example 2 Convert to radians.

(a) $29°\ 40'$

Since $40' = 40/60 = 2/3$ of a degree,

$$\begin{aligned} 29°\ 40' &= 29\frac{2}{3}° \\ &= \frac{89°}{3} \\ &= \frac{89°}{3}\left(\frac{\pi}{180°}\right) \text{ radians} \\ &= \frac{89\pi}{540} \text{ radians.} \end{aligned}$$

The formula method was used here. This answer is exact. If π is replaced with the approximation 3.14159,

$$29°\ 40' \approx \frac{89(\mathbf{3.14159})}{540} \approx .518 \text{ radians.}$$

(b) $74.9162°$

Using the formula method gives

$$\begin{aligned} 74.9162° &= 74.9162°\left(\frac{\pi}{180°}\right) && \textbf{Multiply by } \frac{\pi}{180°} \\ &\approx 74.9162°\left(\frac{3.14159}{180°}\right) \\ &\approx 1.30753 \text{ radians.} \end{aligned}$$ ✥

Example 3 Convert both of the following radian measures to degrees.

(a) $\frac{9\pi}{4}$

Using the proportion method,

$$\begin{aligned} \frac{\frac{9\pi}{4}}{\pi} &= \frac{x}{180°} \\ \frac{9}{4} &= \frac{x}{180°} \\ x &= \frac{9}{4}(180°) = 405°. \end{aligned}$$

(b) $\frac{11\pi}{3}$

$$\frac{\frac{11\pi}{3}}{\pi} = \frac{x}{180°}$$

$$\frac{11}{3} = \frac{x}{180°}$$

$$x = \frac{11}{3}(180°) = 660°$$

Example 4 Convert 4.2 radians to degrees. Write the result to the nearest minute.

By the formula,

$$\begin{aligned} 4.2 \text{ radians} &= 4.2\left(\frac{180°}{\pi}\right) \\ &\approx \frac{4.2(180°)}{3.14159} \\ &\approx 240.64° \\ &\approx 240° + .64° \\ &= 240° + .64(60') \\ &\approx 240° + 38' \\ &= 240°\ 38'. \end{aligned}$$

Many calculators have a $\boxed{\pi}$ key. This key could be used for the conversion in Example 4. A $\boxed{\pi}$ key often gives the value of π to fifteen decimal places, even though the calculator displays only eight or ten places. The extra digits are stored internally and used in calculations. Because of all these decimal places, be sure to round your results to the proper number of significant digits when using the $\boxed{\pi}$ key.

Some calculators feature a key that automatically converts between degrees and radians. With such a key, 84.9076° could be converted to radians as follows.

Enter: 84.9076	Use appropriate key	**Display:** 1.4819172

Converting from radians to degrees can also be done on these calculators.

Enter: 4.2	Use appropriate key	**Display:** 240.64227
30		1718.8734

This last result shows that 30 radians is equivalent to about 1719°. Figure 3.2 shows angles of measure of 30 radians and 30°; as the figure shows, these angle measures are not at all close. Be careful not to confuse these two measures.

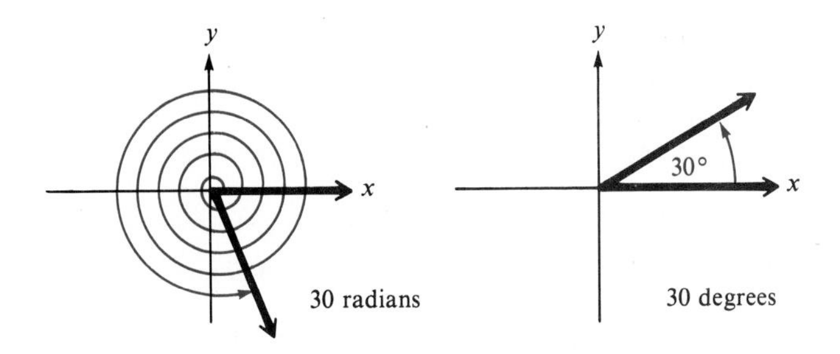

Figure 3.2

Trigonometric function values for angles measured in radians can be found by first converting the radian measure to degrees. (You should try to skip this intermediate step as soon as possible, and find the function values directly from the radian measure.)*

Example 5 Find $\tan \dfrac{2\pi}{3}$.

First convert $2\pi/3$ radians to degrees.

$$\begin{aligned}\tan \frac{2\pi}{3} &= \tan \frac{2\pi}{3} \cdot \frac{180^\circ}{\pi}\\ &= \tan 120^\circ\\ &= -\sqrt{3}\end{aligned}$$

The following table and Figure 3.3 give some equivalent angles measured in degrees and radians. It will be useful to memorize these equivalent values. Keep

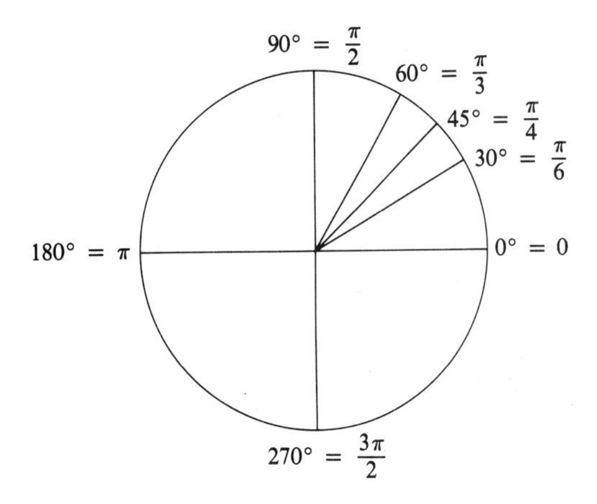

Figure 3.3

*A table giving the values of the trigonometric functions for common radian and degree measures is given inside the front of this book.

in mind that $180° = \pi$ radians. Then it will be easy to reproduce the rest of the table.

Equivalent Angle Measures in Degrees and Radians

Degrees	*Radians*	
	exact	*approximate*
0°	0	0
30°	$\pi/6$	.52
45°	$\pi/4$	.79
60°	$\pi/3$	1.05
90°	$\pi/2$	1.57
180°	π	3.14
270°	$3\pi/2$	4.71
360°	2π	6.28

Example 6 Find $\sin 3\pi/2$.

From the table, $3\pi/2 = 270°$, so

$$\sin \frac{3\pi}{2} = \sin 270° = -1.$$

3.1 Exercises

Convert each of the following degree measures to radians. Leave answers as multiples of π. See Example 1.

1. 60° **2.** 30° **3.** 90° **4.** 120°
5. 150° **6.** 135° **7.** 210° **8.** 270°
9. 300° **10.** 315° **11.** 450° **12.** 480°
13. 20° **14.** 80° **15.** 140° **16.** 320°

Convert each of the following degree measures to radians. See Example 2.

17. 39° **18.** 74° **19.** 42° 30′ **20.** 53° 40′
21. 139° 10′ **22.** 174° 50′ **23.** 64.29° **24.** 85.04°
25. 56° 25′ **26.** 122° 37′ **27.** 47.6925° **28.** 23.0143°
29. −29° 42′ 36″ **30.** −157° 11′ 9″ **31.** −209° 46′ 15″ **32.** −387° 05′ 09″

Convert each of the following radian measures to degrees. See Example 3.

33. $\frac{\pi}{3}$ **34.** $\frac{8\pi}{3}$ **35.** $\frac{7\pi}{4}$ **36.** $\frac{2\pi}{3}$

37. $\frac{11\pi}{6}$ **38.** $\frac{15\pi}{4}$ **39.** $-\frac{\pi}{6}$ **40.** $-\frac{\pi}{4}$

41. $\frac{8\pi}{5}$ **42.** $\frac{7\pi}{10}$ **43.** $\frac{11\pi}{15}$ **44.** $\frac{4\pi}{15}$

45. $\frac{7\pi}{20}$ **46.** $\frac{17\pi}{20}$ **47.** $\frac{11\pi}{30}$ **48.** $\frac{15\pi}{32}$

Convert each of the following radian measures to degrees. Write answers with four decimal places and also to the nearest minute. See Example 4.

49. 2 **50.** 5 **51.** 1.74 **52.** 3.06

53. .0912 **54.** .3417 **55.** 9.84763 **56.** 5.01095

57. -3.47189 **58.** -1.28306 **59.** -4.19806 **60.** -11.5138

Find each of the following without using a calculator or a table. See Examples 5 and 6.

61. $\sin\frac{\pi}{3}$ **62.** $\cos\frac{\pi}{6}$ **63.** $\tan\frac{\pi}{4}$ **64.** $\cot\frac{\pi}{3}$

65. $\sec\frac{\pi}{6}$ **66.** $\csc\frac{\pi}{4}$ **67.** $\sin\frac{\pi}{2}$ **68.** $\csc\frac{\pi}{2}$

69. $\tan\frac{2\pi}{3}$ **70.** $\cot\frac{2\pi}{3}$ **71.** $\sin\frac{5\pi}{6}$ **72.** $\tan\frac{5\pi}{6}$

73. $\cos 3\pi$ **74.** $\sec \pi$ **75.** $\sin\frac{4\pi}{3}$ **76.** $\cot\frac{4\pi}{3}$

77. $\tan\frac{5\pi}{4}$ **78.** $\csc\frac{5\pi}{4}$ **79.** $\sin 3\pi$ **80.** $\cos 5\pi$

81. $\tan\left(-\frac{\pi}{3}\right)$ **82.** $\cot\left(-\frac{2\pi}{3}\right)$ **83.** $\sin\left(-\frac{7\pi}{6}\right)$ **84.** $\cos\left(-\frac{\pi}{6}\right)$

85. The figure shows the same angles measured in both degrees and radians. Complete the missing measures.

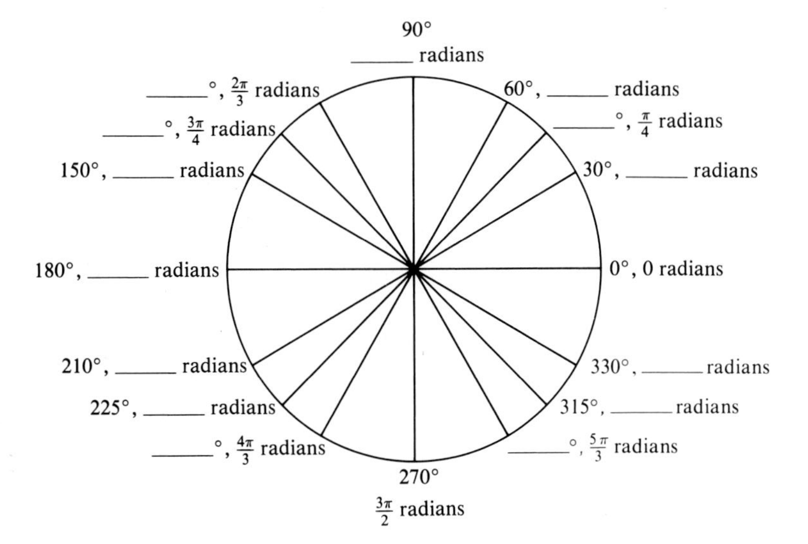

86. A circular pulley is rotating about its center. Through how many radians would it turn in (a) 7 rotations, and (b) 25 rotations?

87. Through how many radians will the hour hand on a clock rotate in (a) 24 hours, and (b) 4 hours?

88. A space vehicle is orbiting the earth in a circular orbit. What radian measure corresponds to (a) 1.5 orbits, and (b) 2/3 of an orbit?

Find the radian measure (as a multiple of π) of the smallest angle between the hands of a clock at each of the following times. Assume that the hour hand is exactly on the given hour.

89. 12:15 **90.** 12:30 **91.** 2:40

92. 3:35 **93.** 9:25 **94.** 10:40

3.2 Applications of Radian Measure

As mentioned in the previous section, radian measure is used to simplify certain formulas. Two of these formulas are discussed in this section. Both would be more complicated if expressed in degrees.

The first of these formulas is used to find the length of an arc of a circle. The formula comes from the fact (proven in plane geometry) that the length of an arc is proportional to the measure of its central angle.

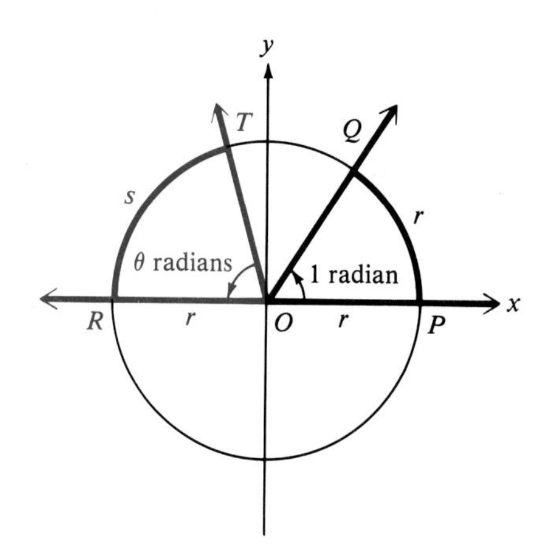

Figure 3.4

In Figure 3.4, angle QOP has a measure of 1 radian and cuts an arc of length r on the circle. Angle ROT has a measure of θ radians and cuts an arc of length s on the circle. Since the lengths of the arcs are proportional to the measure of their central angles,

$$\frac{s}{r} = \frac{\theta}{1}.$$

Multiplying both sides by r gives the following result.

Length of Arc

The length s of the arc cut on a circle of radius r by a central angle of measure θ radians is given by the product of the radius and the radian measure of the angle, or

$$s = r\theta, \qquad \theta \text{ in radians.}$$

This formula is a good example of the usefulness of radian measure. To see why, try to write the equivalent formula for an angle measured in degrees.

Example 1 A circle has a radius of 18.2 cm. Find the length of the arc cut by a central angle having the following measures.

(a) $\frac{3\pi}{8}$ radians

Here $r = 18.2$ cm and $\theta = 3\pi/8$. Since $s = r\theta$,

$$s = 18.2\left(\frac{3\pi}{8}\right) \text{ cm}$$

$$s = \frac{54.6\pi}{8} \text{ cm} \qquad \text{The exact answer}$$

or

$$s \approx 21.4 \text{ cm.}$$

(b) $144°$

The formula $s = r\theta$ requires that θ be measured in radians. First, convert θ to radians by the methods of the previous section.

$$144° = 144°\left(\frac{\pi}{180°}\right) \text{ radians}$$

$$144° = \frac{4\pi}{5} \text{ radians}$$

Now

$$s = 18.2\left(\frac{4\pi}{5}\right) \text{ cm}$$

$$s = \frac{72.8\pi}{5} \text{ cm,}$$

or

$$s \approx 45.7 \text{ cm.}$$ ✥

Example 2 Reno, Nevada, is approximately due north of Los Angeles. The latitude of Reno is 40° N, while that of Los Angeles is 34° N. (The N in 34° N means *north* of the equator.) If the radius of the earth is 6400 km, find the north-south distance between the two cities.

Latitude gives the measure of a central angle with vertex at the earth's center whose initial side goes through the earth's equator and whose terminal side goes through the given location. As shown in Figure 3.5, the central angle for Reno and Los Angeles is 6°. The distance between the two cities can thus be found by the formula $s = r\theta$, after 6° is first converted to radians.

$$6° = 6°\left(\frac{\pi}{180°}\right) = \frac{\pi}{30} \text{ radians}$$

The distance between the two cities is

$$s = r\theta$$

$$s = 6400\left(\frac{\pi}{30}\right) \text{ km}$$

$$= 670 \text{ km.}$$

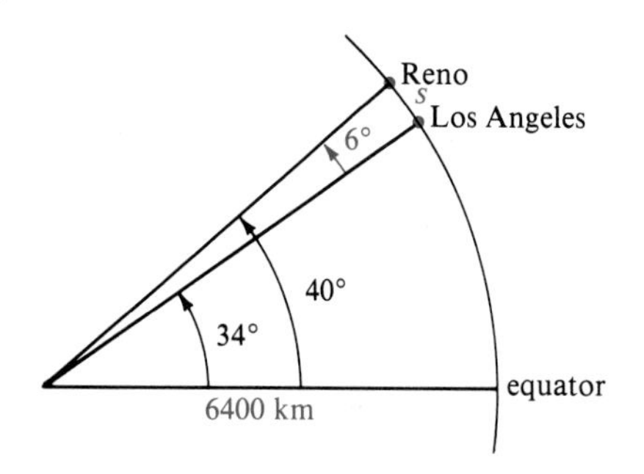

Figure 3.5

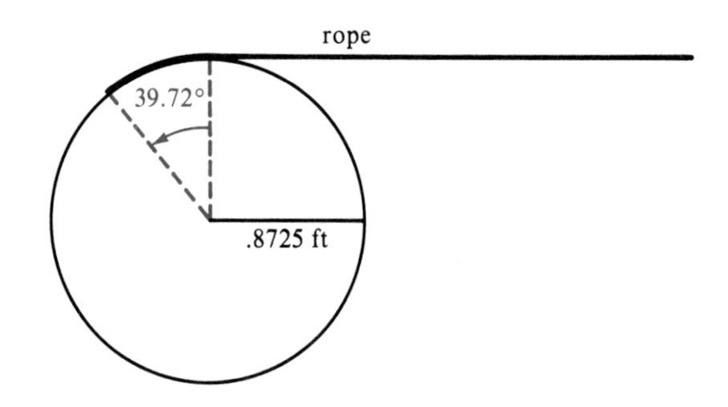

Figure 3.6

Example 3 A rope is being wound around a drum with radius .8725 ft. (See Figure 3.6.) How much rope will be wound around the drum if the drum is rotated through an angle of 39.72°?

The length of rope wound around the drum is just the arc length for a circle of radius .8725 ft and a central angle of 39.72°. Use the formula $s = r\theta$, with the angle converted to radian measure.

$$s = r\theta$$

$$s = (.8725)(39.72°)\left(\frac{\pi}{180°}\right)$$

$$= .6049$$

Rope of length .6049 ft will be wound around the drum.

Example 4 Two gears are adjusted so that the smaller gear drives the larger one as shown in Figure 3.7. If the smaller gear rotates through 225°, through how many degrees will the larger gear rotate?

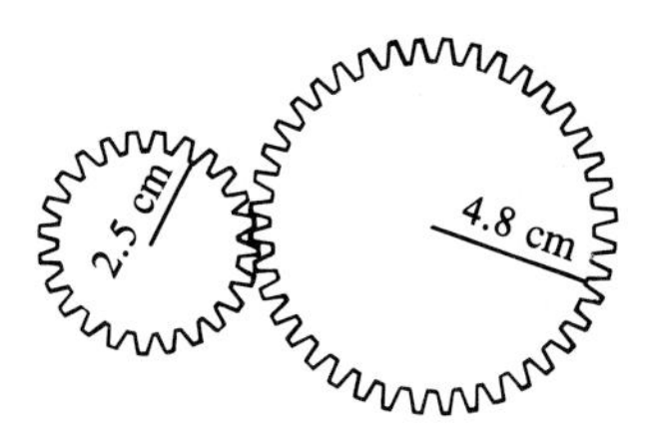

Figure 3.7

First find the radian measure of the angle, which will give the arc length on the smaller gear that determines the motion of the larger gear. Since $225° = 5\pi/4$ radians, for the smaller gear,

$$s = r\theta = 2.5\left(\frac{5\pi}{4}\right) \approx 9.8 \text{ cm}.$$

An arc length of 9.8 cm on the larger gear corresponds to an angle measure θ, in radians, of

$$s = r\theta$$
$$9.8 = 4.8\theta$$
$$2.1 = \theta.$$

Changing back to degrees shows that the larger gear rotates through

$$2.1\left(\frac{180°}{\pi}\right) \approx 120°,$$

to two significant figures.

Sector of a Circle The other useful formula given in this section is used to find the area of a "piece of pie," or sector. A **sector of a circle** is the portion of the interior of a circle cut by a central angle. See Figure 3.8.

To find the area of a sector, assume that the radius of the circle is r. A complete circle can be thought of as an angle with a measure of 2π radians. If a

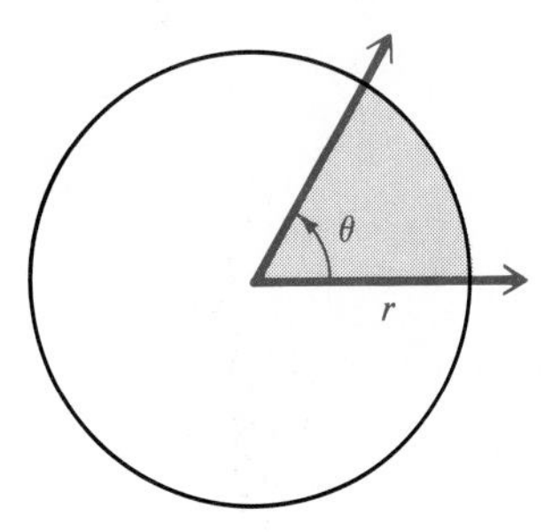

Figure 3.8

central angle for the sector has measure θ radians, then the sector makes up a fraction $\theta/(2\pi)$ of a complete circle. The area of a complete circle is $A = \pi r^2$. Therefore, the area of the sector is given by the product of the fraction $\theta/(2\pi)$ and the total area, πr^2, or

$$\text{area of sector} = \frac{\theta}{2\pi} \cdot \pi r^2 = \frac{1}{2} r^2 \theta, \qquad \theta \text{ in radians.}$$

This discussion is summarized as follows.

Area of Sector

The area of a sector of a circle of radius r and central angle θ in radians is given by

$$A = \frac{1}{2} r^2 \theta, \qquad \theta \text{ in radians.}$$

Example 5 Figure 3.9 shows a field in the shape of a sector of a circle. The central angle is 15° and the radius of the circle is 321 m. Find the area of the field.

First, convert 15° to radians.

$$15^\circ = 15^\circ \left(\frac{\pi}{180^\circ}\right) = \frac{\pi}{12} \text{ radians}$$

Figure 3.9 *This field is a sector of a circle.*

Now, use the formula

$$\text{area of sector} = \frac{1}{2}r^2\theta$$

$$\text{to get} = \frac{1}{2}(321)^2\left(\frac{\pi}{12}\right)$$

$$\approx \frac{1}{2}(103{,}041)\left(\frac{3.14159}{12}\right) \approx 13{,}500 \text{ m}^2.$$

3.2 Exercises

Find the arc length cut by each of the following angles. See Example 1.

1. $r = 8.00$ in, $\theta = \pi$ radians

2. $r = 72.0$ ft, $\theta = \pi/8$ radians

3. $r = 12.3$ cm, $\theta = 2\pi/3$ radians

4. $r = .892$ cm, $\theta = 11\pi/10$ radians

5. $r = 253$ m, $\theta = 2\pi/5$ radians

6. $r = 120$ mm, $\theta = \pi/9$ radians

7. $r = 4.82$ m, $\theta = 60°$

8. $r = 71.9$ cm, $\theta = 135°$

9. $r = 58.402$ m, $\theta = 52.417°$

10. $r = 39.4$ cm, $\theta = 68.059°$

Find the distance in kilometers between each of the following pairs of cities whose latitudes are given. Assume that the cities are on a north-south line and that the radius of the earth is 6400 *km. See Example 2.*

11. Grand Portage, Minnesota, 44° N, and New Orleans, Louisiana, 30° N

12. St. Petersburg, Florida, 28° N, and Detroit, Michigan, 42° N

13. Madison, South Dakota, 44° N, and Dallas, Texas, 33° N

14. Charleston, South Carolina, 33° N, and Toronto, Ontario, 43° N

15. Panama City, Panama, 9° N, and Pittsburgh, Pennsylvania, 40° N

16. Farmersville, California, 36° N, and Penticton, British Columbia, 49° N

17. New York City, New York, 41° N, and Lima, Peru, 12° S

18. Halifax, Nova Scotia, 45° N, and Buenos Aires, Argentina, 34° S

Longitude *gives the measure of a central angle with vertex at the center of the earth measured east or west of the prime meridian in Greenwich, England. See the figure.*

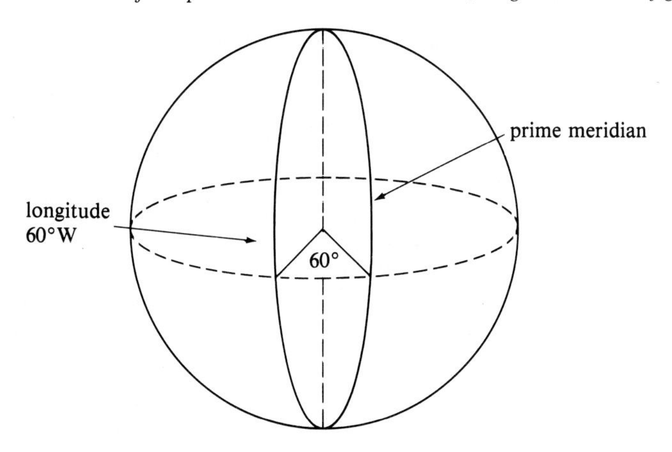

Find the distance between the following pairs of cities located along the equator.

19. Nairobi, Kenya, longitude 40° E, and Singapore, Republic of Singapore, 105° E

20. Quito, Ecuador, longitude 80° W, and Libreville, Gabon, 10° E

Work the following exercises. See Examples 3 and 4.

21. **(a)** How many inches will the weight on the left rise if the pulley is rotated through an angle of 71° 50′?
(b) Through what angle, to the nearest minute, must the pulley be rotated to raise the weight 6 in?

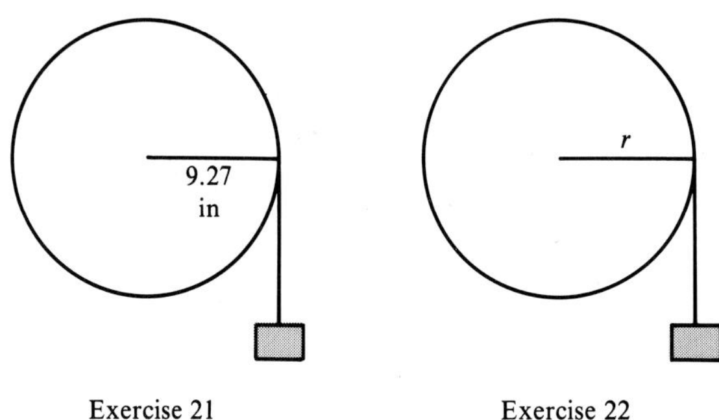

Exercise 21 Exercise 22

22. Find the radius of the pulley on the right if a rotation of 51.6° raises the weight 11.4 cm.

23. The rotation of the smaller wheel in the figure on the left causes the larger wheel to rotate. Through how many degrees will the larger wheel rotate if the smaller one rotates through 60.0°?

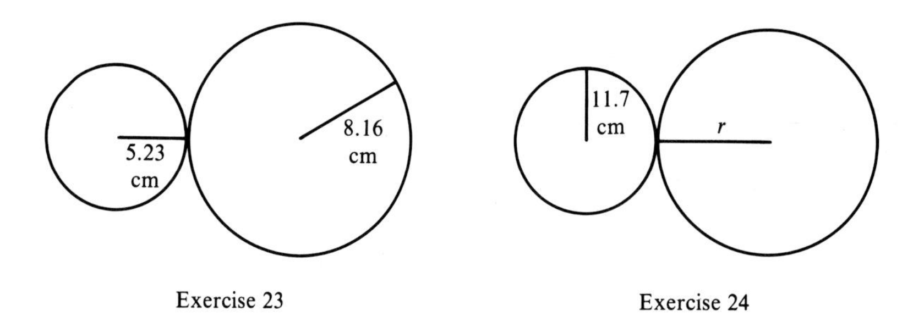

Exercise 23 Exercise 24

24. Find the radius of the larger wheel in the figure on the right if the smaller wheel rotates 80° when the larger wheel rotates 50°.

25. The following figure shows the chain drive of a bicycle. How far will the bicycle move if the pedals are rotated through 180°? Assume that the radius of the bicycle wheel is 13.6 in.

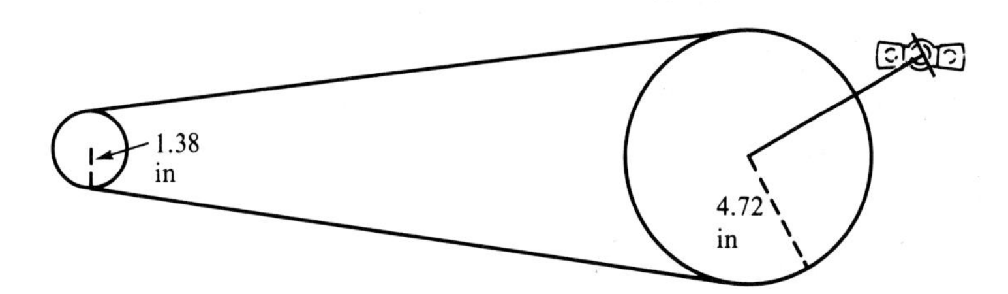

26. The speedometer of a small pickup truck is designed to be accurate with tires of radius 14 in.
(a) Find the number of rotations of a tire in 1 hr if the truck is driven at 55 mph.
(b) Suppose that oversize tires of radius 16 in are placed on the truck. If the truck is now driven for 1 hr with the speedometer reading 55 mph, how far has the truck gone? If the speed limit is 55 mph, does the driver deserve a speeding ticket?

If a central angle is very small, there is little difference in length between an arc and the inscribed chord. See the figure. Approximate each of the following lengths by finding the necessary arc length.

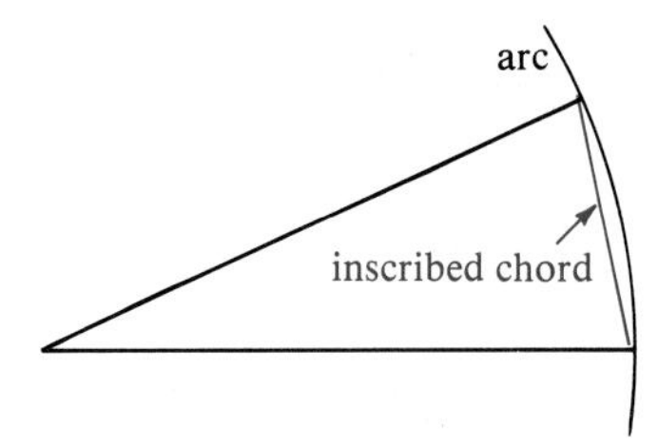

27. A tree 600 m away intercepts an angle of 2°. Find the height of the tree.

28. A building 500 m away intercepts an angle of 3°. Find the height of the building.

29. A railroad track in the desert is 3.5 km away. A train on the track intercepts (horizontally) an angle of 3° 20′. Find the length of the train.

30. An oil tanker 2.3 km at sea intercepts a 1° 30′ angle horizontally. Find the length of the ship.

31. The full moon intercepts an angle of 1/2°. The moon is 240,000 mi away. Find the diameter of the moon.

32. A building with a height of 58 m intercepts an angle of 1° 20′. How far away is the building?

33. The mast of Gale Stockdale's boat is 32 ft high. If it intercepts an angle of 2° 10′, how far away is it?

34. A television tower 530 m high intercepts an angle of 2° 40′. How far away is the tower?

Find the areas of each of the following sectors of a circle. See Example 5.

35. $r = 9.0$ m, $\theta = \pi/3$ radians

36. $r = 15$ ft, $\theta = \pi/5$ radians

37. $r = 29.2$ m, $\theta = 5\pi/6$ radians

38. $r = 59.8$ km, $\theta = 2\pi/3$ radians

39. $r = 52$ cm, $\theta = 3\pi/10$ radians

40. $r = 25$ mm, $\theta = \pi/15$ radians

41. $r = 12.7$ cm, $\theta = 81°$

42. $r = 18.3$ m, $\theta = 125°$

43. $r = 32.6$ m, $\theta = 38° 40'$

44. $r = 59.8$ ft, $\theta = 74° 30'$

45. $r = 86.243$ m, $\theta = 11.7142°$

46. $r = 111.976$ cm, $\theta = 29.8477°$

47. The figure shows Medicine Wheel, an Indian structure in northern Wyoming. This circular structure is perhaps 200 years old. There are 32 spokes in the wheel, all equally spaced.

(a) Find the measure of each central angle in degrees and in radians.

(b) If the radius of the wheel is 76 ft, find the circumference.

(c) Find the length of each arc intercepted by consecutive pairs of spokes.

(d) Find the area of each sector formed by consecutive spokes.

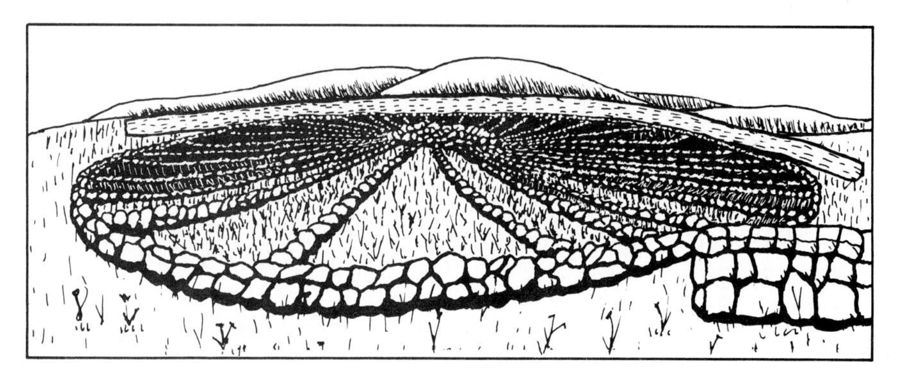

48. The unusual corral in the photograph is separated into 26 areas, many of which approximate sectors of a circle. Assume that the corral has a diameter of 50 m.

(a) Find the central angle for each region, assuming that the 26 regions are all equal sectors, with the fences meeting at the center.
(b) What is the area of each sector?

49. The area of a sector is given approximately by

$$A \approx 0.0087270r^2,$$

where r is the radius of the circle and θ is measured in degrees. Show how this formula was obtained.

Use the formula in Exercise 49 to find the areas of the following sectors.

50. $\theta = 27.472°$, $r = 11.946$ m

51. $\theta = 51.0382°$, $r = 25.1837$ cm

52. $\theta = 38° \ 42' \ 15''$, $r = 42.89734$ cm

53. $\theta = 246° \ 15' \ 37''$, $r = 8.76122$ mm

54. Eratosthenes (*ca.* 230 B.C.) made a famous measurement of the earth. He observed at Syene (the modern Aswan) at noon and at the summer solstice that a vertical stick had no shadow, while at Alexandria (on the same meridian as Syene) the sun's rays were inclined 1/50 of a complete circle to the vertical. See the figure. He then calculated the circumference of the earth from the known distance of 5000 stades between Alexandria and Syene. Obtain Eratosthenes' result of 250,000 stades for the circumference of the earth. There is reason to suppose that a stade is about equal to 516.7 ft. Assuming this, use Eratosthenes' result to calculate the polar diameter of the earth in miles. (The actual polar diameter of the earth, to the nearest mile, is 7900 mi.)*

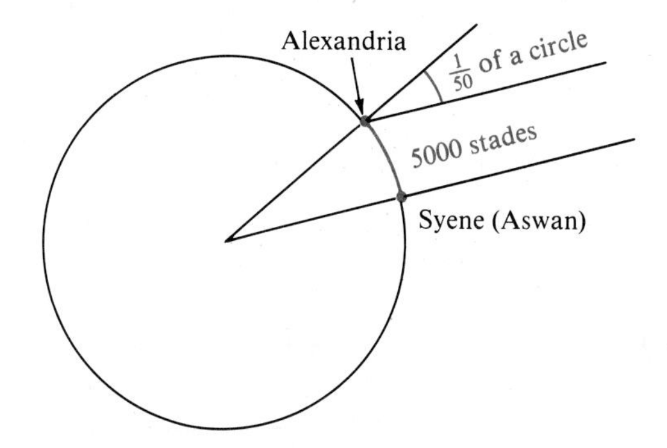

Multiply the area of the base times the height to find the volume of each solid.

55.

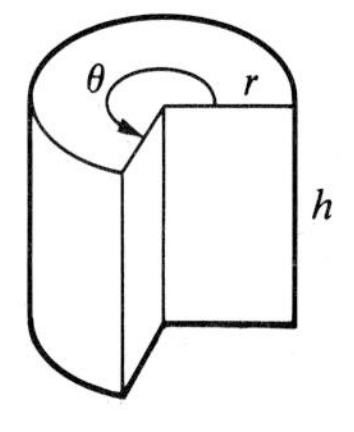

56.

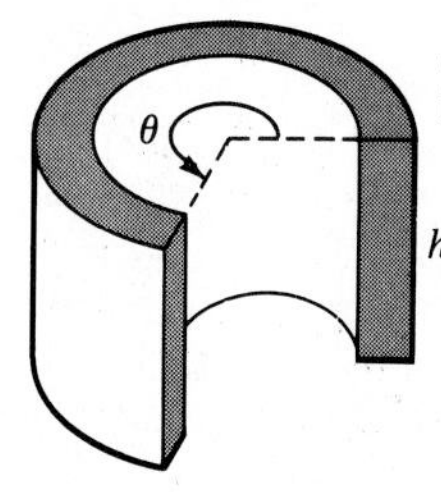

outside radius is r_1,
inside radius is r_2

*Exercise reproduced with permission from *A Survey of Geometry,* v. 1, by Howard Eves, Boston: Allyn and Bacon, 1963.

3.3 Circular Functions of Real Numbers

So far we have defined the six trigonometric functions for *angles*. The angles can be measured either in degrees or in radians. While the domain of the trigonometric functions is a set of angles, the range is a set of real numbers. In advanced work, such as calculus, it is necessary to modify the trigonometric functions so that the domain contains not angles, but real numbers. Trigonometric functions having a domain of real numbers are called **circular functions.**

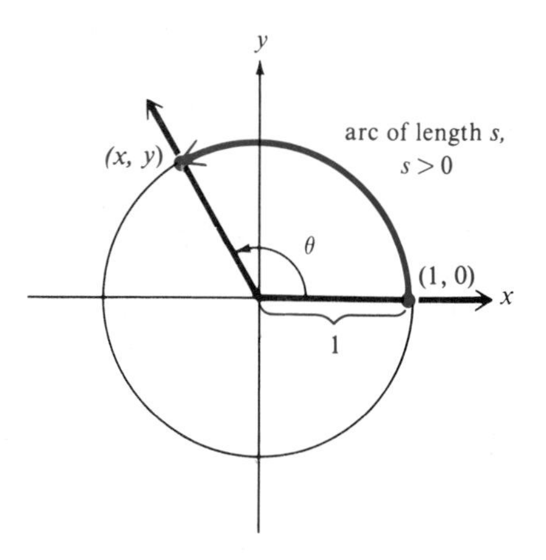

Figure 3.10

To find values of the circular functions for any real number s, use a unit circle, as shown in Figure 3.10. (A **unit circle** has a radius of 1 unit.) Start at the point (1, 0) and lay off an arc of length s along the circle. Go counterclockwise if s is positive, and clockwise if s is negative. Let the endpoint of the arc be at the point (x, y). The six **circular functions** are then defined as follows. (Assume that no denominators are zero.)

Circular Functions

$$\sin s = y \qquad \tan s = \frac{y}{x} \qquad \sec s = \frac{1}{x}$$

$$\cos s = x \qquad \cot s = \frac{x}{y} \qquad \csc s = \frac{1}{y}$$

Since the ordered pair (x, y) is used to locate a point on the unit circle,

$$-1 \leq x \leq 1 \quad \text{and} \quad -1 \leq y \leq 1,$$

making

$$-1 \leq \sin s \leq 1 \quad \text{and} \quad -1 \leq \cos s \leq 1.$$

For any value of s, both $\sin s$ and $\cos s$ exist, so the domain of these functions is the set of all real numbers. For $\tan s$, defined as y/x, $x \neq 0$. The only way x can equal 0 is when the arc length s is $\pi/2$, $3\pi/2$, $-\pi/2$, $-3\pi/2$, and so on. To avoid a zero denominator, the domain of tangent must be restricted to those values of s satisfying

$$s \neq \frac{\pi}{2} + n\pi, \qquad n \text{ any integer.}$$

The definition of secant also has x in the denominator, making the domain of secant the same as the domain of tangent. Both cotangent and cosecant are defined with a denominator of y. To guarantee that $y \neq 0$, the domain of these functions must be the set of all values of s satisfying

$$s \neq 0 + n\pi, \qquad n \text{ any integer.}$$

These circular functions are closely related to the trigonometric functions discussed earlier. To develop this connection, place an angle θ in standard position on a unit circle, as in Figure 3.10. Assume that θ is measured in radians. The arc length cut off by angle θ is s. From the previous section, the arc length is given by $s = r\theta$. Since $r = 1$, the arc length is $s = 1(\theta) = \theta$, making s the radian measure of angle θ. As shown in the figure, (x, y) is a point on the terminal side of θ, with $r = 1$. Using the definitions of the trigonometric functions,

$$\sin \theta = \frac{y}{r} = \frac{y}{1} = y = \sin s$$

$$\cos \theta = \frac{x}{r} = \frac{x}{1} = x = \cos s.$$

Similar results hold for the other four functions.

As shown above, the trigonometric functions and the circular functions lead to the same function values. Because of this, a value such as $\sin \pi/2$ can be found without worrying about whether $\pi/2$ is a real number or the radian measure of an angle. In either case, $\sin \pi/2 = 1$.

All the formulas developed in this book are valid for either angles or real numbers. For example, $\sin \theta = 1/\csc \theta$ is equally valid for θ as the measure of an angle in degrees or radians or for θ as a real number.

Example 1 Find the values of the following.

(a) cos .5149

Since .5149 is in the "radian" column at the left in Table 2, find the word *cosine* across the top.

$$\cos .5149 = .8703$$

To find cos .5149 using a calculator, first make sure the machine is set for radian measure. Then enter .5149 and press the cosine key.

(b) cot 1.3206

The table gives a value of .2555. This value could be found on a calculator using the tangent and reciprocal keys as follows. (Be sure your calculator is set for radians.)

Enter: 1.3206 [TAN] [1/x] **Display:** .255551059 ✣

Reference angles were used to find the values of trigonometric functions for angles larger than $\pi/2$ radians or smaller than 0 radians. To find function values of real numbers larger than $\pi/2$ or smaller than 0, use **reference numbers.**

Reference numbers are found in much the same way that reference angles were found. Use the following rules to find reference numbers or angles. (Here, 3.1416 is used as an approximation for π.)

Reference Numbers or Angles

θ or s in quadrant	*θ' is*	*s' is*
I (0 to 1.5708)	θ	s
II (1.5708 to 3.1416)	$180° - \theta$	$\pi - s$
III (3.1416 to 4.7124)	$\theta - 180°$	$s - \pi$
IV (4.7124 to 6.2832)	$360° - \theta$	$2\pi - s$

Example 2 Find the reference number for each value of s. Use 3.1416 as an approximation for π.

(a) $s = 3.9760$

As shown in Figure 3.11, an angle of measure 3.9760 is in quadrant III. For quadrant III, the reference number is

$$s' = s - \pi = \mathbf{3.9760} - \mathbf{3.1416} = .8344$$

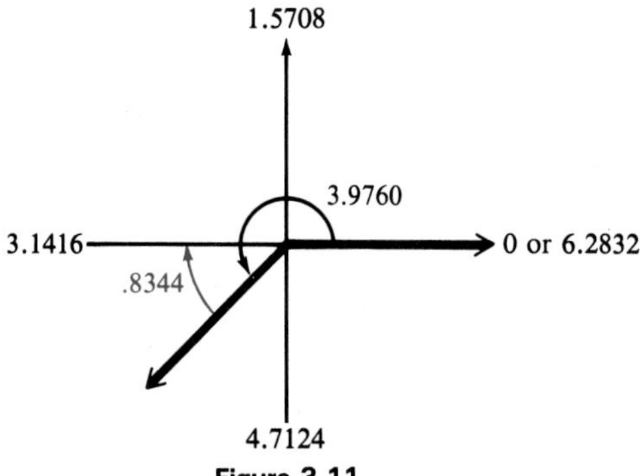

Figure 3.11

(b) $s = -7.36$

This real number is not between 0 and 2π (or 6.2832). Add 2π as many times as needed to get a result that is between 0 and 2π. Here, 2π must be added twice.

$$-7.36 + 2(2\pi) = -7.36 + 2(6.2832) = 5.2064$$

This result is in quadrant IV (see Figure 3.12), and has reference number

$$s' = 2\pi - s = 6.2832 - 5.2064 = 1.0768.$$

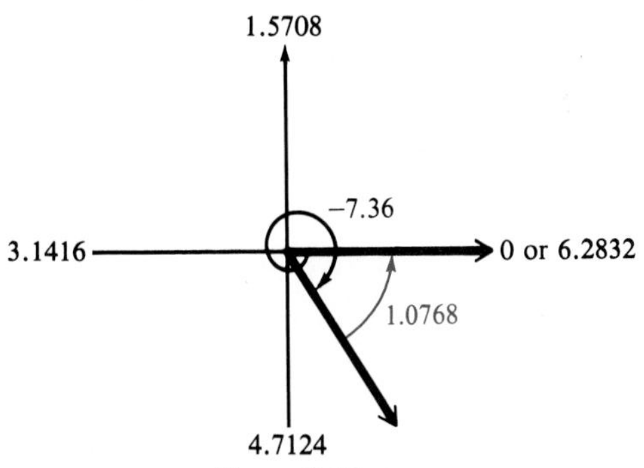

Figure 3.12

(c) $s = \dfrac{2\pi}{3}$

Figure 3.13 shows that $s = 2\pi/3$ is in quadrant II with reference number

$$\pi - \frac{2}{3}\pi = \frac{1}{3}\pi \text{ or } \frac{\pi}{3}.$$

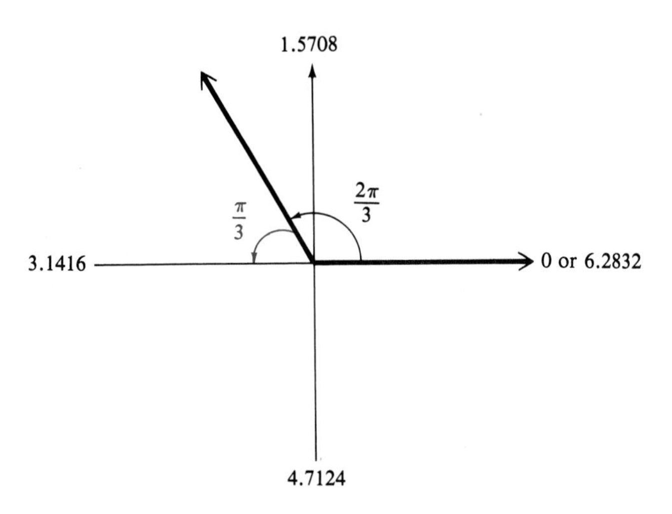

Figure 3.13

(d) $s = -\pi/4$

This quadrant IV angle has a reference number of $\pi/4$. ✜

Example 3 Find the value of each of the following. Use the number 3.1416 as an approximation for π.

(a) sin 6.6759

To use tables, first find the reference number for 6.6759.

$$6.6759 - 6.2832 = .3927$$

From Table 2,

$$\sin 6.6759 = \sin .3927 = .3827.$$

Reference numbers are not needed with most calculators. First set the calculator to radian measure, and then press the sine key.

Enter: 6.6759 [SIN] **Display:** .38269786

(b) cot 2.1031

The reference number is

$$3.1416 - 2.1031 = 1.0385.$$

Since 2.1031 is in quadrant II, cot 2.1031 is negative. From Table 2,

$$\cot 2.1031 = -.5890.$$

Find this value on a calculator using tangent and reciprocal keys.

Enter: 2.1031 [TAN] [1/x] **Display:** $-.58901572$

(c) cos (-2.9234)

Adding 6.2832 gives $-2.9234 + 6.2832 = 3.3598$, with reference number $3.3598 - 3.1416 = .2182$. Since -2.9234 is in quadrant III (draw a sketch),

$$\cos(-2.9234) = -.9763.$$

With a calculator, enter 2.9234, then use the appropriate key to change the sign, and press the cosine key. ✜

Example 4 Find a value of s, $0 \leq s \leq 1.5708$, that makes $\cos s = .0581$.

Look in the body of Table 2 (in a column having *cosine* at top or bottom). The number .0581 is found in a column with cosine at the bottom. Read s from the "radian" column on the right.

$$s = 1.5126$$

The value of s can also be found with a calculator (set for radians) as follows.

Enter: .0581 [INV] [COS] **Display:** 1.51266359 ✜

(Recall that other keys such as [COS^{-1}] or [ARCCOS] may be needed instead of [INV] [COS].)

Interpolation, discussed in the appendix at the end of chapter 2, can also be used with circular functions. Work as in the next example.

Example 5 Find sin .9000 using interpolation.

Proceed as follows.

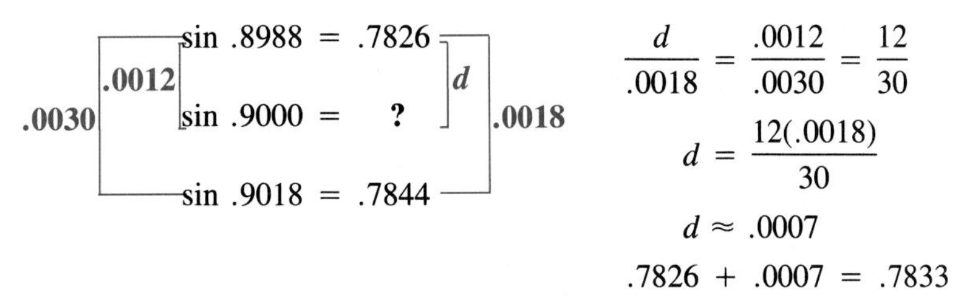

Finally, sin .9000 = .7833. ✚

3.3 Exercises

Use Table 2 or a calculator to find each of the following. See Example 1.

1. tan .4538 **2.** sin .6109 **3.** cot 1.0821 **4.** cos 1.1519

5. sin .8203 **6.** cot .6632 **7.** cos .6429 **8.** tan .9047

9. sin 1.5097 **10.** cot .0465 **11.** csc 1.3875 **12.** tan 1.3032

Use reference numbers or a calculator to find the value of each of the following. Use 3.1416 *as an approximation of* π. *See Example 3.*

13. sin 7.5835 **14.** tan 6.4752 **15.** cot 7.4526 **16.** cos 6.6701

17. tan 4.0230 **18.** cot 3.8426 **19.** cos 4.2528 **20.** sin 3.4645

21. sin (−2.2864) **22.** cot (−2.4871) **23.** cos (−3.0602) **24.** tan (−1.7861)

25. cot 6.0301 **26.** cos 5.2825 **27.** sin 5.9690 **28.** tan 5.4513

29. cos 13.8143 **30.** sin 13.6572 **31.** cot 12.9795 **32.** tan 11.0392

33. csc (−9.4946) **34.** cos (−13.7881) **35.** tan (−23.7744) **36.** sin (−17.5784)

Find a value of s, where $0 \le s \le 1.5708$, *that makes each of the following true. See Example 4.*

37. tan s = .2126 **38.** cos s = .7826 **39.** sin s = .9918 **40.** cot s = .2994

41. cot s = .6208 **42.** tan s = 2.605 **43.** cos s = .5783 **44.** sin s = .9877

45. cot s = .0963 **46.** csc s = 1.021 **47.** tan s = 1.621 **48.** cos s = .9272

Use a calculator, or Table 2 and interpolation, to find each of the following. Give results with four significant digits.

49. cos 1.0000 **50.** sin 1.0000 **51.** tan .3 **52.** cot .5

53. sin 1.2 **54.** cot 1.1 **55.** cos 1.3 **56.** tan 1.4

57. sin 2.0 **58.** cos (−1.3) **59.** tan (−2.4) **60.** sin (−3.1)

Identify the quadrant in which arcs on a unit circle having the following lengths would terminate.

61. $s = 10$ **62.** $s = 18$ **63.** $s = -89.19$ **64.** $s = -104.27$

65. The values of the circular functions repeat every 2π. For this reason, circular functions are used to describe things that repeat periodically. For example, the maximum afternoon temperature in a given city might be approximated by

$$t = 60 - 30 \cos x\pi/6,$$

where t represents the maximum afternoon temperature in month x, with $x = 0$ representing January, $x = 1$ representing February, and so on. Find the maximum afternoon temperature for each of the following months.
(a) January **(b)** April **(c)** May
(d) June **(e)** August **(f)** October

66. The temperature in Fairbanks is approximated by

$$T(x) = 37 \sin\left[\frac{2\pi}{365}(x - 101)\right] + 25,$$

where $T(x)$ is the temperature in degrees Fahrenheit on day x, with $x = 1$ corresponding to January 1 and $x = 365$ corresponding to December 31.* Use a calculator to estimate the temperature on the following days.
(a) March 1 (day 60) **(b)** April 1 (day 91) **(c)** Day 150
(d) June 15 **(e)** September 1 **(f)** October 31

Show that the following points lie (approximately) on a unit circle.

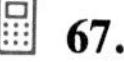

67. (.39852144, .91715902) **68.** (−.81745602, −.57599102)

In calculus courses sin x *and* cos x *are expressed as infinite sums, where x is measured in radians. For all real numbers x and all natural numbers n, the following hold.*

$$\cos x = 1 - \frac{x^2}{2!} + \frac{x^4}{4!} - \cdots + (-1)^{n-1}\frac{x^{2n-2}}{(2n-2)!} + \cdots$$

and

$$\sin x = x - \frac{x^3}{3!} + \frac{x^5}{5!} - \cdots + (-1)^{n-1}\frac{x^{2n-1}}{(2n-1)!} + \cdots$$

where

$$2! = 2 \cdot 1; \quad 3! = 3 \cdot 2 \cdot 1; \quad 4! = 4 \cdot 3 \cdot 2 \cdot 1;$$

and

$$n! = n(n-1)(n-2) \ldots (2)(1).$$

Using these formulas, sin x *and* cos x *can be evaluated to any desired degree of accuracy. For example, let us find* cos .5 *using the first three terms given above and compare this with the result using the first four terms. Using the first three terms,*

---

*Barbara Lando and Clifton Lando, "Is the Graph of Temperature Variation a Sine Curve?" *The Mathematics Teacher,* 70 (September, 1977): 534–37.

$$\cos .5 = 1 - \frac{(.5)^2}{2!} + \frac{(.5)^4}{4!}$$
$$= 1 - \frac{.25}{2} + \frac{.0625}{24}$$
$$= 1 - .125 + .00260$$
$$= .87760.$$

If we use four terms, the result is

$$\cos .5 = 1 - \frac{(.5)^2}{2!} + \frac{(.5)^4}{4!} - \frac{(.5)^6}{6!}$$
$$= 1 - .125 + .00260 - .00002$$
$$= .87758.$$

The two results differ by only .00002, so that in this example the first three terms give a result that is accurate to the fourth decimal place. Using a calculator, or tables more complete than the one in this book, cos .5 = .87758.

Use the first three terms of the results given above for sin x *and* cos x *to find the following. Compare your results with the values in Table 2.*

69. sin 1 **70.** sin .1 **71.** cos 1.4

72. cos .8 **73.** sin .01 **74.** cos .01

3.4 Linear and Angular Velocity

Suppose that point P moves at a constant speed along a circle of radius r and center O. See Figure 3.14. The measure of how fast the position of P is changing is called **linear velocity.** If v represents linear velocity, then

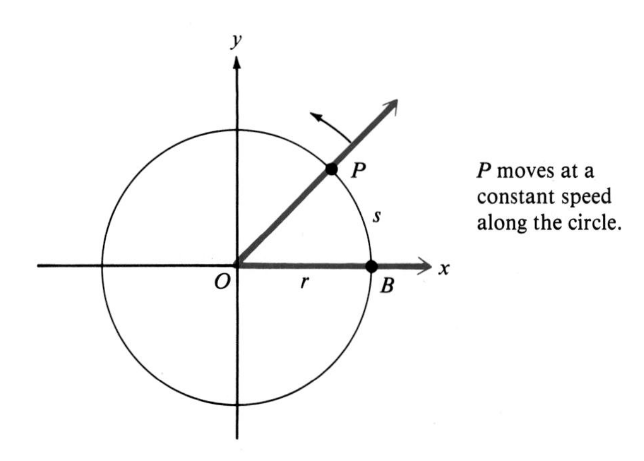

Figure 3.14

$$v = \frac{s}{t},$$

where s is the length of the arc cut by point P at time t. (This formula is just a restatement of the familiar result $d = rt$ with s as distance, v as the rate, and t as time.)

Look at Figure 3.14 again. As point P moves along the circle, ray OP rotates around the origin. Since the ray OP is the terminal side of angle POB, the measure of the angle changes as P moves along the circle. The measure of how fast angle POB is changing is called **angular velocity.** Angular velocity, written ω, can be given as

$$\omega = \frac{\theta}{t}, \qquad \theta \text{ in radians},$$

where θ is the measure of angle POB at time t. As with the earlier formulas in this chapter, θ must be measured in radians, with ω expressed as radians per unit of time. Angular velocity is used in physics and engineering, among other applications.

In Section 3.2 the length s of the arc cut on a circle of radius r by a central angle of measure θ radians was found to be $s = r\theta$. Using this formula, the formula for linear velocity, $v = s/t$, becomes

$$v = \frac{r\theta}{t}$$

or

$$v = r\omega$$

This last formula relates linear and angular velocity.

As mentioned in Section 3.1, a radian is a "pure number," with no units associated with it. This is why the product of the length r, measured in units such as centimeters, and ω, measured in units such as radians per second, is velocity, v, measured in units such as centimeters per second.

All the formulas given in this section are summarized below.

Angular and Linear Velocity

Angular velocity $\quad \omega = \dfrac{\theta}{t}$

(ω in radians per unit time, θ in radians)

Linear velocity $\quad v = \dfrac{s}{t}$

$$v = \frac{r\theta}{t}$$

$$v = r\omega$$

Example 1 Suppose that point P is on a circle with a radius of 10 cm, and ray OP is rotating with angular velocity of $\pi/18$ radians per sec.

(a) Find the angle generated by P in 6 sec.

The velocity of ray OP is $\omega = \pi/18$ radians per sec. Since $\omega = \theta/t$, then in 6 sec

$$\frac{\pi}{18} = \frac{\theta}{6},$$

or $\theta = 6(\pi/18) = \pi/3$ radians.

(b) Find the distance traveled by P along the circle in 6 sec.

In 6 sec P generates an angle of $\pi/3$ radians. Since $s = r\theta$,

$$s = 10\left(\frac{\pi}{3}\right) = \frac{10\pi}{3}\text{cm.}$$

(c) Find the linear velocity of P.

Since $v = s/t$, then in 6 sec

$$v = \frac{10\pi/3}{6} = \frac{5\pi}{9}\text{ cm per sec.}$$

In practical applications, angular velocity is often given as revolutions per unit of time, which must be converted to radians per unit of time before using the formulas given in this section.

Example 2 A belt runs a pulley of radius 6 cm at 80 revolutions per min.

(a) Find the angular velocity of the pulley in radians per sec.

In one minute, the pulley makes 80 revolutions. Each revolution is 2π radians, for a total of

$$80(2\pi) = 160\pi \text{ radians per min.}$$

Since there are 60 seconds in a minute, ω, the angular velocity in radians per second, is given by

$$\omega = \frac{160\pi}{60} = \frac{8\pi}{3}\text{ radians per sec.}$$

(b) Find the linear velocity of the belt in cm per sec.

The linear velocity of the belt will be the same as that of a point on the circumference of the pulley. Thus,

$$v = r\omega$$

$$v = 6\left(\frac{8\pi}{3}\right)$$

$$v = 16\pi \text{ cm per sec}$$

$$v \approx 50.3 \text{ cm per sec.}$$

Example 3 A satellite traveling in a circular orbit 1600 km above Earth takes two hours to make an orbit. Assume that the radius of Earth is 6400 km.

(a) Find the linear velocity of the satellite.

The distance of the satellite from the center of the earth is

$$r = 1600 + 6400 = 8000 \text{ km}.$$

Since it takes 2 hours to complete an orbit, for one orbit $\theta = 2\pi$, and

$$s = r\theta = 8000(2\pi) \text{ km}.$$

Then the linear velocity is

$$\begin{aligned} v = \frac{s}{t} &= \frac{8000(2\pi)}{2} \\ &= 8000\pi \\ &\approx 25{,}000 \text{ km/hr}. \end{aligned}$$

(b) Find the distance traveled in 4.5 hours.

$$\begin{aligned} s = vt &= (8000\pi)(4.5) \\ &= 36{,}000\pi \\ &\approx 110{,}000 \text{ km}. \end{aligned}$$

3.4 Exercises

Use the formula $\omega = \theta/t$ to find the value of the missing variable in each of the following.

1. $\omega = \pi/4$ radians per min, $t = 5$ min

2. $\omega = 2\pi/3$ radians per sec, $t = 3$ sec

3. $\theta = 2\pi/5$ radians, $t = 10$ sec

4. $\theta = 3\pi/4$ radians, $t = 8$ sec

5. $\theta = 3\pi/8$ radians, $\omega = \pi/24$ radians per min

6. $\theta = 2\pi/9$ radians, $\omega = 5\pi/27$ radians per min

7. $\omega = .90674$ radians per min, $t = 11.876$ min

8. $\theta = 3.871142$ radians, $t = 21.4693$ sec

Use the formula $v = r\omega$ to find the value of the missing variable in each of the following.

9. $r = 8$ cm, $\omega = 9\pi/5$ radians per sec

10. $r = 12$ m, $\omega = 2\pi/3$ radians per sec

11. $v = 18$ ft per sec, $r = 3$ ft

12. $v = 9$ m per sec, $r = 5$ m

13. $r = 24.93215$ cm, $\omega = .372914$ radians per sec

14. $v = 107.692$ m per sec, $r = 58.7413$ m

The formula $\omega = \theta/t$ can be rewritten as $\theta = \omega t$. Using ωt for θ changes $s = r\theta$ to $s = r\omega t$. Use this formula to find the values of the missing variables in each of the following.

15. $r = 6$ cm, $\omega = \pi/3$ radians per sec, $t = 9$ sec

16. $r = 9$ yd, $\omega = 2\pi/5$ radians per sec, $t = 12$ sec

17. $s = 6\pi$ cm, $r = 2$ cm, $\omega = \pi/4$ radians per sec

18. $s = 12\pi/5$ m, $r = 3/2$ m, $\omega = 2\pi/5$ radians per sec

19. $s = 3\pi/4$ km, $r = 2$ km, $t = 4$ sec

20. $s = 8\pi/9$ m, $r = 4/3$ m, $t = 12$ sec

21. $r = 37.6584$ cm, $\omega = .714213$ radians per sec, $t = .924473$ sec

22. $s = 5.70201$ m, $r = 8.92399$ m, $\omega = .614277$ radians per sec

Find ω for each of the following.

23. The hour hand of a clock

24. The minute hand of a clock

25. The second hand of a clock

26. A line from the center to the edge of a phonograph record revolving 33 1/3 times per minute

Find v for each of the following.

27. The tip of the minute hand of a clock, if the hand is 7 cm long

28. The tip of the second hand of a clock, if the hand is 28 mm long

29. A point on the edge of a flywheel of radius 2 m rotating 42 times per min

30. A point on the tread of a tire of radius 18 cm, rotating 35 times per min

31. The tip of an airplane propeller 3 m long, rotating 500 times per min (*Hint:* $r = 1.5$)

32. A point on the edge of a gyroscope of radius 83 cm, rotating 680 times per minute

Solve the following problems, which review the ideas of this chapter

33. A railroad track is laid along the arc of a circle of radius 1800 ft. The circular part of the track intercepts a central angle of 40°. How long (in seconds) will it take a point on the front of a train traveling 30 mph to go around this portion of the track?

34. Two pulleys of diameter 4 m and 2 m, respectively, are connected by a belt. The larger pulley rotates 80 times per min. Find the speed of the belt in meters per second and the angular velocity of the smaller pulley.

35. The earth revolves on its axis once every 24 hr. Assuming that the earth's radius is 6400 km, find the following.
(a) Angular velocity of the earth in radians per day and radians per hr
(b) Linear velocity at the North Pole or South Pole
(c) Linear velocity at Quito, Equador, a city on the equator
(d) Linear velocity at Salem, Oregon (halfway from the equator to the North Pole)

36. The earth travels about the sun in an orbit that is almost circular. Assume that the orbit is a circle, with a radius of 93,000,000 mi. See the figure.

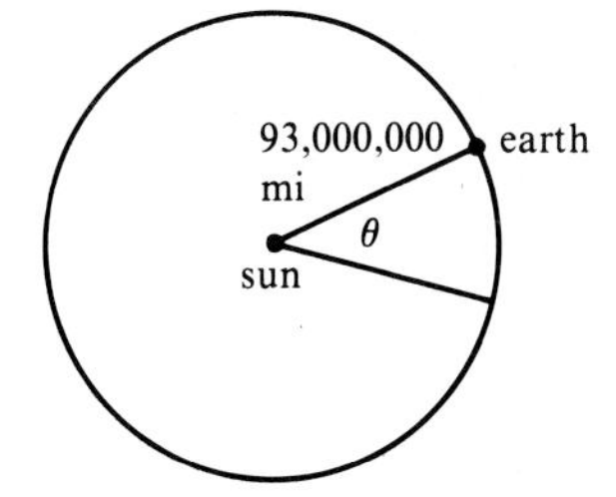

(a) Assume that a year is 356 days, and find θ, the angle formed by the earth's movement in one day.
(b) Give the angular velocity in radians per hour.
(c) Find the linear velocity of the earth in miles per hour.

37. The pulley shown on the left has a radius of 12.96 cm. Suppose that it takes 18 sec for 56 cm of belt to go around the pulley. Find the angular velocity of the pulley in radians per second.

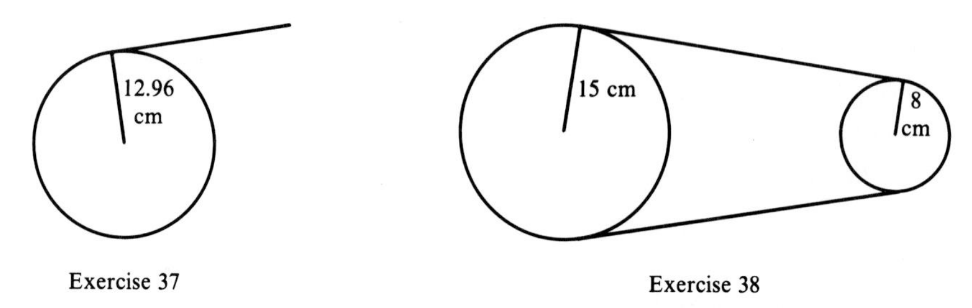

Exercise 37

Exercise 38

38. The two pulleys in the figure on the right have radii of 15 cm and 8 cm, respectively. The larger pulley rotates 25 times in 36 sec. Find the angular velocity of each pulley in radians per sec.

39. A gear is driven by a chain that travels 1.46 m per sec. Find the radius of the gear if it makes 46 revolutions per min.

40. A thread is being pulled off a spool at the rate of 59.4 cm per sec. Find the radius of the spool if it makes 152 revolutions per min.

Chapter 3 *Key Concepts*

Radian An angle that has its vertex at the center of a circle and that cuts an arc on the circle equal in length to the radius of the circle has a measure of **one radian.**

$$\mathbf{180° = \pi \text{ radians}}$$

Converting Between Degrees and Radians

1. Proportion: $\dfrac{\text{radian measure}}{\pi} = \dfrac{\text{degree measure}}{180}$

2. Formulas

From	*To*	*Multiply by*
Radians	Degrees	$\dfrac{180°}{\pi}$
Degrees	Radians	$\dfrac{\pi}{180°}$

Length of Arc The length s of the arc cut on a circle of radius r by a central angle of measure θ radians is given by the product of the radius and the radian measure of the angle, or

$$s = r\theta, \qquad \theta \text{ in radians.}$$

Area of Sector The area of a sector of a circle of radius r and central angle θ in radians is given by

$$A = \frac{1}{2}r^2\theta, \qquad \theta \text{ in radians.}$$

Circular Functions

$$\sin s = y \qquad \tan s = \frac{y}{x} \qquad \sec s = \frac{1}{x}$$

$$\cos s = x \qquad \cot s = \frac{x}{y} \qquad \csc s = \frac{1}{y}$$

Chapter 3 *Key Concepts (continued)*

Reference Numbers or Angles

θ *or* s *in quadrant*	θ' *is*	s' *is*
I (0 to 1.5708)	θ	s
II (1.5708 to 3.1416)	$180° - \theta$	$\pi - s$
III (3.1416 to 4.7124)	$\theta - 180°$	$s - \pi$.
IV (4.7124 to 6.2832)	$360° - \theta$	$2\pi - s$

Angular and Linear Velocity

Angular velocity $\omega = \dfrac{\theta}{t}$

(ω in radians per unit time, θ in radians)

Linear velocity $v = \dfrac{s}{t}$

$$v = \frac{r\theta}{t}$$

$$v = r\omega$$

Chapter 3 *Review Exercises*

Convert each of the following degree measures to radians. Leave answers as multiples of π.

1. $45°$ **2.** $120°$ **3.** $80°$ **4.** $175°$

5. $330°$ **6.** $800°$ **7.** $1020°$ **8.** $2000°$

Convert each of the following radian measures to degrees.

9. $\frac{5\pi}{4}$ **10.** $\frac{9\pi}{10}$ **11.** $\frac{8\pi}{3}$ **12.** $\frac{-6\pi}{5}$

13. $\frac{-11\pi}{18}$ **14.** $\frac{21\pi}{5}$ **15.** $\frac{14\pi}{15}$ **16.** $\frac{33\pi}{5}$

Evaluate each of the following without using a table or a calculator.

17. $\tan \frac{\pi}{3}$ **18.** $\cos \frac{2\pi}{3}$ **19.** $\sin \left(-\frac{5\pi}{6}\right)$ **20.** $\cot \frac{11\pi}{6}$

21. $\tan \left(-\frac{7\pi}{3}\right)$ **22.** $\sec \frac{\pi}{3}$ **23.** $\csc \left(-\frac{11\pi}{6}\right)$ **24.** $\cot \left(-\frac{17\pi}{3}\right)$

Solve each of the following problems.

25. The radius of a circle is 15.2 cm. Find the length of an arc of the circle cut by a central angle of $3\pi/4$ radians.

26. Find the length of arc cut by a central angle of .769 radians on a circle with a radius of 11.4 cm.

27. A circle has a radius of 8.973 cm. Find the length of arc on this circle cut by a central angle of 49.06°.

28. A central angle of $7\pi/4$ radians forms a sector of a circle. Find the area of the sector if the radius of the circle is 28.69 in.

29. Find the area of a sector of a circle cut by a central angle of 21° 40′ in a circle of radius 38.0 m.

30. A tree 2000 yd away intercepts an angle of 1° 10′. Find the height of the tree to two significant digits.

Assume that the radius of Earth is 6400 *km in the next two exercises.*

31. Find the distance in kilometers between cities on a north-south line that are on latitudes 28° N and 12° S, respectively.

32. Two cities on the equator have longitudes of 72° E and 35° W, respectively. Find the distance between the cities.

Find reference numbers for each of the following. Use 3.1416 *as an approximation for* π.

33. 2.0417 **34.** -1.8962 **35.** 7.0101 **36.** 12

Use Table 2 or a calculator to find each of the following. Use 3.1416 *as an approximation for* π.

37. sin 1.0472 **38.** tan 1.2275 **39.** cos ($-$.2443) **40.** cot 3.0543

41. tan 7.3159 **42.** sin 4.8386 **43.** cos 2.1 **44.** tan 3.8

Find s in each of the following. Assume that $0 \le s \le \pi/2$.

45. $\cos s = .9250$

46. $\tan s = 4.011$

47. $\sin s = .4924$

48. $\csc s = 1.236$

49. $\cot s = .5022$

50. $\sec s = 4.560$

Solve each of the following problems.

51. Find t if $\theta = 5\pi/12$ radians and $\omega = 8\pi/9$ radians per sec.

52. Find θ if $t = 12$ sec and $\omega = 9$ radians per sec.

53. Find ω if $t = 8$ sec and $\theta = 2\pi/5$ radians.

54. Find ω if $s = 12\pi/25$ ft, $r = 3/5$ ft, and $t = 15$ sec.

55. Find s if $r = 11.46$ cm, $\omega = 4.283$ radians per sec, and $t = 5.813$ sec.

56. Find the linear velocity of a point on the edge of a flywheel of radius 7 m if the flywheel is rotating 90 times per sec.

Graphs of Trigonometric Functions

Many things in daily life repeat with a predictable pattern: the daily newspaper is delivered at the same time each morning, in warm areas electricity use goes up in the summer and down in the winter, the price of fresh fruit goes down in the summer and up in the winter, and attendance at amusement parks increases in the summer and declines in autumn. There are many examples of these *periodic* phenomena (see Figure 4.1). As we shall see in this chapter, the trigonometric functions are periodic and very useful for describing periodic activities.

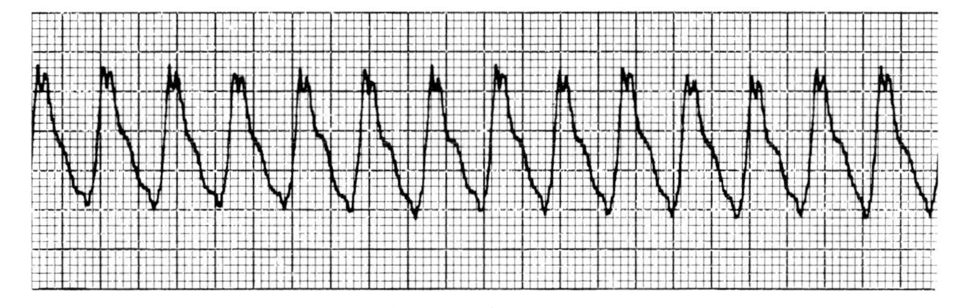

Figure 4.1 *Periodic phenomena are very common in nature. One of the most common is the human heartbeat, as shown in this electrocardiogram (EKG). The EKG shows electrical impulses from the heart. Each small square represents .04 seconds, so that each large square represents .2 seconds.*

4.1 Graphs of Sine and Cosine

By the identities for coterminal angles,

$$\sin 0 = \sin 2\pi$$
$$\sin \pi/2 = \sin (\pi/2 + 2\pi)$$

and
$$\sin \pi = \sin (\pi + 2\pi),$$

with $\sin x = \sin (x + 2\pi)$ for any real number x. The value of the sine function is the same for x and for $x + 2\pi$, making $y = \sin x$ a periodic function with period 2π.

Periodic Function

A **periodic function** is a function f with the property that

$$f(x) = f(x + p),$$

for every real number x in the domain of f and for some positive real number p. The smallest possible positive value of p is the **period** of the function.

While it is true that $\sin x = \sin (x + 4\pi)$ and $\sin x = \sin (x + 6\pi)$, the *smallest* positive value of p making $\sin x = \sin (x + p)$ is $p = 2\pi$, so 2π is the period for sine.

We saw earlier that the sine function $y = \sin x$ has the set of all real numbers as domain, with range $-1 \le y \le 1$, or $-1 \le \sin x \le 1$.* Since the sine function has period 2π, one period of $y = \sin x$ can be graphed using values of x from 0 to 2π. To graph this period, look at Figure 4.2, which shows a unit circle with a point (p, q) marked on it. Based on the definitions for circular functions given in Section 3.3, for any angle x, $p = \cos x$ and $q = \sin x$. As x increases from 0 to $\pi/2$ (or 90°), q (or $\sin x$) increases from 0 to 1, while p (or $\cos x$) decreases from 1 to 0.

As x increases from $\pi/2$ to π (or 180°), q decreases from 1 to 0, while p decreases from 0 to -1. Similar results can be found for the other quadrants, as shown in the following table.

As x increases from	$\sin x$	$\cos x$
0 to $\pi/2$	Increases from 0 to 1	Decreases from 1 to 0
$\pi/2$ to π	Decreases from 1 to 0	Decreases from 0 to -1
π to $3\pi/2$	Decreases from 0 to -1	Increases from -1 to 0
$3\pi/2$ to 2π	Increases from -1 to 0	Increases from 0 to 1

Selecting key values of x and finding the corresponding values of $\sin x$ give the following results. (Decimals are rounded to the nearest tenth.)

*In this chapter, x is used as the independent variable, as in $y = \sin x$, instead of θ or s because such use allows drawing graphs on the familiar xy-coordinate system.

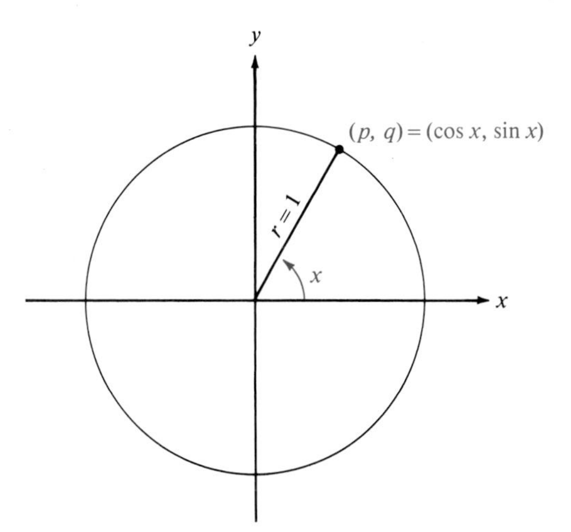

Figure 4.2

x	0	$\pi/4$	$\pi/2$	$3\pi/4$	π	$5\pi/4$	$3\pi/2$	$7\pi/4$	2π
$\sin x$	0	.7	1	.7	0	$-.7$	-1	$-.7$	0

Plotting the points from the table of values and connecting them with a smooth curve gives the solid portion of the graph in Figure 4.3. Since $y = \sin x$ is periodic and has all real numbers as its domain, the graph continues in both directions indefinitely, as indicated by the arrows. This graph is sometimes called a **sine wave** or **sinusoid.** You should memorize the shape of this graph and be able to sketch it quickly. The main points of the graph are $(0, 0)$, $(\pi/2, 1)$, $(\pi, 0)$, $(3\pi/2, -1)$, and $(2\pi, 0)$. By plotting these five points and connecting them with the characteristic sine wave, you can quickly sketch the graph.

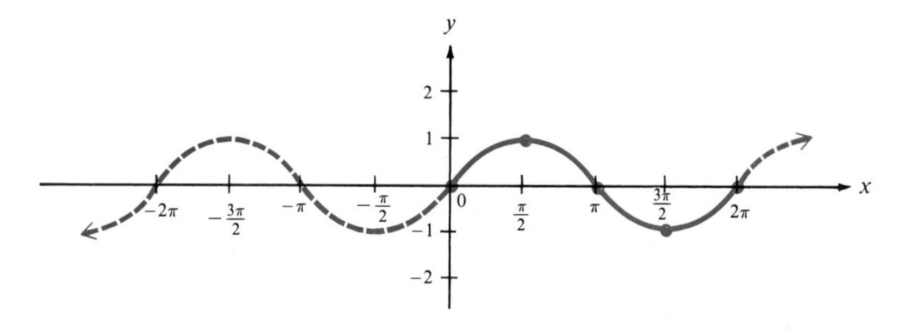

$y = \sin x$ sine wave

Figure 4.3

The same scales are used on both the x and y axes of Figure 4.3 so as not to distort the graph. Since the period of $y = \sin x$ is 2π, it is convenient to use subdivisions of 2π on the x-axis. The more familiar x-values, 1, 2, 3, 4, and so on, are still present, but are usually not shown to avoid cluttering the graph. These values are shown in Figure 4.4.

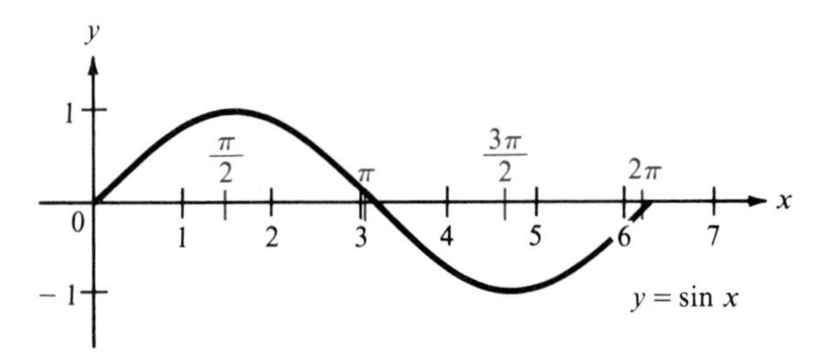

Figure 4.4

Sine graphs occur in many different practical applications. For one application, look back at Figure 4.2 and assume that the line from the origin to the point (p, q) is part of the pedal of a bicycle wheel, with a foot placed at (p, q). As mentioned above, q is equal to $\sin x$, showing that the height of the pedal from the horizontal axis in Figure 4.2 is given by $\sin x$. By choosing various angles for the pedal and calculating q for each angle, the height of the pedal leads to the sine curve shown in Figure 4.5.

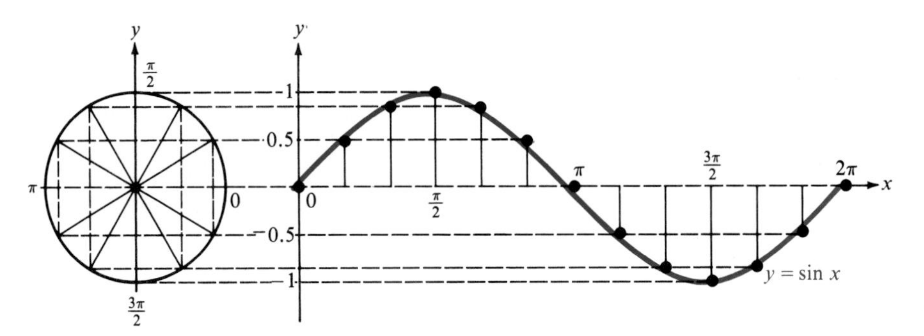

Figure 4.5

The graph of $y = \cos x$ can be found in much the same way as the graph of $y = \sin x$ was found. The domain of cosine is the set of all real numbers, and the range of $y = \cos x$ is $-1 \leq \cos x \leq 1$. A table of values is shown below for $y = \cos x$.

x	0	$\pi/4$	$\pi/2$	$3\pi/4$	π	$5\pi/4$	$3\pi/2$	$7\pi/4$	2π
$\cos x$	1	.7	0	−.7	−1	−.7	0	.7	1

Here the key points are $(0, 1)$, $(\pi/2, 0)$, $(\pi, -1)$, $(3\pi/2, 0)$, and $(2\pi, 1)$.

The graph of $y = \cos x$, in Figure 4.6, has the same shape as the graph of $y = \sin x$. In fact, it is the sine wave, shifted $\pi/2$ units to the left.

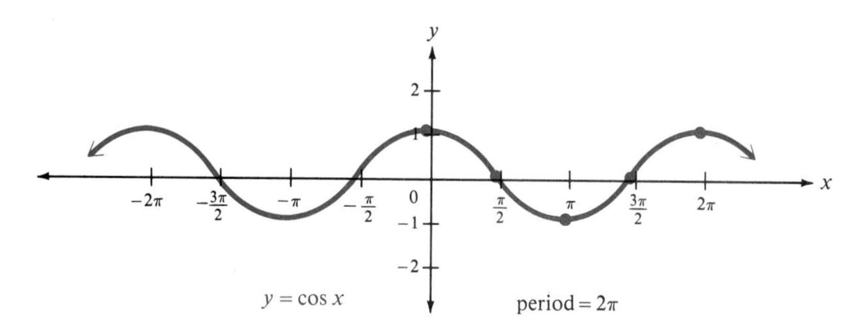

Figure 4.6

The examples in the rest of this section show graphs that are "stretched" either vertically or horizontally, or both when compared with the graphs of $y = \sin x$ or $y = \cos x$.

Example 1 Graph $y = 2 \sin x$.

For a given value of x, the value of y is twice as large as it would be for $y = \sin x$, as shown in the table of values. The only change in the graph is the range, which becomes $-2 \leq y \leq 2$. See Figure 4.7, which also shows a graph of $y = \sin x$ for comparison. ✥

x	0	$\pi/2$	π	$3\pi/2$	2π
$\sin x$	0	1	0	−1	0
$2 \sin x$	0	2	0	−2	0

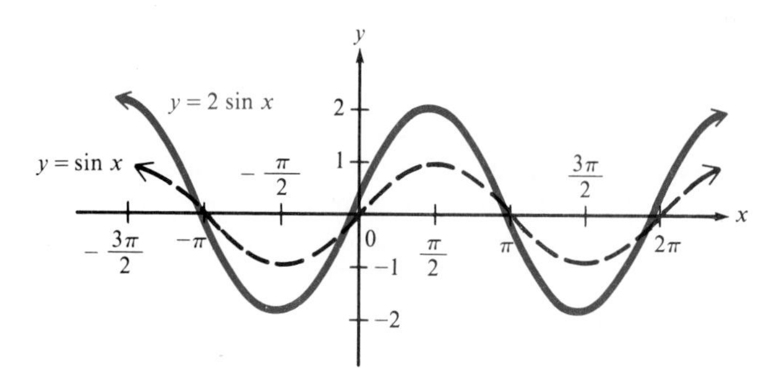

Figure 4.7

Generalizing from the example and assuming $a \neq 0$ gives the following.

Amplitude of Sine and Cosine

> The graph of $y = a \sin x$ or $y = a \cos x$ will have the same shape as $y = \sin x$ or $y = \cos x$, respectively, except with range $-|a| \leq y \leq |a|$. The number $|a|$ is called the **amplitude.**

No matter what the value of the amplitude, the period of $y = a \sin x$ and $y = a \cos x$ is still 2π.

Example 2 Graph $y = \sin 2x$.

Start with a table of values.

x	0	$\pi/4$	$\pi/2$	$3\pi/4$	π	$5\pi/4$	$3\pi/2$	$7\pi/4$
$2x$	0	$\pi/2$	π	$3\pi/2$	2π	$5\pi/2$	3π	$7\pi/2$
$\sin 2x$	0	1	0	-1	0	1	0	-1

As the table shows, multiplying x by 2 shortens the period to half. The amplitude is not changed. Figure 4.8 shows the graph of $y = \sin 2x$. Again, the graph of $y = \sin x$ is included for comparison. ✥

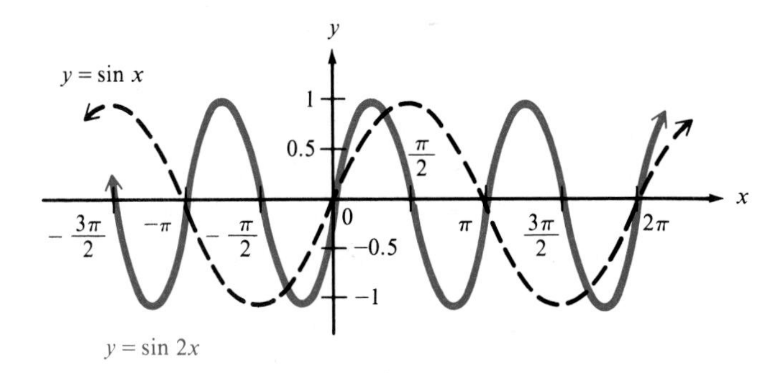

Figure 4.8

Generalizing from Example 2 leads to the following result.

Period of Sine and Cosine

> The graph of $y = \sin bx$ will look like that of $\sin x$, but with a period of $|2\pi/b|$. Also, the graph of $y = \cos bx$ looks like that of $y = \cos x$, but with a period of $|2\pi/b|$.

Exercise 47 asks for a proof of this change in the period.

Example 3 Graph $y = \cos \dfrac{2}{3}x$.

The period here is $|2\pi/(2/3)| = 3\pi$. The amplitude is 1. The key points for one complete period will start at 0 and end at 3π. The midpoint will be at $3\pi/2$ and the quarter points will be located at $3\pi/4$ and $3 \cdot 3\pi/4 = 9\pi/4$. Thus, the key points are $(0, 1)$, $(3\pi/4, 0)$, $(3\pi/2, -1)$, $(9\pi/4, 0)$, and $(3\pi, 1)$. Plot these points and complete the graph as shown in Figure 4.9. ✥

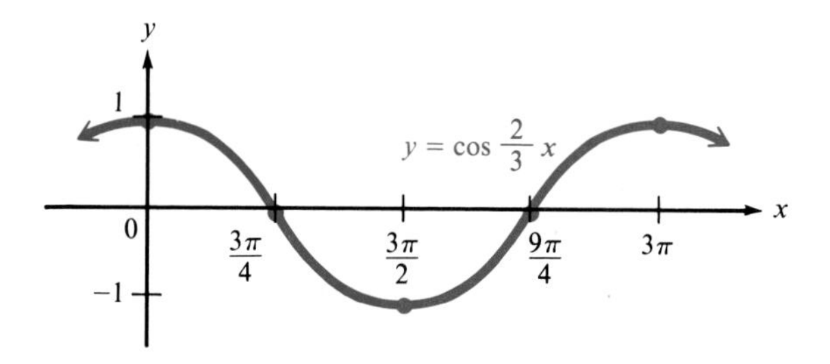

Figure 4.9

Throughout this chapter we assume $b > 0$. If a function has $b < 0$, then the identities given in the next chapter can be used to change the function to one in which $b > 0$. The steps used to graph $y = a \sin bx$ or $y = a \cos bx$, where $b > 0$, are given below.

Graphing the Sine and Cosine Functions

To graph $y = a \sin bx$ or $y = a \cos bx$, with $b > 0$:

1. Find the period, $2\pi/b$. Start at 0 on the x-axis and lay off a distance of $2\pi/b$.
2. Divide the interval from 0 to $2\pi/b$ into four equal parts.
3. Locate the points at which the graph crosses the x-axis.

Function	*Graph crosses x-axis at:*
$y = a \sin bx$	$0, \frac{\pi}{b}, \frac{2\pi}{b}$ (beginning, midpoint, and end of interval)
$y = a \cos bx$	$\frac{\pi}{2b}, \frac{3\pi}{2b}$ (one-fourth and three-fourths points of interval)

4. Find the amplitude, $|a|$.
5. Locate the points at which the graph reaches maximum and minimum values.

Function	*Graph has a maximum when x is:*
$y = a \sin bx$	$\frac{\pi}{2b}$ (for $a > 0$) or $\frac{3\pi}{2b}$ (for $a < 0$)
$y = a \cos bx$	0 and $\frac{2\pi}{b}$ (for $a > 0$) or $\frac{\pi}{b}$ (for $a < 0$)

6. Sketch the graph through the key points.
7. Draw additional periods of the graph, to the right and to the left, as needed.

Example 4 Graph $y = -2 \sin 3x$.

Here $a = -2$ and $b = 3$. The amplitude is $|a| = |-2| = 2$. The period is $2\pi/b = 2\pi/3$, so the key points are located at

$$0, \quad \frac{1}{4}\left(\frac{2\pi}{3}\right) = \frac{\pi}{6}, \quad \frac{1}{2}\left(\frac{2\pi}{3}\right) = \frac{\pi}{3},$$

$$\frac{3}{4}\left(\frac{2\pi}{3}\right) = \frac{\pi}{2}, \quad \text{and} \quad \frac{2\pi}{3}.$$

Sketch the graph using the steps shown in Figure 4.10. Notice how the minus sign affects the location of the maximum and minimum points. ✜

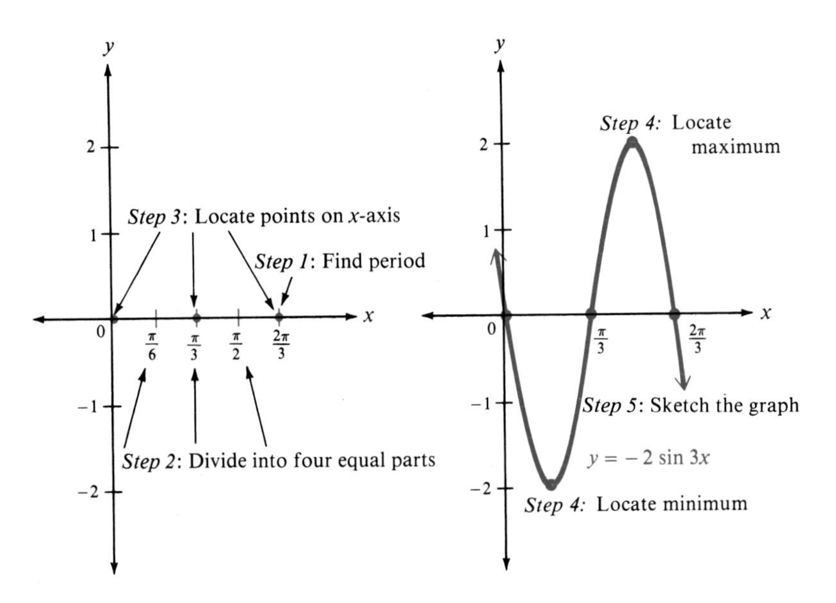

Figure 4.10

Example 5 Graph $y = 3 \cos \frac{1}{2}x$.

The period is $2\pi/(1/2) = 4\pi$. The key points have x-values of

$$0, \quad \frac{1}{4}(4\pi) = \pi, \quad \frac{1}{2}(4\pi) = 2\pi,$$

$$\frac{3}{4}(4\pi) = 3\pi, \quad \text{and} \quad 4\pi.$$

The amplitude is 3. Complete the graph by following the steps shown in Figure 4.11. ✜

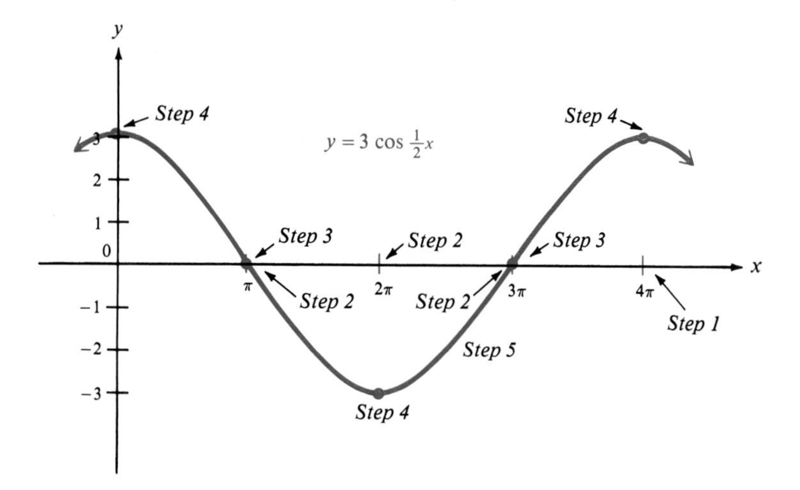

Figure 4.11

Vertical Translations The graph of a function of the form $y = c + f(x)$ is shifted vertically as compared with the graph of $y = f(x)$. See Figure 4.12. The function $y = c + f(x)$ is called a **vertical translation** of $y = f(x)$. The next example illustrates a vertical translation of a trigonometric function.

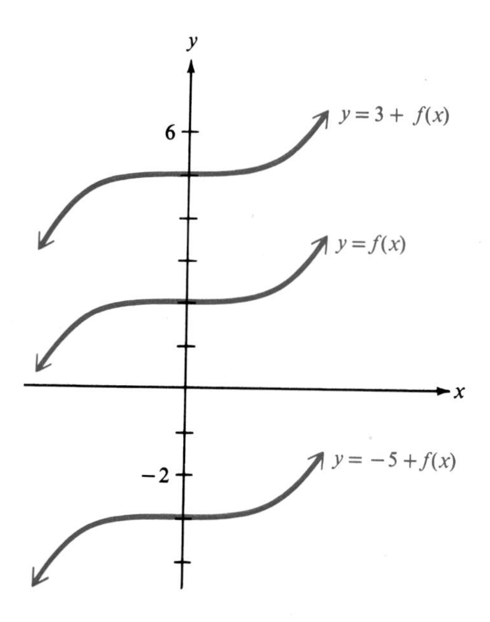

Figure 4.12

Example 6 Graph $y = 3 - 2 \cos 3x$.

The values of y will be 3 greater than the corresponding values of y in $y = -2 \cos 3x$. This means that the graph of $y = 3 - 2 \cos 3x$ is the same as the graph of $y = -2 \cos 3x$, except with a vertical translation of 3 units upward. Since the period of $y = -2 \cos 3x$ is $2\pi/3$, the key points have x-values of

$$0, \quad \frac{\pi}{6}, \quad \frac{\pi}{3}, \quad \frac{\pi}{2}, \quad \text{and} \quad \frac{2\pi}{3}.$$

The graph of $y = -2 \cos 3x$ has amplitude of 2. Because of the negative, the *maximum* value occurs at the midpoint of the key points and the minimum values occur at the endpoints. The key points are shown on the graph in Figure 4.13. ✥

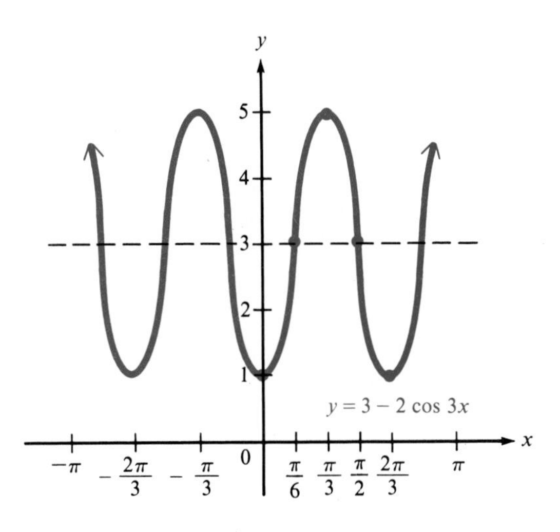

Figure 4.13

Radio stations send out a carrier signal in the form of a sine wave having equation

$$y = A_0 \sin (2\pi\omega_0 t),$$

where A_0 is the amplitude of the carrier signal, ω_0 is the number of periods the signal oscillates through in one second (its **frequency**), and t is time. A carrier signal received by a radio would be a pure tone. To transmit music and voices, a station might change or **modulate** A_0 according to the function

$$A_0(t) = A_0 + mA_0 \sin (2\pi\omega t),$$

where ω is the frequency of a pure tone and m is a constant called the **degree of modulation.** The transmitted signal has equation

$$\begin{aligned} y &= A_0 \sin (2\pi\omega_0 t) + A_0 m \sin (2\pi\omega t) \sin (2\pi\omega_0 t) \\ &= A_0[1 + m \sin (2\pi\omega t)] \sin (2\pi\omega_0 t). \end{aligned}$$

A typical carrier signal and a typical graph of y are shown in Figure 4.14. This process of sending out a radio signal is called **amplitude modulation,** or AM, radio.

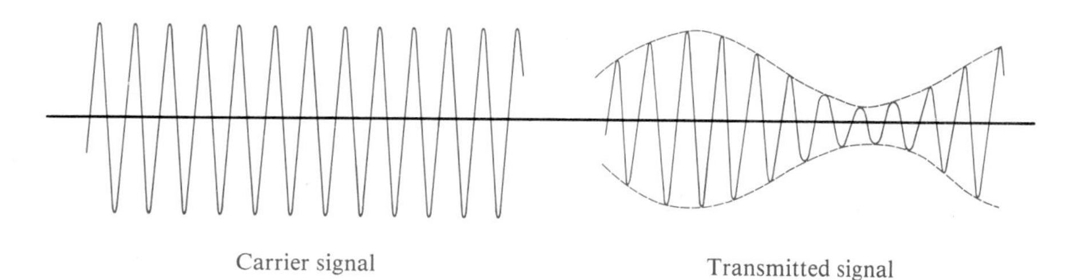

Figure 4.14

Frequency modulation, or FM, radio involves altering the frequency of the carrier signal, rather than its amplitude. A typical graph is shown in Figure 4.15.

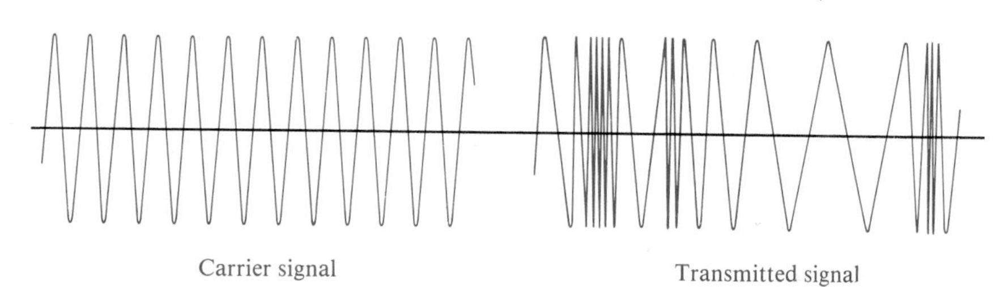

Figure 4.15

4.1 Exercises

Graph the following functions over the interval $-2\pi \le x \le 2\pi$. Identify the amplitude. See Example 1.

1. $y = 2 \cos x$ **2.** $y = 3 \sin x$ **3.** $y = \frac{2}{3} \sin x$ **4.** $y = \frac{3}{4} \cos x$

5. $y = -\cos x$ **6.** $y = -\sin x$ **7.** $y = -2 \sin x$ **8.** $y = -3 \cos x$

Graph each of the following functions over a two-period interval. Give the period, the amplitude, and any vertical translations. See Examples 2–6.

9. $y = \sin \frac{1}{2}x$ **10.** $y = \sin \frac{2}{3}x$ **11.** $y = \cos \frac{1}{3}x$ **12.** $y = \cos \frac{3}{4}x$

13. $y = \sin 3x$ **14.** $y = \sin 2x$ **15.** $y = \cos 2x$ **16.** $y = \cos 3x$

17. $y = -\sin 4x$ **18.** $y = -\cos 6x$ **19.** $y = 2 \sin \frac{1}{4}x$ **20.** $y = 3 \sin 2x$

21. $y = -2 \cos 3x$ **22.** $y = -5 \cos 2x$ **23.** $y = -3 + 2 \sin x$ **24.** $y = 2 - 3 \cos x$

25. $y = 1 - \frac{2}{3} \sin \frac{3}{4}x$ **26.** $y = -1 - 2 \cos 5x$ **27.** $y = 2 - \cos x$ **28.** $y = 1 + \sin x$

29. $y = 1 - 2\cos\frac{1}{2}x$

30. $y = -3 + 3\sin\frac{1}{2}x$

31. $y = -2 + \frac{1}{2}\sin 3x$

32. $y = 1 + \frac{2}{3}\cos\frac{1}{2}x$

33. $y = \cos \pi x$

34. $y = -\sin \pi x$

Graph each of the following functions over two periods.

35. $y = (\sin x)^2$ [*Hint:* $(\sin x)^2 = \sin x \cdot \sin x$]

36. $y = (\cos x)^2$

37. $y = (\sin 2x)^2$

38. $y = (\cos 2x)^2$

39. The graph shown below gives the variation in blood pressure for a typical person. Systolic and diastolic pressures are the upper and lower limits of the periodic changes in pressure that produce the pulse. The length of time between peaks is called the period of the pulse.

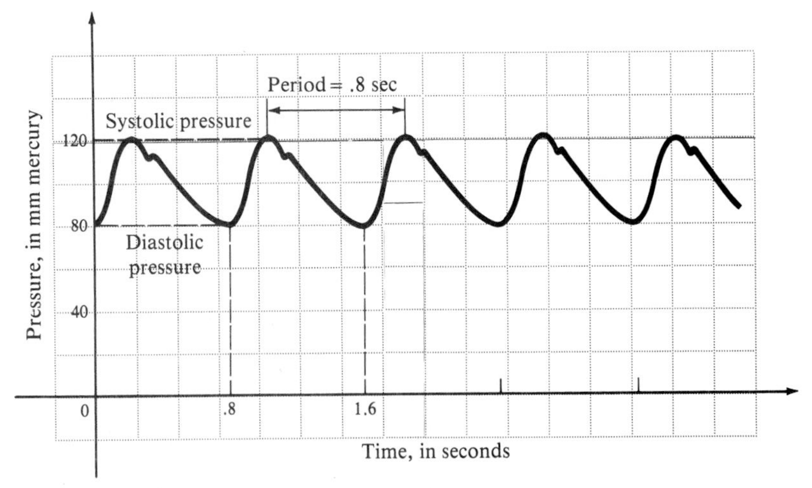

(a) Find the amplitude of the graph.
(b) Find the pulse rate (the number of pulse beats in one minute) for this person.

40. Scientists believe that the average annual temperature in a given location is periodic. The overall temperature at a given place during a given season fluctuates as time goes on, from colder to warmer, and back to colder. The graph shows an idealized description of the temperature for the last few thousand years of a location at the same latitude as Anchorage.

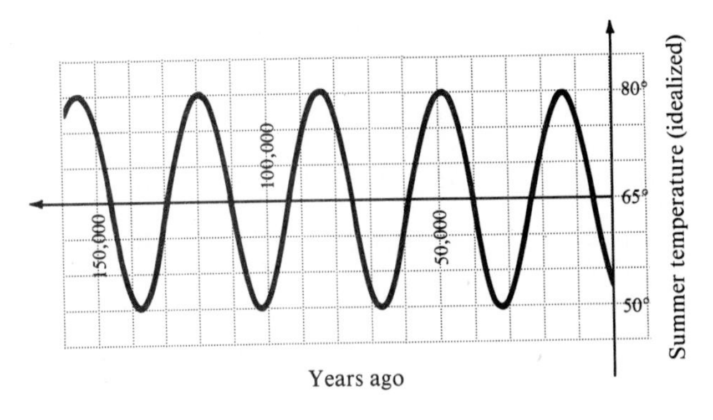

(a) Find the highest and lowest temperatures recorded.
(b) Use these two numbers to find the amplitude (*Hint:* An alternative definition of the amplitude is half the difference between the highest and lowest points on the graph.)
(c) Find the period of the graph.
(d) What is the trend of the temperature now?

41. Many of the activities of living organisms are periodic. For example, the graph below shows the time that flying squirrels begin their evening activity.
(a) Find the amplitude of this graph. (b) Find the period.

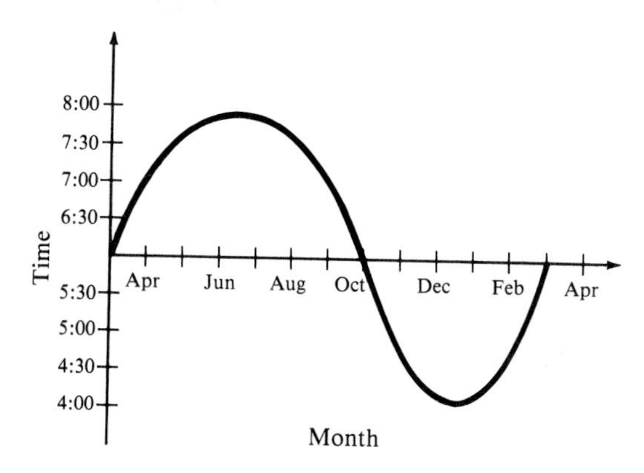

42. The figure shows schematic diagrams of a rhythmically moving arm. The upper arm RO rotates back and forth about the point R; the position of the arm is measured by the angle y between the actual position and the downward vertical position.*
(a) Find an equation of the form $y = a \sin kt$ for the graph at the right.
(b) How long does it take for a complete movement of the arm?

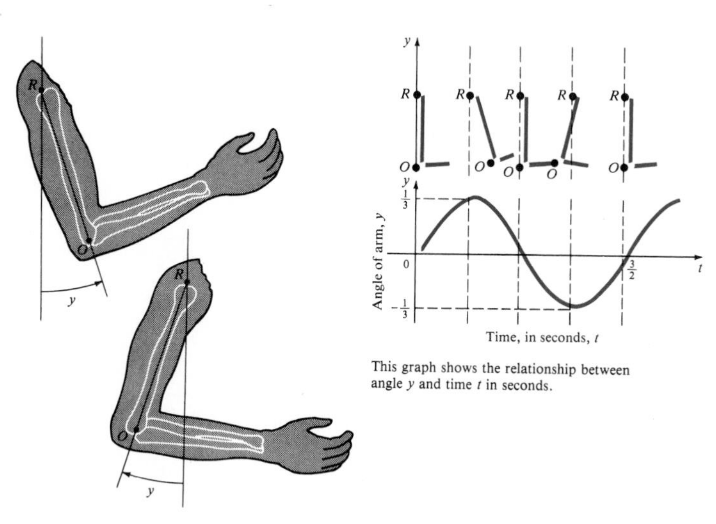

This graph shows the relationship between angle y and time t in seconds.

*From *Calculus for the Life Sciences,* by Rodolfo De Sapio, W. H. Freeman and Company.

Pure sounds produce single sine waves on an oscilloscope. Find the amplitude and period of each sine wave in the following photographs. On the vertical scale, each square represents .5, and on the horizontal scale each square represents 30°.

43.

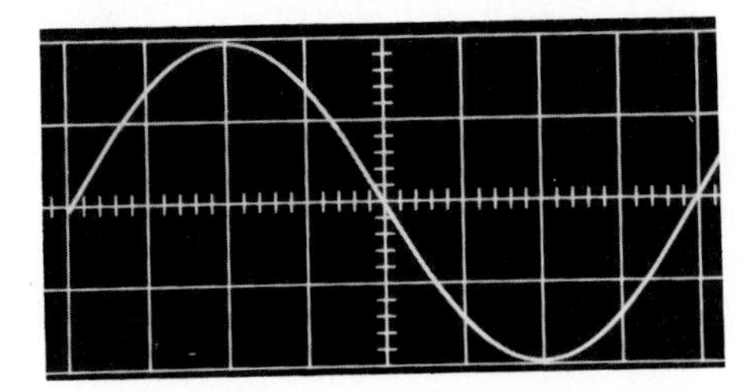

44.

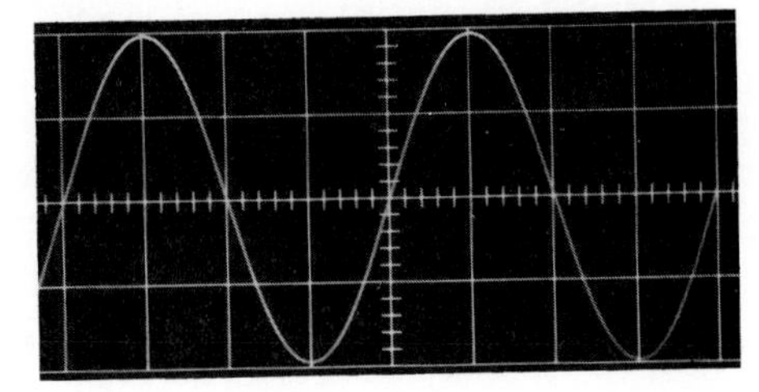

45. The voltage E in an electrical circuit is given by

$$E = 5 \cos 120\pi t,$$

where t is time measured in seconds.

(a) Find the amplitude and the period.

(b) How many cycles are completed in one second? [Recall that the number of cycles (periods) completed in one second is the **frequency** of the function.]

(c) Find E when $t = 0, .03, .06, .09, .12$.

(d) Graph E, for $0 \leq t \leq 1/30$.

46. For another electrical circuit, the voltage E is given by

$$E\ 3.8 \cos 40\pi t,$$

where t is time measured in seconds.

(a) Find the amplitude and the period.

(b) Find the frequency. See Exercise 45(b).

(c) Find E when $t = .02, .04, .08, .12, .14$.

(d) Graph one period of E.

47. To find the period of $y = \sin bx$, where $b > 0$, first observe that as bx varies from 0 to 2π, we get one period of the graph of $y = \sin bx$. Show that x therefore must vary from 0 to $2\pi/b$, so that the period of $y = \sin bx$ is $2\pi/b$.

48. Sketch the graph of $y = \sin x$ for real number values of x from 0 to .2 in increments of .02. On the same axes, draw $y = x$. Use this sketch to argue that for small values of x, $\sin x \approx x$.

4.2 Phase Shifts; Graphs of Cosecant and Secant

In Section 4.1 we discussed the roles played by the real numbers a, b, and c in the graph of a trigonometric function such as

$$y = c + a \sin bx \quad \text{or} \quad y = c + a \cos bx.$$

The value of $|a|$, the amplitude, determines the maximum distance of the graph from the line $y = c$. The number b affects the period, so that if the usual period is p, the new period becomes p/b, assuming $b > 0$. The number c determines the vertical translations; the graph is translated c units up from the x-axis if $c > 0$ and translated $|c|$ units down if $c < 0$.

In general, the graph of the function $y = f(x - d)$ is translated *horizontally* when compared to the graph of $y = f(x)$. The translation is d units to right if $d > 0$ and $|d|$ units to the left if $d < 0$. See Figure 4.16. With trigonometric functions, a horizontal translation is called a **phase shift.** In the function $y = f(x - d)$, the expression $x - d$ is called the **argument.**

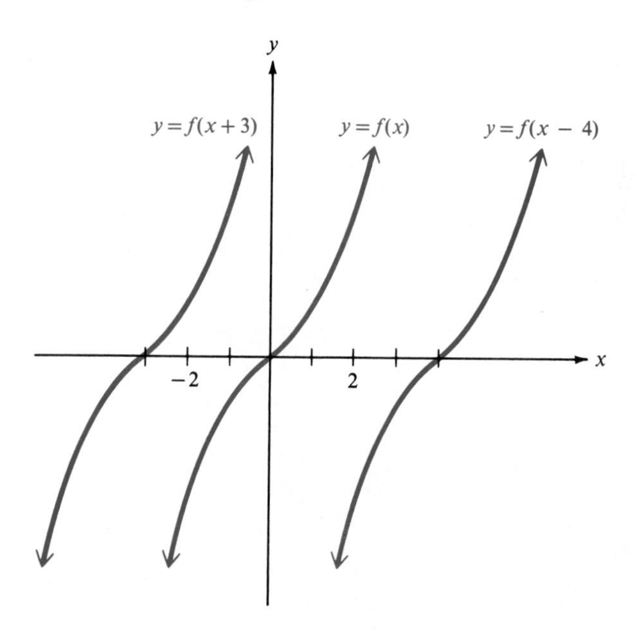

Figure 4.16

Example 1 Graph $y = \sin\left(x - \frac{\pi}{3}\right)$.

The argument $x - \pi/3$ indicates that the graph will be translated $\pi/3$ units to the right. This means that the x-values of the usual key points are each increased by $\pi/3$, as follows.

first point	$0 + \frac{\pi}{3} = \frac{\pi}{3}$
quarter point	$\frac{\pi}{2} + \frac{\pi}{3} = \frac{5\pi}{6}$

midpoint	$\pi + \dfrac{\pi}{3} = \dfrac{4\pi}{3}$
three-quarter point	$\dfrac{3\pi}{2} + \dfrac{\pi}{3} = \dfrac{11\pi}{6}$
last point	$2\pi + \dfrac{\pi}{3} = \dfrac{7\pi}{3}$

Thus, the key points for $y = \sin(x - \pi/3)$ are $(\pi/3, 0)$, $(5\pi/6, 1)$, $(4\pi/3, 0)$, $(11\pi/6, -1)$, and $(7\pi/3, 0)$, as shown in Figure 4.17. The graph of $y = \sin x$ is shown for comparison. ✥

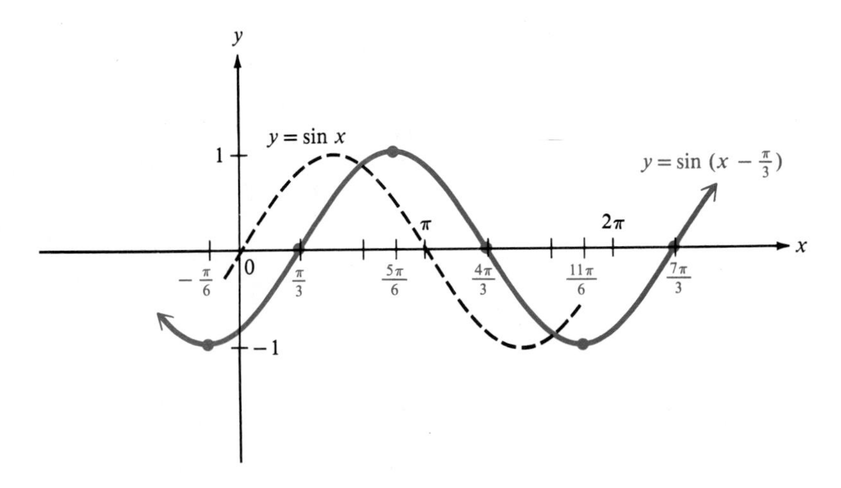

Figure 4.17

Example 2 Graph $y = 3\cos\left(x + \dfrac{\pi}{4}\right)$.

Start by writing $3\cos(x + \pi/4)$ in the form $a\cos(x - d)$.

$$3\cos\left(x + \frac{\pi}{4}\right) = 3\cos\left[x - \left(-\frac{\pi}{4}\right)\right]$$

This result shows that $d = -\pi/4$. Since $-\pi/4$ is negative, the phase shift is $|-\pi/4| = \pi/4$ to the left. The period is 2π and the amplitude is 3. A quick way to find the key points is by setting the argument equal to 0, $\pi/2$, π, $3\pi/2$, and 2π and solving these equations. For example,

$$x + \frac{\pi}{4} = 0 \qquad x + \frac{\pi}{4} = \frac{\pi}{2}$$

$$x = -\frac{\pi}{4}, \qquad x = \frac{\pi}{2} - \frac{\pi}{4}$$

$$x = \frac{\pi}{4},$$

and so on. The graph is shown in Figure 4.18. ✥

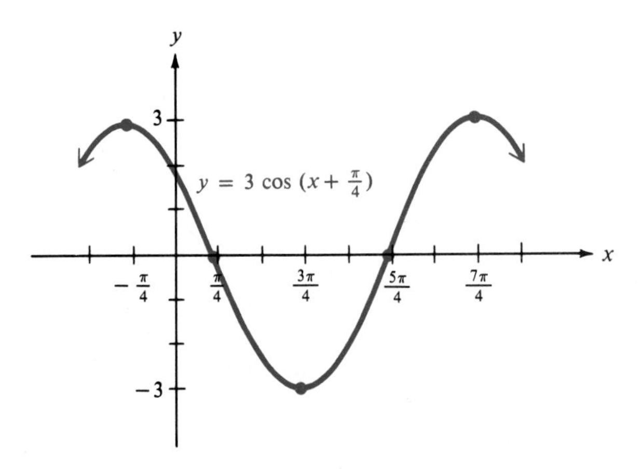

Figure 4.18

The next example shows a function of the form $y = a \cos b(x - d)$. Such functions have both a phase shift (if $d \neq 0$) and a period different from 2π (if $b \neq 1$).

Example 3 Graph $y = -2 \cos (3x + \pi)$.

First write the expression in the form $a \cos b(x - d)$ by factoring 3 out of the parentheses as follows.

$$y = -2 \cos (3x + \pi) = -2 \cos 3\left(x + \frac{\pi}{3}\right)$$

Then $a = -2$, $b = 3$, and $d = -\pi/3$. The amplitude is $|-2| = 2$, and the period is $2\pi/3$ (since the coefficient of x is 3).The phase shift is $|-\pi/3| = \pi/3$ to the left. Thus, the first key point is at $-\pi/3$. Since the period is $2\pi/3$, the last key point will have x-value $-\pi/3 + 2\pi/3 = \pi/3$ and the midpoint will be at $x = 0$. The graph is shown in Figure 4.19 on the following page. ✥

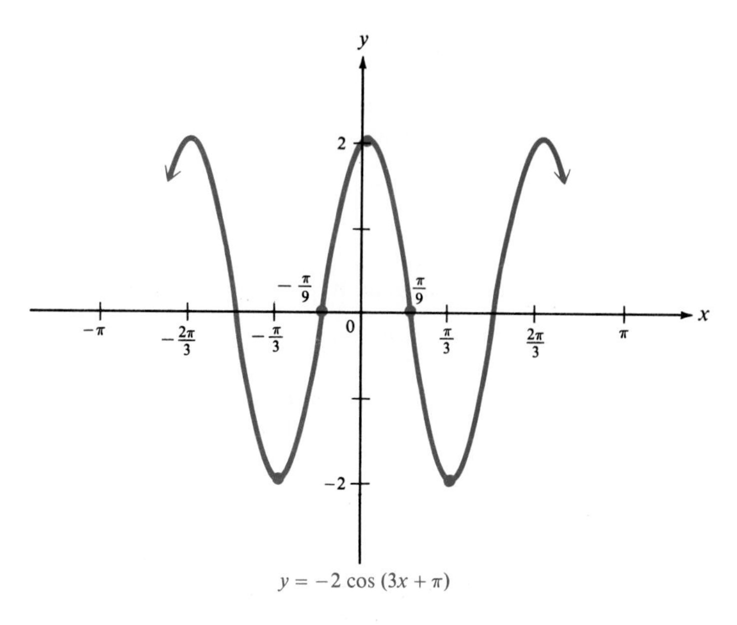

Figure 4.19

The work with sine and cosine graphs is summarized below. Assume $b > 0$.

Sine and Cosine Graphs

Function $y = c + a \sin b(x - d)$

or

$y = c + a \cos b(x - d)$

Amplitude $|a|$

Period $\dfrac{2\pi}{b}$

Vertical translation Up c units if $c > 0$

Down $|c|$ units if $c < 0$

Phase shift (horizontal translation)

d units to the right if $d > 0$

$|d|$ units to the left if $d < 0$

Graphs of Cosecant and Secant Since the function $\csc x = 1/(\sin x)$, the period is 2π, the same as for $\sin x$. When $\sin x = 1$, the value of $\csc x$ is also 1, and when $0 < \sin x < 1$, then $\csc x > 1$. Also, if $-1 < \sin x < 0$, then $\csc x < -1$. As x approaches 0, $\sin x$ approaches 0, and $\csc x$ gets larger and larger. The graph of $\csc x$ approaches the vertical line $x = 0$ but never touches it. The line $x = 0$ is called a **vertical asymptote.** In fact, the lines $x = n\pi$, where n is any integer, are all vertical asymptotes. Using this information and plotting a few points shows that the graph takes the shape of the solid curve shown in Figure 4.20. To show how the two graphs are related, the graph of $y = \sin x$ is also shown, as a dashed curve.

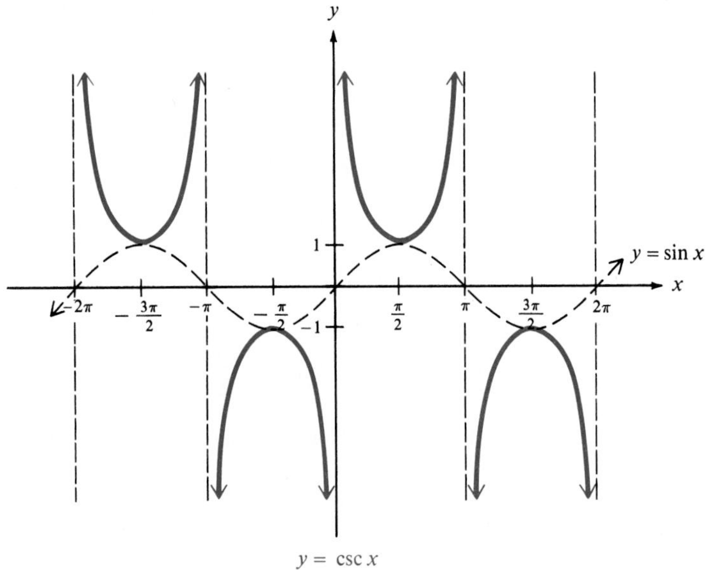

Figure 4.20

The graph of $y = \sec x$, shown in Figure 4.21, is related to the cosine graph in the same way that the graph of $y = \csc x$ is related to the sine graph, because $\sec x = 1/\cos x$.

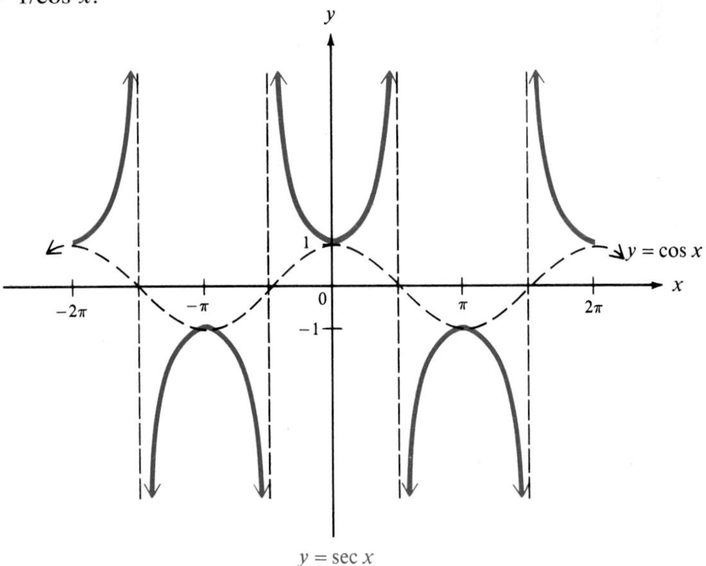

Figure 4.21

The following list summarizes the steps used to graph the cosecant and secant functions.

Graphing the Cosecant and Secant Functions

To graph $y = a \csc bx$ or $y = a \sec bx$, with $b > 0$, follow these steps.

1. Find the period, $2\pi/b$. Start at 0 on the x-axis and lay off an interval of $2\pi/b$.
2. Divide the interval from 0 to $2\pi/b$ into four equal parts.
3. Locate the asymptotes.

Function	*Asymptotes*
$y = a \csc bx$	$x = 0, \quad x = \frac{\pi}{b}, \quad x = \frac{2\pi}{b}$
$y = a \sec bx$	$x = \frac{\pi}{2b}, \quad x = \frac{3\pi}{2b}$

4. Find the amplitude for the corresponding graph, $y = a \sin bx$ or $y = a \cos bx$.
5. Locate the points where the graph has maximum or minimum values.

Function	*Maximum*	*Minimum*
$y = a \csc bx$	$\frac{3\pi}{2b}$	$\frac{\pi}{2b}$
$y = a \sec bx$	$\frac{\pi}{b}$	0

6. Draw additional periods of the graph, to the right and to the left, as needed.

Like the sine and cosine functions, the cosecant and secant functions may be translated vertically and horizontally.

Example 4 Graph $y = \frac{3}{2} \csc\left(x - \frac{\pi}{2}\right)$.

Compared with the graph of $y = \csc x$, this graph has a phase shift of $\pi/2$ to the right. Thus, the asymptotes are the lines $x = \pi/2, 3\pi/2$, and so on. Also, there are no values of y between $-3/2$ and $3/2$. As shown in Figure 4.22, this is related to the increased amplitude of $y = (3/2) \sin x$ compared with $y = \sin x$. (Amplitude does not apply to the secant or cosecant functions; it enters only indirectly from the corresponding cosine or sine graphs.) This means that the graph goes through the points $(\pi, 3/2)$, $(2\pi, -3/2)$, and so on. Two periods are shown in Figure 4.22. ✤

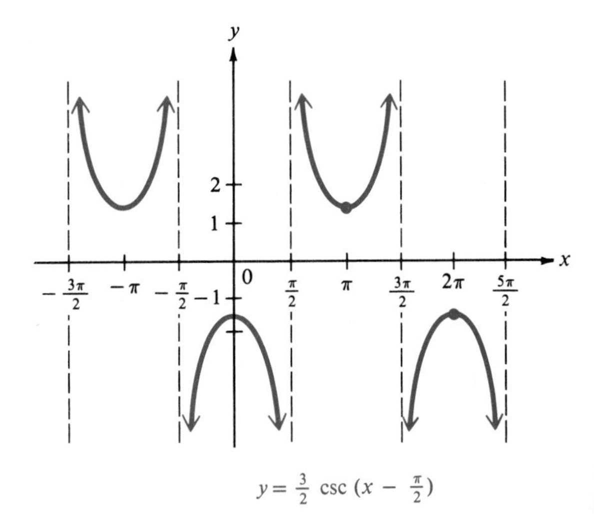

Figure 4.22

Example 5 One example of a phase shift comes up in electrical work. A simple alternating current circuit is shown in Figure 4.23. The relationship between voltage V and current I in the circuit is also shown in the figure.

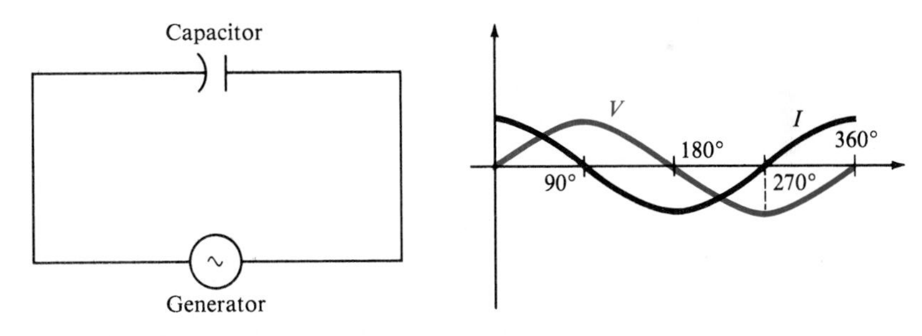

Figure 4.23

As this graph shows, current and voltage are *out of phase* by 90°. In this example, current *leads* the voltage by 90°, or voltage *lags* by 90°. ✤

4.2 Exercises

For each of the following, find the amplitude, the period, any vertical translation, and any phase shift. See Examples 1–3.

1. $y = 2 \sin (x - \pi)$

2. $y = \dfrac{2}{3} \sin \left(x + \dfrac{\pi}{2}\right)$

3. $y = 4 \cos \left(\dfrac{x}{2} + \dfrac{\pi}{2}\right)$

4. $y = -\cos \dfrac{2}{3}\left(x - \dfrac{\pi}{3}\right)$

5. $y = 3 \cos 2\left(x - \frac{\pi}{4}\right)$

6. $y = \frac{1}{2} \sin \left(\frac{x}{2} + \pi\right)$

7. $y = 2 - \sin \left(3x - \frac{\pi}{5}\right)$

8. $y = -1 + \frac{1}{2} \cos (2x - 3\pi)$

Graph each of the following functions over a one-period interval. See Examples 1–4.

9. $y = \cos \left(x - \frac{\pi}{2}\right)$

10. $y = \sin \left(x + \frac{\pi}{4}\right)$

11. $y = \sin \left(x + \frac{\pi}{4}\right)$

12. $y = \cos \left(x - \frac{\pi}{3}\right)$

13. $y = 2 \cos \left(x - \frac{\pi}{3}\right)$

14. $y = 3 \sin \left(x - \frac{3\pi}{2}\right)$

15. $y = \frac{3}{2} \sin 2 \left(x + \frac{\pi}{4}\right)$

16. $y = -\frac{1}{2} \cos 4 \left(x + \frac{\pi}{2}\right)$

17. $y = -4 \sin (2x - \pi)$

18. $y = 3 \cos (4x + \pi)$

19. $y = \frac{1}{2} \cos \left(\frac{1}{2} x - \frac{\pi}{4}\right)$

20. $y = -\frac{1}{4} \sin \left(\frac{3}{4} x + \frac{\pi}{8}\right)$

21. $y = -3 + 2 \sin \left(x + \frac{\pi}{2}\right)$

22. $y = 4 - 3 \cos (x - \pi)$

23. $y = \frac{1}{2} + \sin 2 \left(x + \frac{\pi}{4}\right)$

24. $y = -\frac{5}{2} + \cos 3 \left(x - \frac{\pi}{6}\right)$

25. $y = \csc \left(x - \frac{\pi}{4}\right)$

26. $y = \sec \left(x + \frac{3\pi}{4}\right)$

27. $y = \sec \left(x + \frac{\pi}{4}\right)$

28. $y = \csc \left(x + \frac{\pi}{3}\right)$

29. $y = \sec \left(\frac{1}{2} x + \frac{\pi}{3}\right)$

30. $y = \csc \left(\frac{1}{2} x - \frac{\pi}{4}\right)$

31. $y = 2 + 3 \sec (2x - \pi)$

32. $y = 1 - 2 \csc \left(x + \frac{\pi}{2}\right)$

33. $y = 1 - \frac{1}{2} \csc \left(x - \frac{3\pi}{4}\right)$

34. $y = 2 + \frac{1}{4} \sec \left(\frac{1}{2} x - \pi\right)$

35. $y = \frac{2}{3} \sec \left(\frac{3}{4} x + \pi\right) - 2$

4.3 Graphs of Tangent and Cotangent

Since the values of $y = \tan x$ are positive in quadrants I and III, and negative in quadrants II and IV,

$$\tan (x + \pi) = \tan x,$$

so the period of $y = \tan x$ is π. Thus, the tangent function need be investigated only within an interval of π units. A convenient interval for this purpose is $-\pi/2 < x < \pi/2$, because although the endpoints $-\pi/2$ and $\pi/2$ are not in the domain of $y = \tan x$ (why?), $\tan x$ exists for all other values in the interval. In the interval $0 < x < \pi/2$ $\tan x$ is positive. As x goes from 0 to $\pi/2$, a calculator or Table 2 shows that $\tan x$ gets larger and larger without bound. As x goes from $-\pi/2$ to 0, the values of $\tan x$ approach 0. These results are summarized in the following table.

As x increases from	tan x
0 to $\pi/2$	Increases from 0, without bound
$-\pi/2$ to 0	Increases to 0

Based on these results, the graph of $y = \tan x$ will approach the vertical line $x = \pi/2$ but never touch it, so the line $x = \pi/2$ is a vertical asymptote. The lines $x = \pi/2 + n\pi$, where n is any integer, are all vertical asymptotes. These asymptotes are indicated with light dashed lines on the graph in Figure 4.24. In the interval $-\pi/2 < x < 0$, which corresponds to quadrant IV, tan x is negative, and as x goes from 0 to $-\pi/2$, tan x gets smaller and smaller. A table of values for tan x, where $-\pi/2 < x < \pi/2$, is given below.

x	$-\pi/3$	$-\pi/4$	$-\pi/6$	0	$\pi/6$	$\pi/4$	$\pi/3$
tan x	-1.7	-1	$-.6$	0	.6	1	1.7

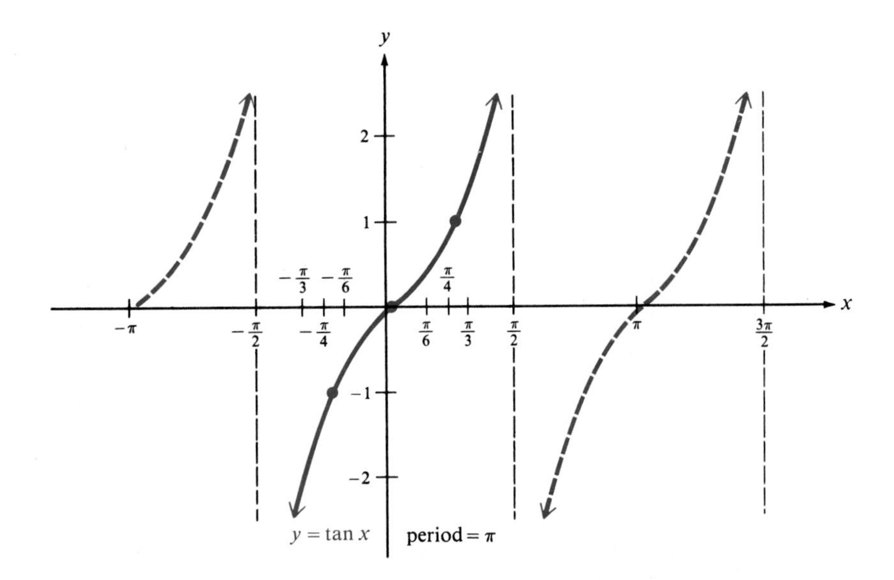

Figure 4.24

Plotting the points from the table and letting the graph approach the asymptotes at $x = \pi/2$ and $x = -\pi/2$ gives the portion of the graph shown with a solid curve in Figure 4.24. More of the graph can be sketched by repeating the same curve, also as shown in the figure. This graph, like the graphs for the sine and cosine functions, should be memorized. Convenient main points are $(-\pi/4, -1)$,

(0, 0), and ($\pi/4$, 1). These points are shown in Figure 4.24. The lines $x = \pi/2$ and $x = -\pi/2$ are vertical asymptotes. The idea of *amplitude,* discussed earlier, applies only to the sine and cosine functions, and so is not used here.)

Example 1 Graph $y = \tan 2x$.

Multiplying x by 2 changes the period from π to $\pi/2$. (Make a table of values to see this.) The effect on the graph is shown in Figure 4.25, where two periods of the function are graphed. Because of this change in period, the asymptotes are the lines $x = -\dfrac{\pi}{4}$ and $x = \dfrac{\pi}{4}$, and the key points are $(-\pi/8, -1)$, (0, 0), and $(\pi/8, 1)$. ✣

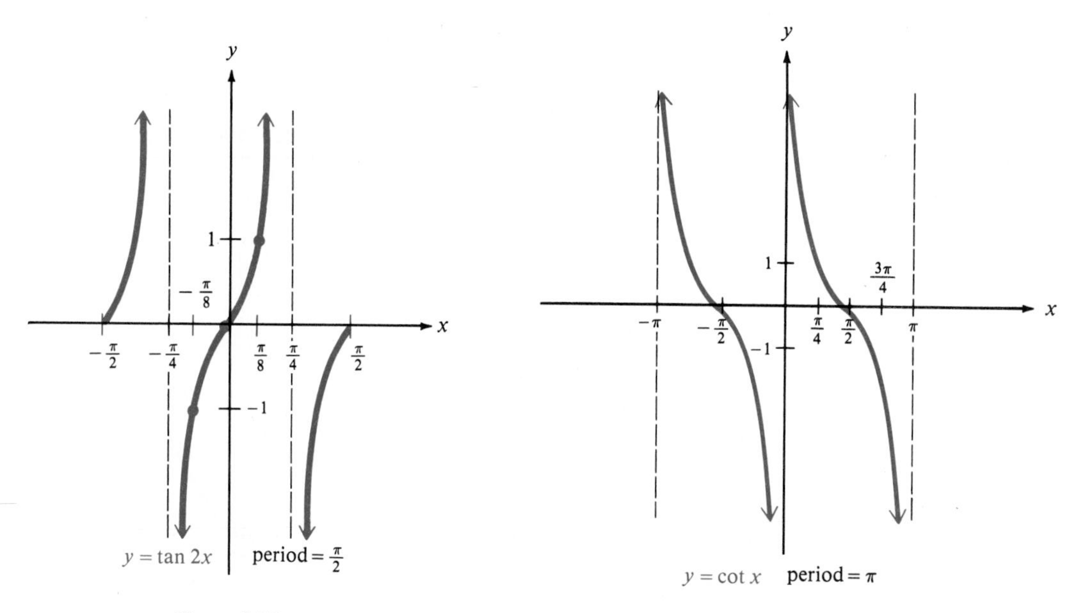

Figure 4.25

Figure 4.26

The fact that $\cot x = 1/(\tan x)$ can be used to find the graph of $y = \cot x$. The period of cotangent, like that of tangent, is π. The domain of $y = \cot x$ excludes $0 + n\pi$, where n is any integer since $1/\tan x$ is undefined for these values of x. Thus, the vertical lines $x = n\pi$ are asymptotes. Values of x that lead to asymptotes for $\tan x$ will make $\cot x = 0$, so $\cot(-\pi/2) = 0$, $\cot \pi/2 = 0$, $\cot 3\pi/2 = 0$, and so on. The values of $\tan x$ increase as x goes from $-\pi/2$ to $\pi/2$, so the values of $\cot x$ will *decrease* as x goes from $-\pi/2$ to $\pi/2$. A table of values for $\cot x$, where $0 < x < \pi$, is shown below.

x	$\pi/6$	$\pi/4$	$\pi/3$	$\pi/2$	$2\pi/3$	$3\pi/4$	$5\pi/6$
$\cot x$	1.7	1	.6	0	−.6	−1	−1.7

Plotting these points and using the information discussed above gives the graph of $y = \cot x$ shown in Figure 4.26. (The graph shows two periods.)

Graphing the Tangent and Cotangent Functions

To graph $y = a \tan bx$ or $y = a \cot bx$, with $b > 0$:

1. Find the period, π/b.
2. On the x-axis, lay off two intervals, each with length half the period, one interval to the left and the other to the right of the midpoint.

Function	*Midpoint*
$y = a \tan bx$	$x = 0$
$y = a \cot bx$	$x = \dfrac{\pi}{2b}$.

3. Draw the asymptotes as vertical dashed lines at the endpoints of the intervals of Step 2.
4. Locate the key points.

Function	*Key Points*
$y = a \tan bx$	$\left(-\dfrac{\pi}{4b}, -a\right)$, $(0, 0)$, $\left(\dfrac{\pi}{4b}, a\right)$
$y = a \cot bx$	$\left(\dfrac{\pi}{4b}, a\right)$, $\left(\dfrac{\pi}{2b}, 0\right)$, $\left(\dfrac{3\pi}{4b}, -a\right)$

5. Sketch the graph.
6. Draw additional periods, both to the right and to the left, as needed.

Example 2 Graph $y = -3 \tan \dfrac{1}{2} x$.

The period is $\pi/(1/2) = 2\pi$. Proceed as shown in Figure 4.27 on the following page. ✥

Example 3 Graph one period of $y = \dfrac{1}{2} \cot 2x$.

The period is $\pi/2$. Lay off an interval on the x-axis $\pi/4$ units on either side of $\pi/2b = \pi/(2 \cdot 2) = \pi/4$. One endpoint will be at 0, and the other will be at $\pi/2$. Thus, the asymptotes are $x = 0$ and $x = \pi/2$. Here, $a = 1/2$, so locate the key points at $(\pi/8, 1/2)$, $(\pi/4, 0)$, and $(3\pi/8, -1/2)$. Complete the graph as shown in Figure 4.28 on the following page. ✥

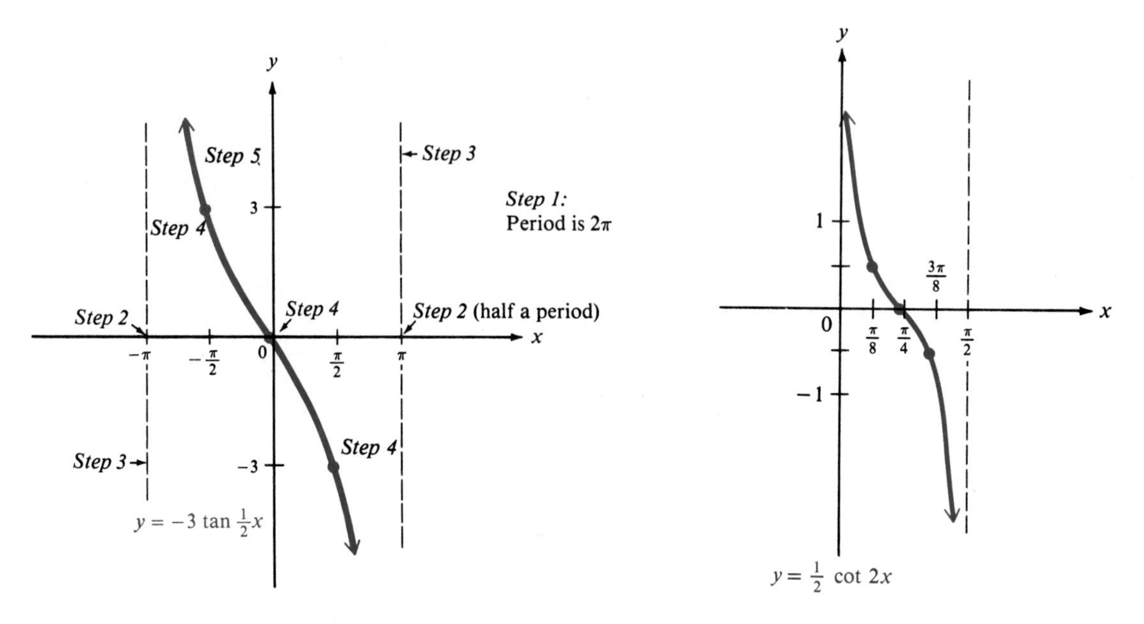

Figure 4.27

Figure 4.28

Tangent and cotangent graphs may also be translated vertically, horizontally, or both, as the next two examples show.

Example 4 Graph $y = 2 + \tan x$.

Every value of y for this function will be 2 units more than the corresponding value of y in $y = \tan x$, causing the graph of $y = 2 + \tan x$ to be translated 2 units vertically compared with the graph of $y = \tan x$. See Figure 4.29. ✜

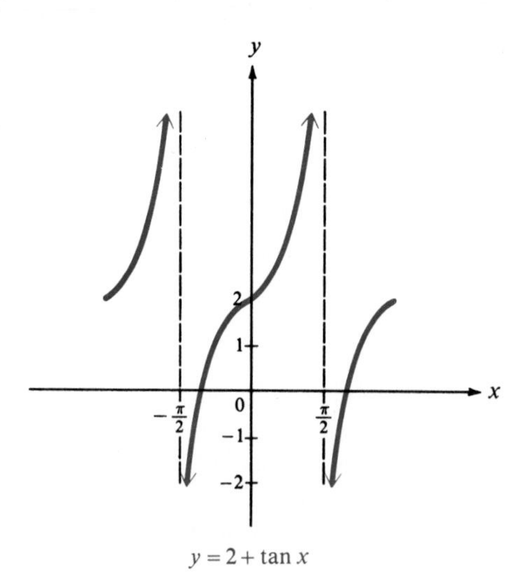

Figure 4.29

Example 5 Graph two periods of $y = -2 - \cot\left(x - \frac{\pi}{4}\right)$.

Here $b = 1$, so the period is π. The graph will be translated down 2 units and $\pi/4$ units to the right. Lay off intervals of $\pi/2$ on either side of $\pi/2 + \pi/4 = 3\pi/4$. (Add $\pi/4$ because of the phase shift.) The endpoints of these intervals give the asymptotes,

$$x = \frac{3\pi}{4} - \frac{\pi}{2} = \frac{\pi}{4} \quad \text{and} \quad x = \frac{3\pi}{4} + \frac{\pi}{2} = \frac{5\pi}{4}.$$

The key points, $(\pi/2, -3)$, $(3\pi/4, -2)$, and $(\pi, -1)$, have been adjusted for both translations. The graph is shown in Figure 4.30. ✣

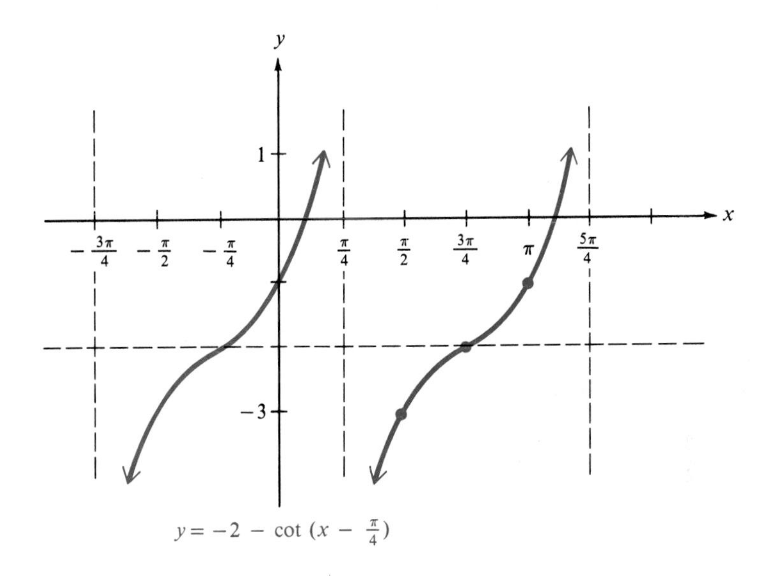

Figure 4.30

4.3 Exercises

Graph the following functions over a one-period interval. See Examples 1–3.

1. $y = 2 \tan x$ **2.** $y = -\tan x$ **3.** $y = -\cot x$ **4.** $y = \frac{1}{2} \cot x$

5. $y = \tan 2x$ **6.** $y = 2 \tan \frac{1}{4} x$ **7.** $y = \cot 3x$ **8.** $y = -\cot \frac{1}{2} x$

Graph the following functions over a two-period interval. Identify the period. See Examples 4 and 5.

9. $y = 1 + \tan x$ **10.** $y = -2 + \tan x$ **11.** $y = -1 + 2 \tan x$ **12.** $y = 3 + \frac{1}{2} \tan x$

13. $y = 1 - \cot x$ **14.** $y = -2 - \cot x$ **15.** $y = \tan (2x - \pi)$ **16.** $y = \tan\left(\frac{x}{2} + \pi\right)$

17. $y = \cot\left(3x + \frac{\pi}{4}\right)$

18. $y = \cot\left(2x - \frac{3\pi}{2}\right)$

19. $y = -1 + \frac{1}{2}\cot(2x - 3\pi)$

20. $y = -2 + 3\tan(4x + \pi)$

21. $y = \frac{2}{3}\tan\left(\frac{3}{4}x - \pi\right) - 2$

22. $y = 1 - 2\cot 2\left(x + \frac{\pi}{2}\right)$

23. A rotating beacon is located at point A next to a long wall. (See the figure.) The beacon is 4 m from the wall. The distance d is given by

$$d = 4\tan 2\pi t,$$

where t is time measured in seconds since the beacon started rotating. (When $t = 0$, the beacon is aimed at point R. When the beacon is aimed to the right of R, the value of d is positive; d is negative if the beacon is aimed to the left of R.) Find d for the following times.

(a) $t = 0$ **(b)** $t = .4$
(c) $t = .8$ **(d)** $t = 1.2$
(e) Why is .25 a meaningless value for t?
(f) What is a meaningful domain for t?

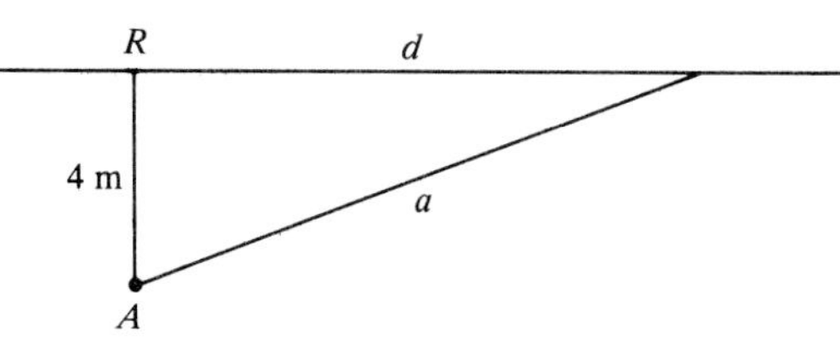

24. In the figure for Exercise 23, the distance a is given by

$$a = 4|\sec 2\pi t|.$$

Find a for the following times.

(a) $t = 0$ **(b)** $t = .86$ **(c)** $t = 1.24$
(d) Why are the absolute value bars needed here, but not in the function giving d?

25. Let a person h_1 ft tall stand d ft from an object h_2 ft tall, where $h_2 > h_1$. Let θ be the angle of elevation to the top of the object. (See the figure.)

(a) Show that $d = (h_2 - h_1)\cot\theta$.
(b) Let $h_2 = 55$ and $h_1 = 5$. Graph d for $0° < \theta < \frac{\pi}{2}$.

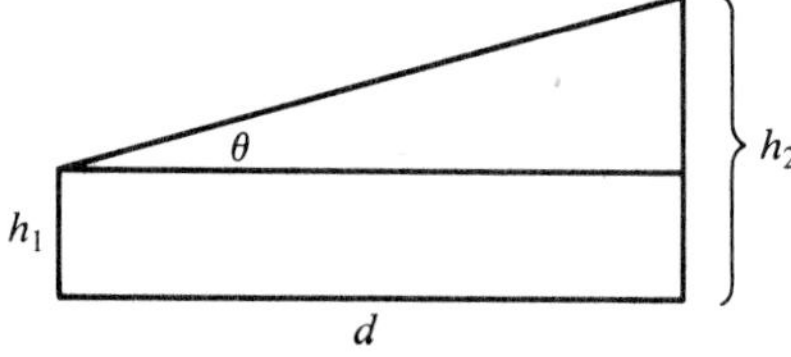

26. The quotient $y = \frac{\sin x}{\cos x}$ does not exist if $x = -\pi/2$ or if $x = \pi/2$. Start at $x = -1.4$ and evaluate the quotient with x increasing by .2 until $x = 1.4$ is reached. Plot the values obtained. What graph is suggested?

4.4 Graphing by Combining Functions (Optional)

A quick method for graphing trigonometric functions that are the sum of two or more functions is by *addition of ordinates*. An **ordinate** is the y-value of an ordered pair. For example, in the ordered pair $(\pi, -1)$, the number -1 is the ordinate. This graphing method is best described by examples.

Example 1 Graph $y = x + \sin x$.

Begin by graphing the functions $y = x$ and $y = \sin x$ separately on the same coordinate axes. Figure 4.31 shows the two graphs. (The graphs of $y = x$ and $y = \sin x$ are given with dashed curves.) Then select some x-values, and for these values add the two corresponding ordinates to get the ordinate of the sum, $x + \sin x$. For example, when $x = 0$, both ordinates are 0, so that $P_1 = (0, 0)$ is a point on the graph of $y = x + \sin x$. When $x = \pi/2$, the ordinates are $\pi/2$ and 1. Their sum is $\pi/2 + 1$, which is approximately 2.6, and $P_2 = (\pi/2, \pi/2 + 1)$, or approximately (1.6, 2.6), is on the graph. At $x = 3\pi/2$, the sum $x + \sin x$ is $3\pi/2 + (-1)$, or approximately 3.7, with $P_3 = (3\pi/2, 3\pi/2 - 1)$, or approximately (4.7, 3.7), on the graph. As many points as necessary can be located in this way. The graph is then completed by drawing a smooth curve through the points. The graph of $y = x + \sin x$ is shown in color in Figure 4.31. ✣

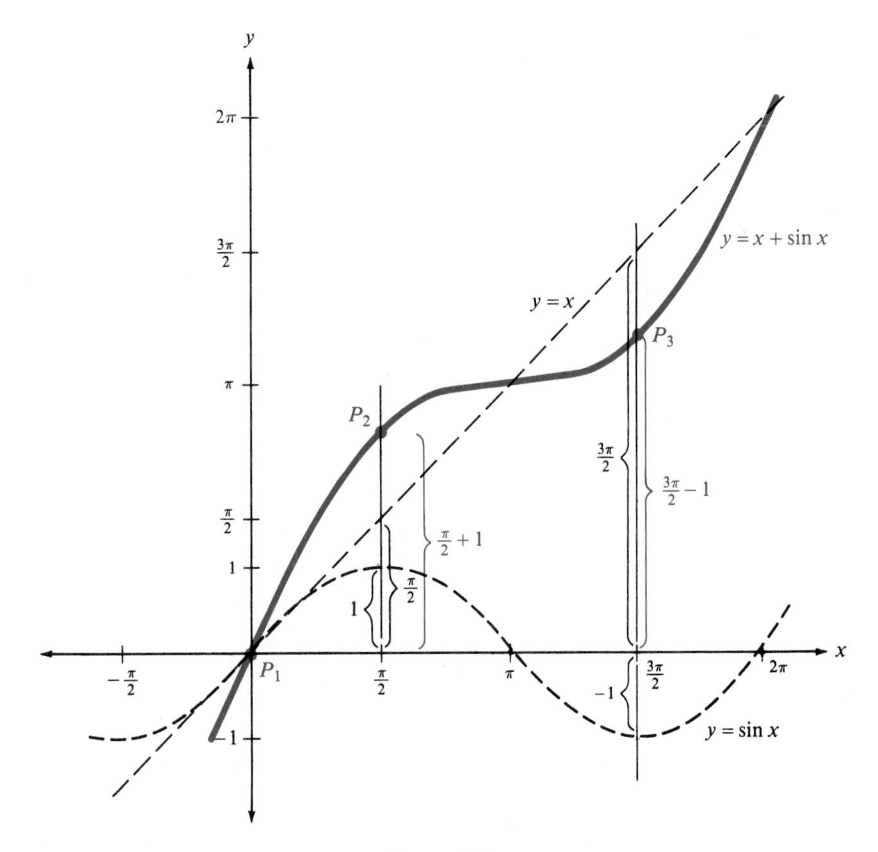

Figure 4.31

As shown on the graph in Figure 4.31, to get the graph of $y = x + \sin x$ the ordinates are actually treated as line segments. For example, the ordinate of P_2 is found by adding the lengths of the two line segments that represent the ordinates of x and $\sin x$ at $\pi/2$. The same is true for the ordinate of P_3 as well as for each of the other ordinates.

Example 2 Graph $y = \cos x - \tan x$.

Think of $y = \cos x - \tan x$ as $y = \cos x + (-\tan x)$. Start by graphing $y = \cos x$ and $y = -\tan x$ on the same axes, as in Figure 4.31. At $x = 0$, the ordinates are 1 and 0, so the ordinate of $y = \cos x + (-\tan x)$ is $1 + 0 = 1$ and the point (0, 1) is on the graph. At any point where the graphs of $\cos x$ and $-\tan x$ cross, the ordinate is doubled. See, for example, P_1 and P_2 on the graph in Figure 4.32.

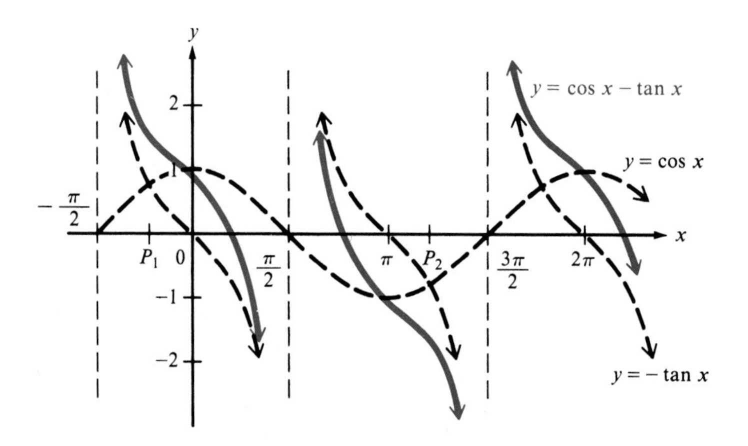

Figure 4.32

The domain of the function $y = \cos x - \tan x$ does not contain $\pi/2 + n\pi$, where n is any integer, because these numbers are not in the domain of $\tan x$. Thus, the lines $x = \pi/2 + n\pi$ are asymptotes, so that as x approaches $\pi/2$ and $3\pi/2$ from the right, the values of y get larger and larger. Also, when x approaches $\pi/2$ and $3\pi/2$ from the left, the values of y get smaller and smaller. A portion of this graph is shown in color in Figure 4.32. ✥

Figures 4.33 and 4.34 on the facing page and Figure 4.35 on page 166 illustrate some naturally occurring examples of ordinate addition.

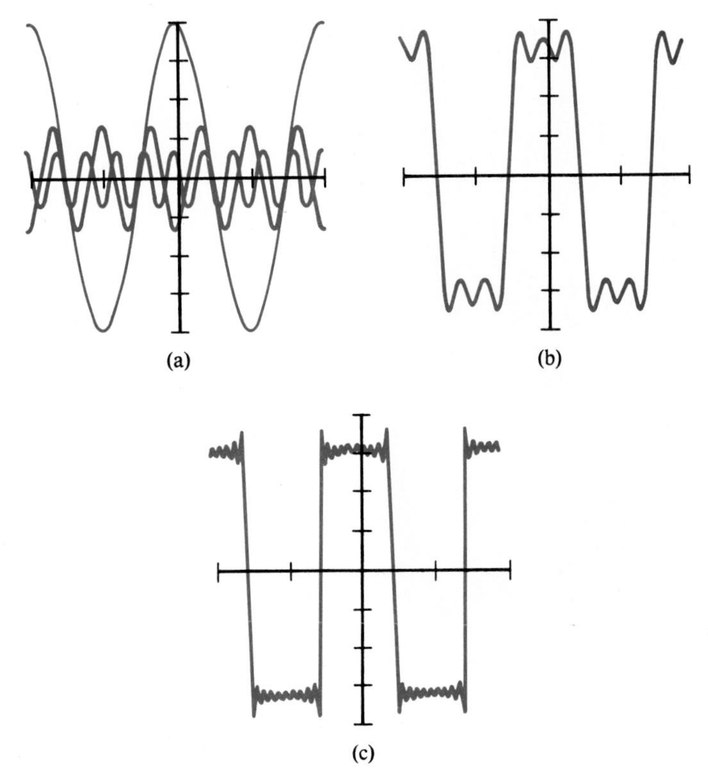

Figure 4.33 *Figure (a) shows three sine waves, of amplitudes 1, $\frac{1}{3}$, and $\frac{1}{5}$, respectively. Figure (b) shows the wave that results from adding together the three waves in Figure (a). Finally, Figure (c) shows the approximation to a square wave produced by adding together 19 sine waves similar to those in Figure (a)—sine waves with amplitudes 1, $\frac{1}{3}$, $\frac{1}{5}$, $\frac{1}{7}$, . . . , $\frac{1}{37}$.**

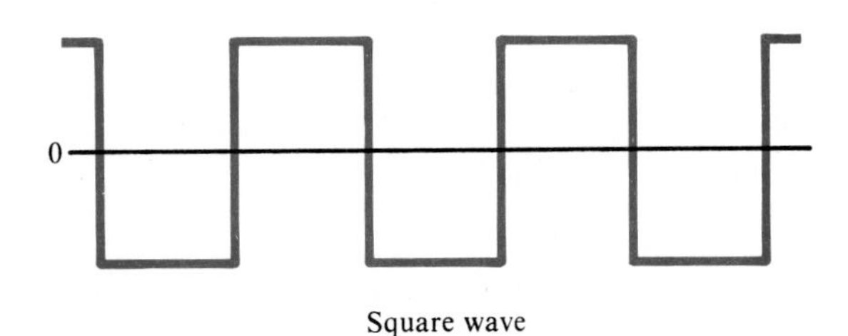

Figure 4.34 *A branch of advanced mathematics called* Fourier analysis *shows how functions can be expressed as the sum of an infinite number of sine waves. For example, the figure shows a square wave, an idealized representation of a repetitive sound such as striking a piano key over and over.*

---

*From *The Science of Musical Sound,* by John R. Pierce, W. H. Freeman and Company. Copyright © 1983 by Scientific American Books, Reprinted by permission.

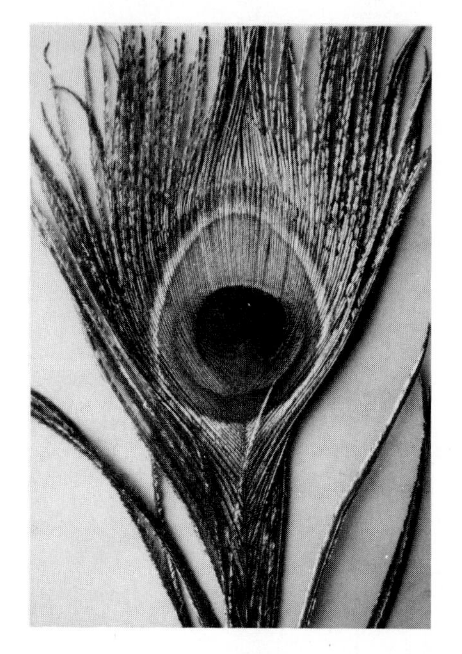

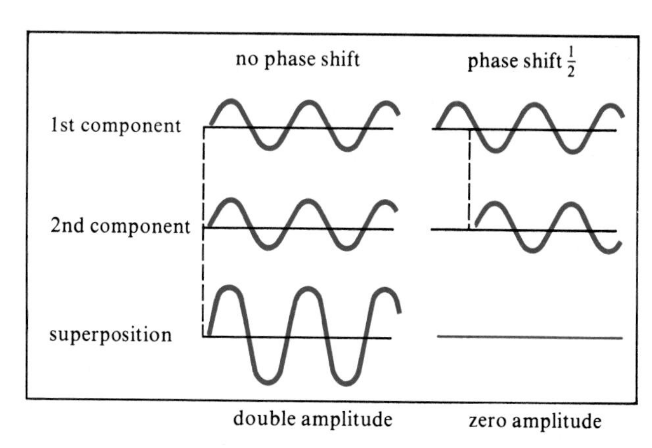

Figure 4.35 *A biological application of the addition of ordinates of sine curves with equal frequency and amplitude is found in the colors of a peacock's feathers. Pure light waves (which are sine waves) are combined by interference to produce colors. There are no pigments in the feathers, which look clear under a microscope.**

4.4 Exercises

Use the method of addition of ordinates to graph each of the following. See Examples 1 and 2.

1. $y = x + \cos x$

2. $y = \sin x - 2x$

3. $y = 3x - \cos 2x$

4. $y = x + 2 - \sin x$

5. $y = \sin x + \sin 2x$

6. $y = \cos x - \cos \frac{1}{2}x$

7. $y = \sin x + \tan x$

8. $y = \sin x + \csc x$

9. $y = 2 \cos x - \sec x$

10. $y = 2 \sec x + \sin x$

11. $y = \cos x + \cot x$

12. $y = \sin x - 2 \cos x$

13. $y = -x + \sec x$

14. $y = x + \csc x$

Use methods similar to those in this section to graph the following functions.

15. $y = x \sin x$

16. $y = x \cos x$

17. $y = 2^{-x} \sin x$

18. $y = x^2 \sin x$

19. The function $y = (6 \cos x)(\cos 8x)$ can occur in AM radio transmissions, as mentioned in Section 4.1. Graph this function on an interval from 0 to 2π. (*Hint:* First graph $y = 6 \cos x$ and $y = \cos 8x$ as dashed lines.)

---

*From *Introduction to Mathematics for Life Scientists,* Third Edition, by Edward Batschelet.

20. Graph $y = \sin 8x + \sin 9x$. The period of this function is very long. By placing two engines that are running at almost the same speed side by side (as suggested by the $8x$ and $9x$), we get an effect of "beats." See the figure.

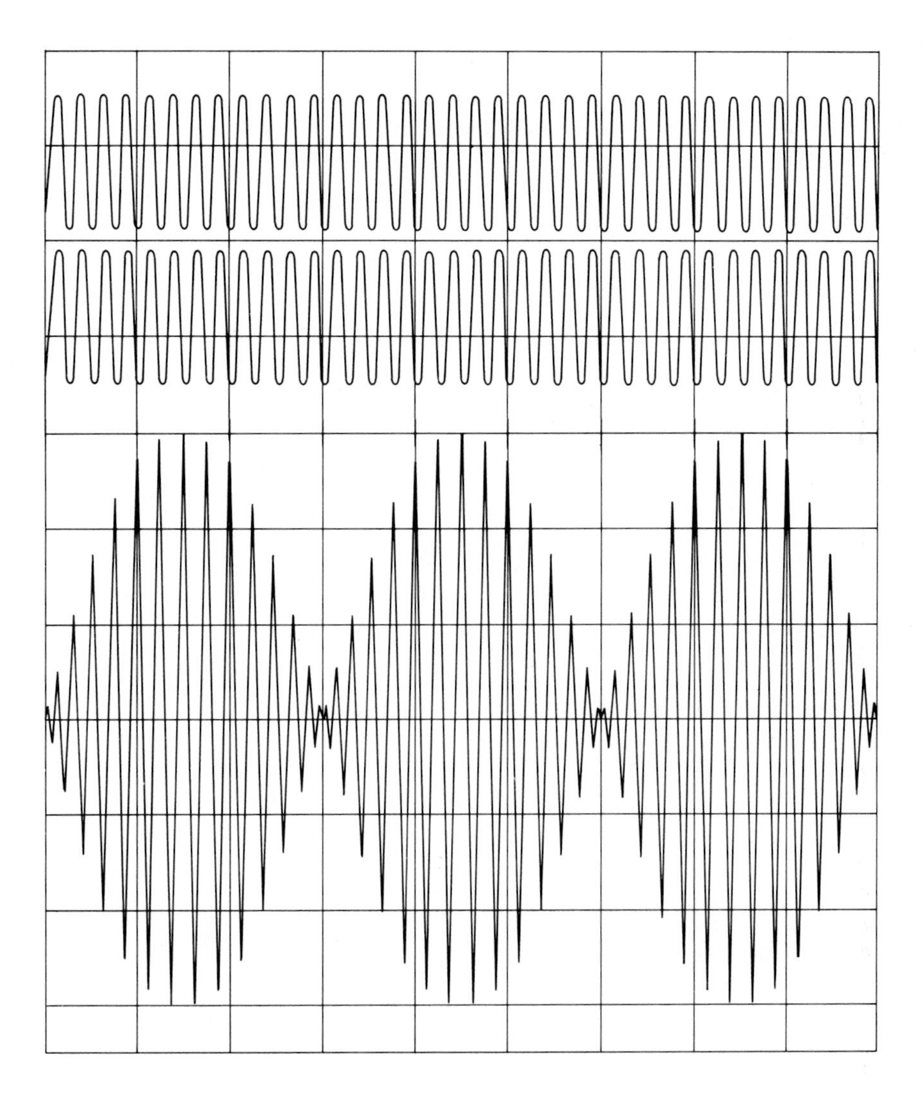

The top two sine waves represent pure tones, such as those put out by a tuning fork or an electronic oscillator. When two such pure tones, having slightly different periods, are played side by side, the amplitudes add algebraically, instant by instant, producing a result such as that shown in the bottom graph. The peaks here are called beats. *Beats result, for example, from engines on an airplane that are running at almost, but not quite, the same speeds. The blowers in air-conditioning or heating units in different apartments also can cause such beats; these can be quite annoying.*

21. If a bumper of a car is given a firm downward push and then released, the car's shock absorbers cause it to bounce back and then quickly return to stable position, producing an example of *damped oscillatory motion*. Such motion often can be represented by a function of the form

$$y = a \cdot e^{-kt} \cdot \sin b(t + c),$$

where a, k, b, and c are constants, with $k > 0$, t represents time, and e is the base of natural logarithms. (To six decimal places, $e = 2.718282$.) Graphs of $y = e^{-t}$ and $y = -e^{-t}$ for $t > 0$ are shown in the figure below. Use these graphs to obtain the graph of $y = e^{-t} \cdot \sin t$.

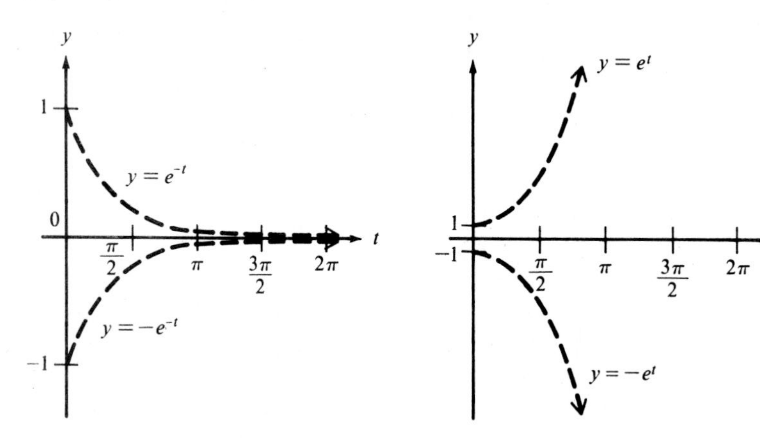

Exercise 21 Exercise 22

22. Motion that gets out of control can be represented by a function of the form

$$y = a \cdot e^{kt} \cdot \sin b(t + c),$$

with variables as given in Exercise 21. Use the graphs of $y = e^t$ and $y = -e^t$ in the figure above to graph $y = e^t \cdot \sin t$.

4.5 Simple Harmonic Motion (Optional)*

Case I: *Point $P(x, y)$ is considered to move around the unit circle counterclockwise at a uniform speed.* This is the simplest case of harmonic motion, as illustrated in Figure 4.36. As in section 3.4, let v be its **linear velocity** or **speed** (number of units P moves along the circle per unit of time.) If the point P is at $(1, 0)$ when $t = 0$, then the arc length s is given by the product of the rate and the time, or $s = vt$ after t units of time. (Recall that for a circle of radius 1, the arc length and the angle have the same measure.) So $y = \sin s = \sin vt$. If a horizontal line through P intersects the y-axis at Q with coordinates $(0, y)$, then

*Adapted from Floyd F. Helton and Margaret L. Lial, *Precalculus Mathematics: A Functions Approach* (Glenview, Ill.: Scott, Foresman and Company, 1983), pp. 211–15.

the formula $y = \sin vt$ describes the up-and-down motion of Q along the y-axis as a function of time t. This oscillatory motion is called **simple harmonic motion.**

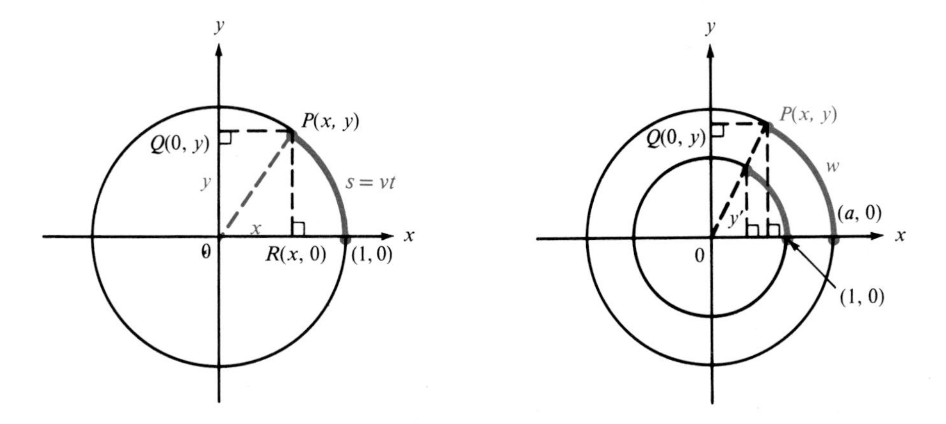

Figure 4.36

Figure 4.37

The amplitude of the motion is 1 and its period is $2\pi/v$. The moving points P and Q complete one cycle per period. The number of cycles per unit of time (called the **frequency**) is the reciprocal of the period, $v/2\pi$. The back-and-forth motion of a point R along the x-axis is another instance of simple harmonic motion, with equation $x = \cos vt$.

The basic notion is extended in Cases II and III below, to give a more general view of simple harmonic motion.

Case II: *The radius of the circle is allowed to be any number.* As in Figure 4.37, suppose that the outer circle has radius a and that P is at $(a, 0)$ when $t = 0$ and is moving counterclockwise at a uniform speed of v units per unit of time. After t time the arc length w is given by $w = vt$. Using geometric similarities in the figure, we have

$$\frac{w}{s} = \frac{a}{1} \quad \text{and} \quad \frac{y}{a} = \frac{y'}{1},$$

$$w = as \quad \text{and} \quad y = ay'.$$

From $y = ay'$, we get

$$\begin{aligned} y &= a \sin s \\ &= a \sin \frac{w}{a} && \text{Since } w = as \\ &= a \sin \frac{vt}{a} && \text{Since } w = vt \\ &= a \sin \left(\frac{v}{a}\right)t. \end{aligned}$$

Letting $b = v/a$, we write this as $y = a \sin bt$.

Case III: *The initial position of the point is anywhere on the circle*. Suppose that at time $t = 0$ the point P is at a distance w_0 from $(a, 0)$, measured along the circle counterclockwise. Then t units of time later it will have moved a distance vt from w_0 and so is along the circle at a distance $w = vt + w_0$ from $(a, 0)$. See Figure 4.38. Now we have

$$y = a \sin \frac{w}{a} \qquad \text{As in Case II}$$

$$= a \sin \frac{vt + w_0}{a}$$

$$= a \sin \left(\frac{vt}{a} + \frac{w_0}{a}\right)$$

$$= a \sin \frac{v}{a}\left(t + \frac{w_0}{v}\right).$$

The final equation may be written as $y = a \sin b(t + d)$, where $b = v/a$ and $d = w_0/v$. Note that d = distance w_0 per speed v = time for the particle to go from $(a, 0)$ to w_0 along the circle.

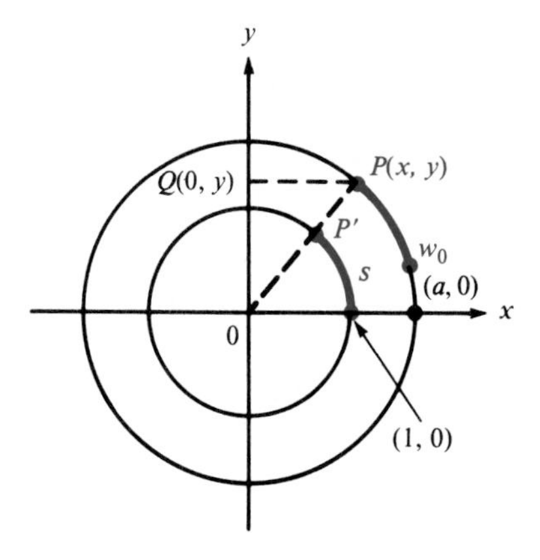

Figure 4.38

The three cases we have just described are represented by the following three formulas:

Simple Harmonic Motion

Case I: $y = \sin vt$, for motion on the unit circle, starting at the point $(1, 0)$ and moving v units per unit of time

Case II: $y = a \sin bt$, where $b = v/a$, for motion on a circle of radius a, starting at the point $(a, 0)$ and moving v units per unit of time

Case III: $y = a \sin b(t + d)$, where $b = v/a$ and $d = w_0/v$, for motion on a circle of radius a, starting w_0 units from $(a, 0)$

Example 1 Write the equation of motion for a point moving at a linear speed of 5 around each of the following circles, and state the amplitude, period, and frequency.

(a) The unit circle, starting at the point (1, 0)

In this case, $y = \sin vt = \sin 5t$. The amplitude is 1, the period is $2\pi/5$, and the frequency is $5/2\pi$.

(b) A circle of radius 3, starting at the point (3, 0)

Here, the function is $y = a \sin bt = 3 \sin (5/3)t$. The amplitude is 3, the period is $2\pi/(5/3) = 6\pi/5$, and the frequency is $5/6\pi$. ✥

The sine-wave function $y = a \sin b(t + d)$ defines the general case of simple harmonic motion. Its graph is a sine wave of amplitude a (the radius of the circle), and its period is $2\pi/b$, or $2\pi a/v$ (the time for P to go once around the circle).

From the calculus it can be shown that for a pendulum, as shown in Figure 4.39, the angle θ is given by $\theta(t) = a \sin (\sqrt{g/l})t$, where a is the maximum angle of displacement of the pendulum arm from the vertical, g is the constant of gravitation ($\approx$ 32), and l is the length of the pendulum arm. (Here, l is measured in feet and the angle θ in radians, and t is time in seconds.) This formula is actually an approximation for small values of θ. The period of motion is $2\pi/\sqrt{32/l}$, or $\pi\sqrt{2l}/4$.

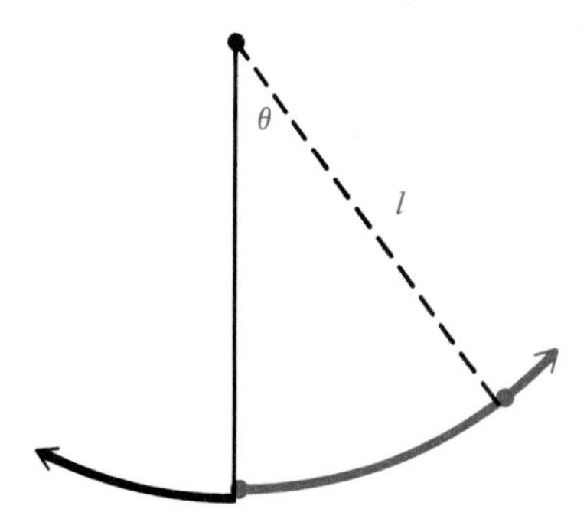

Figure 4.39

Example 2 Suppose that a pendulum arm of length 2 ft is displaced 2 radians from the vertical and released. Write the formula for the motion of the pendulum and state the period and frequency.

In this case, $a = 2$ and $l = 2$.

$$\theta(t) = a \sin \left(\sqrt{\frac{g}{l}}\right)t = 2 \sin \left(\sqrt{\frac{32}{2}}\right)t = 2 \sin 4t$$

The period is $2\pi/4 = \pi/2 \approx 1.6$ (sec) and the frequency is $2/\pi \approx 0.6$ (cycles per sec) ✥

Suppose that a weight is placed on the end of a suspended spring and allowed to come to rest (see Figure 4.40). If it is then pulled down, stretching the spring, and released (neglecting friction), it oscillates up and down periodically in simple harmonic motion. The equation for this motion is $s(t) = a \sin (\sqrt{k/m})t$, where a is the maximum vertical displacement from the position of rest, k is the spring constant, and m is the mass of the weight. Here the spring constant k is determined by the fact (Hooke's law) that the force f required to stretch the spring a distance s is given by $f = ks$.

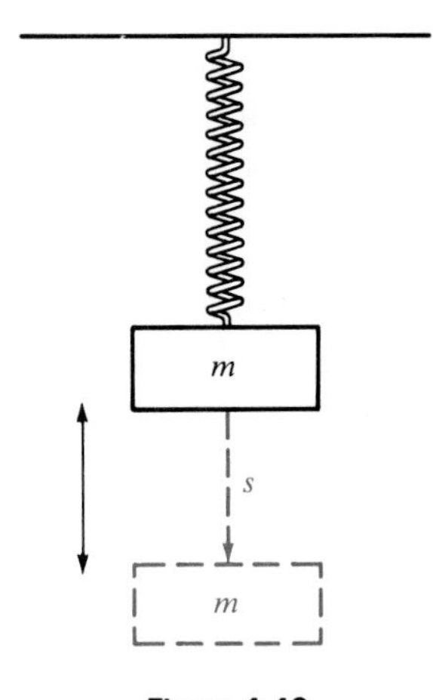

Figure 4.40

Mathematically, the formula for the motion of a weight on a spring is identical to that for the simple pendulum. Only the interpretations of the variables and constants are different. This situation occurs quite frequently in mathematics.

4.5 Exercises

Solve the following problems.

1. Write the equation and then determine the amplitude, period, and frequency of the simple harmonic motion determined by the uniform circular motion of a particle around a unit circle with each of the following linear speeds.
(a) 2 units per sec **(b)** 3 units per sec **(c)** 4 units per sec

2. Repeat Exercise 1 for a circle of radius 2 units.

3. Suppose that in Exercise 2 the point P is initially (when $t = 0$) at the point $(1, \sqrt{3})$ and has a linear speed of 2 units per unit of time. Write the equation of the corresponding simple harmonic motion and determine its amplitude, period, and frequency.

4. What is the period and frequency of oscillation of a pendulum of length 1/2 ft?

5. How long should a pendulum be to have a period of one sec?

6. Suppose that 4 lb of force are required to stretch a spring 2 ft. (Since force $= ks$, $4 = k(2)$, so the spring constant is $k = 2$.) Let a mass of one unit be placed on the spring and allowed to come to rest. If the spring is then stretched 1/2 ft and released, what is the amplitude, period, and frequency of the resulting oscillatory motion?

7. The formula for the up-and-down motion of a weight on a spring is given by $s(t) = a \sin (\sqrt{k/m})t$.

(a) Write the formulas for the period and frequency of the motion.

(b) If the spring constant is $k = 4$, what mass m must be used to produce a period of 1 sec?

Chapter 4 *Key Concepts*

The Trigonometric Functions

Function	*Domain*	*Range*	*Period*
$y = \sin x$	all real numbers	$-1 \leq y \leq 1$	2π
$y = \cos x$	all real numbers	$-1 \leq y \leq 1$	2π
$y = \csc x$	all real $x \neq n\pi$	$y \leq -1$ or $y \geq 1$	2π
$y = \sec x$	all real $x \neq \frac{\pi}{2} + n\pi$	$y \leq -1$ or $y \geq 1$	2π
$y = \tan x$	all real $x \neq \frac{\pi}{2} + n\pi$	all real numbers	π
$y = \cot x$	all real $x \neq n\pi$	all real numbers	π

Graphs of Trigonometric Functions

If $f(x)$ is a trigonometric function with period p, and if $y = c + a \cdot f[b(x - d)]$:

Amplitude $|a|$ (for $f(x) = \sin x$ or $f(x) = \cos x$)

Period $\frac{p}{b}$ $(b > 0)$

Vertical translation Up c units if $c > 0$
Down $|c|$ units if $c < 0$

Phase shift (horizontal translation)

d units to the right if $d > 0$

$|d|$ units to the left if $d < 0$

Chapter 4 *Review Exercises*

For each of the following trigonometric functions, give the amplitude, period, vertical translation, and phase shift, as applicable.

1. $y = 2 \sin x$

2. $y = \tan 3x$

3. $y = -\frac{1}{2} \cos 3x$

4. $y = 2 \sin 5x$

5. $y = 1 + 2 \sin \frac{1}{4} x$

6. $y = 3 - \frac{1}{4} \cos \frac{2}{3} x$

7. $y = 3 \cos \left(x + \frac{\pi}{2}\right)$

8. $y = -\sin \left(x - \frac{3\pi}{4}\right)$

9. $y = \frac{1}{2} \csc \left(2x - \frac{\pi}{4}\right)$

10. $y = 2 \sec (\pi x - 2\pi)$

11. $y = \frac{1}{3} \cos \left(3x - \frac{\pi}{3}\right)$

12. $y = \cot \left(\frac{x}{2} + \frac{3\pi}{4}\right)$

Graph the following functions over a one-period interval.

13. $y = 3 \sin x$

14. $y = \frac{1}{2} \sec x$

15. $y = -\tan x$

16. $y = -2 \cos x$

17. $y = 2 + \cot x$

18. $y = -1 + \csc x$

19. $y = \sin 2x$

20. $y = \tan 3x$

21. $y = 3 \cos 2x$

22. $y = \frac{1}{2} \cot 3x$

23. $y = \cos \left(x - \frac{\pi}{4}\right)$

24. $y = \tan \left(x - \frac{\pi}{2}\right)$

25. $y = \sec \left(2x + \frac{\pi}{3}\right)$

26. $y = \sin \left(3x + \frac{\pi}{2}\right)$

27. $y = 1 + 2 \cos 3x$

28. $y = -1 - 3 \sin 2x$

29. $y = 2 \sin \pi x$

30. $y = -\frac{1}{2} \cos (\pi x - \pi)$

31. $y = 1 - 2 \sec \left(x - \frac{\pi}{4}\right)$

32. $y = -\csc (2x - \pi) + 1$

Graph each of the following using the method of addition of ordinates.

33. $y = \tan x - x$

34. $y = \cos x + \frac{1}{2} x$

35. $y = \sin x + \cos x$

36. $y = \tan x + \cot x$

37. The figure on the following page shows the population of lynx and hares in Canada for the years 1847–1903. The hares are food for the lynx. An increase in hare population causes an increase in lynx population some time later. The increasing lynx population then causes a decline in hare population.

(a) Estimate the length of one period.
(b) Estimate maximum and minimum hare poulation.

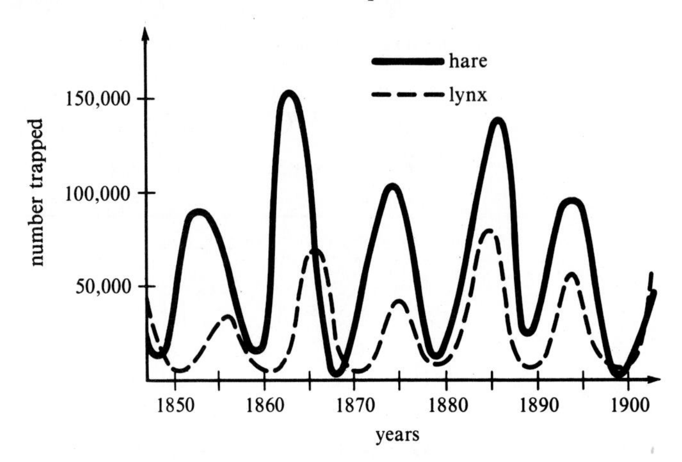

38. The amount of pollution in the air fluctuates with the seasons. It is lower after heavy spring rains and higher after periods of little rain. In addition to this seasonal fluctuation, the long-term trend is upward. An idealized graph of this situation is shown in the figure. Trigonometric functions can be used to describe the fluctuating part of the pollution levels. Powers of the number e (e is the base of natural logarithms; to six decimal places, $e = 2.718282$) can be used to show the long-term growth. In fact, the pollution level in a certain area might be given by

$$P(t) = 7(1 - \cos 2\pi t)(t + 10) + 100e^{.2t},$$

where t is time in years, with $t = 0$ representing January 1 of the base year. Thus, July 1 of the same year would be represented by $t = .5$, and October 1 of the following year would be represented by $t = 1.75$. Find the pollution levels on the following dates.

(a) January 1, base year
(b) July 1, base year
(c) January 1, following year
(d) July 1, following year

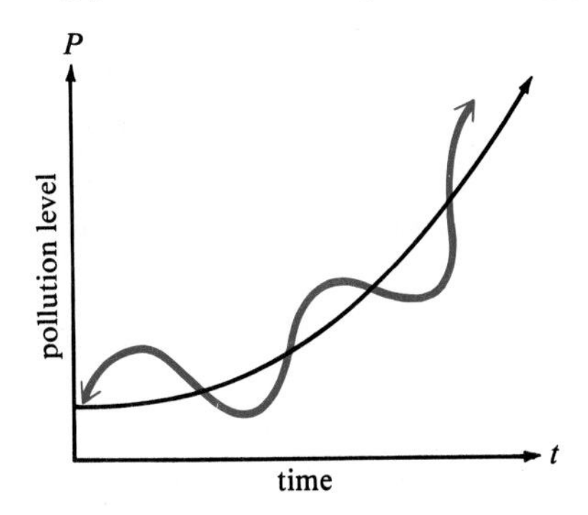

5

Trigonometric Identities

A **conditional equation,** such as $2x + 1 = 9$ or $m^2 - 2m = 3$, is true for only certain values in the domain of its variable. For example, $2x + 1 = 9$ is true only for $x = 4$, and $m^2 - 2m = 3$ is true only for $m = 3$ and $m = -1$. On the other hand, an **identity** is an equation that is true for *every* value in the domain of its variable. Examples of identities include

$$5(x + 3) = 5x + 15 \qquad \text{and} \qquad (a + b)^2 = a^2 + 2ab + b^2.$$

This chapter discusses identities involving trigonometric functions. The variables in the trigonometric functions represent either angles or real numbers. The domain of the variable is assumed to be all values for which a given function is defined.

5.1 Fundamental Identities

This section reviews the fundamental trigonometric identities and discusses some of their uses.* For convenience, the definitions of the six trigonometric functions are restated at the top of the following page. If x, y, r, and θ are defined as shown in Figure 5.1, then

*All the identities given in this chapter are summarized at the end of the chapter and inside the back cover.

$$\sin\theta = \frac{y}{r} \qquad \cos\theta = \frac{x}{r} \qquad \tan\theta = \frac{y}{x}$$

$$\csc\theta = \frac{r}{y} \qquad \sec\theta = \frac{r}{x} \qquad \cot\theta = \frac{x}{y}.$$

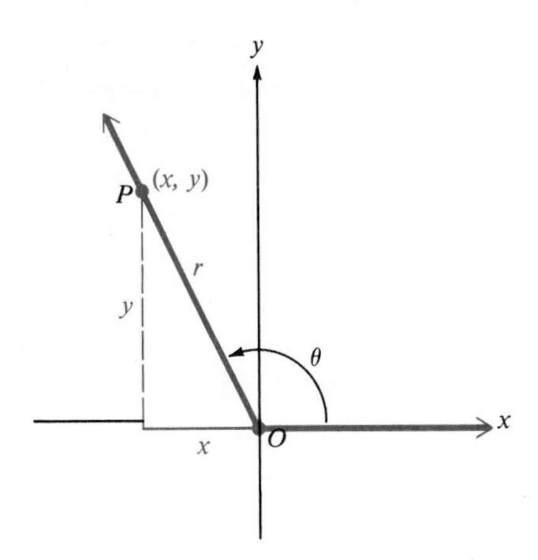

Figure 5.1

In chapter 1, these definitions were used to derive the following **reciprocal identities,** which are true for all suitable replacements of the variable.

$$\cot\theta = \frac{1}{\tan\theta} \qquad \csc\theta = \frac{1}{\sin\theta} \qquad \sec\theta = \frac{1}{\cos\theta}$$

Each of these reciprocal identities leads to another identity. For example, $\csc\theta = 1/\sin\theta$ gives $\sin\theta = 1/\csc\theta$.

From the definitions of the trigonometric functions,

$$\frac{\sin\theta}{\cos\theta} = \frac{y/r}{x/r} = \frac{y}{x} = \tan\theta.$$

or

$$\tan\theta = \frac{\sin\theta}{\cos\theta}.$$

In a similar manner,

$$\cot\theta = \frac{\cos\theta}{\sin\theta}.$$

These last two identities are sometimes called the **quotient identities.**

In chapter 1 the definitions of the trigonometric functions were used to derive the identity

$$\sin^2 \theta + \cos^2 \theta = 1.$$

Dividing both sides by $\cos^2 \theta$ then leads to

$$\tan^2 \theta + 1 = \sec^2 \theta,$$

while dividing through by $\sin^2 \theta$ gives

$$1 + \cot^2 \theta = \csc^2 \theta.$$

These last three identities are the **Pythagorean identities.**

As suggested by the circle shown in Figure 5.2, an angle θ having the point (x, y) on its terminal side has a corresponding angle $-\theta$ with a point $(x, -y)$ on its terminal side. From the definition of sine,

$$\sin(-\theta) = \frac{-y}{r} \quad \text{and} \quad \sin \theta = \frac{y}{r},$$

so that $\sin(-\theta)$ and $\sin \theta$ are negatives of each other, or

$$\sin(-\theta) = -\sin \theta.$$

Figure 5.2 shows an angle θ in quadrant II, but the same result holds for θ in any quadrant.

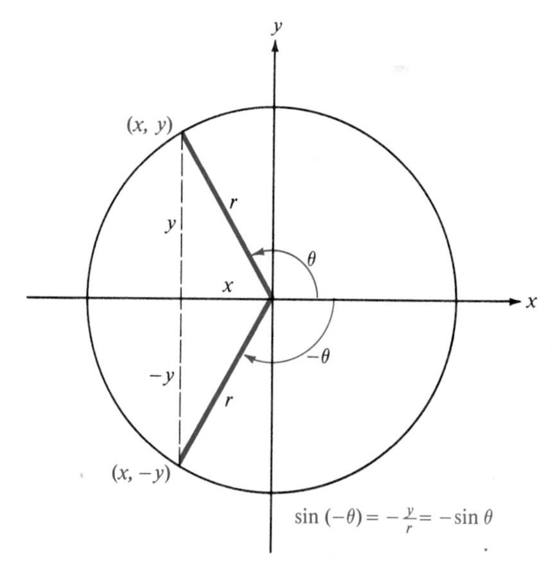

Figure 5.2

Also, by definition,

$$\cos(-\theta) = \frac{x}{r} \quad \text{and} \quad \cos \theta = \frac{x}{r},$$

so that

$$\cos(-\theta) = \cos \theta.$$

These formulas for $\sin(-\theta)$ and $\cos(-\theta)$ can be used to find $\tan(-\theta)$ in terms of $\tan\theta$:

$$\tan(-\theta) = \frac{\sin(-\theta)}{\cos(-\theta)} = \frac{-\sin\theta}{\cos\theta},$$

or

$$\tan(-\theta) = -\tan\theta.$$

The preceding three identities are **negative-angle identities.**

The identities given in this section are summarized below. As a group, these are called the **fundamental identities.**

Fundamental Identities

Reciprocal identities

$$\cot\theta = \frac{1}{\tan\theta} \qquad \sec\theta = \frac{1}{\cos\theta} \qquad \csc\theta = \frac{1}{\sin\theta}$$

Quotient identities

$$\tan\theta = \frac{\sin\theta}{\cos\theta} \qquad \cot\theta = \frac{\cos\theta}{\sin\theta}$$

Pythagorean identities

$$\sin^2\theta + \cos^2\theta = 1 \qquad \tan^2\theta + 1 = \sec^2\theta \qquad 1 + \cot^2\theta = \csc^2\theta$$

Negative-angle identities

$$\sin(-\theta) = -\sin\theta \qquad \cos(-\theta) = \cos\theta \qquad \tan(-\theta) = -\tan\theta$$

One use for trigonometric identities is to find the values of other trigonometric functions from the value of a given trigonometric function. For example, given a value of $\tan\theta$, the value of $\cot\theta$ can be found from the identity $\cot\theta = 1/\tan\theta$. In fact, given any trigonometric function value and the quadrant in which θ lies, the values of all the other trigonometric functions can be found by using identities, as in the following example.

Example 1 If $\tan\theta = -5/3$ and θ is in quadrant II, find the values of the other trigonometric functions using fundamental identities.

The identity $\cot\theta = 1/\tan\theta$ leads to $\cot\theta = -3/5$. Next, find $\sec\theta$ from the identity $\tan^2\theta + 1 = \sec^2\theta$.

$$\left(-\frac{5}{3}\right)^2 + 1 = \sec^2\theta$$

$$\frac{25}{9} + 1 = \sec^2\theta$$

$$\frac{34}{9} = \sec^2\theta$$

$$-\sqrt{\frac{34}{9}} = \sec\theta \qquad \text{Take the square root of both sides}$$

$$-\frac{\sqrt{34}}{3} = \sec\theta$$

Choose the negative square root since $\sec\theta$ is negative in quadrant II. Now find $\cos\theta$:

$$\cos\theta = \frac{1}{\sec\theta} = \frac{-3}{\sqrt{34}} = -\frac{3\sqrt{34}}{34},$$

after rationalizing the denominator. Find $\sin\theta$ by using the identity $\sin^2\theta + \cos^2\theta = 1$, with $\cos\theta = -3/\sqrt{34}$.

$$\sin^2\theta + \left(\frac{-3}{\sqrt{34}}\right)^2 = 1$$

$$\sin^2\theta = 1 - \frac{9}{34}$$

$$\sin^2\theta = \frac{25}{34}$$

$$\sin\theta = \frac{5}{\sqrt{34}}$$

$$\sin\theta = \frac{5\sqrt{34}}{34} \qquad \text{Rationalize}$$

The positive square root is used since $\sin\theta$ is positive in quadrant II. Finally, since $\csc\theta$ is the reciprocal of $\sin\theta$,

$$\csc\theta = \frac{\sqrt{34}}{5}.$$ ✤

Example 2 Express $\cos x$ in terms of $\tan x$.

Since $\sec x$ is related to both $\cos x$ and $\tan x$ by identities, start with $\tan^2 x + 1 = \sec^2 x$. Then take reciprocals to get

$$\frac{1}{\tan^2 x + 1} = \frac{1}{\sec^2 x}$$

or

$$\frac{1}{\tan^2 x + 1} = \cos^2 x$$

$$\pm\sqrt{\frac{1}{\tan^2 x + 1}} = \cos x. \qquad \text{Take the square root of both sides}$$

Finally,
$$\cos x = \frac{\pm 1}{\sqrt{\tan^2 x + 1}}.$$

Rationalize the denominator to get
$$\cos x = \frac{\pm\sqrt{\tan^2 x + 1}}{\tan^2 x + 1}.$$

Choose the + sign or the − sign, depending on the quadrant of x. ✤

Another use of identities is to simplify trigonometric expressions by substituting one side of an identity for the other side. For example, the expression $\sin^2\theta + \cos^2\theta$ can be replaced by 1, as in the following example.

Example 3 Use the fundamental identities to write $\tan\theta + \cot\theta$ in terms of $\sin\theta$ and $\cos\theta$, and then simplify the expression.

From the fundamental identities,
$$\tan\theta + \cot\theta = \frac{\sin\theta}{\cos\theta} + \frac{\cos\theta}{\sin\theta}.$$

Simplify this expression by adding the two fractions on the right side, using the common denominator $\cos\theta\sin\theta$.
$$\begin{aligned}\tan\theta + \cot\theta &= \frac{\sin^2\theta}{\cos\theta\sin\theta} + \frac{\cos^2\theta}{\cos\theta\sin\theta}\\ &= \frac{\sin^2\theta + \cos^2\theta}{\cos\theta\sin\theta}\end{aligned}$$

Now substitute 1 for $\sin^2\theta + \cos^2\theta$.
$$\tan\theta + \cot\theta = \frac{1}{\cos\theta\sin\theta}$$ ✤

Some problems in calculus are simplified by making an appropriate trigonometric substitution, as in the next example.

Example 4 Remove the radical in the expression $\sqrt{9 + x^2}$ by replacing x with $3\tan\theta$, where $0 < \theta < \pi/2$.

Letting $x = 3\tan\theta$ gives
$$\begin{aligned}\sqrt{9 + x^2} &= \sqrt{9 + (3\tan\theta)^2}\\ &= \sqrt{9 + 9\tan^2\theta}\\ &= \sqrt{9(1 + \tan^2\theta)}\\ &= 3\sqrt{1 + \tan^2\theta}\\ &= 3\sqrt{\sec^2\theta}.\end{aligned}$$

On the interval $0 < \theta < \pi/2$, the value of $\sec \theta$ is positive, giving

$$\sqrt{9 + x^2} = 3 \sec \theta.$$

The result of Example 4 could be written as

$$\sec \theta = \frac{\sqrt{9 + x^2}}{3}.$$

In a right triangle, $\sec \theta$ is the ratio of the length of the hypotenuse to the side adjacent to the angle. This definition was used to label the right triangle in Figure 5.3, and then the Pythagorean theorem was used to find the length of the side opposite angle θ.

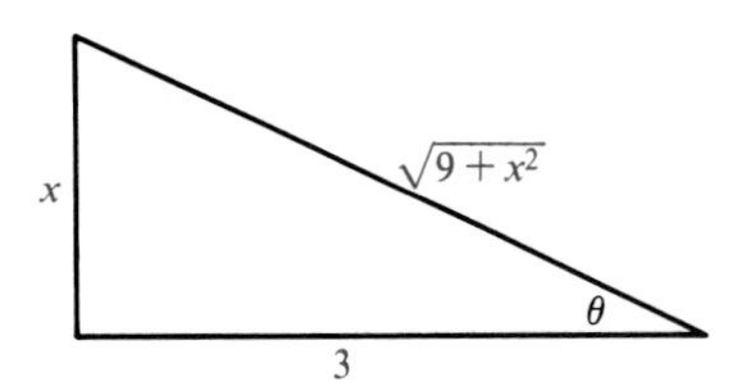

Figure 5.3

From the right triangle in Figure 5.3,

$$\sin \theta = \frac{x}{\sqrt{9 + x^2}}, \qquad \cos \theta = \frac{3}{\sqrt{9 + x^2}},$$

and so on.

5.1 Exercises

Find sin *s for each of Exercises 1–4. See Example 1.*

1. $\cos s = 3/4$, s in quadrant I

2. $\cot s = -1/3$, s in quadrant IV

3. $\cos s = \sqrt{5}/5$, $\tan s < 0$

4. $\tan s = -\sqrt{7}/2$, $\sec s > 0$

5. Find $\tan \theta$ if $\cos \theta = -2/5$, and $\sin \theta < 0$.

6. Find $\csc \alpha$ if $\tan \alpha = 6$, and $\cos \alpha > 0$.

Use the fundamental identities to find the remaining five trigonometric functions of θ. See Example 1.

7. $\sin \theta = \frac{2}{3}$, θ in quadrant II

8. $\cos \theta = \frac{1}{5}$, θ in quadrant I

9. $\tan \theta = -\frac{1}{4}$, θ in quadrant IV

10. $\tan \theta = \frac{2}{3}$, θ in quadrant III

11. $\sec \theta = -3$, θ in quadrant II

12. $\csc \theta = -\frac{5}{2}$, θ in quadrant III

13. $\cot \theta = \frac{4}{3}$, $\sin \theta > 0$

14. $\sin \theta = -\frac{4}{5}$, $\cos \theta < 0$

15. $\sec \theta = \frac{4}{3}$, $\sin \theta < 0$

16. $\cos \theta = -\frac{1}{4}$, $\sin \theta > 0$

For each trigonometric expression in Column I, choose the expression from Column II that completes a fundamental identity.

Column I	*Column II*
17. $\dfrac{\cos x}{\sin x}$	**(a)** $\sin^2 x + \cos^2 x$
18. $\tan x$	**(b)** $\cot x$
19. $\cos(-x)$	**(c)** $\sec^2 x$
20. $\tan^2 x + 1$	**(d)** $\dfrac{\sin x}{\cos x}$
21. 1	**(e)** $\cos x$

For each expression in Column I, choose the expression from Column II that completes an identity. You will have to rewrite one or both expressions, using a fundamental identity, to recognize the matches.

Column I	*Column II*
22. $-\tan x \cos x$	**(a)** $\dfrac{\sin^2 x}{\cos^2 x}$
23. $\sec^2 x - 1$	**(b)** $\dfrac{1}{\sec^2 x}$
24. $\dfrac{\sec x}{\csc x}$	**(c)** $\sin(-x)$
25. $1 + \sin^2 x$	**(d)** $\csc^2 x - \cot^2 x + \sin^2 x$
26. $\cos^2 x$	**(e)** $\tan x$

In each of the following, use the fundamental identities to get an equivalent expression involving only sines and cosines, and then simplify it. See Example 3.

27. $\csc^2 \beta - \cot^2 \beta$

28. $\dfrac{\tan(-\theta)}{\sec \theta}$

29. $\tan(-\alpha)\cos(-\alpha)$

30. $\cot^2 x(1 + \tan^2 x)$

31. $\tan^2\theta - \dfrac{\sec^2\theta}{\csc^2\theta}$

32. $\dfrac{\tan x \csc x}{\sec x}$

33. $\sec\theta + \tan\theta$

34. $\dfrac{\sec\alpha}{\tan\alpha + \cot\alpha}$

35. $\sec^2 t - \tan^2 t$

36. $\csc^2\gamma + \sec^2\gamma$

37. $\cot^2\beta - \csc^2\beta$

38. $1 + \cot^2\alpha$

39. $\dfrac{1 + \tan^2\theta}{\cot^2\theta}$

40. $\dfrac{1 - \sin^2 t}{\csc^2 t}$

41. $\cot^2\beta\sin^2\beta + \tan^2\beta\cos^2\beta$

42. $\sec^2 x + \cos^2 x$

43. $\dfrac{\cot^2\alpha + \csc^2\alpha}{\cos^2\alpha}$

44. $1 - \tan^4\theta$

45. $1 - \cot^4 s$

46. $\tan^4\gamma - \cot^4\gamma$

Complete this chart, so that each trigonometric function in the column at the left is expressed in terms of the functions given across the top. See Example 2.

		$\sin\theta$	$\cos\theta$	$\tan\theta$	$\cot\theta$	$\sec\theta$	$\csc\theta$
47.	$\sin\theta$	$\sin\theta$	$\pm\sqrt{1 - \cos^2\theta}$	$\dfrac{\pm\tan\theta}{\sqrt{1 + \tan^2\theta}}$			$\dfrac{1}{\csc\theta}$
48.	$\cos\theta$		$\cos\theta$	$\dfrac{\pm\sqrt{\tan^2\theta + 1}}{\tan^2\theta + 1}$		$\dfrac{1}{\sec\theta}$	
49.	$\tan\theta$			$\tan\theta$	$\dfrac{1}{\cot\theta}$		
50.	$\cot\theta$			$\dfrac{1}{\tan\theta}$	$\cot\theta$	$\dfrac{\pm\sqrt{\sec^2\theta - 1}}{\sec^2\theta - 1}$	
51.	$\sec\theta$		$\dfrac{1}{\cos\theta}$			$\sec\theta$	
52.	$\csc\theta$	$\dfrac{1}{\sin\theta}$					$\csc\theta$

53. Suppose that $\cos\theta = x/(x + 1)$. Find $\sin\theta$.

54. Find $\tan\alpha$ if $\sec\alpha = (p + 4)/p$.

Show that each of the following is not an identity by replacing the variables with numbers that show the result to be false.

55. $(\sin s + \cos s)^2 = 1$

56. $(\tan s + 1)^2 = \sec^2 s$

57. $2\sin s = \sin 2s$

58. $\sin x = \sqrt{1 - \cos^2 x}$

59. $\sin^3 x + \cos^3 x = 1$

60. $\sin x + \sin y = \sin(x + y)$

Use the indicated substitution to remove the radical in the given expression in Exercises 61–66. Assume $0 < \theta < \pi/2$. Then find the indicated functions. See Example 4.

61. $\sqrt{16 + 9x^2}$, let $x = \frac{4}{3}\tan\theta$; find $\sin\theta$ and $\cos\theta$

62. $\sqrt{x^2 - 25}$, let $x = 5\sec\theta$; find $\sin\theta$ and $\tan\theta$

63. $\sqrt{(1 - x^2)^3}$, let $x = \cos\theta$; find $\sin\theta$ and $\tan\theta$

64. $\dfrac{\sqrt{x^2 - 9}}{x}$, let $x = 3\sec\theta$; find $\sin\theta$ and $\tan\theta$

65. $x^2\sqrt{1 + 16x^2}$, let $x = \frac{1}{4}\tan\theta$; find $\sin\theta$ and $\cos\theta$

66. $x^2\sqrt{9 + x^2}$, let $x = 3\tan\theta$; find $\sin\theta$ and $\cos\theta$

67. Let $\cos x = 1/5$. Find all possible values for

$$\frac{\sec x - \tan x}{\sin x}.$$

68. Let $\csc x = -3$. Find all possible values for

$$\frac{\sin x + \cos x}{\sec x}.$$

For students who have studied logarithms: Prove each of the following identities for first-quadrant values of s.

69. $\log \sin s = -\log \csc s$

70. $\log \tan s = \log \sin s - \log \cos s$

71. $\log \sec s = \frac{1}{2}\log(\tan^2 s + 1)$

72. $\log \csc s = -\log \sin s$

5.2 Verifying Trigonometric Identities

One of the skills required for more advanced work in mathematics (and especially in calculus) is the ability to use the trigonometric identities to write trigonometric expressions in alternate forms. This skill is developed by using the fundamental identities to verify that a trigonometric equation is an identity (for those values of the variable for which it is defined). Here are some hints that may help you get started.

Verifying Identities

1. Memorize the fundamental identities given in the last section. Whenever you see either side of a fundamental identity, the other side should come to mind. Also, be aware of equivalent forms of the fundamental identities. For example $\sin^2 \theta = 1 - \cos^2 \theta$ is an alternate form of $\sin^2 \theta + \cos^2 \theta = 1$.
2. Try to rewrite the more complicated side of the equation so that it is identical to the simpler side.
3. It is often helpful to express all trigonometric functions in the equation in terms of sine and cosine and then simplify the result.
4. Usually any factoring or indicated algebraic operations should be performed. For example, the expression $\sin^2 x + 2 \sin x + 1$ can be factored as follows: $(\sin x + 1)^2$. The sum or difference of two trigonometric expressions, such as
$$\frac{1}{\sin \theta} + \frac{1}{\cos \theta},$$
can be added or subtracted in the same way as any other rational expression:
$$\frac{1}{\sin \theta} + \frac{1}{\cos \theta} = \frac{\cos \theta}{\sin \theta \cos \theta} + \frac{\sin \theta}{\sin \theta \cos \theta}$$
$$= \frac{\cos \theta + \sin \theta}{\sin \theta \cos \theta}.$$
5. As you select substitutions, keep in mind the side you are not changing, because it represents your goal. For example, to verify the identity
$$\tan^2 x + 1 = \frac{1}{\cos^2 x},$$
try to think of an identity that relates $\tan x$ to $\cos x$. Here, since $\sec x = 1/\cos x$ and $\sec^2 x = \tan^2 x + 1$, the secant function is the best link between the two sides.
6. If an expression contains $1 + \sin x$, multiplying both numerator and denominator by $1 - \sin x$ would give $1 - \sin^2 x$, which could be replaced with $\cos^2 x$. Similar results for $1 - \sin x$, $1 + \cos x$, and $1 - \cos x$ may be useful.

These hints are used in the following examples. A word of warning: Verifying identities is not the same as solving equations. Techniques used in solving equations, such as adding the same terms to both sides, or multiplying both sides by the same term, are not valid when working with identities since you are starting with a statement (to be verified) that may not be true.

Example 1 Verify that

$$\cot s + 1 = \csc s(\cos s + \sin s)$$

is an identity.

Use the fundamental identities to rewrite one side of the equation so that it is identical to the other side. Since the right side is more complicated, its is probably a good idea to work with it. Here we use the method of changing all the trigonometric functions to sine or cosine.

Steps	*Reasons*
$\csc s\,(\cos s + \sin s) = \dfrac{1}{\sin s}(\cos s + \sin s)$	$\csc s = \dfrac{1}{\sin s}$
$= \dfrac{\cos s}{\sin s} + \dfrac{\sin s}{\sin s}$	Distributive property
$= \cot s + 1$	$\dfrac{\cos s}{\sin s} = \cot s$; $\dfrac{\sin s}{\sin s} = 1$

The given equation is an identity since the right side equals the left side. ✣

Example 2 Verify that

$$\tan^2 \alpha\,(1 + \cot^2 \alpha) = \frac{1}{1 - \sin^2 \alpha}$$

is an identity.

Working with the left side gives the following.

$\tan^2 \alpha\,(1 + \cot^2 \alpha) = \tan^2 \alpha + \tan^2 \alpha \cot^2 \alpha$	Distributive property
$= \tan^2 \alpha + \tan^2 \alpha \cdot \dfrac{1}{\tan^2 \alpha}$	$\cot^2 \alpha = \dfrac{1}{\tan^2 \alpha}$
$= \tan^2 \alpha + 1$	$\tan^2 \alpha \cdot \dfrac{1}{\tan^2 \alpha} = 1$
$= \sec^2 \alpha$	$\tan^2 \alpha + 1 = \sec^2 \alpha$
$= \dfrac{1}{\cos^2 \alpha}$	$\sec^2 \alpha = \dfrac{1}{\cos^2 \alpha}$
$= \dfrac{1}{1 - \sin^2 \alpha}$	$\cos^2 \alpha = 1 - \sin^2 \alpha$

Since the left side equals the right side, the given equation is an identity. ✣

Example 3 Show that

$$\frac{\tan t - \cot t}{\sin t \cos t} = \sec^2 t - \csc^2 t.$$

Work with the left side.

$$\frac{\tan t - \cot t}{\sin t \cos t}$$

$$= \frac{\tan t}{\sin t \cos t} - \frac{\cot t}{\sin t \cos t}$$

$$= \tan t \cdot \frac{1}{\sin t \cos t} - \cot t \cdot \frac{1}{\sin t \cos t}$$

$$= \frac{\sin t}{\cos t} \cdot \frac{1}{\sin t \cdot \cos t} - \frac{\cos t}{\sin t} \cdot \frac{1}{\sin t \cos t} \qquad \tan t = \frac{\sin t}{\cos t};\ \cot t = \frac{\cos t}{\sin t}$$

$$= \frac{1}{\cos^2 t} - \frac{1}{\sin^2 t}$$

$$= \sec^2 t - \csc^2 t \qquad \frac{1}{\cos^2 t} = \sec^2 t;\ \frac{1}{\sin^2 t} = \csc^2 t$$

✤

Example 4 Show that

$$\frac{\cos x}{1 - \sin x} = \frac{1 + \sin x}{\cos x}$$

is an identity.

This time we will work on the right side. Use the suggestion given at the beginning of the section to multiply the numerator and denominator on the right by $1 - \sin x$.

$$\frac{1 + \sin x}{\cos x} = \frac{(1 + \sin x)(1 - \sin x)}{\cos x\,(1 - \sin x)}$$

$$= \frac{1 - \sin^2 x}{\cos x\,(1 - \sin x)}$$

$$= \frac{\cos^2 x}{\cos x\,(1 - \sin x)} \qquad 1 - \sin^2 x = \cos^2 x$$

$$= \frac{\cos x}{1 - \sin x}$$ ✤

If both sides of an identity appear to be equally complex, the identity can be verified by working independently on the left side and on the right side, until each side is changed into some common third result. *Each step, on each side, must be reversible.* With all steps reversible, the procedure is as follows.

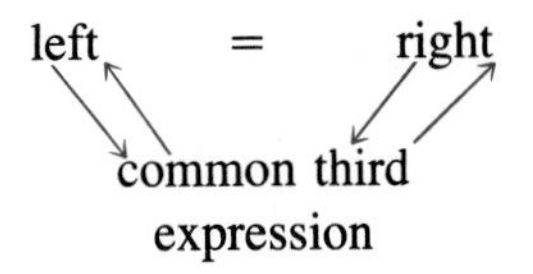

The left hand side leads to the third expression, which leads back to the right-hand side. This procedure is just a shortcut for the procedure used in the first examples of this section: the left side is changed into the right side, but by going through an intermediate step.

Example 5 Show that

$$\frac{\sec \alpha + \tan \alpha}{\sec \alpha - \tan \alpha} = \frac{1 + 2 \sin \alpha + \sin^2 \alpha}{\cos^2 \alpha}$$

is an identity.

Both sides appear equally complex, so prove the identity by changing each side into a common third expression. Work first on the left, multiplying numerator and denominator by $\cos \alpha$.

$$\frac{\sec \alpha + \tan \alpha}{\sec \alpha - \tan \alpha} = \frac{(\sec \alpha + \tan \alpha)\cos \alpha}{(\sec \alpha - \tan \alpha)\cos \alpha}$$

$$= \frac{\sec \alpha \cos \alpha + \tan \alpha \cos \alpha}{\sec \alpha \cos \alpha - \tan \alpha \cos \alpha}$$

$$= \frac{1 + \tan \alpha \cos \alpha}{1 - \tan \alpha \cos \alpha} \qquad \sec \alpha \cos \alpha = 1$$

$$= \frac{1 + \dfrac{\sin \alpha}{\cos \alpha} \cdot \cos \alpha}{1 - \dfrac{\sin \alpha}{\cos \alpha} \cdot \cos \alpha} \qquad \tan \alpha = \frac{\sin \alpha}{\cos \alpha}$$

$$= \frac{1 + \sin \alpha}{1 - \sin \alpha}$$

On the right-hand side of the original statement, start by factoring.

$$\frac{1 + 2 \sin \alpha + \sin^2 \alpha}{\cos^2 \alpha} = \frac{(1 + \sin \alpha)^2}{\cos^2 \alpha}$$

$$= \frac{(1 + \sin \alpha)^2}{1 - \sin^2 \alpha} \qquad \cos^2 \alpha = 1 - \sin^2 \alpha$$

$$= \frac{(1 + \sin \alpha)^2}{(1 + \sin \alpha)(1 - \sin \alpha)}$$

$$= \frac{1 + \sin \alpha}{1 - \sin \alpha}$$

We now have shown that

$$\frac{\sec \alpha + \tan \alpha}{\sec \alpha - \tan \alpha} = \frac{1 + \sin \alpha}{1 - \sin \alpha} = \frac{1 + 2 \sin \alpha + \sin^2 \alpha}{\cos^2 \alpha},$$

proving that the original equation is an identity. ✣

5.2 Exercises

For each of the following, perform the indicated operations and simplify the result.

1. $\tan \theta + \dfrac{1}{\tan \theta}$
2. $\dfrac{\cos s}{\sin x} + \dfrac{\sin x}{\cos x}$
3. $\cot s (\tan s + \sin s)$
4. $\sec \beta(\cos \beta + \sin \beta)$
5. $\dfrac{1}{\csc^2 \theta} + \dfrac{1}{\sec^2 \theta}$
6. $\dfrac{1}{\sin \alpha - 1} - \dfrac{1}{\sin \alpha + 1}$
7. $\dfrac{\cos x}{\sec x} + \dfrac{\sin x}{\csc x}$
8. $\dfrac{\cos \gamma}{\sin \gamma} + \dfrac{\sin \gamma}{1 + \cos \gamma}$
9. $(1 + \sin t)^2 + \cos^2 t$
10. $(1 + \tan s)^2 - 2 \tan s$
11. $\dfrac{1}{1 + \cos x} - \dfrac{1}{1 - \cos x}$
12. $(\sin \alpha - \cos \alpha)^2$

Factor each of the following trigonometric expressions.

13. $\sin^2 \gamma - 1$
14. $\sec^2 \theta - 1$
15. $(\sin x + 1)^2 - (\sin x - 1)^2$
16. $(\tan x + \cot x)^2 - (\tan x - \cot x)^2$
17. $2 \sin^2 x + 3 \sin x + 1$
18. $4 \tan^2 \beta + \tan \beta - 3$
19. $4 \sec^2 x + 3 \sec x - 1$
20. $2 \csc^2 x + 7 \csc x - 30$
21. $\cos^4 x + 2 \cos^2 x + 1$
22. $\cot^4 x + 3 \cot^2 x + 2$
23. $\sin^3 x - \cos^3 x$
24. $\sin^3 \alpha + \cos^3 \alpha$

Use the fundamental identities to simplify each of the given expressions.

25. $\tan \theta \cos \theta$
26. $\cot \alpha \sin \alpha$
27. $\sec r \cos r$
28. $\cot t \tan t$
29. $\dfrac{\sin \beta \tan \beta}{\cos \beta}$
30. $\dfrac{\csc \theta \sec \theta}{\cot \theta}$
31. $\sec^2 x - 1$
32. $\csc^2 t - 1$
33. $\dfrac{\sin^2 x}{\cos^2 x} + \sin x \csc x$
34. $\dfrac{1}{\tan^2 \alpha} + \cot \alpha \tan \alpha$

Verify each of the following trigonometric identities. See Examples 1–5.

35. $\dfrac{\cot \theta}{\csc \theta} = \cos \theta$
36. $\dfrac{\tan \alpha}{\sec \alpha} = \sin \alpha$
37. $\dfrac{1 - \sin^2 \beta}{\cos \beta} = \cos \beta$
38. $\dfrac{\tan^2 \gamma + 1}{\sec \gamma} = \sec \gamma$
39. $\cos^2 \theta (\tan^2 \theta + 1) = 1$
40. $\sin^2 \beta (1 + \cot^2 \beta) = 1$
41. $\sin^2 \alpha + \tan^2 \alpha + \cos^2 \alpha = \sec^2 \alpha$
42. $\cot s + \tan s = \sec s \csc s$
43. $\dfrac{\sin^2 \gamma}{\cos \gamma} = \sec \gamma - \cos \gamma$
44. $\dfrac{\cos \alpha}{\sec \alpha} + \dfrac{\sin \alpha}{\csc \alpha} = \sec^2 \alpha - \tan^2 \alpha$
45. $\dfrac{\cos \theta}{\sin \theta \cot \theta} = 1$
46. $\sin^4 \theta - \cos^4 \theta = 2 \sin^2 \theta - 1$
47. $\tan^2 \gamma \sin^2 \gamma = \tan^2 \gamma + \cos^2 \gamma - 1$
48. $(1 - \cos^2 \alpha)(1 + \cos^2 \alpha) = 2 \sin^2 \alpha - \sin^4 \alpha$
49. $\dfrac{(\sec \theta - \tan \theta)^2 + 1}{\sec \theta \csc \theta - \tan \theta \csc \theta} = 2 \tan \theta$
50. $\dfrac{\cos \theta + 1}{\tan^2 \theta} = \dfrac{\cos \theta}{\sec \theta - 1}$

51. $\dfrac{1}{\sec \alpha - \tan \alpha} = \sec \alpha + \tan \alpha$

52. $\dfrac{1}{1 - \sin \theta} + \dfrac{1}{1 + \sin \theta} = 2 \sec^2 \theta$

53. $\dfrac{1 - \cos x}{1 + \cos x} = (\cot x - \csc x)^2$

54. $\dfrac{\tan s}{1 + \cos s} + \dfrac{\sin s}{1 - \cos s} = \cot s + \sec s \csc s$

55. $\dfrac{1}{\tan \alpha - \sec \alpha} + \dfrac{1}{\tan \alpha + \sec \alpha} = -2 \tan \alpha$

56. $\dfrac{\cot \alpha + 1}{\cot \alpha - 1} = \dfrac{1 + \tan \alpha}{1 - \tan \alpha}$

57. $\dfrac{\csc \theta + \cot \theta}{\tan \theta + \sin \theta} = \cot \theta \csc \theta$

58. $\sin^2 \alpha \sec^2 \alpha + \sin^2 \alpha \csc^2 \alpha = \sec^2 \alpha$

59. $\sec^4 x - \sec^2 x = \tan^4 x + \tan^2 x$

60. $\dfrac{1 - \sin \theta}{1 + \sin \theta} = \sec^2 \theta - 2 \sec \theta \tan \theta + \tan^2 \theta$

61. $\sin \theta + \cos \theta = \dfrac{\sin \theta}{1 - \dfrac{\cos \theta}{\sin \theta}} + \dfrac{\cos \theta}{1 - \dfrac{\sin \theta}{\cos \theta}}$

62. $\dfrac{\sin \theta}{1 - \cos \theta} - \dfrac{\sin \theta \cos \theta}{1 + \cos \theta} = \csc \theta(1 + \cos^2 \theta)$

63. $\dfrac{\sec^4 s - \tan^4 s}{\sec^2 s + \tan^2 s} = \sec^2 s - \tan^2 s$

64. $\dfrac{\cot^2 t - 1}{1 + \cot^2 t} = 1 - 2 \sin^2 t$

65. $\dfrac{\tan^2 t - 1}{\sec^2 t} = \dfrac{\tan t - \cot t}{\tan t + \cot t}$

66. $(1 + \sin x + \cos x)^2 = 2(1 + \sin x)(1 + \cos x)$

67. $(\sin s + \cos s)^2 \cdot \csc s = 2 \cos s + \dfrac{1}{\sin s}$

68. $\dfrac{\sin^3 t - \cos^3 t}{\sin t - \cos t} = 1 + \sin t \cos t$

69. $\dfrac{1 + \cos x}{1 - \cos x} - \dfrac{1 - \cos x}{1 + \cos x} = 4 \cot x \csc x$

70. $(\sec \alpha - \tan \alpha)^2 = \dfrac{1 - \sin \alpha}{1 + \sin \alpha}$

71. $(\sec \alpha + \csc \alpha)(\cos \alpha - \sin \alpha) = \cot \alpha - \tan \alpha$

72. $\dfrac{\sin^4 \alpha - \cos^4 \alpha}{\sin^2 \alpha - \cos^2 \alpha} = 1$

73. $\dfrac{\cot^2 x + \sec^2 x + 1}{\cot^2 x} = \sec^4 x$

74. $\dfrac{\cos x - (\sin x - 1)}{\cos x + (\sin x - 1)} = \dfrac{\sin x}{1 - \cos x}$ [*Hint:* Multiply numerator and denominator on the left by $\cos x - (\sin x - 1)$.]

Given a complicated equation involving trigonometric functions, it is a good idea to decide whether it really is an identity before trying to prove that it is. Substitute $s = 1$ and $s = 2$ into each of the following (with the calculator set for radian measure). If you get the same results on both sides of the equation, it may *be an identity. Then prove that it is.*

75. $\dfrac{2 + 5 \cos s}{\sin s} = 2 \csc s + 5 \cot s$

76. $1 + \cot^2 s = \dfrac{\sec^2 s}{\sec^2 s - 1}$

77. $\dfrac{\tan s - \cot s}{\tan s + \cot s} = 2 \sin^2 s$

78. $\dfrac{1}{1 + \sin s} + \dfrac{1}{1 - \sin s} = \sec^2 s$

79. $\dfrac{1 - \tan^2 s}{1 + \tan^2 s} = \cos^2 s - \sin s$

80. $\dfrac{\sin^3 s - \cos^3 s}{\sin s - \cos s} = \sin^2 s + 2 \sin s \cos s + \cos^2 s$

81. $\sin^2 s + \cos^2 s = \dfrac{1}{2}(1 - \cos 4s)$

82. $\cos 3s = 3 \cos s + 4 \cos^3 s$

Show that the following are not identities for real numbers s and t.

83. $\sin(\csc s) = 1$

84. $\sqrt{\cos^2 s} = \cos s$

85. $\csc t = \sqrt{1 + \cot^2 t}$

86. $\sin t = \sqrt{1 - \cos^2 t}$

87. Let $\tan \theta = t$ and show that

$$\sin \theta \cos \theta = \frac{t}{t^2 + 1}.$$

88. When does $\sin x = \sqrt{1 - \cos^2 x}$?

For students who have studied natural logarithms: Verify each identity.

89. $\ln e^{|\sin x|} = |\sin x|$

90. $\ln |\tan x| = -\ln |\cot x|$

91. $\ln |\sec x - \tan x| = -\ln |\sec x + \tan x|$

92. $-\ln |\csc t - \cot t| = \ln |\csc t + \cot t|$

5.3 Sum and Difference Identities for Cosine

Several examples presented throughout this book should have convinced you by now that $\cos(A - B)$ does *not* equal $\cos A - \cos B$. For example, if $A = \pi/2$ and $B = 0$,

$$\cos(A - B) = \cos(\pi/2 - 0) = \cos \pi/2 = 0,$$

while

$$\cos A - \cos B = \cos \pi/2 - \cos 0 = 0 - 1 = -1.$$

The actual formula for $\cos(A - B)$ is derived in this section. Start by locating angles A and B in standard position on a unit circle, with $B < A$. Let S and Q be the points where angles A and B, respectively, cut the circle. Locate point R on the unit circle so that angle POR equals the difference $A - B$. See Figure 5.4 on the following page.

Point Q is on the unit circle, so by the work with circular functions in chapter 3, the x-coordinate of Q is given by the cosine of angle B, while the y-coordinate of Q is given by the sine of angle B:

$$Q \text{ has coordinates } (\cos B, \sin B).$$

In the same way,

$$S \text{ has coordinates } (\cos A, \sin A),$$

and

$$R \text{ has coordinates } (\cos(A - B), \sin(A - B)).$$

Angle SOQ also equals $A - B$. Since the central angles SOQ and POR are equal, chords PR and SQ are equal. By the distance formula, since $PR = SQ$,

$$\sqrt{[\cos(A - B) - 1]^2 + [\sin(A - B)]^2} = \sqrt{(\cos A - \cos B)^2 + (\sin A - \sin B)^2}.$$

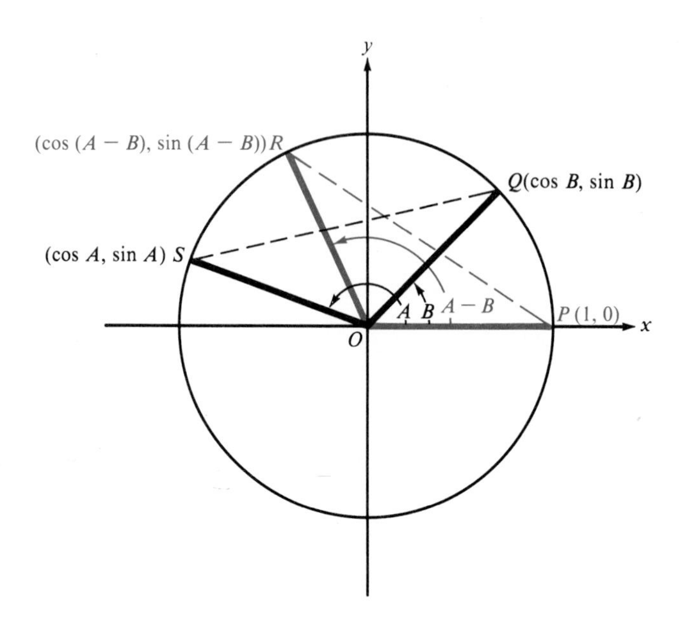

Figure 5.4

Squaring both sides and clearing parentheses gives

$$\cos^2 (A - B) - 2 \cos (A - B) + 1 + \sin^2 (A - B)$$
$$= \cos^2 A - 2 \cos A \cos B + \cos^2 B + \sin^2 A - 2 \sin A \sin B + \sin^2 B.$$

Since $\sin^2 x + \cos^2 x = 1$ for any value of x, rewrite the equation as

$$2 - 2 \cos (A - B) = 2 - 2 \cos A \cos B - 2 \sin A \sin B$$
$$\mathbf{\cos (A - B) = \cos A \cos B + \sin A \sin B.}$$

This is the identity for $\cos (A - B)$. Although Figure 5.4 shows angles A and B in the second and first quadrants, respectively, it can be shown that this result is the same for any values of these angles.

To find a similar expression for $\cos (A + B)$, rewrite $A + B$ as $A - (-B)$ and use the identity for $\cos (A - B)$ found above, along with the fact that $\cos (-B) = \cos B$ and $\sin (-B) = -\sin B$.

$$\begin{aligned} \cos (A + B) &= \cos [A - (-B)] \\ &= \cos A \cos (-B) + \sin A \sin (-B) \\ &= \cos A \cos B + \sin A(- \sin B) \\ \mathbf{\cos (A + B)} &= \mathbf{\cos A \cos B - \sin A \sin B} \end{aligned}$$

The two formulas we have just derived are summarized at the top of the next page.

Cosine of Sum or Difference

$$\cos(A - B) = \cos A \cos B + \sin A \sin B$$
$$\cos(A + B) = \cos A \cos B - \sin A \sin B$$

These identities are important in calculus and other areas of mathematics and in certain applications. Although a calculator or table can be used to find cos 15°, for example, the method shown below is used to give practice using the sum and difference identities, as well as to get an exact value.

Example 1 Find the exact value of the following without using tables or calculators.

(a) cos 15°

To find cos 15°, write 15° as the sum or difference of two angles with known function values. Since we know the trigonometric function values of both 45° and 30°, write 15° as 45° − 30°. Then use the identity for the cosine of the difference of two angles.

$$\begin{aligned}\cos 15^\circ &= \cos(45^\circ - 30^\circ)\\ &= \cos 45^\circ \cos 30^\circ + \sin 45^\circ \sin 30^\circ\\ &= \frac{\sqrt{2}}{2}\cdot\frac{\sqrt{3}}{2} + \frac{\sqrt{2}}{2}\cdot\frac{1}{2}\\ &= \frac{\sqrt{6}+\sqrt{2}}{4}\end{aligned}$$

(b) $$\begin{aligned}\cos\frac{5}{12}\pi &= \cos\left(\frac{\pi}{6}+\frac{\pi}{4}\right)\\ &= \cos\frac{\pi}{6}\cos\frac{\pi}{4} - \sin\frac{\pi}{6}\sin\frac{\pi}{4}\\ &= \frac{\sqrt{3}}{2}\cdot\frac{\sqrt{2}}{2} - \frac{1}{2}\cdot\frac{\sqrt{2}}{2}\\ &= \frac{\sqrt{6}-\sqrt{2}}{4}\end{aligned}$$

(c) cos 87° cos 93° − sin 87° sin 93°

From the identity for the cosine of the sum of two angles,

$$\begin{aligned}\cos 87^\circ \cos 93^\circ - \sin 87^\circ \sin 93^\circ &= \cos(87^\circ + 93^\circ)\\ &= \cos(180^\circ)\\ &= -1.\end{aligned}$$

The identities for the cosine of the sum and difference of two angles can be used to derive other identities. Recall the *cofunction identities*, which were presented earlier for values of θ where $0^\circ \leq \theta \leq 90^\circ$.

$$\cos(90° - \theta) = \sin\theta \qquad \cot(90° - \theta) = \tan\theta$$
$$\sin(90° - \theta) = \cos\theta \qquad \sec(90° - \theta) = \csc\theta$$
$$\tan(90° - \theta) = \cot\theta \qquad \csc(90° - \theta) = \sec\theta$$

These identities now can be generalized for any angle θ, not just those between 0° and 90°. For example, substituting 90° for A and θ for B in the identity given above for $\cos(A - B)$ gives

$$\begin{aligned}\cos(90° - \theta) &= \cos 90° \cos\theta + \sin 90° \sin\theta \\ &= 0 \cdot \cos\theta + 1 \cdot \sin\theta \\ &= \sin\theta.\end{aligned}$$

This result is true for *any* value of θ since the identity for $\cos(A - B)$ is true for any values of A and B. For the derivations of other cofunction identities, see Exercises 72 and 73.

Example 2 Find an angle θ that satisfies each of the following.

(a) $\cot\theta = \tan 25°$

Since tangent and cotangent are cofunctions,

$$\cot\theta = \tan(90° - \theta).$$

This means that

$$\tan(90° - \theta) = \tan 25°,$$

or

$$\begin{aligned}90° - \theta &= 25° \\ \theta &= 65°.\end{aligned}$$

(b) $\sin\theta = \cos(-30°)$

In the same way,

$$\sin\theta = \cos(90° - \theta) = \cos(-30°),$$

giving

$$\begin{aligned}90° - \theta &= -30° \\ \theta &= 120°.\end{aligned}$$

(c) $\csc\dfrac{3\pi}{4} = \sec\theta$

Cosecant and secant are cofunctions, so

$$\csc\frac{3\pi}{4} = \sec\left(\frac{\pi}{2} - \frac{3\pi}{4}\right) = \sec\theta$$

$$\sec\left(-\frac{\pi}{4}\right) = \sec\theta$$

$$-\frac{\pi}{4} = \theta.$$

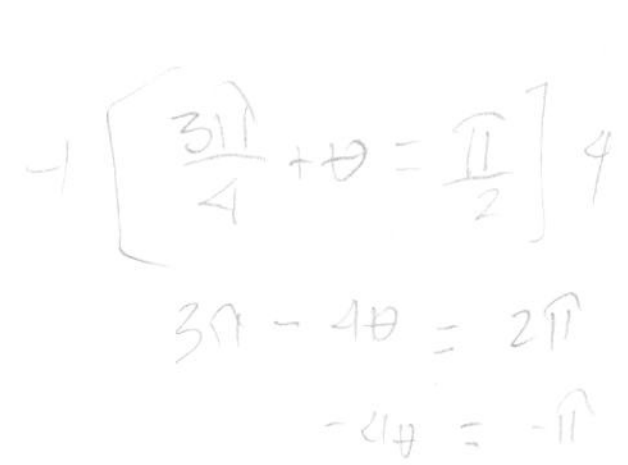

Example 3 Write $\cos(180° - \theta)$ as a trigonometric function of θ.

Use the difference identity. Replace A with 180° and B with θ.

$$\begin{aligned}\cos(180° - \theta) &= \cos 180° \cos\theta + \sin 180° \sin\theta \\ &= (-1)\cos\theta + (0)\sin\theta \\ &= -\cos\theta\end{aligned}$$

Example 4 Suppose that $\sin x = 1/2$, $\cos y = -12/13$, and both x and y are in quadrant II. Find $\cos(x + y)$.

By the identity above, $\cos(x + y) = \cos x \cos y - \sin x \sin y$. The values of $\sin x$ and $\cos y$ are given, so that $\cos(x + y)$ can be found if $\cos x$ and $\sin y$ are known. To find $\cos x$, use the identity $\sin^2 x + \cos^2 x = 1$, and substitute 1/2 for $\sin x$.

$$\begin{aligned}\sin^2 x + \cos^2 x &= 1 \\ \left(\frac{1}{2}\right)^2 + \cos^2 x &= 1 \\ \frac{1}{4} + \cos^2 x &= 1 \\ \cos^2 x &= \frac{3}{4} \\ \cos x &= \pm\frac{\sqrt{3}}{2}\end{aligned}$$

Since x is in quadrant II, $\cos x$ is negative, so

$$\cos x = -\frac{\sqrt{3}}{2}.$$

Find $\sin y$ as follows.

$$\begin{aligned}\sin^2 y + \cos^2 y &= 1 \\ \sin^2 y + \left(-\frac{12}{13}\right)^2 &= 1 \\ \sin^2 y + \frac{144}{169} &= 1 \\ \sin y &= \pm\frac{5}{13}\end{aligned}$$

Since y is in quadrant II,

$$\sin y = \frac{5}{13}.$$

(These values of $\cos x$ and $\sin y$ could also be found by sketching right triangles in quadrant II, labeling the known sides, and using the Pythagorean theorem to find the missing sides.) Now find $\cos(x + y)$.

$$\begin{aligned}\cos(x+y) &= \cos x \cos y - \sin x \sin y \\ &= -\frac{\sqrt{3}}{2}\cdot\left(-\frac{12}{13}\right) - \frac{1}{2}\cdot\frac{5}{13} \\ &= \frac{12\sqrt{3}}{26} - \frac{5}{26} \\ &= \frac{12\sqrt{3}-5}{26}\end{aligned}$$

5.3 Exercises

Write each of the following in terms of the cofunction of a complementary angle. See Example 2.

1. $\tan 87°$ **2.** $\sin 15°$ **3.** $\cos \pi/12$ **4.** $\sin 2\pi/5$

5. $\csc(-14° 24')$ **6.** $\sin 142° 14'$ **7.** $\sin 5\pi/8$ **8.** $\cot 9\pi/10$

9. $\sec 146° 42'$ **10.** $\tan 174° 3'$ **11.** $\cot 176.9814°$ **12.** $\sin 98.0142°$

Use the cofunction identities to fill in each of the following blanks with the appropriate trigonometric function name. See Example 2.

13. $\cot \frac{\pi}{3} =$ ______ $\frac{\pi}{6}$ **14.** $\sin \frac{2\pi}{3} =$ ______ $\left(-\frac{\pi}{6}\right)$ **15.** ______ $33° = \sin 57°$

16. ______ $72° = \cot 18°$ **17.** $\cos 70° = \dfrac{1}{\_\_\_\_\_\_\ 20°}$ **18.** $\tan 24° = \dfrac{1}{\_\_\_\_\_\_\ 66°}$

Tell whether each of the following is true *or* false.

19. $\cos 42° = \cos(30° + 12°)$

20. $\cos(-24°) = \cos 16° - \cos 40°$

21. $\cos 74° = \cos 60° \cos 14° + \sin 60° \sin 14°$

22. $\cos 140° = \cos 60° \cos 80° - \sin 60° \sin 80°$

23. $\cos \pi/3 = \cos \pi/12 \cos \pi/4 - \sin \pi/12 \sin \pi/4$

24. $\cos 2\pi/3 = \cos 11\pi/12 \cos \pi/4 + \sin 11\pi/12 \sin \pi/4$

25. $\cos 70° \cos 20° - \sin 70° \sin 20° = 0$

26. $\cos 85° \cos 40° + \sin 85° \sin 40° = \sqrt{2}/2$

Use the cofunction identities to find the angle θ that makes each of the following true. See Example 2.

27. $\tan \theta = \cot(45° + 2\theta)$

28. $\sin \theta = \cos(2\theta - 10°)$

29. $\sec \theta = \csc(\theta/2 + 20°)$

30. $\cos \theta = \sin(3\theta + 10°)$

31. $\sin(3\theta - 15°) = \cos(\theta + 25°)$

32. $\cot(\theta - 10°) = \tan(2\theta + 20°)$

Use the sum and difference identities for cosine to find the exact value of each of the following. See Example 1.

33. $\cos 285°$

34. $\cos(-15°)$

35. $\cos(-105°)$

36. $\cos 75°$

37. $\cos 7\pi/12$

38. $\cos(-5\pi/12)$

39. $\cos 40° \cos 50° - \sin 40° \sin 50°$

40. $\cos 80° \cos 35° + \sin 80° \sin 35°$

41. $\cos(-10°) \cos 35° + \sin(-10°) \sin 35°$

42. $\cos 112.146° \cos 67.854° - \sin 112.146° \sin 67.854°$

43. $\cos 174.983° \cos 95.017° - \sin 174.983° \sin 95.017°$

44. $\cos 348.502° \cos 78.502° + \sin 348.502° \sin 78.502°$

45. $\cos 2\pi/5 \cos \pi/10 - \sin 2\pi/5 \sin \pi/10$

46. $\cos 7\pi/9 \cos 2\pi/9 - \sin 7\pi/9 \sin 2\pi/9$

Write each of the following as a function of θ or x. See Example 3.

47. $\cos(30° + \theta)$

48. $\cos(45° - \theta)$

49. $\cos(60° + \theta)$

50. $\cos(\theta - 30°)$

51. $\cos(3\pi/2 - x)$

52. $\cos(x + \pi/4)$

For each of the following, find $\cos(s + t)$ and $\cos(s - t)$. See Example 4.

53. $\cos s = -1/5$ and $\sin t = 3/5$, s and t in quadrant II

54. $\sin s = 2/3$ and $\sin t = -1/3$, s in quadrant II and t in quadrant IV

55. $\sin s = 3/5$ and $\sin t = -12/13$, s in quadrant I and t in quadrant III

56. $\cos s = -8/17$ and $\cos t = -3/5$, s and t in quadrant III

57. $\cos s = -15/17$ and $\sin t = 4/5$, s in quadrant II and t in quadrant I

58. $\sin s = -8/17$ and $\cos t = -8/17$, s and t in quadrant III

59. $\sin s = \sqrt{5}/7$ and $\sin t = \sqrt{6}/8$, s and t in quadrant I

60. $\cos s = \sqrt{2}/4$ and $\sin t = -\sqrt{5}/6$, s and t in quadrant IV

Verify each of the following identities.

61. $\cos(\pi/2 + x) = -\sin x$

62. $\sec(\pi - x) = -\sec x$

63. $\cos 2x = \cos^2 x - \sin^2 x$ [*Hint:* $\cos 2x = \cos(x + x)$.]

64. $\cos(x + y) + \cos(x - y) = 2 \cos x \cos y$

65. $\dfrac{\cos(\alpha - \theta) - \cos(\alpha + \theta)}{\cos(\alpha - \theta) + \cos(\alpha + \theta)} = \tan\theta \tan\alpha$

66. $1 + \cos 2x - \cos^2 x = \cos^2 x$ (*Hint:* Use the result in Exercise 63.)

67. $\cos(\pi + s - t) = -\sin s \sin t - \cos s \cos t$

68. $\cos(\pi/2 + s - t) = \sin(t - s)$

69. $\cos(\alpha + \beta) \cos(\alpha - \beta) = 1 - \sin^2\alpha - \sin^2\beta$

70. $\cos 4x \cos 7x - \sin 4x \sin 7x = \cos 11x$

71. Use the identities for the cosine of the sum and difference of two angles to complete each of the following. See Example 3.

(a) $\cos(0° - \theta) =$

(b) $\cos(90° - \theta) =$

(c) $\cos(180° - \theta) =$

(d) $\cos(270° - \theta) =$

(e) $\cos(0° + \theta) =$

(f) $\cos(90° + \theta) =$

(g) $\cos(180° + \theta) =$

(h) $\cos(270° + \theta) =$

72. Use the identity $\cos(90° - \theta) = \sin\theta$, and replace θ with $90° - A$, to derive the identity $\cos A = \sin(90° - A)$.

73. Use the results of Exercise 72 to derive the identity $\tan\theta = \cot(90° - \theta)$.

Let $\sin s = -0.09463$ *and* $\cos t = 0.83499$, *where* s *terminates in quadrant III and* t *in quadrant IV. Find each of the following.*

74. $\cos(s - t)$ **75.** $\cos(s + t)$ **76.** $\cos 2s$ **77.** $\cos 2t$

78. Let $f(x) = \cos x$. Prove that $\dfrac{f(x + h) - f(x)}{h} = \cos x\left(\dfrac{\cos h - 1}{h}\right) - \sin x\left(\dfrac{\sin h}{h}\right)$.

5.4 Sum and Difference Identities for Sine and Tangent

Formulas for $\sin(A + B)$ and $\sin(A - B)$ can be developed from the results in Section 5.3. Start with the cofunction relationship

$$\sin\theta = \cos(90° - \theta).$$

Replace θ with $A + B$.

$$\begin{aligned}\sin(A + B) &= \cos[90° - (A + B)]\\ &= \cos[(90° - A) - B]\end{aligned}$$

Using the formula for $\cos(A - B)$ from the previous section gives

$$\sin(A + B) = \cos(90° - A)\cos B + \sin(90° - A)\sin B$$
$$\mathbf{\sin(A + B) = \sin A\cos B + \cos A\sin B.}$$

(The cofunction relationships were used in the last step.)

Now write $\sin(A - B)$ as $\sin[A + (-B)]$ and use the identity for $\sin(A + B)$ to get

$$\begin{aligned}\sin(A - B) &= \sin[A + (-B)]\\ &= \sin A\cos(-B) + \cos A\sin(-B)\\ &= \sin A\cos B - \cos A\sin B,\end{aligned}$$

or, in summary,

$$\mathbf{\sin(A - B) = \sin A\cos B - \cos A\sin B.}$$

Using the identities for $\sin(A + B)$, $\cos(A + B)$, $\sin(A - B)$, and $\cos(A - B)$, and the identity $\tan\theta = \sin\theta/\cos\theta$, gives the following identities.

$$\mathbf{\tan(A + B) = \frac{\tan A + \tan B}{1 - \tan A\tan B}}$$

$$\mathbf{\tan(A - B) = \frac{\tan A - \tan B}{1 + \tan A\tan B}}$$

We show the proof for the first of these two identities. The proof for the other is very similar. Start with

$$\tan (A + B) = \frac{\sin (A + B)}{\cos (A + B)}$$

$$= \frac{\sin A \cos B + \cos A \sin B}{\cos A \cos B - \sin A \sin B}.$$

To express this result in terms of the tangent function, multiply both numerator and denominator by $1/(\cos A \cos B)$.

$$\tan (A + B) = \frac{\dfrac{\sin A \cos B + \cos A \sin B}{1}}{\dfrac{\cos A \cos B - \sin A \sin B}{1}} \cdot \frac{\dfrac{1}{\cos A \cos B}}{\dfrac{1}{\cos A \cos B}}$$

$$= \frac{\dfrac{\sin A \cos B}{\cos A \cos B} + \dfrac{\cos A \sin B}{\cos A \cos B}}{\dfrac{\cos A \cos B}{\cos A \cos B} - \dfrac{\sin A \sin B}{\cos A \cos B}}$$

$$= \frac{\dfrac{\sin A}{\cos A} + \dfrac{\sin B}{\cos B}}{1 - \dfrac{\sin A}{\cos A} \cdot \dfrac{\sin B}{\cos B}}$$

Using the identity $\tan \theta = \sin \theta / \cos \theta$,

$$\tan (A + B) = \frac{\tan A + \tan B}{1 - \tan A \tan B}.$$

The identities given in this section are summarized below.

Sine and Tangent of Sum and Difference

$$\sin (A + B) = \sin A \cos B + \cos A \sin B$$

$$\sin (A - B) = \sin A \cos B - \cos A \sin B$$

$$\tan (A + B) = \frac{\tan A + \tan B}{1 - \tan A \tan B}$$

$$\tan (A - B) = \frac{\tan A - \tan B}{1 + \tan A \tan B}$$

Again, the following example and the corresponding exercises are given primarily to offer practice in using these new identities.

Example 1 Find the *exact* value of the following.

(a) $\sin 75° = \sin(45° + 30°)$

$$= \sin 45° \cos 30° + \cos 45° \sin 30°$$

$$= \frac{\sqrt{2}}{2} \cdot \frac{\sqrt{3}}{2} + \frac{\sqrt{2}}{2} \cdot \frac{1}{2}$$

$$= \frac{\sqrt{6}}{4} + \frac{\sqrt{2}}{4} = \frac{\sqrt{6} + \sqrt{2}}{4}$$

(b) $\tan \frac{7\pi}{12} = \tan\left(\frac{\pi}{3} + \frac{\pi}{4}\right)$

$$= \frac{\tan \frac{\pi}{3} + \tan \frac{\pi}{r}}{1 - \tan \frac{\pi}{3} \tan \frac{\pi}{4}}$$

$$= \frac{\sqrt{3} + 1}{1 - \sqrt{3} \cdot 1}$$

$$= \frac{\sqrt{3} + 1}{1 - \sqrt{3}} \cdot \frac{1 + \sqrt{3}}{1 + \sqrt{3}}$$

$$= \frac{\sqrt{3} + 3 + 1 + \sqrt{3}}{1 - 3}$$

$$= \frac{4 + 2\sqrt{3}}{-2} = -2 - \sqrt{3}$$

(c) $\sin 40° \cos 160° - \cos 40° \sin 160° = \sin(40° - 160°)$

$$= \sin(-120°)$$

$$= -\sin 120°$$

$$= -\frac{\sqrt{3}}{2}$$ ✥

Example 2 Write each of the following as a function of θ.

(a) $\sin(30° + \theta)$

Using the identity for $\sin(A + B)$,

$$\sin(30° + \theta) = \sin 30° \cos \theta + \cos 30° \sin \theta$$

$$= \frac{1}{2} \cos \theta + \frac{\sqrt{3}}{2} \sin \theta.$$

(b) $\tan(45° - \theta) = \dfrac{\tan 45° - \tan \theta}{1 + \tan 45° \tan \theta} = \dfrac{1 - \tan \theta}{1 + \tan \theta}$ ✥

Example 3 If $\sin A = 4/5$ and $\cos B = -5/13$, where A is in quadrant II and B is in quadrant III, find each of the following.

(a) $\sin (A + B)$

The identity for $\sin (A + B)$ requires $\sin A$, $\cos A$, $\sin B$, and $\cos B$. Two of these values are given. The two missing values, $\cos A$ and $\sin B$, must be found first. These values can be found with the identity $\sin^2 x + \cos^2 x = 1$. To find $\cos A$, use

$$\sin^2 A + \cos^2 A = 1$$
$$\frac{16}{25} + \cos^2 A = 1$$
$$\cos^2 A = \frac{9}{25}$$
$$\cos A = -\frac{3}{5}. \qquad A \text{ is in quadrant II}$$

In the same way, $\sin B = -12/13$. Now use the formula for $\sin (A + B)$.

$$\sin (A + B) = \frac{4}{5}\left(-\frac{5}{13}\right) + \left(-\frac{3}{5}\right)\left(-\frac{12}{13}\right)$$
$$= \frac{-20}{65} + \frac{36}{65} = \frac{16}{65}$$

(b) $\tan (A + B)$

Use the values of sine and cosine from part (a) to get $\tan A = -4/3$ and $\tan B = 12/5$. Then

$$\tan (A + B) = \frac{-\frac{4}{3} + \frac{12}{5}}{1 - \left(-\frac{4}{3}\right)\left(\frac{12}{5}\right)} = \frac{\frac{16}{15}}{1 + \frac{48}{15}} = \frac{\frac{16}{15}}{\frac{63}{15}} = \frac{16}{63}.$$ ✦

The next example shows how to verify identities using the results of this section and the last.

Example 4 Show that

$$\sin\left(\frac{\pi}{6} + s\right) + \cos\left(\frac{\pi}{3} + s\right) = \cos s$$

is an identity.

Work on the left side, using the identities for $\sin(A + B)$ and $\cos(A + B)$.

$$\sin\left(\frac{\pi}{6} + s\right) + \cos\left(\frac{\pi}{3} + s\right)$$

$$= \left(\sin\frac{\pi}{6}\cos s + \cos\frac{\pi}{6}\sin s\right)$$

$\sin(A + B) = \sin A\cos B + \cos A\sin B$

$$+ \left(\cos\frac{\pi}{3}\cos s - \sin\frac{\pi}{3}\sin s\right)$$

$\cos(A + B) = \cos A\cos B - \sin A\sin B$

$$= \left(\frac{1}{2}\cos s + \frac{\sqrt{3}}{2}\sin s\right)$$

$\sin\pi/6 = 1/2$; $\cos\pi/6 = \sqrt{3}/2$

$$+ \left(\frac{1}{2}\cos s - \frac{\sqrt{3}}{2}\sin s\right)$$

$\cos\pi/3 = 1/2$; $\sin\pi/3 = \sqrt{3}/2$

$$= \frac{1}{2}\cos s + \frac{1}{2}\cos s$$

$$= \cos s$$ ✣

5.4 Exercises

Use the identities in this section to find the exact value of each of the following. See Example 1.

1. $\sin 15^\circ$
2. $\sin 105^\circ$
3. $\tan 15^\circ$
4. $\tan(-105^\circ)$
5. $\sin(-105^\circ)$
6. $\tan 5\pi/12$
7. $\sin 5\pi/12$
8. $\sin 285^\circ$
9. $\sin 76^\circ \cos 31^\circ - \cos 76^\circ \sin 31^\circ$
10. $\sin 40^\circ \cos 50^\circ + \cos 40^\circ \sin 50^\circ$
11. $\dfrac{\tan 80^\circ + \tan 55^\circ}{1 - \tan 80^\circ \tan 55^\circ}$
12. $\dfrac{\tan 80^\circ - \tan(-55^\circ)}{1 + \tan 80^\circ \tan(-55^\circ)}$
13. $\dfrac{\tan 100^\circ + \tan 80^\circ}{1 - \tan 100^\circ \tan 80^\circ}$
14. $\sin 100^\circ \cos 10^\circ - \cos 100^\circ \sin 10^\circ$
15. $\sin \pi/5 \cos 3\pi/10 + \cos \pi/5 \sin 3\pi/10$
16. $\dfrac{\tan 5\pi/12 + \tan \pi/4}{1 - \tan 5\pi/12 \tan \pi/4}$
17. $\sin 79.802^\circ \cos 100.198^\circ + \cos 79.802^\circ \sin 100.198^\circ$
18. $\sin 296.4372^\circ \cos 26.4372^\circ - \cos 296.4372^\circ \sin 26.4372^\circ$
19. $\dfrac{\tan 151.9063^\circ + \tan 28.0937^\circ}{1 - \tan 151.9063^\circ \tan 28.0937^\circ}$
20. $\dfrac{\tan 214.91^\circ + \tan 145.09^\circ}{1 - \tan 214.91^\circ \tan 145.09^\circ}$

Use identities to write each of the following as a function of θ or s. See Example 2.

21. $\sin(45° + \theta)$ **22.** $\sin(\theta - 30°)$ **23.** $\tan(\theta + 30°)$ **24.** $\tan(60° - \theta)$

25. $\tan(\pi/4 + s)$ **26.** $\sin(\pi/4 + s)$ **27.** $\sin(180° - \theta)$ **28.** $\sin(270° - \theta)$

29. $\tan(180° + \theta)$ **30.** $\tan(360° - \theta)$ **31.** $\sin(\pi + \theta)$ **32.** $\tan(\pi - \theta)$

For each of the following, find $\sin(s + t)$, $\sin(s - t)$, $\tan(s + t)$, *and* $\tan(s - t)$. *See Example 3.*

33. $\cos s = 3/5$ and $\sin t = 5/13$, s and t in quadrant I

34. $\cos s = -1/5$ and $\sin t = 3/5$, s and t in quadrant II

35. $\sin s = 2/3$ and $\sin t = -1/3$, s in quadrant II and t in quadrant IV

36. $\sin s = 3/5$ and $\sin t = -12/13$, s in quadrant I and t in quadrant III

37. $\cos s = -8/17$ and $\cos t = -3/5$, s and t in quadrant III

38. $\cos s = -15/17$ and $\sin t = 4/5$, s in quadrant II and t in quadrant I

39. $\sin s = -4/5$ and $\cos t = 12/13$, s in quadrant III and t in quadrant IV

40. $\sin s = -5/13$ and $\sin t = 3/5$, s in quadrant III and t in quadrant II

41. $\sin s = -8/17$ and $\cos t = -8/17$, s and t in quadrant III

42. $\sin s = 2/3$ and $\sin t = 2/5$, s and t in quadrant I

43. $\cos s = -\sqrt{7}/4$ and $\sin t = \sqrt{3}/5$, s and t in quadrant II

44. $\cos s = \sqrt{11}/5$ and $\cos t = \sqrt{2}/6$, s and t in quadrant IV

Verify that each of the following are identities. See Example 4.

45. $\sin\left(\frac{\pi}{2} + x\right) = \cos x$

46. $\sin\left(\frac{3\pi}{2} + x\right) = -\cos x$

47. $\tan\left(\frac{\pi}{2} + x\right) = -\cot x \quad \left(\textit{Hint: } \tan\theta = \frac{\sin\theta}{\cos\theta}\right)$

48. $\tan\left(\frac{\pi}{4} + x\right) = \frac{1 + \tan x}{1 - \tan x}$

49. $\sin 2x = 2\sin x \cos x \quad [\textit{Hint: } \sin 2x = \sin(x + x)]$

50. $\sin(x + y) + \sin(x - y) = 2\sin x \cos y$

51. $\tan(x - y) - \tan(y - x) = \frac{2(\tan x - \tan y)}{1 + \tan x \tan y}$

52. $\sin(210° + x) - \cos(120° + x) = 0$

53. $\frac{\cos(\alpha - \beta)}{\cos\alpha \sin\beta} = \tan\alpha + \cot\beta$

54. $\frac{\sin(s + t)}{\cos s \cos t} = \tan s + \tan t$

55. $\frac{\sin(x - y)}{\sin(x + y)} = \frac{\tan x - \tan y}{\tan x + \tan y}$

56. $\frac{\sin(x + y)}{\cos(x - y)} = \frac{\cot x + \cot y}{1 + \cot x \cot y}$

57. $\frac{\sin(s - t)}{\sin t} + \frac{\cos(s - t)}{\cos t} = \frac{\sin s}{\sin t \cos t}$

58. $\frac{\tan(\alpha + \beta) - \tan\beta}{1 + \tan(\alpha + \beta)\tan\beta} = \tan\alpha$

Let $\sin s = 0.599832$, *where s terminates in quadrant II. Let* $\sin t = -0.845992$, *where t terminates in quadrant III. Find each of the following.*

59. $\sin (s + t)$ **60.** $\sin (s - t)$ **61.** $\tan (s + t)$

62. $\sin 2s$ **63.** $\tan 2t$ **64.** $\sin 2t$

65. Why is it not possible to follow Example 2 and find a formula for $\tan (270° - \theta)$?

66. What happens when you try to evaluate

$$\frac{\tan 65.902° + \tan 24.098°}{1 - \tan 65.902° \tan 24.098°}?$$

Derive a formula for each of the following.

67. $\sin (A + B + C)$

68. $\cos (A + B + C)$

69. Let $f(x) = \sin x$. Show that

$$\frac{f(x + h) - f(x)}{h} = \sin x \left(\frac{\cos h - 1}{h}\right) + \cos x \left(\frac{\sin h}{h}\right).$$

70. The slope of a line is defined as the ratio of the vertical change and the horizontal change. As shown in the sketch on the left, the tangent of the *angle of inclination* θ is given by the ratio of the side opposite and the side adjacent. This ratio is the same as that used in finding the slope, m, so that $m = \tan \theta$. In the figure on the right, let the two lines have angles of inclination α and β, and slopes m_1 and m_2, respectively. Let θ be the smallest positive angle between the lines. Show that

$$\tan \theta = \frac{m_2 - m_1}{1 + m_1 m_2}.$$

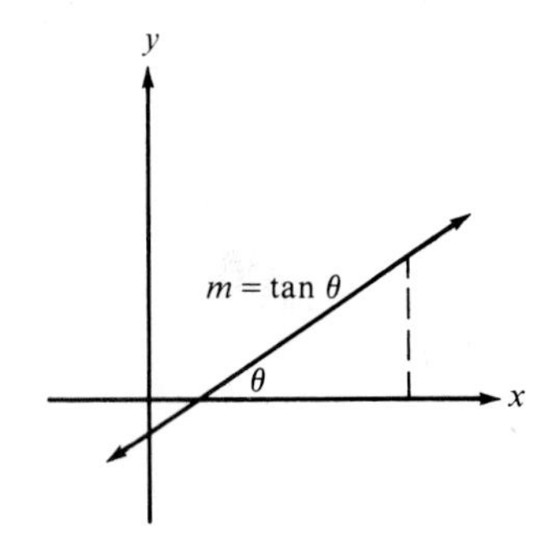

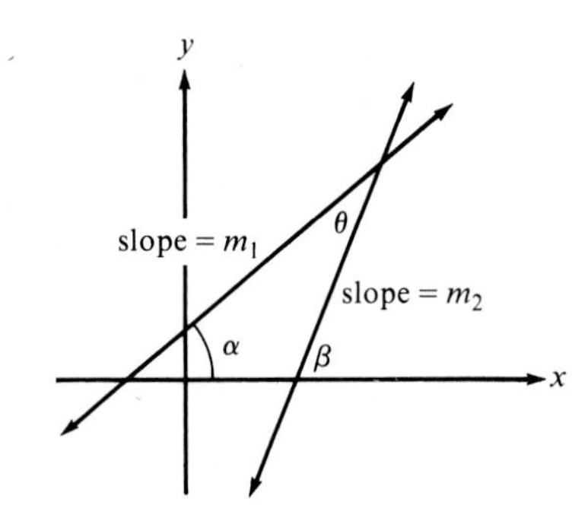

Use the results from Exercise 70 to find the angle between the following pairs of lines. Round to the nearest tenth of a degree.

71. $x + y = 9, \quad 2x + y = -1$

72. $5x - 2y + 4 = 0, \quad 3x + 5y = 6$

5.5 Double-Angle Identities

Some special cases of the identities for the sum of two angles are used often enough to be expressed as separate identities. These are the identities that result from the addition identities when $A = B$, so that $A + B = 2A$. These identities, called the **double-angle identities,** are derived in this section.

In the identity $\cos(A + B) = \cos A \cos B - \sin A \sin B$, let $B = A$ to derive an expression for $\cos 2A$.

$$\begin{aligned} \cos 2A &= \cos(A + A) \\ &= \cos A \cos A - \sin A \sin A \\ \mathbf{\cos 2A} &= \mathbf{\cos^2 A - \sin^2 A} \end{aligned}$$

Two other useful forms of this identity can be obtained by substituting either $\cos^2 A = 1 - \sin^2 A$ or $\sin^2 A = 1 - \cos^2 A$. Replace $\cos^2 A$ with $1 - \sin^2 A$ to get

$$\begin{aligned} \cos 2A &= \cos^2 A - \sin^2 A \\ &= (1 - \sin^2 A) - \sin^2 A \\ \mathbf{\cos 2A} &= \mathbf{1 - 2\sin^2 A}, \end{aligned}$$

and replace $\sin^2 A$ with $1 - \cos^2 A$ to get

$$\begin{aligned} \cos 2A &= \cos^2 A - (1 - \cos^2 A) \\ &= \cos^2 A - 1 + \cos^2 A \\ \mathbf{\cos 2A} &= \mathbf{2\cos^2 A - 1}. \end{aligned}$$

Find $\sin 2A$ with the identity $\sin(A + B) = \sin A \cos B + \cos A \sin B$, letting $B = A$.

$$\begin{aligned} \sin 2A &= \sin(A + A) \\ &= \sin A \cos A + \cos A \sin A \\ \mathbf{\sin 2A} &= \mathbf{2\sin A \cos A} \end{aligned}$$

Use the identity for $\tan(A + B)$ to find $\tan 2A$.

$$\begin{aligned} \tan 2A &= \tan(A + A) \\ &= \frac{\tan A + \tan A}{1 - \tan A \tan A} \\ \mathbf{\tan 2A} &= \mathbf{\frac{2\tan A}{1 - \tan^2 A}} \end{aligned}$$

These identities are summarized on the following page.

Double-Angle Identities

$$\cos 2A = \cos^2 A - \sin^2 A$$
$$\cos 2A = 1 - 2\sin^2 A$$
$$\cos 2A = 2\cos^2 A - 1$$
$$\sin 2A = 2\sin A \cos A$$
$$\tan 2A = \frac{2\tan A}{1 - \tan^2 A}$$

Example 1 Simplify each of the following using identities.

(a) $\sin 15° \cos 15°$

The product of the sine and cosine of the same angle suggests the identity for $\sin 2A$: $\sin 2A = 2\sin A \cos A$. With this identity,

$$\begin{aligned}\sin 15° \cos 15° &= \left(\frac{1}{2}\right)(2) \sin 15° \cos 15° \\ &= \frac{1}{2}(2 \sin 15° \cos 15°) \\ &= \frac{1}{2}(\sin 2 \cdot 15°) \\ &= \frac{1}{2} \sin 30° \\ &= \frac{1}{2} \cdot \frac{1}{2} \\ &= \frac{1}{4}.\end{aligned}$$

(b) $\cos^2 7x - \sin^2 7x$

This expression suggests one of the identities for $\cos 2A$: $\cos 2A = \cos^2 A - \sin^2 A$. Substituting $7x$ for A gives

$$\cos^2 7x - \sin^2 7x = \cos 2(7x) = \cos 14x.$$ ✣

Example 2 Find the values of the six trigonometric functions of θ if $\cos 2\theta = 4/5$ and θ terminates in quadrant II.

Use one of the double-angle identities for cosine to get a trigonometric function of θ.

$$\cos 2\theta = 1 - 2\sin^2 \theta$$
$$\frac{4}{5} = 1 - 2\sin^2 \theta$$

$$-\frac{1}{5} = -2\sin^2\theta$$

$$\frac{1}{10} = \sin^2\theta$$

$$\sin\theta = \frac{\sqrt{10}}{10}$$

Choose the positive square root since θ terminates in quadrant II. Values of $\cos\theta$ and $\tan\theta$ can now be found by using the fundamental identities or by sketching and labeling a right triangle in quadrant II. Using identities gives

$$\sin^2\theta + \cos^2\theta = 1$$

$$\frac{1}{10} + \cos^2\theta = 1$$

$$\cos^2\theta = \frac{9}{10}$$

$$\cos\theta = \frac{-3}{\sqrt{10}} = -\frac{3\sqrt{10}}{10}.$$

Verify that $\tan\theta = \sin\theta/\cos\theta = -1/3$. Find the other three functions using reciprocals.

$$\csc\theta = \frac{1}{\sin\theta} = \sqrt{10}, \quad \sec\theta = \frac{1}{\cos\theta} = -\frac{\sqrt{10}}{3}, \quad \cot\theta = \frac{1}{\tan\theta} = -3$$ ✤

Example 3 Given $\cos\theta = 3/5$, where $3\pi/2 < \theta < 2\pi$, find $\sin 2\theta$, $\cos 2\theta$, and $\tan 2\theta$.

From $\cos\theta = 3/5$, the identity $\sin^2\theta + \cos^2\theta = 1$ leads to $\sin\theta = \pm 4/5$. Since θ terminates in quadrant IV, $\sin\theta = -4/5$. Then, using the double-angle identities,

$$\sin 2\theta = 2\sin\theta\cos\theta = 2\left(-\frac{4}{5}\right)\left(\frac{3}{5}\right) = -\frac{24}{25};$$

$$\cos 2\theta = \cos^2\theta - \sin^2\theta = \frac{9}{25} - \frac{16}{25} = -\frac{7}{25}; \text{ and}$$

$$\tan 2\theta = \frac{\sin 2\theta}{\cos 2\theta} = \frac{-24/5}{-7/25} = \frac{24}{7}.$$

As an alternative way of finding $\tan 2\theta$, start with $\sin\theta = -4/5$ and $\cos\theta = 3/5$, to get $\tan\theta = -4/3$, with

$$\tan 2\theta = \frac{2\tan\theta}{1 - \tan^2\theta} = \frac{2\left(-\frac{4}{3}\right)}{1 - \frac{16}{9}} = \frac{-\frac{8}{3}}{-\frac{7}{9}} = \frac{24}{7}.$$ ✤

The methods used above to derive the identities for double angles can also be used to find identities for expressions such as $\sin 3s$, as in the next example.

Example 4 Write $\sin 3s$ in terms of $\sin s$.

$$\begin{aligned}
\sin 3s &= \sin (2s + s) \\
&= \sin 2s \cos s + \cos 2s \sin s \\
&= (2 \sin s \cos s)\cos s + (\cos^2 s - \sin^2 s)\sin s \\
&= 2 \sin s \cos^2 s + \cos^2 s \sin s - \sin^3 s \\
&= 2 \sin s(1 - \sin^2 s) + (1 - \sin^2 s)\sin s - \sin^3 s \\
&= 2 \sin s - 2 \sin^3 s + \sin s - \sin^3 s - \sin^3 s \\
&= 3 \sin s - 4 \sin^3 s
\end{aligned}$$

✥

Example 5 Show that

$$\cot x \sin 2x = 1 + \cos 2x$$

is an identity.

Work on the left side.

$$\begin{aligned}
\cot x \sin 2x &= \frac{\cos x}{\sin x} \cdot \sin 2x && \cot x = \frac{\cos x}{\sin x} \\
&= \frac{\cos x}{\sin x}(2 \sin x \cos x) && \sin 2x = 2 \sin x \cos x \\
&= 2 \cos^2 x \\
&= 1 + \cos 2x && 2 \cos^2 x - 1 = \cos 2x
\end{aligned}$$

✥

5.5 Exercises

Use an identity to write each of the following as a single trigonometric function or as a single number. See Example 1.

1. $2 \cos^2 15° - 1$

2. $\cos^2 15° - \sin^2 15°$

3. $\dfrac{2 \tan 15°}{1 - \tan^2 15°}$

4. $1 - 2 \sin^2 15°$

5. $2 \sin \pi/3 \cos \pi/3$

6. $\dfrac{2 \tan \pi/3}{1 - \tan^2 \pi/3}$

7. $1 - 2 \sin^2 22\frac{1}{2}°$

8. $2 \sin 22\frac{1}{2}° \cos 22\frac{1}{2}°$

9. $2 \cos^2 67\frac{1}{2}° - 1$

10. $\dfrac{2 \tan 67\frac{1}{2}°}{1 - \tan^2 67\frac{1}{2}°}$

11. $\sin \pi/8 \cos \pi/8$

12. $\cos^2 \pi/8 - 1/2$

13. $\dfrac{\tan 51°}{1 - \tan^2 51°}$

14. $\dfrac{\tan 34°}{2(1 - \tan^2 34°)}$

15. $\dfrac{1}{4} - \dfrac{1}{2}\sin^2 47.1°$

16. $\dfrac{1}{8}\sin 29.5° \cos 29.5°$

17. $\sin^2 2\pi/5 - \cos^2 2\pi/5$

18. $\dfrac{2 \tan \pi/5}{\tan^2 \pi/5 - 1}$

19. $2 \sin 5x \cos 5x$

20. $2 \cos^2 6\alpha - 1$

21. $\cos^2 2\alpha - \sin^2 2\alpha$

22. $\dfrac{2 \tan 3r}{1 - \tan^2 3r}$

23. $\dfrac{2 \tan x/9}{1 - \tan^2 x/9}$

24. $\dfrac{\tan 2y/5}{1 - \tan^2 2y/5}$

Find the exact value of each of the following in two ways: evaluate the expression directly, and use an appropriate identity. Do not use tables or a calculator.

25. $\sin 2(45°)$

26. $\cos 2(45°)$

27. $\cos 2(60°)$

28. $\tan 2(60°)$

29. $\cos 2\left(\dfrac{5\pi}{3}\right)$

30. $\sin 2\left(\dfrac{\pi}{3}\right)$

31. $\tan 2\left(-\dfrac{\pi}{3}\right)$

32. $\cos 2\left(-\dfrac{9\pi}{4}\right)$

33. $\tan 2\left(-\dfrac{4\pi}{3}\right)$

34. $\tan 2\left(\dfrac{-13\pi}{6}\right)$

35. $\sin 2\left(\dfrac{-11\pi}{2}\right)$

36. $\sin 2\left(\dfrac{-17\pi}{2}\right)$

Use the identities in this section to find values of the six trigonometric functions for each of the following. See Examples 2 and 3.

37. θ, given $\cos 2\theta = 3/5$ and θ terminates in quadrant I

38. α, given $\cos 2\alpha = 3/4$ and α terminates in quadrant III

39. x, given $\cos 2x = -5/12$ and $\pi/2 < x < \pi$

40. t, given $\cos 2t = 2/3$ and $\pi/2 < t < \pi$

41. 2θ, given $\sin \theta = 2/5$ and $\cos \theta < 0$

42. 2β, given $\cos \beta = -12/13$ and $\sin \beta > 0$

43. $2x$, given $\tan x = 2$ and $\cos x > 0$

44. $2x$, given $\tan x = 5/3$ and $\sin x < 0$

45. 2α, given $\sin \alpha = -\sqrt{5}/7$ and $\cos \alpha > 0$

46. 2α, given $\cos \alpha = \sqrt{3}/5$ and $\sin \alpha > 0$

Express each of the following as a trigonometric function of x. See Example 4.

47. $\tan^2 2x$

48. $\cos^2 2x$

49. $\cos 3x$

50. $\sin 4x$

51. $\tan 3x$

52. $\cos 4x$

53. $\tan 4x$

54. $\sin 5x$

Verify each of the following identities. See Example 5.

55. $(\sin \gamma + \cos \gamma)^2 = \sin 2\gamma + 1$

56. $\cos 2s = \cos^4 s - \sin^4 s$

57. $\sec 2x = \dfrac{\sec^2 x + \sec^4 x}{2 + \sec^2 x - \sec^4 x}$

58. $\sin 2\theta = \dfrac{4 \tan \theta \cos^2 \theta - 2 \tan \theta}{1 - \tan^2 \theta}$

59. $\cot 2\beta = \dfrac{\cot^2 \beta - 1}{2 \cot \beta}$

60. $\tan 8k - \tan 8k \tan^2 4k = 2 \tan 4k$

61. $\sin 2\gamma = \dfrac{2 \tan \gamma}{1 + \tan^2 \gamma}$

62. $-\tan 2\theta = \dfrac{2 \tan \theta}{\sec^2 \theta - 2}$

63. $\cos 2y = \dfrac{2 - \sec^2 y}{\sec^2 y}$

64. $\cot s + \tan s = 2 \csc 2s$

65. $\sin 4\alpha = 4 \sin \alpha \cos \alpha \cos 2\alpha$

66. $\dfrac{1 + \cos 2x}{\sin 2x} = \cot x$

67. $\tan (\theta - 45°) + \tan (\theta + 45°) = 2 \tan 2\theta$

68. $\cot 4\theta = \dfrac{1 - \tan^2 2\theta}{2 \tan 2\theta}$

69. $\dfrac{2 \cos 2\alpha}{\sin 2\alpha} = \cot \alpha - \tan \alpha$

70. $\sin 4\gamma = 4 \sin \gamma \cos \gamma - 8 \sin^3 \gamma \cos \gamma$

71. $\sin 2\alpha \cos 2\alpha = \sin 2\alpha - 4 \sin^3 \alpha \cos \alpha$

72. $\dfrac{\sin^3 t - \cos^3 t}{\sin t - \cos t} = \dfrac{2 + \sin 2t}{2}$

73. $\cos 2x = \dfrac{1 - \tan^2 x}{1 + \tan^2 x}$

74. $\dfrac{\sin 2\theta}{\sin \theta} - \dfrac{\cos 2\theta}{\cos \theta} = \sec \theta$

75. $\tan s + \cot s = 2 \csc 2s$

76. $\dfrac{\cot \alpha - \tan \alpha}{\cot \alpha + \tan \alpha} = \cos 2\alpha$

77. $1 + \tan x \tan 2x = \sec 2x$

78. $\cot \theta \tan (\theta + \pi) - \sin (\pi - \theta) \cos \left(\dfrac{\pi}{2} - \theta\right) = \cos^2 \theta$

Let $\sin s = -0.481143$, with $3\pi/2 < s < 2\pi$. Find each of the following.

79. $\sin 2s$

80. $\cos 2s$

81. $\tan 2s$

82. $\sec 2s$

Use a fundamental identity to simplify each of the following.

83. $\sin^2 2x + \cos^2 2x$

84. $1 + \tan^2 4\alpha$

85. $\cot^2 3r + 1$

86. $\sin^2 11\alpha + \cos^2 11\alpha$

If an object is dropped in a vacuum, then the distance, d, the object falls in t seconds is given by

$$d = \frac{1}{2} gt^2,$$

where g is the acceleration due to gravity. At any particular point on the earth's surface, the value of g is a constant, roughly 978 cm per sec per sec. A more exact value of g at any point on the earth's surface is given by

$$g = 978.0524(1 + 0.005297 \sin^2 \phi - 0.0000059 \sin^2 2\phi) - 0.000094h$$

in cm per second per second, where ϕ is the latitude of the point and h is the altitude of the point in feet. Find g, rounding to the nearest thousandth, given the following.

87. $\phi = 47° 12'$, $h = 387.0$ ft

88. $\phi = 68°47'$, $h = 1145$ ft

5.6 Half-Angle Identities

From the alternative forms of the double-angle identity for cosine, three additional identities can be derived. These **half-angle identities,** listed below, are useful in calculus.

Half-Angle Identities

$$\cos\frac{A}{2} = \pm\sqrt{\frac{1+\cos A}{2}} \qquad \sin\frac{A}{2} = \pm\sqrt{\frac{1-\cos A}{2}}$$

$$\tan\frac{A}{2} = \pm\sqrt{\frac{1-\cos A}{1+\cos A}} \qquad \tan\frac{A}{2} = \frac{\sin A}{1+\cos A}$$

$$\tan\frac{A}{2} = \frac{1-\cos A}{\sin A}$$

In these identities, the plus or minus sign is selected according to the quadrant in which $A/2$ terminates. For example, if A represents an angle of 324°, then $A/2 =$ 162°, which lies in quadrant II. In quadrant II, $\cos A/2$ and $\tan A/2$ are negative, while $\sin A/2$ is positive.

To derive the identity for $\sin A/2$, start with the following double-angle identity for cosine.

$$\cos 2x = 1 - 2\sin^2 x$$

Then solve for $\sin x$.

$$2\sin^2 x = 1 - \cos 2x$$

$$\sin x = \pm\sqrt{\frac{1-\cos 2x}{2}}$$

Now let $2x = A$, so that $x = A/2$, and substitute into this last expression.

$$\sin\frac{A}{2} = \pm\sqrt{\frac{1-\cos A}{2}}$$

The identity for $\cos A/2$ is derived in a very similar way, starting with the double-angle identity $\cos 2x = 2\cos^2 x - 1$. Finally, the identity for $\tan A/2$ comes from the half-angle identities for sine and cosine.

$$\tan\frac{A}{2} = \frac{\pm\sqrt{\dfrac{1-\cos A}{2}}}{\pm\sqrt{\dfrac{1+\cos A}{2}}} = \pm\sqrt{\frac{1-\cos A}{1+\cos A}}$$

The other two identities for $\tan A/2$ given above are more difficult to derive. See Exercise 62 at the end of this section.

Example 1 Find tan 22.5°.

Since 22.5° = 45°/2, find tan 22.5° with a half-angle identity for tangent. The result is positive since 22.5° is in the first quadrant.

$$\tan 22.5° = \tan \frac{45°}{2}$$
$$= \sqrt{\frac{1 - \cos 45°}{1 + \cos 45°}}$$
$$= \sqrt{\frac{1 - \frac{\sqrt{2}}{2}}{1 + \frac{\sqrt{2}}{2}}}$$
$$= \sqrt{\frac{2 - \sqrt{2}}{2 + \sqrt{2}}}$$

Now rationalize the denominator.

$$\tan 22.5° = \sqrt{\frac{2 - \sqrt{2}}{2 + \sqrt{2}}}$$
$$= \sqrt{\frac{(2 - \sqrt{2})(2 - \sqrt{2})}{(2 + \sqrt{2})(2 - \sqrt{2})}}$$
$$= \sqrt{\frac{4 - 4\sqrt{2} + 2}{4 - 2}} = \sqrt{\frac{6 - 4\sqrt{2}}{2}} = \sqrt{3 - 2\sqrt{2}}$$

Tan 22.5° could also be found with one of the other identities given above, such as

$$\tan \frac{A}{2} = \frac{\sin A}{1 + \cos A}.$$

Replacing A with 45° gives

$$\tan 22.5° = \tan \frac{45°}{2} = \frac{\sin 45°}{1 + \cos 45°} = \frac{\frac{\sqrt{2}}{2}}{1 + \frac{\sqrt{2}}{2}}.$$

Now multiply numerator and denominator by 2.

$$\tan 22.5° = \frac{\sqrt{2}}{2 + \sqrt{2}}$$

Verify that rationalizing the denominator gives $\sqrt{2} - 1$. Two values have now been found for tan 22.5°, $\sqrt{3 - 2\sqrt{2}}$ and $\sqrt{2} - 1$. In Exercise 61 you are asked to show that these values are equal. ✣

Example 2 Simplify each of the following.

(a) $\pm\sqrt{\dfrac{1 + \cos 12x}{2}}$

Start with the identity for $\cos A/2$,

$$\cos \frac{A}{2} = \pm\sqrt{\frac{1 + \cos A}{2}},$$

and replace A with $12x$ to get

$$\pm\sqrt{\frac{1 + \cos 12x}{2}} = \cos \frac{12x}{2} = \cos 6x.$$

(b) $\dfrac{1 - \cos 5\alpha}{\sin 5\alpha}$

Use the third identity for $\tan A/2$ given above to get

$$\frac{1 - \cos 5\alpha}{\sin 5\alpha} = \tan \frac{5\alpha}{2}.$$ ✣

Example 3 Given $\cos s = 2/3$, with s terminating in quadrant IV, find $\cos s/2$, $\sin s/2$, and $\tan s/2$.

Since s terminates in quadrant IV,

$$\frac{3\pi}{2} < s < 2\pi.$$

Dividing through by 2 gives

$$\frac{3\pi}{4} < \frac{s}{2} < \pi,$$

showing that $s/2$ terminates in quadrant II. In this quadrant the values of $\cos s/2$ and $\tan s/2$ are negative, and the value of $\sin s/2$ is positive. Use the appropriate half-angle identities to get

$$\sin \frac{s}{2} = \sqrt{\frac{1 - \frac{2}{3}}{2}} = \sqrt{\frac{1}{6}} = \frac{\sqrt{6}}{6};$$

$$\cos \frac{s}{2} = -\sqrt{\frac{1 + \frac{2}{3}}{2}} = -\sqrt{\frac{5}{6}} = -\frac{\sqrt{30}}{6}; \text{ and}$$

$$\tan \frac{s}{2} = \frac{\frac{\sqrt{6}}{6}}{-\frac{\sqrt{30}}{6}} = -\frac{\sqrt{5}}{5}.$$ ✣

Example 4 Show that

$$\left(\sin\frac{x}{2} + \cos\frac{x}{2}\right)^2 = 1 + \sin x$$

is an identity.

Work on the left.

$$\left(\sin\frac{x}{2} + \cos\frac{x}{2}\right)^2$$

$$= \sin^2\frac{x}{2} + 2\sin\frac{x}{2}\cos\frac{x}{2} + \cos^2\frac{x}{2}$$

$$= 1 + 2\sin\frac{x}{2}\cos\frac{x}{2} \qquad \sin^2\frac{x}{2} + \cos^2\frac{x}{2} = 1$$

$$= 1 + \sin 2\left(\frac{x}{2}\right) \qquad 2\sin\frac{x}{2}\cos\frac{x}{2} = \sin 2\left(\frac{x}{2}\right)$$

$$= 1 + \sin x$$ ✥

5.6 Exercises

Use an identity to write each of the following as a single trigonometric function. See Example 2.

1. $\sqrt{\dfrac{1 - \cos 40°}{2}}$
2. $\sqrt{\dfrac{1 + \cos 76°}{2}}$
3. $\sqrt{\dfrac{1 - \cos 147°}{1 + \cos 147°}}$
4. $\sqrt{\dfrac{1 + \cos 165°}{1 - \cos 165°}}$
5. $\dfrac{1 - \cos 59.74°}{\sin 59.74°}$
6. $\dfrac{\sin 158.2°}{1 + \cos 158.2°}$
7. $\pm\sqrt{\dfrac{1 + \cos 18x}{2}}$
8. $\pm\sqrt{\dfrac{1 + \cos 20\alpha}{2}}$
9. $\pm\sqrt{\dfrac{1 - \cos 8\theta}{1 + \cos 8\theta}}$
10. $\pm\sqrt{\dfrac{1 - \cos 5A}{1 + \cos 5A}}$
11. $\pm\sqrt{\dfrac{1 + \cos x/4}{2}}$
12. $\pm\sqrt{\dfrac{1 - \cos 3\theta/5}{2}}$

Determine whether the positive or negative square root should be selected for each of the following.

13. $\sin 195° = \pm\sqrt{\dfrac{1 - \cos 390°}{2}}$
14. $\cos 58° = \pm\sqrt{\dfrac{1 + \cos 116°}{2}}$
15. $\tan 225° = \pm\sqrt{\dfrac{1 - \cos 450°}{1 + \cos 450°}}$
16. $\sin(-10°) = \pm\sqrt{\dfrac{1 - \cos(-20°)}{2}}$

Use the identities in this section to find the sine, cosine, and tangent for each of the following. See Examples 1 and 3.

17. $\theta = 22.5°$

18. $\theta = 15°$

19. $\theta = 195°$

20. $\theta = 67.5°$

21. $x = -\pi/8$

22. $x = -5\pi/6$

23. $x = 5\pi/2$

24. $x = 3\pi/2$

Find each of the following. See Example 3.

25. $\cos \theta/2$, given $\cos \theta = 1/4$ and θ terminates in quadrant I

26. $\sin \theta/2$, given $\cos \theta = -5/8$ and θ terminates in quadrant II

27. $\tan \theta/2$, given $\sin \theta = 3/5$ and θ terminates in quadrant II

28. $\cos \theta/2$, given $\sin \theta = -1/5$ and θ terminates in quadrant III

29. $\sin \alpha/2$, given $\tan \alpha = 2$ and α terminates in quadrant I

30. $\cos \alpha/2$, given $\cot \alpha = -3$ and α terminates in quadrant II

31. $\tan \beta/2$, given $\tan \beta = \sqrt{7}/3$ and β terminates in quadrant III

32. $\cot \beta/2$, given $\tan \beta = -\sqrt{5}/2$ and β terminates in quadrant II

33. $\sin \theta$, given $\cos 2\theta = 3/5$ and θ terminates in quadrant I

34. $\cos \theta$ given $\cos 2\theta = 1/2$ and θ terminates in quadrant II

35. $\cos x$, given $\cos 2x = -5/12$ and $\pi/2 < x < \pi$

36. $\sin x$, given $\cos 2x = 2/3$ and $\pi < x < 3\pi/2$

Verify that each of the following equations is an identity. See Example 4.

37. $\sec^2 \dfrac{x}{2} = \dfrac{2}{1 + \cos x}$

38. $\cot^2 \dfrac{x}{2} = \dfrac{(1 + \cos x)^2}{\sin^2 x}$

39. $\sin^2 \dfrac{x}{2} = \dfrac{\tan x - \sin x}{2 \tan x}$

40. $\dfrac{\sin 2x}{2 \sin x} = \cos^2 \dfrac{x}{2} - \sin^2 \dfrac{x}{2}$

41. $\dfrac{2}{1 + \cos x} - \tan^2 \dfrac{x}{2} = 1$

42. $\tan \dfrac{\theta}{2} = \dfrac{\sin \theta}{1 + \cos \theta}$

43. $\tan \dfrac{\alpha}{2} = \dfrac{1 - \cos \alpha}{\sin \alpha}$

44. $\tan \dfrac{\gamma}{2} = \csc \gamma - \cot \gamma$

45. $\dfrac{\tan \dfrac{x}{2} + \cot \dfrac{x}{2}}{\cot \dfrac{x}{2} - \tan \dfrac{x}{2}} = \sec x$

46. $1 - \tan^2 \dfrac{\theta}{2} = \dfrac{2 \cos \theta}{1 + \cos \theta}$

47. $\cos x = \dfrac{1 - \tan^2 \dfrac{x}{2}}{1 + \tan^2 \dfrac{x}{2}}$

48. $\dfrac{\sin 2\alpha - 2 \sin \alpha}{2 \sin \alpha + \sin 2\alpha} = -\tan^2 \dfrac{\alpha}{2}$

49. $8 \sin^2 \dfrac{\gamma}{2} \cos^2 \dfrac{\gamma}{2} = 1 - \cos 2\gamma$

50. $\cos^2 \dfrac{x}{2} = \dfrac{1 + \sec x}{2 \sec x}$

An airplane flying faster than sound sends out sound waves that form a cone, as shown in the figure. The cone intersects the ground to form a hyperbola. As this hyperbola passes over a particular point on the ground, a sonic boom is heard at that point.

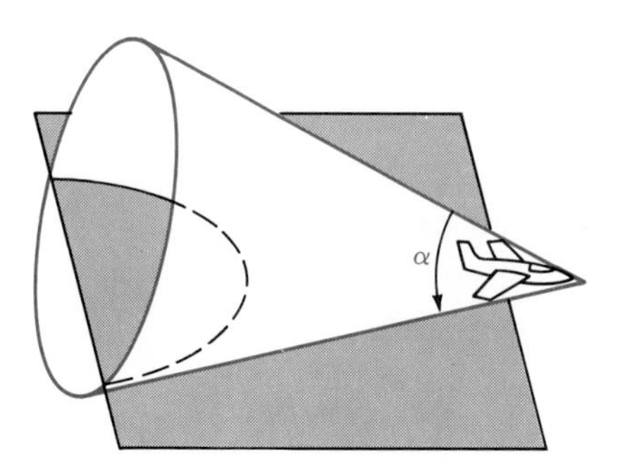

If α *is the angle at the vertex of the cone, then*

$$\sin\frac{\alpha}{2} = \frac{1}{m},$$

where m *is the Mach number for the speed of the plane. (We assume* $m > 1$.*) The Mach number is the ratio of the speed of the plane and the speed of sound. Thus, a speed of Mach 1.4 means that the plane is flying at 1.4 times the speed of sound. Find* α *or* m*, as necessary, for each of the following.*

51. $m = 3/2$ **52.** $m = 5/4$ **53.** $m = 2$ **54.** $m = 5/2$

55. $\alpha = 30°$ **56.** $\alpha = 60°$

Let $\cos s = -0.592147$*, with* $\pi < s < 3\pi/2$*. Use the identities in this section to find each of the following.*

57. $\sin\frac{1}{2}s$ **58.** $\cos\frac{1}{2}s$ **59.** $\tan\frac{1}{2}s$ **60.** $\csc\frac{1}{2}s$

61. In Example 1, the identity

$$\tan\frac{A}{2} = \pm\sqrt{\frac{(1 - \cos A)}{(1 + \cos A)}}$$

was used to find $\tan 22.5° = \sqrt{3 - 2\sqrt{2}}$, and then the identity

$$\tan\frac{A}{2} = \frac{\sin A}{(1 + \cos A)}$$

was used to get $\tan 22.5° = \sqrt{2} - 1$. Show that these answers are the same.

62. Go through the following steps to prove that

$$\tan\frac{A}{2} = \frac{\sin A}{1 + \cos A}.$$

(a) Start with $\tan A/2 = \pm\sqrt{(1 - \cos A)/(1 + \cos A)}$, and multiply numerator and denomintor by $\sqrt{1 + \cos A}$ to show that

$$\tan \frac{A}{2} = \pm \left| \frac{\sin A}{1 + \cos A} \right|.$$

(b) Show that $1 + \cos A \geq 0$, giving

$$\tan \frac{A}{2} = \frac{\pm|\sin A|}{1 + \cos A}.$$

(c) By considering the quadrant in which A lies, show that $\tan A/2$ and $\sin A$ have the same sign, with

$$\tan \frac{A}{2} = \frac{\sin A}{1 + \cos A}.$$

5.7 *Sum and Product Identities*

While the identities in this section are used less frequently than those in the preceding sections, they are important in further work in mathematics. One group of identities can be used to rewrite a product of two functions as a sum or difference. The other group can be used to rewrite a sum or difference of two functions as a product. Some of these identities can also be used to rewrite an expression involving both sine and cosine functions as an expression with only one of these functions. In Sections 6.3 and 6.4 on conditional equations, the need for this kind of change will become clear.

All of the identities in this section result from the sine and cosine of the sum and difference identities. For example, adding the two identities for $\sin (A + B)$ and $\sin (A - B)$ gives

$$\begin{aligned} \sin (A + B) &= \sin A \cos B + \cos A \sin B \\ \sin (A - B) &= \sin A \cos B - \cos A \sin B \\ \hline \sin (A + B) + \sin (A - B) &= 2 \sin A \cos B, \end{aligned}$$

or

$$\sin A \cos B = \frac{1}{2} [\sin (A + B) + \sin (A - B)].$$

Subtract $\sin (A - B)$ from $\sin (A + B)$ to get

$$\cos A \sin B = \frac{1}{2} [\sin (A + B) - \sin (A - B)].$$

Use the identities for $\cos(A + B)$ and $\cos(A - B)$ in a similar manner to get

$$\cos A \cos B = \frac{1}{2}[\cos(A + B) + \cos(A - B)]$$

$$\sin A \sin B = \frac{1}{2}[\cos(A - B) - \cos(A + B)].$$

Example 1 Rewrite $\cos 2\theta \sin \theta$ as the sum or difference of two functions.

Use the identity for $\cos A \sin B$.

$$\begin{aligned} \cos 2\theta \sin \theta &= \frac{1}{2}[\sin(2\theta + \theta) - \sin(2\theta - \theta)] \\ &= \frac{1}{2}(\sin 3\theta - \sin \theta) \\ &= \frac{1}{2}\sin 3\theta - \frac{1}{2}\sin \theta \end{aligned}$$

Example 2 Use an identity to evaluate $\cos 15° \cos 45°$.

Use the identity for $\cos A \cos B$.

$$\begin{aligned} \cos 15° \cos 45° &= \frac{1}{2}[\cos(15° + 45°) + \cos(15° - 45°)] \\ &= \frac{1}{2}[\cos 60° + \cos(-30°)] \\ &= \frac{1}{2}(\cos 60° + \cos 30°) \\ &= \frac{1}{2}\left(\frac{1}{2} + \frac{\sqrt{3}}{2}\right) = \frac{1 + \sqrt{3}}{4} \end{aligned}$$

Now these new identities can be used to obtain identities that are useful in calculus when a sum of trigonometric functions must be written as a product. To begin, let $A + B = x$ and let $A - B = y$. Then adding x and y gives $x + y = (A + B) + (A - B) = 2A$, or

$$A = \frac{x + y}{2}.$$

Finding $x - y$ gives $x - y = (A + B) - (A - B) = 2B$, or

$$B = \frac{x - y}{2}.$$

With these results, the identity

$$\sin A \cos B = \frac{1}{2}[\sin(A + B) + \sin(A - B)]$$

becomes

$$\sin\left(\frac{x + y}{2}\right)\cos\left(\frac{x - y}{2}\right) = \frac{1}{2}(\sin x + \sin y),$$

or

$$\sin x + \sin y = 2\sin\left(\frac{x + y}{2}\right)\cos\left(\frac{x - y}{2}\right).$$

Three other identities are obtained in a very similar way.

$$\sin x - \sin y = 2\cos\left(\frac{x + y}{2}\right)\sin\left(\frac{x - y}{2}\right)$$

$$\cos x + \cos y = 2\cos\left(\frac{x + y}{2}\right)\cos\left(\frac{x - y}{2}\right)$$

$$\cos x - \cos y = -2\sin\left(\frac{x + y}{2}\right)\sin\left(\frac{x - y}{2}\right)$$

Example 3 Write $\sin 2\gamma - \sin 4\gamma$ as a product of two functions.

Use the identity for $\sin x - \sin y$.

$$\begin{aligned}\sin 2\gamma - \sin 4\gamma &= 2\cos\left(\frac{2\gamma + 4\gamma}{2}\right)\sin\left(\frac{2\gamma - 4\gamma}{2}\right)\\ &= 2\cos\frac{6\gamma}{2}\sin\frac{-2\gamma}{2}\\ &= 2\cos 3\gamma\sin(-\gamma)\\ &= -2\cos 3\gamma\sin\gamma\end{aligned}$$

Example 4 Verify that the equation $\dfrac{\sin 3s + \sin s}{\cos s + \cos 3s} = \tan 2s$ is an identity.

Work as follows.

$$\frac{\sin 3s + \sin s}{\cos s + \cos 3s} = \frac{2\sin\left(\frac{3s + s}{2}\right)\cos\left(\frac{3s - s}{2}\right)}{2\cos\left(\frac{s + 3s}{2}\right)\cos\left(\frac{s - 3s}{2}\right)}$$

$$= \frac{\sin 2s \cos s}{\cos 2s \cos(-s)}$$

$$= \frac{\sin 2s}{\cos 2s}$$

$$= \tan 2s$$

5.7 Exercises

Rewrite each of the following as a sum or difference of trigonometric functions. See Examples 1 and 2.

1. $\cos 45° \sin 25°$
2. $2 \sin 74° \cos 114°$
3. $3 \cos 5x \cos 3x$
4. $2 \sin 2x \sin 4x$
5. $\sin(-\theta) \sin(-3\theta)$
6. $4 \cos 8\alpha \sin(-4\alpha)$
7. $-8 \cos 4y \cos 5y$
8. $2 \sin 3k \sin 14k$

Rewrite each of the following as a product of trigonometric functions. See Example 3.

9. $\sin 60° - \sin 30°$
10. $\sin 28° + \sin(-18°)$
11. $\cos 42° + \cos 148°$
12. $\cos 2x - \cos 8x$
13. $\sin 12\beta - \sin 3\beta$
14. $\cos 5x + \cos 10x$
15. $-3 \sin 2x + 3 \sin 5x$
16. $-\cos 8s + \cos 14s$

Verify that each of the following is an identity. See Example 4.

17. $\tan x = \dfrac{\sin 3x - \sin x}{\cos 3x + \cos x}$

18. $\dfrac{\sin 5t + \sin 3t}{\cos 3t - \cos 5t} = \cot t$

19. $\dfrac{\cot 2\theta}{\tan 3\theta} = \dfrac{\cos 5\theta + \cos \theta}{\cos \theta - \cos 5\theta}$

20. $\dfrac{\cos \alpha + \cos \beta}{\cos \alpha - \cos \beta} = -\cot\left(\dfrac{\alpha + \beta}{2}\right) \cot\left(\dfrac{\alpha - \beta}{2}\right)$

21. $\dfrac{1}{\tan 2s} = \dfrac{\sin 3s - \sin s}{\cos s - \cos 3s}$

22. $\dfrac{\sin^2 5\alpha - 2 \sin 5\alpha \sin 3\alpha + \sin^2 3\alpha}{\sin^2 5\alpha - \sin^2 3\alpha} = \dfrac{\tan \alpha}{\tan 4\alpha}$

23. $\sin 6\theta \cos 4\theta - \sin 3\theta \cos 7\theta = \sin 3\theta \cos \theta$

24. $\sin 8\beta \sin 4\beta + \cos 10\beta \cos 2\beta = \cos 6\beta \cos 2\beta$

25. $\sin^2 u - \sin^2 v = \sin(u + v) \sin(u - v)$

26. $\cos^2 u - \cos^2 v = -\sin(u + v) \sin(u - v)$

27. Show that the double-angle identity for sine can be considered a special case of the identity $\sin s \cos t = (1/2)[\sin(s + t) + \sin(s - t)]$.

28. Show that the double-angle identity $\cos 2s = 2 \cos^2 s - 1$ is a special case of the identity $\cos s \cos t = (1/2)[\cos(s + t) + \cos(s - t)]$.

5.8 Reduction of $a \sin \theta \pm b \cos \theta$ to $k \sin(\theta \pm \alpha)$ (Optional)

The expressions $a \sin \theta + b \cos \theta$ and $a \sin \theta - b \cos \theta$ occur so frequently that it is useful to have a way of rewriting them in a simpler form. To find this form, rewrite $a \sin \theta + b \cos \theta$ as follows. (The use of $\sqrt{a^2 + b^2}$ comes from the Pythagorean formula, as we shall see.)

$$a \sin \theta + b \cos \theta = \frac{\sqrt{a^2 + b^2}}{\sqrt{a^2 + b^2}}(a \sin \theta + b \cos \theta)$$

$$= \sqrt{a^2 + b^2}\left(\frac{a}{\sqrt{a^2 + b^2}} \sin \theta + \frac{b}{\sqrt{a^2 + b^2}} \cos \theta\right)$$

$$= \sqrt{a^2 + b^2}\left(\sin \theta \frac{a}{\sqrt{a^2 + b^2}} + \cos \theta \frac{b}{\sqrt{a^2 + b^2}}\right) \quad \textbf{(1)}$$

Now choose an angle α so that

$$\sin \alpha = \frac{b}{\sqrt{a^2 + b^2}} \quad \text{and} \quad \cos \alpha = \frac{a}{\sqrt{a^2 + b^2}}.$$

See Figure 5.5.

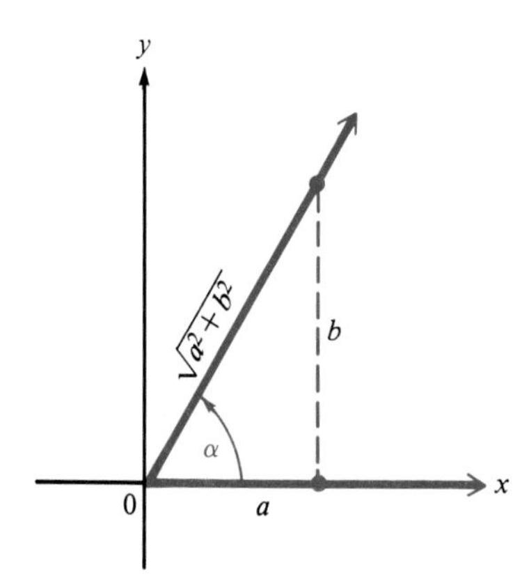

Figure 5.5

Substitute the values of $\sin \alpha$ and $\cos \alpha$ into equation (1).

$$a \sin \theta + b \cos \theta = \sqrt{a^2 + b^2}\,(\sin \theta \,\mathbf{\cos \alpha} + \cos \theta \,\mathbf{\sin \alpha})$$

Using the identity for $\sin (A + B)$,

$$\boldsymbol{a \sin \theta + b \cos \theta = \sqrt{a^2 + b^2} \sin (\theta + \alpha).}$$

This identity is called the **reduction identity.** In summary,

Reduction Identity

$$\boldsymbol{a \sin \theta + b \cos \theta = \sqrt{a^2 + b^2} \sin (\theta + \alpha),}$$

$$\text{where } \sin \alpha = \frac{b}{\sqrt{a^2 + b^2}} \quad \text{and} \quad \cos \alpha = \frac{a}{\sqrt{a^2 + b^2}}.$$

Example 1 Rewrite $\frac{1}{2} \sin \theta + \frac{\sqrt{3}}{2} \cos \theta$ using the reduction identity given above.

From the identity above, $a = \frac{1}{2}$ and $b = \frac{\sqrt{3}}{2}$, so that

$$a \sin \theta + b \cos \theta = \sqrt{a^2 + b^2} \sin (\theta + \alpha)$$

becomes

$$\frac{1}{2} \sin \theta + \frac{\sqrt{3}}{2} \cos \theta = 1 \cdot \sin (\theta + \alpha),$$

where angle α satisfies the conditions

$$\sin \alpha = \frac{b}{\sqrt{a^2 + b^2}} = \frac{\sqrt{3}}{2} \quad \text{and} \quad \cos \alpha = \frac{a}{\sqrt{a^2 + b^2}} = \frac{1}{2}.$$

The smallest possible positive value of α that satisfies both of these conditions is $\alpha = 60°$. Thus,

$$\frac{1}{2} \sin \theta + \frac{\sqrt{3}}{2} \cos \theta = \sin (\theta + 60°).$$ ✥

Example 2 Express $2 \sin (\theta - 60°)$ as $a \sin \theta + b \cos \theta$, where a and b are constants.

First write

$$2 \sin (\theta - 60°) = 2 \sin [\theta + (-60°)].$$

From the reduction identity,

$$\sin (-60°) = \frac{b}{\sqrt{a^2 + b^2}} = \frac{b}{2},$$

and

$$\cos (-60°) = \frac{a}{\sqrt{a^2 + b^2}} = \frac{a}{2}.$$

Since $\sin (-60°) = b/2$, and since $\sin (-60°) = -\sin 60°$,

$$\sin 60° = -\frac{b}{2}$$

$$\frac{\sqrt{3}}{2} = -\frac{b}{2}$$

$$b = -\sqrt{3}.$$

In the same way, verify that $a = 1$. Thus,

$$2 \sin (\theta - 60°) = \sin \theta - \sqrt{3} \cos \theta.$$ ✥

The method used in Example 2 is not the best method for working this problem; using the identity for $\sin (A - B)$ would have been faster. However, we show this method to illustrate the reduction identity.

The reduction identity in this section is useful when graphing functions that are the sums of sine and cosine functions. It can be used instead of the method of addition of ordinates discussed in chapter 4.

Example 3 Graph $y = \sin x + \cos x$.

Rewrite $\sin x + \cos x$ as follows. Since $a = b = 1$, $\sqrt{a^2 + b^2} = \sqrt{2}$, and

$$\sin x + \cos x = \sqrt{2} \sin (x + \alpha).$$

To find α, let

$$\sin \alpha = \frac{b}{\sqrt{a^2 + b^2}} = \frac{1}{\sqrt{2}} \quad \text{and} \quad \cos \alpha = \frac{a}{\sqrt{a^2 + b^2}} = \frac{1}{\sqrt{2}}.$$

The smallest positive angle satisfying these conditions is $\alpha = \pi/4$, with

$$y = \sin x + \cos x = \sqrt{2} \sin \left(x + \frac{\pi}{4}\right).$$

The graph of this function has an amplitude of $\sqrt{2}$, a period of 2π, and a phase shift of $\pi/4$ to the left, as shown in Figure 5.6 ✥

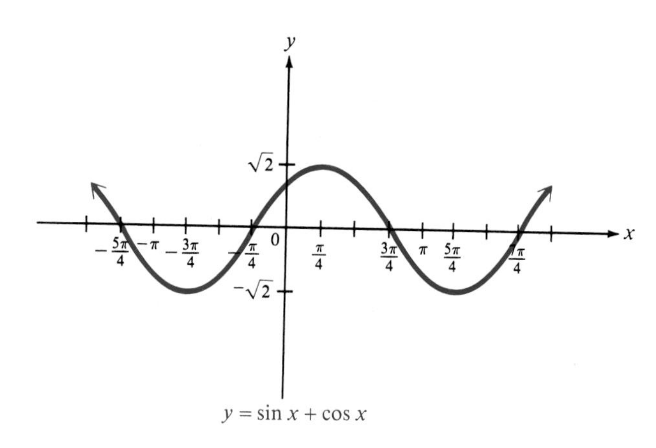

Figure 5.6

5.8 Exercises

Use the reduction identity to simplify each of the following for angles between 0° and 360°. Find angles to the nearest degree. Choose the smallest possible positive value of α. See Example 1.

1. $-\sin x + \cos x$
2. $\sqrt{3} \sin x - \cos x$
3. $5 \sin \theta - 12 \cos \theta$
4. $12 \sin A + 5 \cos A$
5. $-15 \sin x + 8 \cos x$
6. $15 \sin B - 8 \cos B$
7. $-7 \sin \theta - 24 \cos \theta$
8. $24 \sin t - 7 \cos t$
9. $3 \sin x + 4 \cos x$
10. $-4 \sin x + 3 \cos x$
11. $\sqrt{5} \sin x + 2 \cos x$
12. $3 \sin x - \sqrt{7} \cos x$

Use the reduction identity to express each of the following as a $\sin\theta + b\cos\theta$. See Example 2.

13. $2\sin(\theta + 120°)$

14. $2\sin(\theta - 30°)$

15. $\sqrt{2}\sin\left(\theta - \frac{\pi}{4}\right)$

16. $\sqrt{2}\sin\left(\theta + \frac{3\pi}{4}\right)$

17. $\sin(\theta + 90°)$

18. $\sin(\theta - 90°)$

Graph each of the following by first changing the function to a sine function. See Example 3.

19. $y = \sqrt{3}\sin x + \cos x$

20. $y = \sin x - \sqrt{3}\cos x$

21. $y = -\sin x + \cos x$

22. $y = -\sin x - \cos x$

23. $y = \sqrt{3}\sin 2x + \cos 2x$

24. $y = \sqrt{3}\sin 2x - \cos 2x$

Chapter 5 *Key Concepts*

Fundamental Identities

Reciprocal identities

$$\cot \theta = \frac{1}{\tan \theta} \qquad \sec \theta = \frac{1}{\cos \theta} \qquad \csc \theta = \frac{1}{\sin \theta}$$

Quotient identities

$$\tan \theta = \frac{\sin \theta}{\cos \theta} \qquad \cot \theta = \frac{\cos \theta}{\sin \theta}$$

Pythagorean identities

$$\sin^2 \theta + \cos^2 \theta = 1 \qquad \tan^2 \theta + 1 = \sec^2 \theta \qquad 1 + \cot^2 \theta = \csc^2 \theta$$

Negative-angle identities

$$\sin(-\theta) = -\sin \theta \qquad \cos(-\theta) = \cos \theta \qquad \tan(-\theta) = -\tan \theta$$

Double-Angle Identities

$$\cos 2A = \cos^2 A - \sin^2 A$$

$$\cos 2A = 1 - 2\sin^2 A$$

$$\cos 2A = 2\cos^2 A - 1$$

$$\sin 2A = 2 \sin A \cos A$$

$$\tan 2A = \frac{2 \tan A}{1 - \tan^2 A}$$

Half-Angle Identities

$$\sin \frac{A}{2} = \pm\sqrt{\frac{1 - \cos A}{2}}$$

$$\cos \frac{A}{2} = \pm\sqrt{\frac{1 + \cos A}{2}}$$

$$\tan \frac{A}{2} = \pm\sqrt{\frac{1 - \cos A}{1 + \cos A}}$$

$$\tan \frac{A}{2} = \frac{1 - \cos A}{\sin A}$$

$$\tan \frac{A}{2} = \frac{\sin A}{1 + \cos A}$$

Chapter 5 *Key Concepts (continued)*

Sum or Difference Identities

$$\cos(A - B) = \cos A \cos B + \sin A \sin B$$
$$\cos(A + B) = \cos A \cos B - \sin A \sin B$$
$$\sin(A + B) = \sin A \cos B + \cos A \sin B$$
$$\sin(A - B) = \sin A \cos B - \cos A \sin B$$
$$\tan(A + B) = \frac{\tan A + \tan B}{1 - \tan A \tan B}$$
$$\tan(A - B) = \frac{\tan A - \tan B}{1 + \tan A \tan B}$$

Sum and Product Identities

$$\sin A \cos B = \frac{1}{2}[\sin(A + B) + \sin(A - B)]$$
$$\cos A \sin B = \frac{1}{2}[\sin(A + B) - \sin(A - B)]$$
$$\cos A \cos B = \frac{1}{2}[\cos(A + B) + \cos(A - B)]$$
$$\sin A \sin B = \frac{1}{2}[\cos(A - B) - \cos(A + B)]$$
$$\sin x + \sin y = 2 \sin\left(\frac{x + y}{2}\right) \cos\left(\frac{x - y}{2}\right)$$
$$\sin x - \sin y = 2 \cos\left(\frac{x + y}{2}\right) \sin\left(\frac{x - y}{2}\right)$$
$$\cos x + \cos y = 2 \cos\left(\frac{x + y}{2}\right) \cos\left(\frac{x - y}{2}\right)$$
$$\cos x - \cos y = -2 \sin\left(\frac{x + y}{2}\right) \sin\left(\frac{x - y}{2}\right)$$

Chapter 5 *Review Exercises*

1. Use the trigonometric identities to find the remaining five trigonometric functions of x, given that $\cos x = 3/5$ and x is in quadrant IV.
2. Given $\tan x = -5/4$, where $\pi/2 < x < \pi$, use trigonometric identities to find the other trigonometric functions of x.
3. Given $x = \pi/8$, use trigonometric identities to find $\sin x$, $\cos x$, and $\tan x$.
4. Given $\sin 2\theta = \sqrt{3}/2$ and 2θ terminates in quadrant II, use trigonometric identities to find $\tan \theta$.

For each item in Column I, give the letter of the item in Column II that completes an identity.

Column I	*Column II*
5. $\cos 210°$	**(a)** $\sin(-35°)$
6. $\sin 35°$	**(b)** $\cos 55°$
7. $\tan(-35°)$	**(c)** $\sqrt{\dfrac{1 + \cos 150°}{2}}$
8. $-\sin 35°$	**(d)** $2 \sin 150° \cos 150°$
9. $\cos 35°$	**(e)** $\cos 150° \cos 60° - \sin 150° \sin 60°$
10. $\cos 75°$	**(f)** $\cot(-35°)$
11. $\sin 75°$	**(g)** $\cos^2 150° - \sin^2 150°$
12. $\sin 300°$	**(h)** $\sin 15° \cos 60° + \cos 15° \sin 60°$
13. $\cos 300°$	**(i)** $\cos(-35°)$
	(j) $\cot 125°$

For each item in Column I, give the letter of the item in Column II that completes an identity.

Column I	*Column II*
14. $\sec x$	**(a)** $\dfrac{1}{\sin x}$
15. $\csc x$	**(b)** $\dfrac{1}{\cos x}$
16. $\tan x$	**(c)** $\dfrac{\sin x}{\cos x}$
17. $\cot x$	**(d)** $\dfrac{1}{\cot^2 x}$
18. $\sin^2 x$	**(e)** $\dfrac{1}{\cos^2 x}$
19. $\tan^2 x + 1$	**(f)** $\dfrac{\cos x}{\sin x}$
20. $\tan^2 x$	**(g)** $\dfrac{1}{\sin^2 x}$
	(h) $1 - \cos^2 x$

Use identities to express each of the following in terms of $\sin \theta$ *and* $\cos \theta$, *and simplify.*

21. $\sec^2 \theta - \tan^2 \theta$

22. $\dfrac{\cot \theta}{\sec \theta}$

23. $\tan^2 \theta\,(1 + \cot^2 \theta)$

24. $\csc \theta + \cot \theta$

25. $\csc^2 \theta + \sec^2 \theta$

26. $\tan \theta - \sec \theta \csc \theta$

For each of the following find $\sin (x + y)$, $\cos (x - y)$, *and* $\tan (x + y)$.

27. $\sin x = -1/4$, $\cos y = -4/5$, x and y in quadrant III

28. $\sin y = -2/3$, $\cos x = -1/5$, x in quadrant II, y in quadrant III

29. $\sin x = 1/10$, $\cos y = 4/5$, x in quadrant I, y in quadrant IV

30. $\cos x = 2/9$, $\sin y = -1/2$, x in quadrant IV, y in quadrant III

Find sine and cosine of each of the following.

31. θ, given $\cos 2\theta = -3/4$, 2θ in quadrant II

32. B, given $\cos 2B = 1/8$, B in quadrant IV

33. $2x$, given $\tan x = 3$, $\sin x < 0$

34. $2y$, given $\sec y = -5/3$, $\sin y > 0$

Find each of the following.

35. $\cos \theta/2$, given $\cos \theta = -1/2$, θ in quadrant II

36. $\sin A/2$, given $\cos A = -3/4$, A in quadrant II

37. $\tan x$, given $\tan 2x = 2$, $\pi < x < 3\pi/2$

38. $\sin y$, given $\cos 2y = -1/3$, $\pi/2 < y < \pi$

Verify that each of the following equations is an identity.

39. $\sin^2 x - \sin^2 y = \cos^2 y - \cos^2 x$

40. $2 \cos^3 x - \cos x = \dfrac{\cos^2 x - \sin^2 x}{\sec x}$

41. $-\cot \dfrac{x}{2} = \dfrac{\sin 2x + \sin x}{\cos 2x - \cos x}$

42. $\dfrac{\sin^2 x}{2 - 2 \cos x} = \cos^2 \dfrac{x}{2}$

43. $\dfrac{\sin 2x}{\sin x} = \dfrac{2}{\sec x}$

44. $2 \cos A - \sec A = \cos A - \dfrac{\tan A}{\csc A}$

45. $\dfrac{2 \tan B}{\sin 2B} = \sec^2 B$

46. $\tan \beta = \dfrac{1 - \cos 2\beta}{\sin 2\beta}$

47. $1 + \tan^2 \alpha = 2 \tan \alpha \csc 2\alpha$

48. $-\dfrac{\sin (A - B)}{\sin (A + B)} = \dfrac{\cot A - \cot B}{\cot A + \cot B}$

49. $\dfrac{\sin t}{1 - \cos t} = \cot \dfrac{t}{2}$

50. $2 \cos (A + B) \sin (A + B) = \sin 2A \cos 2B + \sin 2B \cos 2A$

51. $\dfrac{2 \cot x}{\tan 2x} = \csc^2 x - 2$

52. $\sin t = \dfrac{\cos t \sin 2t}{1 + \cos 2t}$

53. $\tan \theta \sin 2\theta = 2 - 2 \cos^2 \theta$

54. $\csc A \sin 2A - \sec A = \cos 2A \sec A$

55. $2 \tan x \csc 2x - \tan^2 x = 1$

56. $2 \cos^2 \theta - 1 = \dfrac{1 - \tan^2 \theta}{1 + \tan^2 \theta}$

57. $\tan\theta\cos^2\theta = \dfrac{2\tan\theta\cos^2\theta - \tan\theta}{1 - \tan^2\theta}$

58. $-\cot\dfrac{x}{2} = \dfrac{\sin 2x + \sin x}{\cos 2x - \cos x}$

59. $2\cos^3 x - \cos x = \dfrac{\cos^2 x - \sin^2 x}{\sec x}$

60. $\sin^3\theta = \sin\theta - \cos^2\theta\sin\theta$

61. $\cos^4\theta = \dfrac{3}{8} + \dfrac{1}{2}\cos 2\theta + \dfrac{1}{8}\cos 4\theta$

62. $\tan\dfrac{7}{2}x = \dfrac{2\tan\dfrac{7}{4}x}{1 - \tan^2\dfrac{7}{4}x}$

63. $\sec^2\alpha - 1 = \dfrac{\sec 2\alpha - 1}{\sec 2\alpha + 1}$

64. $\dfrac{\sin 3t + \sin 2t}{\sin 3t - \sin 2t} = \dfrac{\tan\dfrac{5t}{2}}{\tan\dfrac{t}{2}}$

65. $\tan 4\theta = \dfrac{2\tan 2\theta}{2 - \sec^2 2\theta}$

66. $\sin 2\alpha = \dfrac{2(\sin\alpha - \sin^3\alpha)}{\cos\alpha}$

67. $2\cos^2\dfrac{x}{2}\tan x = \tan x + \sin x$

68. $\csc\theta - \cot\theta = \tan\dfrac{\theta}{2}$

69. $\cot\left(x + \dfrac{\pi}{2}\right) = \tan(x + \pi)$

70. $\dfrac{1}{2}\cot\dfrac{x}{2} - \dfrac{1}{2}\tan\dfrac{x}{2} = \cot x$

71. Exact values of the trigonometric functions of 15° can be found by the following method, an alternative to the use of the half-angle formulas. Start with a right triangle *ABC* having a 60° angle at *A* and a 30° angle at *B*. Let the hypotenuse of this triangle have length 2. Extend side *BC* and draw a semicircle with diameter along *BC* extended, center at *B*, and radius *AB*. Draw segment *AE*. (See the figure.) Since any angle inscribed in a semicircle is a right angle, triangle *AED* is a right triangle.

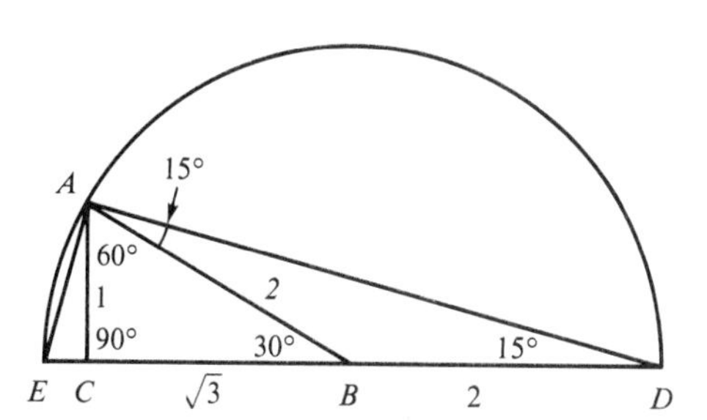

Prove each of the following statements.

(a) Triangle *ABD* is isosceles.

(b) Angle *ABD* is 150°.

(c) Angle *DAB* and angle *ADB* are each 15°.

(d) *DC* has length $2 + \sqrt{3}$.

(e) Since *AC* has length 1, the length of *AD* is $AD = \sqrt{1^2 + (2 + \sqrt{3})^2}$. Reduce this to $\sqrt{8 + 4\sqrt{3}}$, and show that this result equals $\sqrt{6} + \sqrt{2}$.

72. Use angle *ADB* of triangle *ADE* and find cos 15°.

73. Show that *AE* has length $\sqrt{6} - \sqrt{2}$. Then find sin 15°.

74. Use triangle *ACE* and find tan 15°.

Inverse Trigonometric Functions and Trigonometric Equations

In many applications of trigonometry, the sine (or other trigonometric function value) of an angle is known, and the angle itself must be found. This is done with *inverse trigonometric functions,* discussed in this chapter. We begin with a discussion of inverse functions in general, and then look at inverse trigonometric functions. The last part of the chapter shows how to solve *conditional equations* involving trigonometric functions or inverse trigonometric functions.

6.1 Inverse Functions

Addition and subtraction are inverse operations. If 8 is added to a number x and then 8 is subtracted from the sum, the result is x. Some functions are inverses, too. For example, if

$$f(x) = 3x \qquad \text{and} \qquad g(x) = \frac{1}{3}x,$$

then

$$f(12) = 3 \cdot 12 = 36$$

and

$$g(36) = \frac{1}{3}(36) = 12.$$

These steps show that $g(x)$ "undoes" what $f(x)$ did to the number 12. Not all functions have inverse functions. To have an inverse function, a function must be a *one-to-one function*.

Recall that a function is a relation in which each value of x leads to exactly one value of y. In the function $y = 5x + 4$, two different values of x produce two *different* values of y. But in the function $y = x^2$, two different values of x can lead to the *same* value of y. For example, choosing 4 for x leads to $y = 4^2 = 16$, while choosing -4 for x also leads to $y = (-4)^2 = 16$. The two different values of x, 4 and -4, lead to the same value of y, 16. A function such as $y = 5x + 4$, in which different values of x lead to different values of y, is called a **one-to-one function.** The next example shows how to decide if a function is one-to-one.

Example 1 Decide if the following functions are one-to-one functions.

(a) $y = 4x + 12$

Suppose $a \neq b$. Then $4a \neq 4b$, and $4a + 12 \neq 4b + 12$. Thus, if a and b are different values, then corresponding y-values are different, and so $y = 4x + 12$ is a one-to-one function.

(b) $y = \sqrt{25 - x^2}$

If two different x-values lead to the *same* y-value, then the function is not one-to-one. Here, if $x = 3$, then $y = \sqrt{25 - 3^2} = \sqrt{25 - 9} = \sqrt{16} = 4$. Choosing $x = -3$ also produces $y = 4$. Since two different values of x lead to the same value of y, this function is not one-to-one. ✤

As Example 1(b) suggests, functions with x raised to an even power generally are not one-to-one functions.

There is a useful graphical test that tells if a function is one-to-one. Figure 6.1 shows the graph of a function cut by a horizontal line. Each of the points at which the horizontal line cuts the graph has the same value of y, but a different value of x. In this graph three different values of x lead to the same value of y, showing that the function is not one-to-one. This idea suggests the **horizontal line test.**

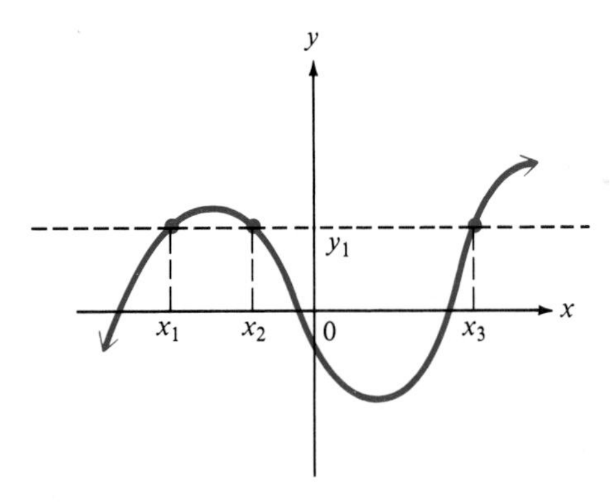

Figure 6.1

Horizontal Line Test

> If it is possible to draw a horizontal line that cuts the graph of a function in more than one point, then the function is not one-to-one.

Example 2 Is the function in Figure 6.2 a one-to-one function?

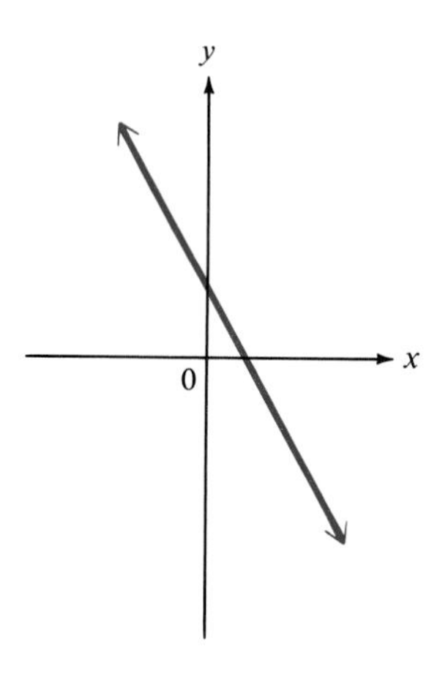

Figure 6.2

No horizontal line will cut this graph in more than one point. Therefore the function is one-to-one. ✥

One-to-one functions are important because a one-to-one function f has an *inverse function*. By definition, the inverse function of the one-to-one function f is found by interchanging the elements of the ordered pairs of f.

Inverse Function

> The **inverse function** of the one-to-one function f, written f^{-1} (read "f inverse"), is defined as
>
> $$f^{-1} = \{(y, x) | (x, y) \text{ belongs to } f\}.$$

Suppose $f = \{(x, y) | y = 3x + 1\}$. If $x = 2$, then $y = 7$, so the ordered pair $(2, 7)$ belongs to f. Also, $(4, 13)$ and $(-2, -5)$ belong to f. By the definition of an inverse function, $(7, 2)$, $(13, 4)$, and $(-5, -2)$ belong to f^{-1}. Since the elements of the ordered pairs of f and f^{-1} are reversed, the domain of f is the range of f^{-1} and the range of f is the domain of f^{-1}.

To get the equation of the inverse function, use the definition and exchange x and y in the equation for f. If $f = \{(x, y) | y = 3x + 1\}$, then f^{-1} can be found by first exchanging x and y in the equation $y = 3x + 1$. Doing this gives $x = 3y + 1$. Now solve for y.

$$x = 3y + 1$$
$$x - 1 = 3y$$
$$\frac{x - 1}{3} = y$$

This is the equation of the inverse function, so $f^{-1}(x)$ can be used to represent y:

$$f^{-1}(x) = \frac{x - 1}{3}.$$

In this example the domain and range of f are the set of real numbers, and so the range and domain of f^{-1} are the set of all real numbers, also.

The procedure for finding $f^{-1}(x)$, given $f(x)$, is summarized below.

Finding $f^{-1}(x)$

Given the one-to-one function $f(x)$, find $f^{-1}(x)$ as follows.

1. Replace $f(x)$ with y.
2. Exchange x and y in $y = f(x)$.
3. Solve the equation from Step 2 for y.
4. Replace y with $f^{-1}(x)$.

Example 3 For each of the following functions that is one-to-one, find the equation of its inverse function.

(a) $f(x) = \dfrac{4x + 6}{5}$

This is a linear function, with a line as its graph. By the horizontal line test, this function is one-to-one and thus has an inverse. To find the inverse, let $f(x) = y$ and then exchange x and y. Finally, solve for y.

$$y = \frac{4x + 6}{5} \qquad \text{Let } f(x) = y$$

$$x = \frac{4y + 6}{5} \qquad \text{Exchange } x \text{ and } y$$

$$5x = 4y + 6$$

$$y = \frac{5x - 6}{4} \qquad \text{Solve for } y$$

Replace y with f^{-1}. Thus, $f^{-1}(x) = \dfrac{5x - 6}{4}$.

(b) $f(x) = x^3 - 1$

This function is one-to-one since $x_1^3 - 1 \neq x_2^3 - 1$ for any two different values of x_1 and x_2. Therefore the function has an inverse. Begin with the equation $y = x^3 - 1$, and exchange x and y to get $x = y^3 - 1$. Solving for y gives $y = \sqrt[3]{x + 1}$, or $f^{-1}(x) = \sqrt[3]{x + 1}$.

(c) $f(x) = x^2$

Here, $2^2 = 4$ and $(-2)^2 = 4$. This function is not one-to-one, so it has no inverse function. ✥

The graph of an inverse can be obtained from the graph of a function. To see how, suppose that a point (a, b) is on the graph of a function f as in Figure 6.3. By the definition of the inverse function, the point (b, a) would belong to f^{-1}. The figure shows that the line segment connecting (a, b) and (b, a) is perpendicular to the line $y = x$ and is cut in half by $y = x$. Because of this, the points (a, b) and (b, a) are "mirror images" of each other with respect to the line $y = x$.

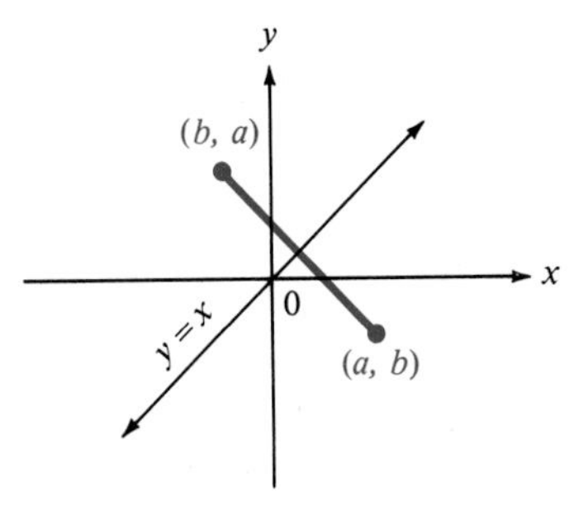

Figure 6.3

Using this idea, the graph of f^{-1} can be found by locating the mirror image of each point of f with respect to the line $y = x$. Figure 6.4 shows the graph of f and f^{-1} from Example 3(a) above. The graph of the function and its inverse from Example 3(b) above are shown in Figure 6.5. In each case, f and f^{-1} are mirror images of each other with respect to the line $y = x$.

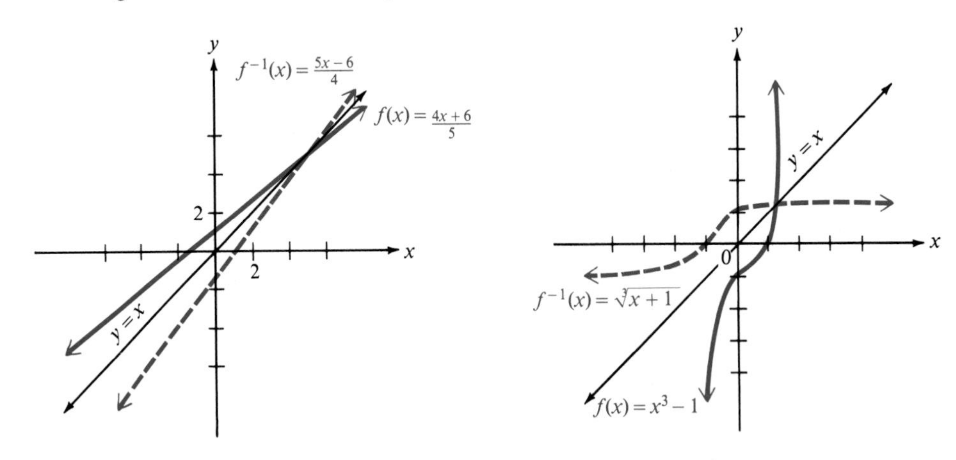

Figure 6.4

Figure 6.5

Given two functions, how can we tell whether or not they are inverses of each other? To find a method, look at Example 3(b) above, which showed that $f(x) = x^3 - 1$ has an inverse function $f^{-1}(x) = \sqrt[3]{x + 1}$. Now find $f[f^{-1}(x)]$.

$$\begin{aligned} f[f^{-1}(x)] &= f[\sqrt[3]{x + 1}] \\ &= (\sqrt[3]{x + 1})^3 - 1 \\ &= x + 1 - 1 = x \end{aligned}$$

Also,

$$\begin{aligned} f^{-1}[f(x)] &= f^{-1}[x^3 - 1] \\ &= \sqrt[3]{(x^3 - 1) + 1} \\ &= \sqrt[3]{x^3} = x. \end{aligned}$$

Since these functions "undo" one another, they are inverses. Generalizing from this example gives the following rule.

Test for Inverse Functions

Two one-to-one functions f and g are inverses of each other if and only if

$$f[g(x)] = x \quad \text{and} \quad g[f(x)] = x.$$

The phrase "if and only if" above means that two things are true: (1) if f and g are inverse functions of each other, then $f[g(x)] = x$ and $g[f(x)] = x$, and (2) if $f[g(x)] = x$ and $g[f(x)] = x$, then f and g are inverses of each other.

This rule can be used to show that two one-to-one functions are inverses.

Example 4 Decide whether $f(x) = 3x - 1$ and $g(x) = (x + 1)/3$ are inverses of each other.

First find $f[g(x)]$.

$$\begin{aligned} f[g(x)] &= f\left(\frac{x + 1}{3}\right) \\ &= 3\left(\frac{x + 1}{3}\right) - 1 \\ &= x + 1 - 1 = x \end{aligned}$$

Then find $g[f(x)]$.

$$\begin{aligned} g[f(x)] &= g(3x - 1) \\ &= \frac{(3x - 1) + 1}{3} \\ &= \frac{3x}{3} = x \end{aligned}$$

Since both $f[g(x)]$ and $g[f(x)]$ equal x, the functions are inverses of each other. ✣

6.1 Exercises

Which of the following functions are one-to-one? See Examples 1 and 2.

1.

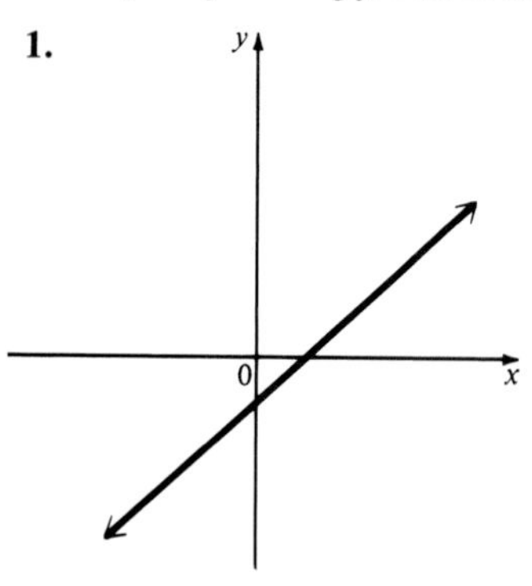

2.

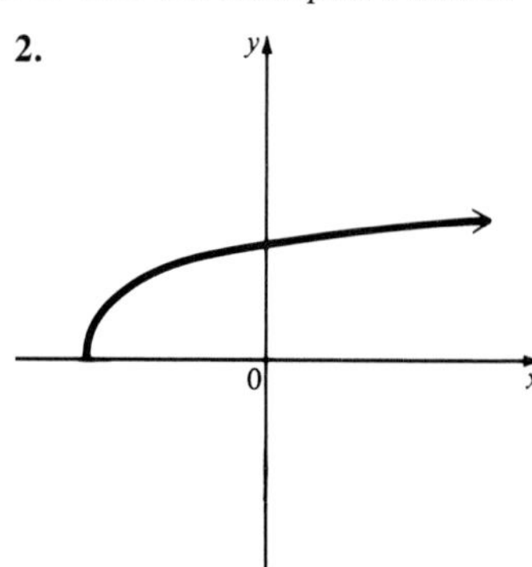

3.

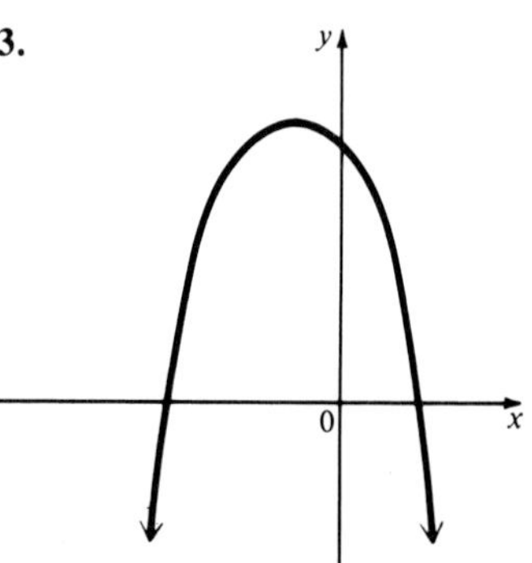

4. $y = 2x + 7$

5. $y = 3x - 4$

6. $y = -x^2$

7. $y = (x - 3)^2$

8. $y = \sqrt{x - 1}$

9. $y = \sqrt{5x + 1}$

10. $y = \dfrac{2}{x}$

11. $y = \dfrac{2}{3 - x}$

12. $y = \dfrac{x + 1}{4}$

In Exercises 13–18, if the function is one-to-one, sketch the graph of its inverse. See Figures 6.4 and 6.5.

13.

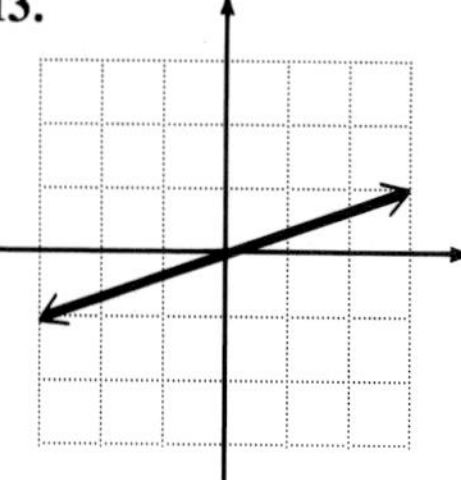

14.

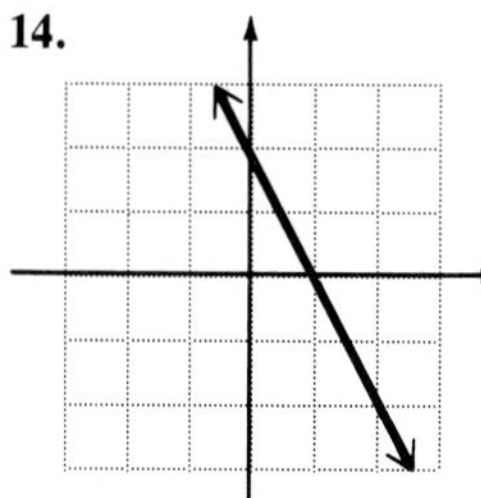

15.

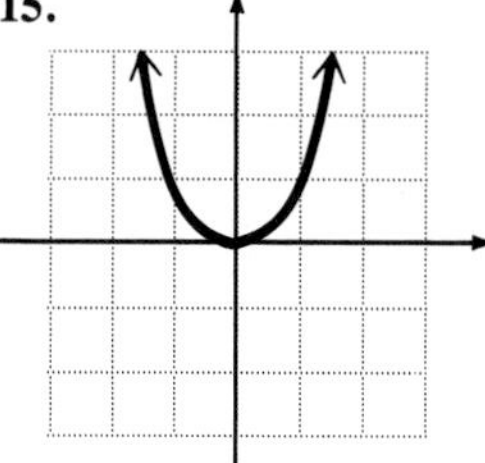

16.

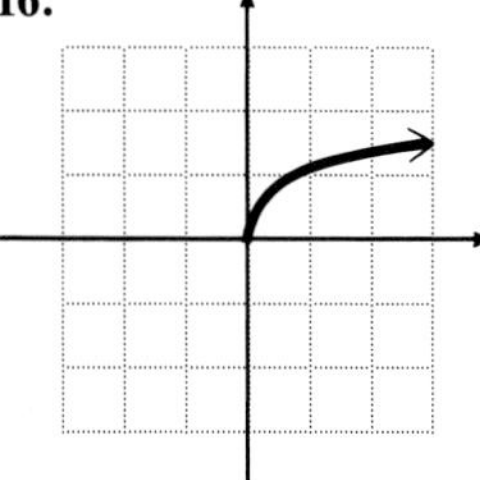

17.

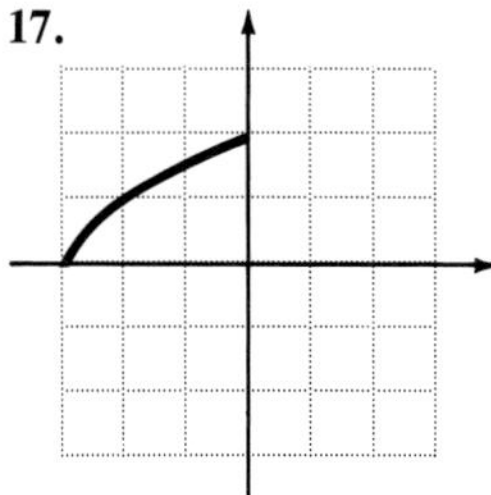

18.

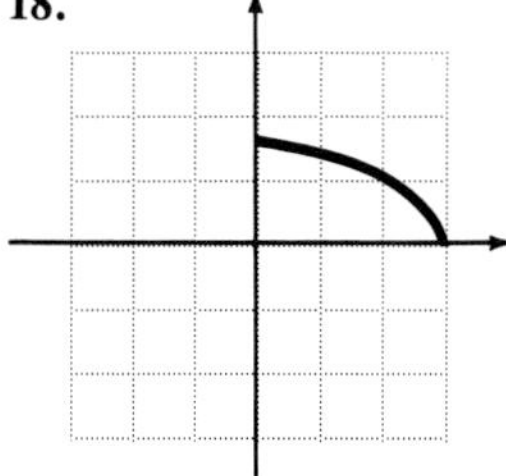

Sketch the graph of each of the following on the indicated interval, using a solid line. Then sketch the graph of the inverse, using a dashed line.

19. $y = \sin \theta, \ -\pi/2 \leq \theta \leq \pi/2$

20. $y = \cos \theta, \ 0 \leq \theta \leq \pi$

21. $y = \tan \theta, \ -\pi/2 < \theta < \pi/2$

22. $y = 2 \sin \theta, \ -\pi/2 \leq \theta \leq \pi/2$

For each of the following functions that are one-to-one, write an equation for the inverse function. See Example 3.

23. $y = x + 3$ **24.** $y + 1 = 3x$ **25.** $2y + 1 = 3x$ **26.** $2y + 15x = 30$

27. $y = 2x^2$ **28.** $2y - 1 = x^2$ **29.** $y + 1 = \sqrt[3]{x}$ **30.** $y = 6$

31. $y = x^4 + 2$ **32.** $y = 3x^6$

Decide whether the following pairs of functions are inverses of each other. See Example 4.

33. $f(x) = -8x$ and $g(x) = -\frac{1}{8}x$ **34.** $f(x) = 2x + 4$ and $g(x) = \frac{1}{2}x - 2$

35. $f(x) = \frac{1}{x + 1}$ and $g(x) = \frac{x - 9}{12}$ **36.** $f(x) = \frac{1}{x + 1}$ and $g(x) = \frac{1 - x}{x}$

37. $f(x) = \frac{1}{x}$ and $g(x) = \frac{1}{x}$ **38.** $f(x) = 4x$ and $g(x) = \frac{4}{x}$

6.2 Inverse Trigonometric Functions

In the last section the inverse of a one-to-one function was defined; this section defines the inverses of the trigonometric functions. A function must be one-to-one to have an inverse, but from the graph of $y = \sin x$ in Figure 6.6 and the horizontal line test, it is clear that $y = \sin x$ is not a one-to-one function. By suitably restricting the domain of the sine function, however, a one-to-one function can be defined. It is common to restrict the domain of $y = \sin x$ to the inverval $-\pi/2 \leq x \leq \pi/2$, which gives the part of the graph shown in color in Figure 6.6. We will call this function with restricted domain $y = \text{Sin } x$ (with a capital S) to distinguish it from $y = \sin x$ (with a lowercase s), which has the real numbers as domain. As Figure 6.6 shows, the range of both $y = \sin x$ and $y = \text{Sin } x$ is $-1 \leq y \leq 1$. Reflecting the graph of $y = \text{Sin } x$ about the line $y = x$ gives the graph of the inverse function, shown in Figure 6.7.

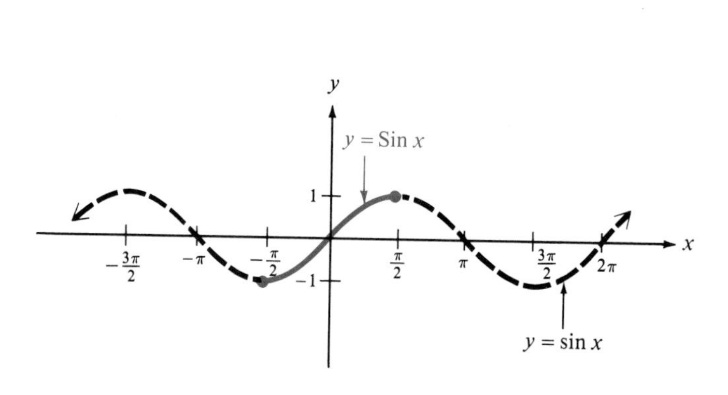

Figure 6.6

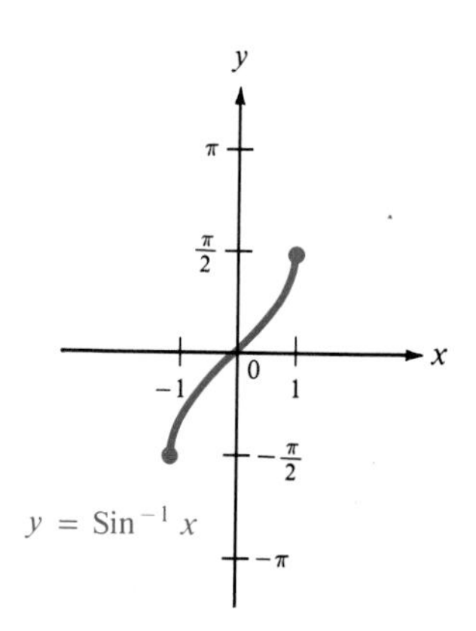

Figure 6.7

The equation of the inverse of $y = \text{Sin } x$ is found by first exchanging x and y to get $x = \text{Sin } y$. This equation then is solved for y by writing $\boldsymbol{y = \textbf{Sin}^{-1} x}$ (read "inverse sine of x". Note that $\text{Sin}^{-1} x$ does not mean $1/\text{Sin } x$.) As Figure 6.7 shows, the domain of $y = \text{Sin}^{-1} x$ is $-1 \le x \le 1$, while the restricted domain of $y = \text{Sin } x$, $-\pi/2 \le y \le \pi/2$, is the range of $y = \text{Sin}^{-1} x$. An alternative notation for $\text{Sin}^{-1} x$ is **Arcsin x.** By definition,

$\text{Sin}^{-1} x$ or Arcsin x

$y = \text{Sin}^{-1} x$ or $y = \text{Arcsin } x$ means $x = \text{Sin } y$, for $-\pi/2 \le y \le \pi/2$.

Think of $y = \text{Sin}^{-1} x$ or $y = \text{Arcsin } x$ as "y is the angle whose sine is x." The notations $\text{Sin}^{-1}x$ and Arcsin x both will be used in the rest of this book.

Example 1 Find y in radians for each of the following.

(a) $y = \text{Arcsin } \dfrac{1}{2}$

Think of $y = \text{Arcsin } (1/2)$ as

y is the angle whose sine is $1/2$.

Rewrite the expression as $\text{Sin } y = 1/2$. Since $\sin \pi/6 = 1/2$ and $\pi/6$ is in the range of the Arcsin function, $y = \pi/6$.

(b) $y = \text{Sin}^{-1} (-1)$

Writing the alternative equation, $\text{Sin } y = -1$, shows that $y = -\pi/2$. The value $y = 3\pi/2$ is incorrect, even though $\sin 3\pi/2 = -1$, because $3\pi/2$ is not in the range of the Sin^{-1} function. ✥

The function $y = \text{Cos}^{-1} x$ (or $y = \text{Arccos } x$) is defined by choosing a function $y = \text{Cos } x$ with the limited domain $0 \le x \le \pi$. This domain becomes the range of $y = \text{Cos}^{-1} x$, $0 \le y \le \pi$. The range of $y = \text{Cos } x$, the interval $-1 \le y \le 1$, becomes the domain of $y = \text{Cos}^{-1} x$, $-1 \le x \le 1$. By definition,

$\text{Cos}^{-1} x$ or Arccos x

$y = \text{Cos}^{-1} x$ or $y = \text{Arccos } x$ means $x = \text{Cos } y$, for $0 \le y \le \pi$.

The graph of $y = \text{Cos}^{-1} x$ is shown in Figure 6.8.

Example 2 Find each of the following.

(a) y in radians if $y = \text{Arccos } 1$

Think of $y = \text{Arccos } 1$ as "y is the angle whose cosine is 1," or $\cos y = 1$. Then $y = 0$, since $\cos 0 = 1$ and 0 is in the range of the Arccos function.

(b) θ in degrees if $\theta = \text{Cos}^{-1}\left(-\frac{1}{\sqrt{2}}\right)$

The values for Cos^{-1} are in quadrants I and II, since the range is $0 \leq y \leq \pi$. Because $-1/\sqrt{2}$ is negative, we are restricted to values in quadrant II. Write the statement as $\text{Cos}\ \theta = -1/\sqrt{2}$. In quadrant II, $\cos 135° = -1/\sqrt{2}$, so $\theta = 135°$. ✦

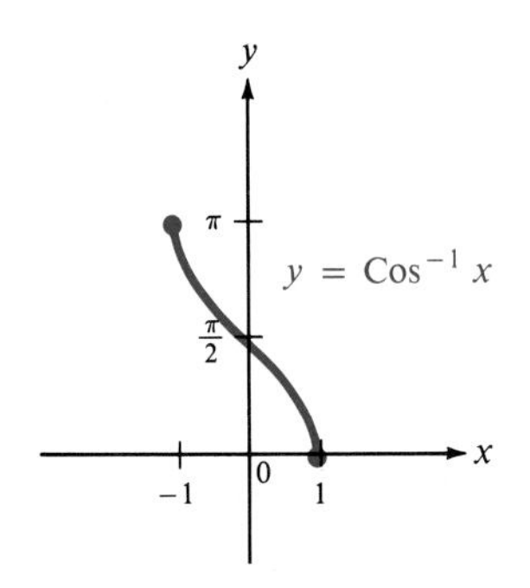

Figure 6.8

For each of the other trigonometric functions, an inverse function is defined by a suitable restriction on the domain, just as with sine and cosine. The six **inverse trigonometric functions,** their domains, and their ranges are given below. The information in this table, particularly the range for each function, should be memorized.

Inverse Trigonometric Functions

Function	*Domain*	*Range*	
		Intervals	*Quadrants*
$y = \text{Sin}^{-1} x$	$-1 \leq x \leq 1$	$-\pi/2 \leq y \leq \pi/2$	I and IV
$y = \text{Cos}^{-1} x$	$-1 \leq x \leq 1$	$0 \leq y \leq \pi$	I and II
$y = \text{Tan}^{-1} x$	Any real number	$-\pi/2 < y < \pi/2$	I and IV
$y = \text{Cot}^{-1} x$	Any real number	$0 < y < \pi$	I and II
$y = \text{Sec}^{-1} x$	$x \leq -1$ or $x \geq 1$	$0 \leq y \leq \pi,\ y \neq \frac{\pi}{2}$*	I and II
$y = \text{Csc}^{-1} x$	$x \leq -1$ or $x \geq 1$	$-\pi/2 \leq y \leq \pi/2,\ y \neq 0$*	I and IV

*Sec^{-1} and Csc^{-1} are sometimes defined with different ranges.

The graph of $y = \text{Tan}^{-1} x$ is shown in Figure 6.9. The graphs of the other three inverse trigonometric functions are left for the exercises.

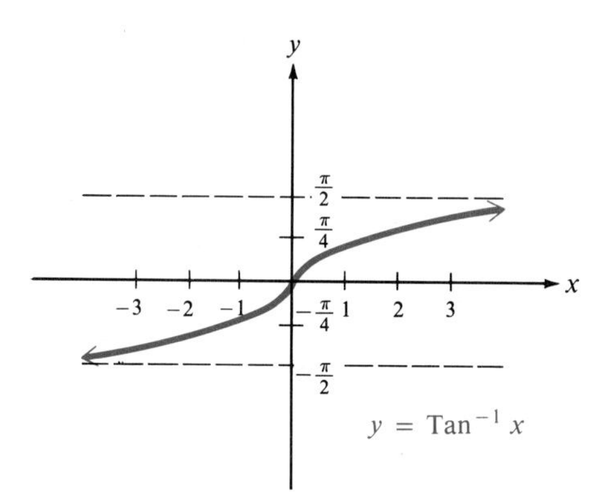

Figure 6.9

The inverse trigonometric function keys on a calculator give results in the proper quadrant, for the Sin^{-1}, Cos^{-1}, and Tan^{-1} functions, according to the definitions of these functions. For example, on a calculator, in degrees,

$$\text{Sin}^{-1} .5 = 30°, \qquad \text{Sin}^{-1} (-.5) = -30°,$$
$$\text{Tan}^{-1} (-1) = -45°, \quad \text{and} \quad \text{Cos}^{-1} (-.5) = 120°.$$

Similar results are found when the calculator is set for radian measure. This is not the case for Cot^{-1}. For example, since the sequence [1/x] [INV] [TAN] is used to find Cot^{-1}, the calculator gives values of Cot^{-1} with the same range as Tan^{-1}, $-\pi/2 < y < \pi/2$, which is not the correct range for Cot^{-1}. For Cot^{-1} the proper range must be considered and the results adjusted accordingly. This is also true for Sec^{-1} and Csc^{-1}.

Example 3 Find each of the following.

(a) θ in degrees if $\theta = \text{Arctan } 1$

Here θ must be in quadrant I. The alternative statement, $\text{Tan } \theta = 1$, leads to $\theta = 45°$.

(b) y in radians if $y = \text{Sec}^{-1} 2$

Write the equation as $\text{Sec } y = 2$. For $\text{Sec}^{-1} x$, y is in quadrant I or II. Since 2 is positive, y is in quadrant I and $y = \pi/3$, since $\text{Sec}(\pi/3) = 2$.

(c) θ in degrees if $\theta = \text{Arccot } (-.3541)$

A calculator or Table 2 shows that $\text{Cot } 70° \, 30' = .3541$. The restriction on the range of Arccot means that θ must be in quadrant II, so that

$$\theta = 180° - 70° \, 30' = 109° \, 30'.$$

Example 4 Evaluate each of the following without a calculator or tables.

(a) $\sin\left(\text{Tan}^{-1}\dfrac{3}{2}\right)$

Let

$$\theta = \text{Tan}^{-1}\, 3/2, \text{ so that } \text{Tan}\,\theta = 3/2.$$

Since Tan^{-1} is defined only in quadrants I and IV and since 3/2 is positive, θ is in quadrant I. Sketch θ in quadrant I, and label a triangle as shown in Figure 6.10. The hypotenuse is $\sqrt{13}$ and the value of sine is the quotient of the side opposite and the hypotenuse, so

$$\sin\left(\text{Tan}^{-1}\frac{3}{2}\right) = \sin\theta = \frac{3}{\sqrt{13}} = \frac{3\sqrt{13}}{13}.$$

To check this result on a calculator, enter 3/2 as 1.5. Then find $\text{Tan}^{-1}\, 1.5$, and finally find $\sin\,(\text{Tan}^{-1}\, 1.5)$. Store this result and calculate $3\sqrt{13}/13$, which should agree with the result for $\sin\,(\text{Tan}^{-1}\, 1.5)$. Since the values are only approximations, this check does not *prove* that the result is correct, but it is highly suggestive that it is correct.

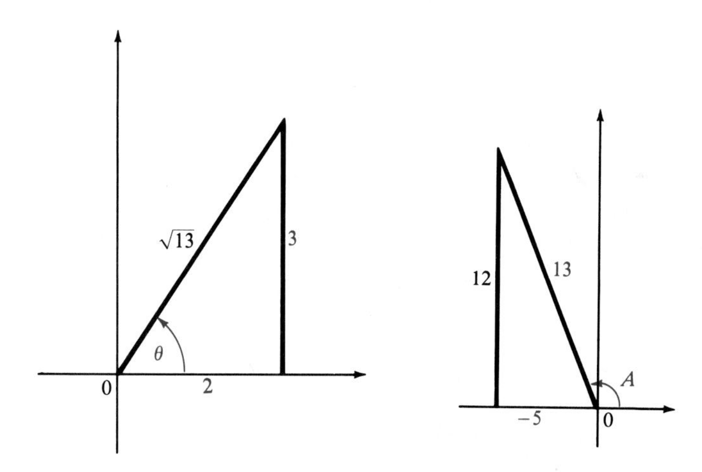

Figure 6.10 **Figure 6.11**

(b) $\tan\left(\text{Cos}^{-1}\dfrac{-5}{13}\right)$

Let $A = \text{Cos}^{-1}\,(-5/13)$. Then $\text{Cos}\,A = -5/13$. Since $\text{Cos}^{-1}\,\theta$ for a negative value of θ is in quadrant II, sketch A in quadrant II, as shown in Figure 6.11. From the triangle in Figure 6.11,

$$\tan\left(\text{Cos}^{-1}\frac{-5}{13}\right) = \tan A = -\frac{12}{5}.$$ ✤

Example 5 Evaluate the following without using tables or a calculator.

(a) $\cos(\text{Arctan}\sqrt{3} + \text{Arcsin } 1/3)$

Let $A = \text{Arctan}\sqrt{3}$ and $B = \text{Arcsin } 1/3$ so that $\text{Tan } A = \sqrt{3}$ and $\text{Sin } B = 1/3$. Sketch both A and B in quadrant I, as shown in Figure 6.12.

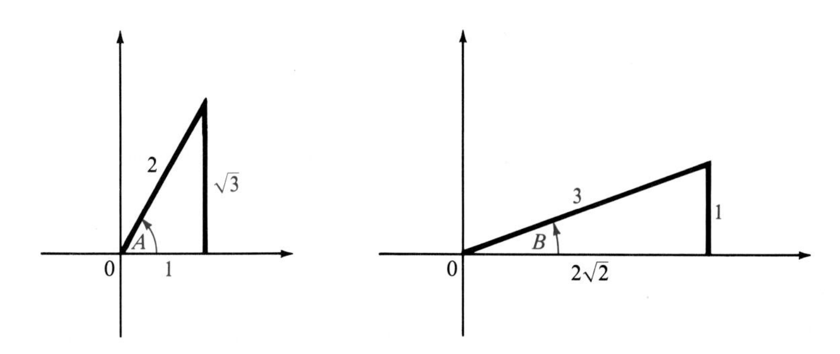

Figure 6.12

Now use the identity for $\cos(A + B)$.

$$\cos(A+B) = \cos A \cos B - \sin A \sin B$$

$$\cos(\text{Arctan}\sqrt{3} + \text{Arcsin } 1/3) = \cos(\textbf{Arctan}\sqrt{\mathbf{3}})\cos(\textbf{Arcsin 1/3}) - \sin(\textbf{Arctan}\sqrt{\mathbf{3}})\sin(\textbf{Arcsin 1/3}) \qquad \textbf{(1)}$$

From the sketch in Figure 6.12,

$$\cos(\text{Arctan}\sqrt{3}) = \cos A = \frac{1}{2},$$

$$\sin(\text{Arctan}\sqrt{3}) = \sin A = \frac{\sqrt{3}}{2},$$

$$\cos\left(\text{Arcsin}\frac{1}{3}\right) = \cos B = \frac{2\sqrt{2}}{3},$$

$$\sin\left(\text{Arcsin}\frac{1}{3}\right) = \sin B = \frac{1}{3}.$$

Substitute these values into equation (1) to get

$$\begin{aligned}\cos\left(\text{Arctan}\sqrt{3} + \text{Arcsin}\frac{1}{3}\right) &= \frac{1}{2}\cdot\frac{2\sqrt{2}}{3} - \frac{\sqrt{3}}{2}\cdot\frac{1}{3}\\ &= \frac{2\sqrt{2}}{6} - \frac{\sqrt{3}}{6}\\ &= \frac{2\sqrt{2} - \sqrt{3}}{6}.\end{aligned}$$

(b) $\tan\left(2 \text{ Arcsin } \dfrac{2}{5}\right)$

Let Arcsin (2/5) = B. Then

$$\tan\left(2 \text{ Arcsin } \frac{2}{5}\right) = \tan (2B)$$
$$= \frac{2 \tan B}{1 - \tan^2 B}.$$

Since Arcsin (2/5) = B, $\sin B = 2/5$. Sketch a triangle in quadrant I, find the length of the third side, and then find $\tan B$. From the triangle in Figure 6.13, $\tan B = 2/\sqrt{21}$, and

$$\tan\left(2 \text{ Arcsin } \frac{2}{5}\right) = \frac{2\left(\dfrac{2}{\sqrt{21}}\right)}{1 - \left(\dfrac{2}{\sqrt{21}}\right)^2}$$
$$= \frac{\dfrac{4}{\sqrt{21}}}{1 - \dfrac{4}{21}}$$
$$= \frac{\dfrac{4}{\sqrt{21}}}{\dfrac{17}{21}}$$
$$= \frac{4\sqrt{21}}{17}.$$

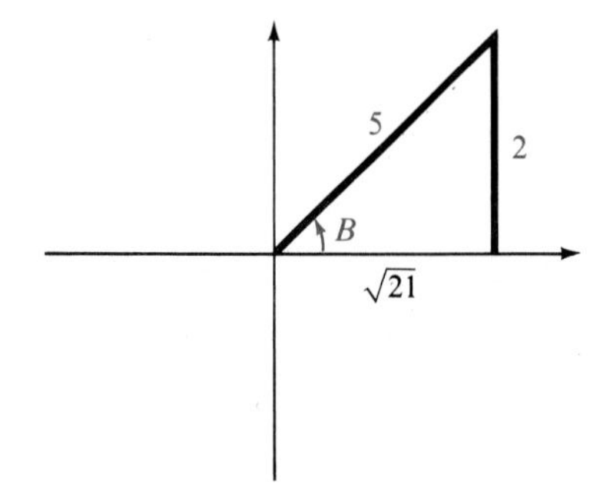

Figure 6.13

Example 6 Write $\sin(\text{Tan}^{-1} u)$ as an expression in u.

Let $\theta = \text{Tan}^{-1} u$, so that $\text{Tan}\ \theta = u$. Here u may be positive or negative. Since $-\pi/2 < \text{Tan}^{-1} u < \pi/2$, sketch θ in quadrants I and IV and label two triangles as shown in Figure 6.14. Since sine is given by the quotient of the side opposite and the hypotenuse,

$$\sin(\text{Tan}^{-1} u) = \sin\theta = \frac{u}{\sqrt{u^2 + 1}}.$$

The result, $u/\sqrt{u^2 + 1}$, is positive when u is positive and negative when u is negative. ✥

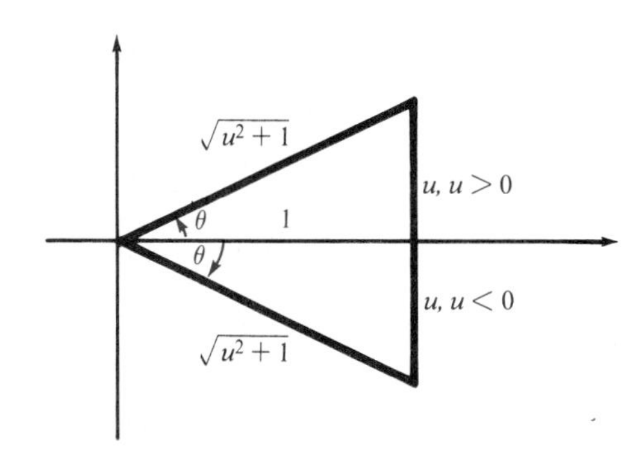

Figure 6.14

6.2 Exercises

For each of the following give the value of y in radians without using a calculator or tables. See Examples 1–3.

1. $y = \text{Arcsin}\left(-\frac{1}{2}\right)$ **2.** $y = \text{Arccos}\ \frac{\sqrt{3}}{2}$ **3.** $y = \text{Tan}^{-1}\ 1$

4. $y = \text{Cot}^{-1}(-1)$ **5.** $y = \text{Arcsin}\ 0$ **6.** $y = \text{Arccos}(-1)$

7. $y = \text{Cos}^{-1}\ \frac{1}{2}$ **8.** $y = \text{Sin}^{-1}\left(-\frac{\sqrt{3}}{2}\right)$ **9.** $y = \text{Sec}^{-1}(-\sqrt{2})$

10. $y = \text{Csc}^{-1}(-2)$ **11.** $y = \text{Arccot}(-\sqrt{3})$ **12.** $y = \text{Arccos}\ 0$

Give the value of each of the following in decimal degrees.

13. $\text{Sin}^{-1}(-.1334)$ **14.** $\text{Cos}^{-1}(-.1334)$ **15.** $\text{Arccos}(-.3987)$

16. Arcsin .7790 **17.** $\text{Csc}^{-1}\ 1.942$ **18.** $\text{Cot}^{-1}\ 1.767$

Give the value of each of the following in radians.

19. Arctan 1.111 **20.** Arcsin .8192 **21.** $\text{Cot}^{-1}(-.9217)$

22. $\text{Sec}^{-1}(-1.287)$ **23.** Arcsin .9283 **24.** Arccos .4462

Give the value of each of the following without using a calculator or tables. See Examples 4 and 5.

25. $\tan\left(\text{Arccos}\,\frac{3}{4}\right)$

26. $\sin\left(\text{Arccos}\,\frac{1}{4}\right)$

27. $\cos\,(\text{Tan}^{-1}\,(-2))$

28. $\sec\left(\text{Sin}^{-1}\left(-\frac{1}{5}\right)\right)$

29. $\cot\left(\text{Arcsin}\left(-\frac{2}{3}\right)\right)$

30. $\cos\left(\text{Arctan}\,\frac{8}{3}\right)$

31. $\sec\left(\text{Arccot}\,\frac{3}{5}\right)$

32. $\cos\left(\text{Arcsin}\,\frac{12}{13}\right)$

33. $\cos\left(\text{Arccos}\,\frac{1}{2}\right)$

34. $\sin\left(\text{Arcsin}\,\frac{\sqrt{3}}{2}\right)$

35. $\tan\,(\text{Tan}^{-1}\,(-1))$

36. $\cot\,(\text{Cot}^{-1}\,(-\sqrt{3}))$

37. $\sec\,(\text{Sec}^{-1}\,2)$

28. $\csc\,(\text{Csc}^{-1}\,\sqrt{2})$

39. $\text{Arccos}\left(\cos\frac{\pi}{4}\right)$

40. $\text{Arctan}\left(\tan\left(-\frac{\pi}{4}\right)\right)$

41. $\text{Arcsin}\left(\sin\frac{\pi}{3}\right)$

42. $\text{Arccos}\,(\cos 0)$

43. $\sin\left(2\,\text{Tan}^{-1}\,\frac{12}{5}\right)$

44. $\cos\left(2\,\text{Sin}^{-1}\,\frac{1}{4}\right)$

45. $\cos\left(2\,\text{Arctan}\,\frac{4}{3}\right)$

46. $\tan\left(2\,\text{Cos}^{-1}\,\frac{1}{4}\right)$

47. $\sin\left(2\,\text{Cos}^{-1}\,\frac{1}{5}\right)$

48. $\cos\,(2\,\text{Arctan}\,(-2))$

49. $\tan\left(2\,\text{Arcsin}\left(-\frac{3}{5}\right)\right)$

50. $\sin\left(2\,\text{Arccos}\,\frac{2}{9}\right)$

51. $\sin\left(\text{Sin}^{-1}\,\frac{1}{2} + \text{Tan}^{-1}\,(-3)\right)$

52. $\cos\left(\text{Tan}^{-1}\,\frac{5}{12} - \text{Cot}^{-1}\,\frac{4}{3}\right)$

53. $\cos\left(\text{Arcsin}\,\frac{3}{5} + \text{Arccos}\,\frac{5}{13}\right)$

54. $\tan\left(\text{Arccos}\,\frac{\sqrt{3}}{2} - \text{Arcsin}\left(-\frac{3}{5}\right)\right)$

Use a calculator to find each of the following to six decimal places.

55. $\cos\,(\text{Tan}^{-1}\,.5)$

56. $\sin\,(\text{Cos}^{-1}\,.25)$

57. $\tan\,(\text{Arcsin}\,.1225)$

58. $\cot\,(\text{Arccos}\,.5823)$

Write each of the following as an expression in u. See Example 6.

59. $\sin\,(\text{Arccos}\,u)$

60. $\tan\,(\text{Arccos}\,u)$

61. $\sec\,(\text{Cot}^{-1}\,u)$

62. $\csc\,(\text{Sec}^{-1}\,u)$

63. $\cot\,(\text{Aresin}\,u)$

64. $\cos\,(\text{Arcsin}\,u)$

65. $\sin\left(\text{Sec}^{-1}\,\frac{u}{2}\right)$

66. $\cos\left(\text{Tan}^{-1}\,\frac{3}{u}\right)$

67. $\tan\left(\text{Arcsin}\,\frac{u}{\sqrt{u^2+2}}\right)$

68. $\cos\left(\text{Arccos}\,\frac{u}{\sqrt{u^2+5}}\right)$

69. $\sec\left(\text{Arccot}\,\frac{\sqrt{4-u^2}}{u}\right)$

70. $\csc\left(\text{Arctan}\,\frac{\sqrt{9-u^2}}{u}\right)$

Graph each of the following, and give the domain and range.

71. $y = \text{Cot}^{-1}\,x$

72. $y = \text{Tan}^{-1}\,2x$

73. $y = \text{Arcsec}\,x$

74. $y = \text{Arccsc}\,x$

75. $y = 2\,\text{Cos}^{-1}\,x$

76. $y = \text{Sin}^{-1}\,\frac{x}{2}$

77. Enter 1.003 in your calculator, and press the keys for inverse sine. The response will indicate that something is wrong. What is wrong?

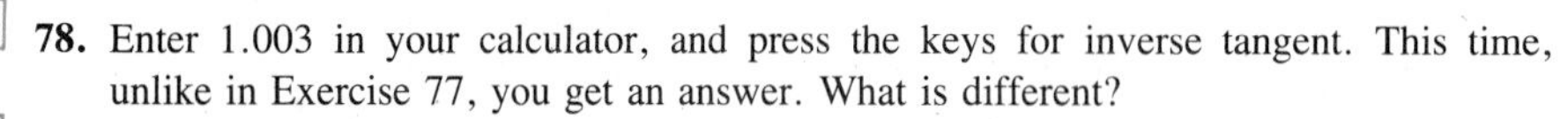

78. Enter 1.003 in your calculator, and press the keys for inverse tangent. This time, unlike in Exercise 77, you get an answer. What is different?

79. Enter 1.74283 in your calculator (set for radians), and press the sine key. Then press the keys for inverse sine. You get 1.398763 instead of 1.74283. What happened?

80. Suppose that an airplane flying faster than sound goes directly over you. Assume that the plane is flying level. At the instant that you feel the sonic boom from the plane, the angle of elevation to the plane is given by

$$\alpha = 2 \text{ Arcsin } \frac{1}{m},$$

where m is the Mach number of the plane's speed. (See the exercises at the end of Section 5.6.) Find α to the nearest degree for each of the following values of m.

(a) $m = 1.2$ **(b)** $m = 1.5$ **(c)** $m = 2$ **(d)** $m = 2.5$

81. A painting 1 m high and 3 m from the floor will cut off an angle θ to an observer, where

$$\theta = \text{Tan}^{-1}\left(\frac{x}{x^2 + 2}\right).$$

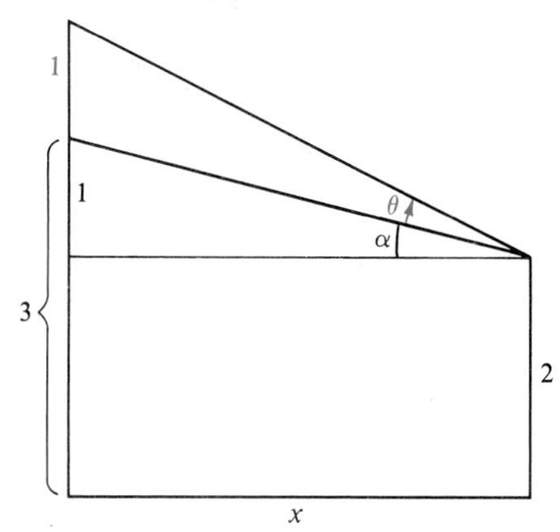

Assume that the observer is x m from the wall where the painting is displayed and that the eyes of the observer are 2 m above the ground. See the figure. Find the value of θ for the following values of x. Round to the nearest degree.

(a) 1 **(b)** 2 **(c)** 3

(d) Derive the formula given above. (*Hint:* Use the identity for tan $(\theta + \alpha)$. Use right triangles.)

The following expressions were used by the mathematicians who computed the value of π *to* 100,000 *decimal places. Use a calculator to verify that each is (approximately) correct.*

82. $\pi = 16 \text{ Tan}^{-1} \frac{1}{5} - 4 \text{ Tan}^{-1} \frac{1}{239}$

83. $\pi = 24 \text{ Tan}^{-1} \frac{1}{8} + 8 \text{ Tan}^{-1} \frac{1}{57} + 4 \text{ Tan}^{-1} \frac{1}{239}$

84. $\pi = 48 \text{ Tan}^{-1} \frac{1}{18} + 32 \text{ Tan}^{-1} \frac{1}{57} - 20 \text{ Tan}^{-1} \frac{1}{239}$

6.3 Trigonometric Equations

As mentioned earlier, a **conditional equation** is an equation in which some replacements for the variable make the statement true and some replacements make it false. For example,

$$2x + 3 = 5, \qquad x^2 - 5x = 10, \qquad \text{and} \qquad 2x = 8$$

are conditional equations. In this section, we discuss conditional equations that involve trigonometric functions.

Conditional equations with trigonometric functions can usually be solved by using algebraic methods and trigonometric identities to simplify the equations. The next examples illustrate the methods used in solving a trigonometric equation.

Example 1 Solve $3 \sin x = \sqrt{3} + \sin x$.

Collect all terms with $\sin x$ on one side of the equation, and solve first for $\sin x$.

$$3 \sin x = \sqrt{3} + \sin x$$
$$2 \sin x = \sqrt{3}$$
$$\sin x = \frac{\sqrt{3}}{2}$$

There are many possible solutions for x. For example, $\sin 60° = \sqrt{3}/2$ as do $\sin 120°$, $\sin 420°$, and so on. There are an infinite number of values of x that satisfy this equation. This infinite number of solutions can be expressed as

$$x = 60° + 360° \cdot n \qquad \text{or} \qquad x = 120° + 360° \cdot n,$$

where n is any integer. Often the solutions of a trigonometric equation are required only for some particular interval. For example, if the solutions of this equation are restricted to the interval $0° \leq x \leq 360°$, only the solutions $60°$ and $120°$ would be given. ✜

When an equation involves more than one trigonometric function, it is often helpful to use a suitable identity to rewrite the equation in terms of just one trigonometric function, as in the following example.

Example 2 Solve $\sin x + \cos x = 0$ for $0° \leq x < 360°$.

One way to solve this equation is to divide both sides by $\cos x$. (Remember that this can be done only if we assume that $\cos x$ does not equal 0.) Then the quotient $\sin x/\cos x$ can be replaced by $\tan x$, and the other trigonometric functions are eliminated.

$$\sin x + \cos x = 0$$

$$\frac{\sin x}{\cos x} + \frac{\cos x}{\cos x} = \frac{0}{\cos x}$$

$$\tan x + 1 = 0$$

$$\tan x = -1$$

The solutions in the given interval are 135° and 315°.

We assumed here that $\cos x \neq 0$. If $\cos x = 0$, the given equation becomes $\sin x + 0 = 0$, or $\sin x = 0$. If $\cos x = 0$, it is not possible for $\sin x = 0$, so no solutions were missed by assuming $\cos x \neq 0$. ✥

Sometimes factoring can be used to isolate different trigonometric functions.

Example 3 Solve $\sin x \tan x = \sin x$ for $0° \leq x < 360°$.

Subtract $\sin x$ from both sides, then factor on the left.

$$\sin x \tan x = \sin x$$

$$\sin x \tan x - \sin x = 0$$

$$\sin x(\tan x - 1) = 0$$

Now set each factor equal to 0.

$$\sin x = 0 \quad \text{or} \quad \tan x - 1 = 0$$

$$\tan x = 1$$

$$x = 0° \quad \text{or} \quad x = 180° \qquad x = 45° \quad \text{or} \quad x = 225°$$ ✥

There are four solutions for Example 3. Trying to solve the equation by dividing both sides by $\sin x$ would give just $\tan x = 1$, which would give $x = 45°$ or $x = 225°$. The other two solutions would not appear. The missing solutions are the ones that make the divisor, $\sin x$, equal 0. For this reason, it is best to avoid dividing by a variable expression if possible. However, in an equation like that in Example 2, dividing both sides by a variable simplified the solution considerably.

It is important to remember that dividing by a variable expression requires checking to see whether the numbers that make that expression 0 are solutions of the original equation.

Sometimes a trigonometric equation can be solved by first squaring both sides, and then using a trigonometric identity. This works for those identities that involve squares, like $\sin^2 x + \cos^2 x = 1$ or $\tan^2 x + 1 = \sec^2 x$. When both sides of an equation are squared, remember to check for any numbers that satisfy the squared equation but not the given equation *(extraneous solutions)* by substituting potential solutions into the given equation.

Example 4 Solve $\tan x + \sqrt{3} = \sec x$ for $0 \le x < 2\pi$.

Square both sides; then express $\sec^2 x$ in terms of $\tan^2 x$.

$$\begin{aligned} \tan x + \sqrt{3} &= \sec x \\ \tan^2 x + 2\sqrt{3}\tan x + 3 &= \sec^2 x \\ \tan^2 x + 2\sqrt{3}\tan x + 3 &= 1 + \tan^2 x \\ 2\sqrt{3}\tan x &= -2 \\ \tan x &= -\frac{1}{\sqrt{3}} \end{aligned}$$

The possible solutions in the given interval are $5\pi/6$ and $11\pi/6$. Now check the possible solutions. Try $5\pi/6$ first.

$$\tan x + \sqrt{3} = \tan\frac{5\pi}{6} + \sqrt{3} = \frac{-\sqrt{3}}{3} + \sqrt{3} = \frac{2\sqrt{3}}{3}$$

$$\sec x = \sec\frac{5\pi}{6} = \frac{-2\sqrt{3}}{3}$$

The check shows that $5\pi/6$ is not a solution. Now check $11\pi/6$.

$$\tan\frac{11\pi}{6} + \sqrt{3} = \frac{-\sqrt{3}}{3} + \sqrt{3} = \frac{2\sqrt{3}}{3}$$

$$\sec\frac{11\pi}{6} = \frac{2\sqrt{3}}{3}$$

This solution satisfies the equation, so $11\pi/6$ is the only solution of the given equation. ✤

Some trigonometric equations are quadratic in form and can be solved by the methods used to solve quadratic equations.

Example 5 Solve $\tan^2 x + \tan x - 2 = 0$ for $0 \le x < 2\pi$.

Let $y = \tan x$, so that the equation becomes $y^2 + y - 2 = 0$, which factors as $(y - 1)(y + 2) = 0$. Substituting $\tan x$ back for y gives.

$$(\tan x - 1)(\tan x + 2) = 0.$$

Set each factor equal to 0.

$$\begin{aligned} \tan x - 1 &= 0 \quad \text{or} \quad \tan x + 2 = 0 \\ \tan x &= 1 \quad \text{or} \quad \tan x = -2 \end{aligned}$$

The solutions for $\tan x = 1$ in the interval $0 \le x < 2\pi$ are $x = \pi/4$ or $5\pi/4$. To solve $\tan x = -2$ in the interval $0 \le x < 2\pi$, use Table 2 or a calculator. If you use a calculator, be sure to set it for radian measure first. The calculator gives $x \approx -1.10715$, a quadrant IV angle. However, we want all the solutions of $\tan x = -2$ in the interval $0 \le x < 2\pi$. To get the solutions in the desired interval, first add π to -1.10715, and then add 2π:

$$x \approx -1.10715 + \pi \approx 2.03444$$

or

$$x \approx -1.10715 + 2\pi \approx 5.17604.$$

The solutions in the required interval (to five decimal places) are

$$\frac{\pi}{4}, \quad \frac{5\pi}{4}, \quad 2.03444, \quad \text{and} \quad 5.17604.$$

When a trigonometric equation that is quadratic in form cannot be factored, the quadratic formula can be used to solve the equation.

Example 6 Solve $\cot^2 x + 3 \cot x = 1$ for $0° \le x < 360°$.

Write the equation in quadratic form, with 0 on one side.

$$\cot^2 x + 3 \cot x - 1 = 0$$

Since this equation cannot be factored, use the quadratic formula, with $a = 1$, $b = 3$, $c = -1$, and $\cot x$ as the variable.

$$\cot x = \frac{-3 \pm \sqrt{9 + 4}}{2} = \frac{-3 \pm \sqrt{13}}{2} = \frac{-3 \pm 3.606}{2}$$

$$\cot x = .303 \qquad \text{or} \qquad \cot x = -3.303$$

Use Table 2 or a calculator to find x to the nearest ten minutes.

$$x = 73° \, 10', \quad 253° \, 10', \quad 163° \, 10', \quad \text{or} \quad 343° \, 10'$$

The methods for solving trigonometric equations illustrated in the examples can be summarized as follows.

Solving Trigonometric Equations

1. If only one trigonometric function is present, first solve the equation for that function.
2. If more than one trigonometric function is present, rearrange the equation so that one side equals 0. Then try to factor and set each factor equal to 0 to solve.
3. If Step 2 does not work, try using identities to change the form of the equation. It may be helpful to square both sides of the equation first.
4. If the equation is quadratic in form, but not factorable, use the quadratic formula.

When squaring both sides, check the answers by pluging them in the original equation, if they equal, that is the answer (s)

6.3 Exercises

Solve each of the following equations for $0 \le x < 2\pi$. Use 3.1416 as an approximation for π when you use values from Table 2. See Examples 1, 4, and *5.*

1. $2 \cot x + 1 = -1$

2. $\sin x + 2 = 3$

3. $2 \sin x + 3 = 4$

4. $2 \sec x + 1 = \sec x + 3$

5. $\tan^2 x - 3 = 0$

6. $\sec^2 x - 2 = -1$

7. $(\cot x - \sqrt{3})(2 \sin x + \sqrt{3}) = 0$

8. $(\tan x - 1)(\cos x - 1) = 0$

9. $(\cot x - 1)(\sqrt{3} \cot x + 1) = 0$

10. $(\csc x + 2)(\csc x - \sqrt{2}) = 0$

11. $\cos^2 x + 2 \cos x + 1 = 0$

12. $2 \cos^2 x - \sqrt{3} \cos x = 0$

13. $-2 \sin^2 x = 3 \sin x + 1$

14. $3 \sin^2 x - \sin x = 2$

15. $\cos^2 x - \sin^2 x = 0$

16. $\dfrac{2 \tan x}{3 - \tan^2 x} = 1$

Solve each of the following equations for $0^\circ \le \theta < 360^\circ$. Find θ to the nearest ten minutes. See Examples 2, 3, and 5.

17. $2 \sin \theta - 1 = \csc \theta$

18. $\tan \theta + 1 = \sqrt{3} + \sqrt{3} \cot \theta$

19. $\tan \theta - \cot \theta = 0$

20. $\sec^2 \theta = 2 \tan \theta + 4$

21. $\cos^2 \theta = \sin^2 \theta + 1$

22. $\csc^2 \theta - 2 \cot \theta = 0$

23. $\tan^3 \theta = 3 \tan \theta$

24. $2 \cos^4 \theta = \cos^2 \theta$

25. $\sin^2 \theta \cos \theta = \cos \theta$

26. $2 \tan^2 \theta \sin \theta - \tan^2 \theta = 0$

27. $5 \sec^2 \theta = 6 \sec \theta$

28. $3 \cot^2 \theta = \cot \theta$

29. $\sin^2 \theta \cos^2 \theta = 0$

30. $\sec^2 \theta \tan \theta = 2 \tan \theta$

31. $4(1 + \sin \theta) = \dfrac{3}{1 - \sin \theta}$

32. $8 \cos \theta = \cot \theta$

33. $\sin \theta + \cos \theta = 1$

34. $\sec \theta - \tan \theta = 1$

To solve the following equations, you will need the quadratic formula. Find all solutions in the interval $0^\circ \le x < 360^\circ$. Give solutions to the nearest ten minutes. See Example 6.

35. $9 \sin^2 x - 6 \sin x = 1$

36. $4 \cos^2 x + 4 \cos x = 1$

37. $\tan^2 x + 4 \tan x + 2 = 0$

38. $3 \cot^2 x - 3 \cot x - 1 = 0$

39. $\sin^2 x - 2 \sin x + 3 = 0$

40. $2 \cos^2 x + 2 \cos x - 1 = 0$

41. $\cot x + 2 \csc x = 3$

42. $2 \sin x = 1 - 2 \cos x$

43. In an electric circuit, let V represent the electromotive force in volts at t seconds. Assume $V = \cos 2\pi t$. Find the smallest positive value of t where $0 \le t \le 1/2$ for each of the following values of V.
(a) $V = 0$ **(b)** $V = .5$ **(c)** $V = .25$

44. A coil of wire rotating in a magnetic field induces a voltage given by

$$e = 20 \sin \left(\frac{\pi t}{4} - \frac{\pi}{2}\right),$$

where t is time in seconds. Find the smallest positive time to produce the following voltages.
(a) 0 **(b)** $10\sqrt{3}$

45. The equation

$$.342D \cos \theta + h \cos^2 \theta = \frac{16D^2}{V_0^2}$$

is used in reconstructing accidents in which a vehicle vaults into the air after hitting an obstruction. V_0 is the velocity in feet per second of the vehicle when it hits, D is the distance (in feet) from the obstruction to the landing point, and h is the difference in height (in feet) between the landing point and the takeoff point. Angle θ is the takeoff angle, the angle between the horizontal and the path of the vehicle. Find θ to the nearest degree if $V_0 = 60$, $D = 80$, and $h = 2$.

6.4 Trigonometric Equations with Multiple Angles

Conditional trigonometric equations in which a half angle or multiple angle is given, such as $2 \sin (x/2) = 1$, often require an additional step to solve. This is shown in the following example.

Example 1 Solve $2 \sin (x/2) = 1$ for $0° \le x < 360°$.

Dividing the inequality $0° \le x < 360°$ by 2 gives

$$0° \le \frac{x}{2} < 180°.$$

To find all values of $x/2$ in the interval 0° to 180°, begin by solving for the trigonometric function.

$$2 \sin \frac{x}{2} = 1$$

$$\sin \frac{x}{2} = \frac{1}{2}$$

Both $\sin 30° = \frac{1}{2}$ and $\sin 150° = \frac{1}{2}$ and 30° and 150° are in the given interval for $x/2$, so

$$\frac{x}{2} = 30° \quad \text{or} \quad \frac{x}{2} = 150°$$

$$x = 60° \quad \text{or} \quad x = 300°.$$

Sometimes equations with multiple angles can be solved by using an appropriate identity, as in the next examples. In Example 2 it is important to notice that $\cos 2x$ cannot be changed to $\cos x$ by dividing by 2:

$$\frac{\cos 2x}{2} \ne \cos x.$$

The only way to change $\cos 2x$ to a trigonometric function of x is by using one of the identities for $\cos 2x$.

Example 2 Solve $\cos 2x = \cos x$ for $0 \leq x < 2\pi$.

The first step is to change $\cos 2x$ to a trigonometric function of x. Use the identity $\cos 2x = 2\cos^2 x - 1$ so that the equation involves only the cosine function.

$$\begin{aligned} \cos 2x &= \cos x \\ 2\cos^2 x - 1 &= \cos x \\ 2\cos^2 x - \cos x - 1 &= 0 \\ (2\cos x + 1)(\cos x - 1) &= 0 \end{aligned}$$

$$2\cos x + 1 = 0 \quad \text{or} \quad \cos x - 1 = 0$$

$$\cos x = -\frac{1}{2} \quad \text{or} \quad \cos x = 1$$

In the required interval,

$$x = \frac{2\pi}{3} \quad \text{or} \quad \frac{4\pi}{3} \quad \text{or} \quad x = 0.$$

The solutions are 0, $2\pi/3$, and $4\pi/3$. ✥

As Example 1 showed, the domain may be affected when the equation simplifies to a trigonometric function of a half angle or a multiple angle. The next example illustrates this further.

Example 3 Solve $4 \sin x \cos x = \sqrt{3}$ for $0° \leq x < 360°$.

The identity $2 \sin x \cos x = \sin 2x$ is useful here.

$$\begin{aligned} 4 \sin x \cos x &= \sqrt{3} \\ 2(2 \sin x \cos x) &= \sqrt{3} \\ 2 \sin 2x &= \sqrt{3} \\ \sin 2x &= \frac{\sqrt{3}}{2} \end{aligned}$$

From the given domain $0° \leq x < 360°$, the domain for $2x$ is $0° \leq 2x < 720°$. Now list all solutions in this interval:

$$2x = 60°, 120°, 420°, 480°$$

or

$$x = 30°, 60°, 210°, 240°.$$

The last two solutions for $2x$ were found by adding 360° to 60° and 120°, respectively. ✥

Example 4 Solve $\tan 3x + \sec 3x = 2$ for $0 \leq x < 2\pi$.

One trigonometric function must be eliminated. Subtract $\sec 3x$ from both sides, and then square both sides.

$$\tan 3x + \sec 3x = 2$$
$$\tan 3x = 2 - \sec 3x$$
$$\tan^2 3x = 4 - 4 \sec 3x + \sec^2 3x$$

On the right, $(a + b)^2 = a^2 + 2ab + b^2$ was used. Since there are two terms involving secant, $\sec^2 3x$ and $\sec 3x$, use the identity $\tan^2 3x + 1 = \sec^2 3x$, or $\tan^2 3x = \sec^2 3x - 1$, to replace $\tan^2 3x$.

$$\sec^2 3x - 1 = 4 - 4 \sec 3x + \sec^2 3x$$
$$0 = 5 - 4 \sec 3x$$
$$4 \sec 3x = 5$$
$$\sec 3x = \frac{5}{4}$$
$$\frac{1}{\cos 3x} = \frac{5}{4}$$
$$\cos 3x = \frac{4}{5}$$

From Table 2 or a calculator, using the domain $0 \le 3x < 6\pi$ and knowing that cosine is positive in quadrants I and IV,

$$3x = .6435, 5.6397, 6.9267, 11.9229, 13.2099, 18.2061$$
$$x = .2145, 1.8799, 2.3089, 3.9743, 4.4033, 6.0687.$$

Since both sides of the equation were squared, each of these proposed solutions must be checked. Verify by substitution in the given equation that the solutions are .2145, 2.3089, and 4.4033. ✣

6.4 Exercises

Solve each of the following equations for $0 \le x < 2\pi$. See Examples 1–4.

1. $\cos 2x = \frac{\sqrt{3}}{2}$

2. $\cos 2x = -\frac{1}{2}$

3. $\sin 3x = -1$

4. $\sin 3x = 0$

5. $3 \tan 3x = \sqrt{3}$

6. $\cot 3x = \sqrt{3}$

7. $\sqrt{2} \cos 2x = -1$

8. $2\sqrt{3} \sin 2x = \sqrt{3}$

9. $\sin \frac{x}{2} = \sqrt{2} - \sin \frac{x}{2}$

10. $\sin x = \sin 2x$

11. $\tan 4x = 0$

12. $\cos 2x - \cos x = 0$

13. $\sec^4 2x = 4$

14. $\sin^2 \frac{x}{2} - 1 = 0$

15. $\sin \frac{x}{2} = \cos \frac{x}{2}$

16. $\sec \frac{x}{2} = \cos \frac{x}{2}$

17. $\cos 2x + \cos x = 0$

18. $\sin x \cos x = \frac{1}{4}$

Solve each of the following equations for $0° \le \theta < 360°$. Use Table 2 or a calculator to find solutions to the nearest ten minutes as necessary. See Examples 1–4.

19. $\sqrt{2} \sin 3\theta - 1 = 0$

20. $-2 \cos 2\theta = \sqrt{3}$

21. $\cos \frac{\theta}{2} = 1$

22. $\sin \frac{\theta}{2} = 1$

23. $2\sqrt{3} \sin \frac{\theta}{2} = 3$

24. $2\sqrt{3} \cos \frac{\theta}{2} = -3$

25. $2 \sin \theta = 2 \cos 2\theta$

26. $\cos \theta - 1 = \cos 2\theta$

27. $1 - \sin \theta = \cos 2\theta$

28. $\sin 2\theta = 2 \cos^2 \theta$

29. $\csc^2 \frac{\theta}{2} = 2 \sec \theta$

30. $\cos \theta = \sin^2 \frac{\theta}{2}$

31. $2 - \sin 2\theta = 4 \sin 2\theta$

32. $4 \cos 2\theta = 8 \sin \theta \cos \theta$

33. $2 \cos^2 2\theta = 1 - \cos 2\theta$

34. $\sin \theta = \cos \frac{\theta}{2}$

35. $\tan 3\theta + \sec 3\theta = 1$

36. $\cot \frac{\theta}{2} = \csc \frac{\theta}{2} + 1$

37. $\sin \frac{\theta}{2} = \cos \frac{\theta}{2} + \frac{1}{4}$

38. $\sin 2\theta + \cos 2\theta = \frac{1}{2}$

For the following equations, use the sum and product identities from Section 5.7. Give all solutions in the interval $0 \le x < 2\pi$.

39. $\sin x + \sin 3x = \cos x$

40. $\cos 4x - \cos 2x = \sin x$

41. $\sin 3x - \sin x = 0$

42. $\cos 2x + \cos x = 0$

43. $\sin 4x + \sin 2x \doteq 2 \cos x$

44. $\cos 5x + \cos 3x = 2 \cos 4x$

45. The seasonal variation in the length of daylight can be represented by a sine function. For example, the daily number of hours of daylight in New Orleans is given by

$$h = \frac{35}{3} + \frac{7}{3} \sin \frac{2\pi x}{365},$$

where x is the number of days after March 21 (disregarding leap year).*

(a) On what date will there be about 14 hours of daylight?

(b) What date has the least number of hours of daylight?

(c) When will there be about 10 hours of daylight?

---

*From *A Sourcebook of Applications of School Mathematics* by Donald Bushaw et al. The material was prepared with the support of National Science Foundation Grant No. SED72-01123 A05. However, any opinions, findings, conclusions, or recommendations expressed herein are those of the authors and do not necessarily reflect the views of NSF.

46. The British nautical mile is defined as the length of a minute of arc of a meridian. Since the earth is flat at its poles, the nautical mile, in feet, is given by

$$L = 6{,}077 - 31 \cos 2\theta,$$

where θ is the latitude in degrees. (See the figure.)

(a) Find the latitude(s) at which the nautical mile is 6,074 feet.

(b) At what latitude(s) is the nautical mile 6,108 feet?

(c) In the United States the nautical mile is defined everywhere as 6,080.2 feet. At what latitude(s) does this agree with the British nautical mile?*

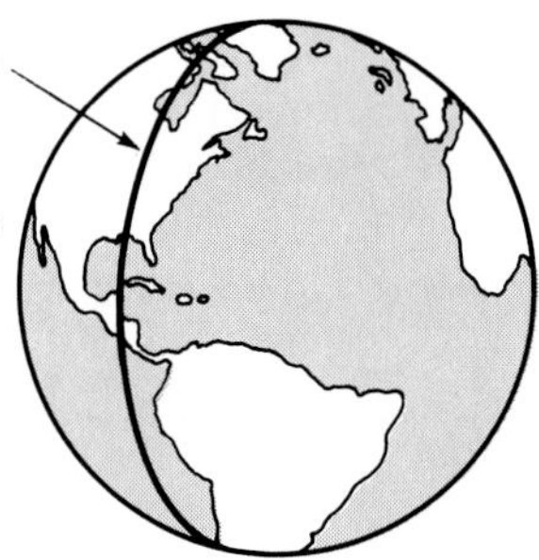

The study of alternating electric current requires the solutions of equations of the form $i = I_{max} \sin 2\pi ft$, for time t in seconds, where i is instantaneous current in amperes, I_{max} is maximum current in amperes, and f is the number of cycles per second.† Find the smallest positive value of t, given the following data.

47. $i = 40, I_{max} = 100, f = 60$

48. $i = 50, I_{max} = 100, f = 120$

49. $i = I_{max}, f = 60$

50. $i = \frac{1}{2} I_{max}, f = 60$

6.5 Inverse Trigonometric Equations

Section 6.2 introduced the inverse trigonometric functions. Recall, for example, that $x = \text{Sin } y$ means the same thing as $y = \text{Arcsin } x$ or $y = \text{Sin}^{-1} x$. Sometimes the solution of a trigonometric equation with more than one variable requires inverse trigonometric functions, as shown in the following examples.

---

*From *A Sourcebook of Applications of School Mathematics* by Donald Bushaw et al. Copyright © 1980 by The Mathematical Association of America. Reprinted by permission. The material was prepared with the support of National Science Foundation Grant No. SED72-01123 A05. However, any opinions, findings, conclusions, or recommendations expressed herein are those of the authors and do not necessarily reflect the views of NSF.

†Ralph H. Hannon, *Basic Technical Mathematics with Calculus* (Philadelphia: W. B. Saunders Co., 1978), pp. 300–302.

Example 1 Solve $y = 3 \cos 2x$ for x.

We want $\cos 2x$ alone on one side of the equation so we can solve for $2x$ and then for x. First, divide both sides of the equation by 3.

$$y = 3 \cos 2x$$

$$\frac{y}{3} = \cos 2x$$

Now write the statement in the alternate form

$$2x = \text{Arccos } \frac{y}{3}.$$

Finally, multiply both sides by 1/2.

$$x = \frac{1}{2} \text{ Arccos } \frac{y}{3}$$ ✣

The next examples show how to solve equations involving inverse trigonometric functions.

Example 2 Solve $2 \text{ Arcsin } x = \pi$.

First solve for Arcsin x.

$$2 \text{ Arcsin } x = \pi$$

$$\text{Arcsin } x = \frac{\pi}{2}$$

Use the definition of Arcsin x to get

$$x = \text{Sin } \frac{\pi}{2}$$

or

$$x = 1.$$

Verify that the solution satisfies the given equation. ✣

Example 3 Solve $\text{Cos}^{-1} x = \text{Sin}^{-1} \frac{1}{2}$.

Let $\text{Sin}^{-1} (1/2) = u$. Then $\text{Sin } u = 1/2$ and the equation becomes

$$\text{Cos}^{-1} x = u,$$

for u in quadrant I. This can be written as

$$\text{Cos } u = x.$$

Sketch a triangle and label it using the facts that u is in quadrant I and $\sin u = 1/2$. See Figure 6.15 on the following page.

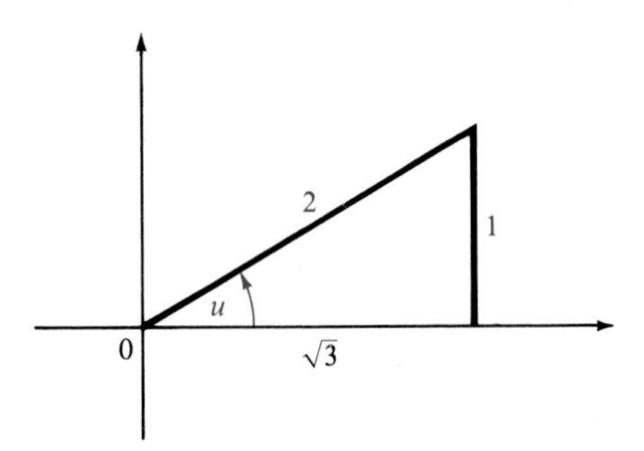

Figure 6.15

Since $x = \text{Cos } u$,

$$x = \frac{\sqrt{3}}{2}.$$

Some equations with inverse trigonometric functions require the use of identities.

Example 4 Solve $\text{Arcsin } x - \text{Arccos } x = \pi/6$.

Begin by adding Arccos x to both sides of the equation so that one inverse function is alone on one side of the equation.

$$\text{Arcsin } x - \text{Arccos } x = \frac{\pi}{6}$$

$$\text{Arcsin } x = \text{Arccos } x + \frac{\pi}{6}$$

Use the definition of Arcsin to write this statement as

$$\sin\left(\text{Arccos } x + \frac{\pi}{6}\right) = x.$$

Let $u = \text{Arccos } x$, so $0 \leq u \leq \pi$ by definition. Then

$$\sin\left(u + \frac{\pi}{6}\right) = x. \quad \textbf{(1)}$$

Using the identity for $\sin(A + B)$,

$$\sin\left(u + \frac{\pi}{6}\right) = \sin u \cos \frac{\pi}{6} + \cos u \sin \frac{\pi}{6}.$$

Substitute this result into equation (1) to get

$$\sin u \cos \frac{\pi}{6} + \cos u \sin \frac{\pi}{6} = x. \quad \textbf{(2)}$$

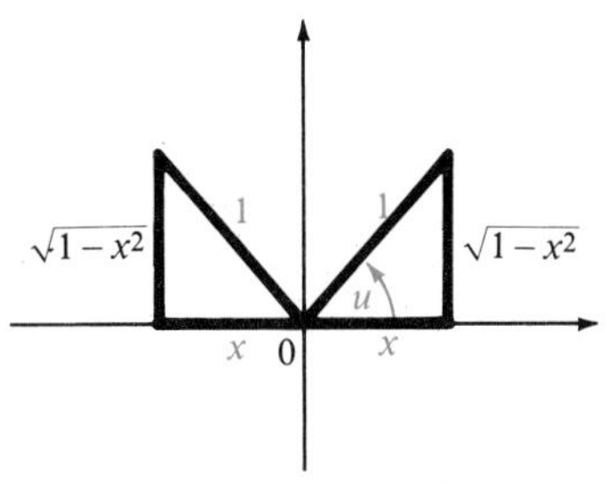

u can be in either quadrant I or II

Figure 6.16

From $u = \text{Arccos } x$, $\text{Cos } u = x$. Since x may be positive or negative, sketch a triangle in both quadrants I and II and label it as shown in Figure 6.16. From the triangle, $\sin u = \sqrt{1 - x^2}$ (for either quadrant.) Now substitute into equation (2) using $\sin u = \sqrt{1 - x^2}$, $\sin \pi/6 = 1/2$, $\cos \pi/6 = \sqrt{3}/2$, and $\cos u = x$.

$$(\sqrt{1 - x^2})\frac{\sqrt{3}}{2} + x \cdot \frac{1}{2} = x$$
$$(\sqrt{1 - x^2})\sqrt{3} + x = 2x$$
$$(\sqrt{3})\sqrt{1 - x^2} = x$$

Squaring both sides gives

$$3(1 - x^2) = x^2$$
$$3 - 3x^2 = x^2$$
$$3 = 4x^2$$
$$x = \pm\sqrt{\frac{3}{4}}$$
$$= \pm\frac{\sqrt{3}}{2}.$$

To check, first replace x with $\sqrt{3}/2$ in the original equation:

$$\text{Arcsin } \frac{\sqrt{3}}{2} - \text{Arccos } \frac{\sqrt{3}}{2} = \frac{\pi}{3} - \frac{\pi}{6} = \frac{\pi}{6},$$

as required. However, if $x = -\sqrt{3}/2$,

$$\text{Arcsin}\left(-\frac{\sqrt{3}}{2}\right) - \text{Arccos}\left(-\frac{\sqrt{3}}{2}\right) = \frac{-\pi}{3} - \frac{5\pi}{6} = \frac{-7\pi}{6} \neq \frac{\pi}{6}.$$

Only $x = \sqrt{3}/2$ satisfies the original equation. Each proposed solution was checked because both sides of the equation were squared.

6.5 Exercises

Solve each of the following equations for x. See Example 1.

1. $y = 5 \cos x$

2. $4y = \sin x$

3. $2y = \cot 3x$

4. $6y = \frac{1}{2} \sec x$

5. $y = 3 \tan 2x$

6. $y = 3 \sin \frac{x}{2}$

7. $y = 6 \cos \frac{x}{4}$

8. $y = -\sin \frac{x}{3}$

9. $y = -2 \cos 5x$

10. $y = 3 \cot 5x$

11. $y = \cos (x + 3)$

12. $y = \tan (2x - 1)$

13. $y = \sin x - 2$

14. $y = \cot x + 1$

15. $y = 2 \sin x - 4$

16. $y = 4 + 3 \cos x$

Solve each of the following equations. See Examples 2 and 3.

17. $\frac{4}{3} \text{Sin}^{-1} \frac{y}{4} = \pi$

18. $4\pi + 4 \text{ Tan}^{-1} y = \pi$

19. $2 \text{ Arccos} \left(\frac{y - \pi}{3}\right) = 2\pi$

20. $\text{Arccos} \left(y - \frac{\pi}{3}\right) = \frac{\pi}{6}$

21. $\text{Arcsin } x = \text{Arctan } \frac{3}{4}$

22. $\text{Arctan } x = \text{Arccos } \frac{5}{13}$

23. $\text{Cos}^{-1} x = \text{Sin}^{-1} \frac{3}{5}$

24. $\text{Cot}^{-1} x = \text{Tan}^{-1} \frac{4}{3}$

Solve each of the following equations. See Example 4.

25. $\text{Sin}^{-1} x - \text{Tan}^{-1} 1 = -\frac{\pi}{4}$

26. $\text{Sin}^{-1} x + \text{Tan}^{-1} \sqrt{3} = \frac{2\pi}{3}$

27. $\text{Arccos } x + 2 \text{ Arcsin } \frac{\sqrt{3}}{2} = \pi$

28. $\text{Arccos } x + 2 \text{ Arcsin } \frac{\sqrt{3}}{2} = \frac{\pi}{3}$

29. $\text{Arcsin } 2x + \text{Arccos } x = \frac{\pi}{6}$

30. $\text{Arcsin } 2x + \text{Arcsin } x = \frac{\pi}{2}$

31. $\text{Cos}^{-1} x + \text{Tan}^{-1} x = \frac{\pi}{2}$

32. $\text{Tan}^{-1} x + \text{Cos}^{-1} x = \frac{\pi}{2}$

33. Solve $d = 550 + 450 \cos \frac{\pi}{50} t$ for t in terms of d.

34. Solve $d = 40 + 60 \cos \frac{\pi}{6} (t - 2)$ for t in terms of d.

35. In the study of alternating electric current, instantaneous voltage is given by

$$e = E_{max} \sin 2\pi f t,$$

where f is the number of cycles per second, E_{max} is the maximum voltage, and t is time in seconds.

(a) Solve the equation for t.

(b) Find the smallest positive value of t if $E_{max} = 12$, $e = 5$, and $f = 100$.

36. When a large-view camera is used to take a picture of an object that is not parallel to the film, the lens board should be tilted so that the planes containing the subject, the lens board, and the film intersect in a line (see the figure). This gives the best "depth of field."*

(a) Write two equations, one relating α, x, and z, and the other relating β, x, y, and z.

(b) Eliminate z from the equations in part (a) to get one equation relating α, β, x, and y.

(c) Solve the equation from part (b) for α.

(d) Solve the equation from part (b) for β.

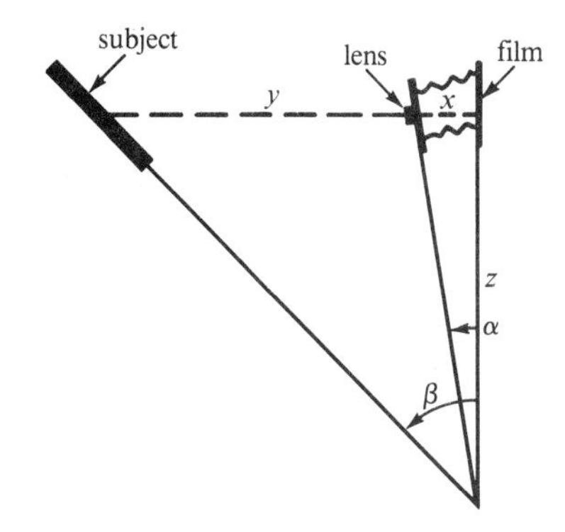

37. In many computer languages, such as BASIC and FORTRAN, only the Arctan function is available. To use the other inverse trigonometric functions, it is necessary to express them in terms of Arctangent. This can be done as follows.

(a) Let u = Arcsin x. Solve the equation for x in terms of u.

(b) Use the result of part (a) to label the three sides of the triangle in the figure in terms of x.

(c) Use the triangle from part (b) to write an equation for tan u in terms of x.

(d) Solve the equation from part (c) for u.

(e) Use the equation from part (d) to calculate Arcsin (1/2). Compare the answer with the actual value of Arcsin (1/2).

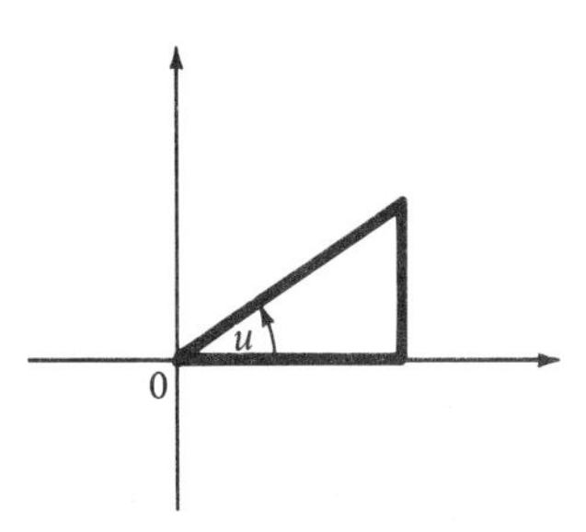

---

*From *A Sourcebook of Applications of School Mathematics* by Donald Bushaw et al. Copyright © 1980 by The Mathematical Association of America. Reprinted by permission. The material was prepared with the support of National Science Foundation Grant No. SED72-01123 A05. However, any opinions, findings, conclusions, or recommendations expressed herein are those of the authors and do not necessarily reflect the views of NSF.

Chapter 6 *Key Concepts*

Horizontal Line Test If it is possible to draw a horizontal line that cuts the graph of a function in more than one point, then the function is not one-to-one.

Inverse Function The **inverse function** of the one-to-one function f, written f^{-1} (read "f inverse"), is defined as

$$f^{-1} = \{(y, x) | (x, y) \textbf{ belongs to } f\}.$$

Finding $f^{-1}(x)$ Given the one-to-one function $f(x)$, find $f^{-1}(x)$ as follows.

1. Replace $f(x)$ with y.
2. Exchange x and y in $y = f(x)$.
3. Solve the equation from Step 2 for y.
4. Replace y with $f^{-1}(x)$.

Testing for Inverses Two one-to-one functions f and g are inverses of each other if and only if

$$f[g(x)] = x \quad \text{and} \quad g[f(x)] = x.$$

Inverse Trigonometric Functions

Function	*Domain*	*Range*
$y = \text{Sin}^{-1} x$	$-1 \le x \le 1$	$-\pi/2 \le y \le \pi/2$
$y = \text{Cos}^{-1} x$	$-1 \le x \le 1$	$0 \le y \le \pi$
$y = \text{Tan}^{-1} x$	Any real number	$-\pi/2 < y < \pi/2$
$y = \text{Cot}^{-1} x$	Any real number	$0 < y < \pi$
$y = \text{Sec}^{-1} x$	$x \le -1$ or $x \ge 1$	$0 \le y \le \pi, \quad y \ne \pi/2$
$y = \text{Csc}^{-1} x$	$x \le -1$ or $x \ge 1$	$-\pi/2 \le y \le \pi/2, \quad y \ne 0$

Chapter 6 *Key Concepts (continued)*

Solving Trigonometric Equations

1. If only one trigonometric function is present, first solve the equation for that function.
2. If more than one trigonometric function is present, rearrange the equation so that one side equals 0. Then try to factor and set each factor equal to 0 to solve.
3. If Step 2 does not work, try using identities to change the form of the equation. It may be helpful to square both sides of the equation first.
4. If the equation is quadratic in form, but not factorable, use the quadratic formula.

Chapter 6 *Review Exercises*

In Exercises 1–3, if the indicated function is one-to-one, sketch the graph of its inverse.

1.

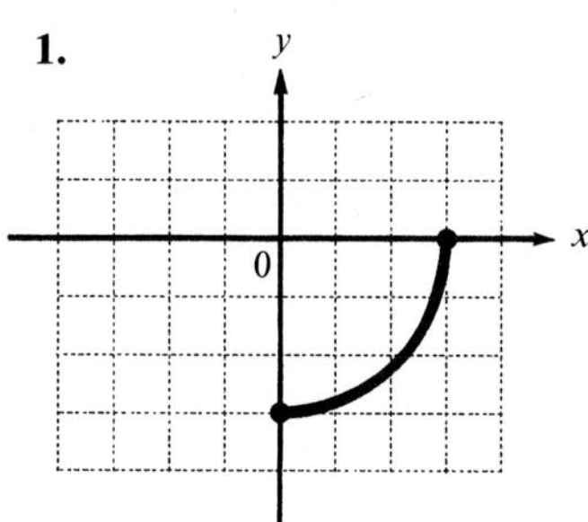

2.

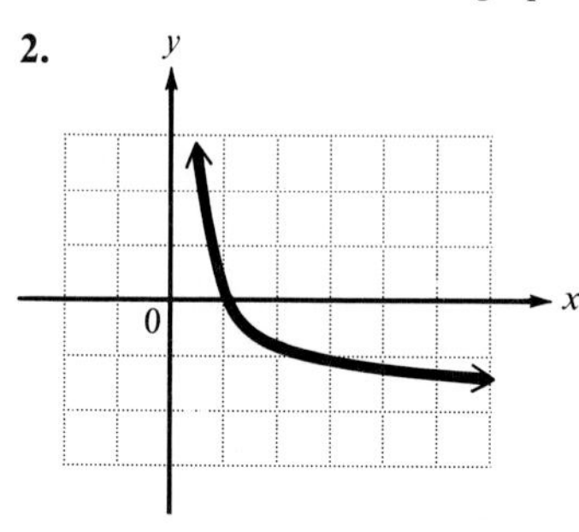

3.

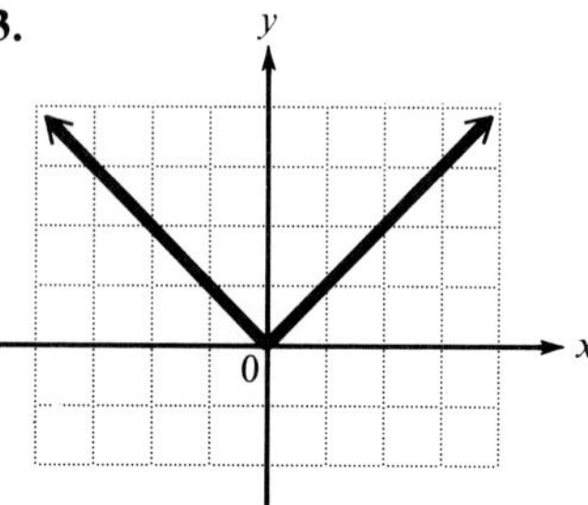

Which of the following functions are one-to-one?

4. $y = 5x - 2$

5. $y = x^2 + 3$

6. $y = \dfrac{-1}{x}$

7. $y = \sqrt{4 - x}$

8. $y = \sin x$

For each of the following functions that are one-to-one, write an equation for the inverse function.

9. $y = 12x + 3$

10. $4x + 3y = 10$

11. $y = \dfrac{2}{x - 9}$

12. $2y + 6 = x^2$

Decide whether the following pairs of functions are inverses of each other.

13. $f(x) = 5x$ and $g(x) = \dfrac{1}{5}x$

14. $f(x) = 3x + 6$ and $g(x) = \dfrac{1}{3}x - 6$

15. $f(x) = \dfrac{1}{x + 2}$ and $g(x) = \dfrac{1}{x - 2}$

16. $f(x) = \dfrac{4}{x}$ and $g(x) = \dfrac{x}{4}$

For each of the following give the value of y in radians without using a calculator or tables.

17. $y = \text{Sin}^{-1} \dfrac{\sqrt{2}}{2}$

18. $y = \text{Arccos}\left(\dfrac{-1}{2}\right)$

19. $y = \text{Tan}^{-1}(-\sqrt{3})$

Find each of the following in degrees to the nearest ten minutes.

20. $y = \text{Arctan } 1.780$

21. $y = \text{Sin}^{-1}(-.6604)$

22. $y = \text{Cos}^{-1} .8039$

Find each of the following without using tables or a calculator.

23. $\sin\left(\text{Sin}^{-1} \dfrac{1}{2}\right)$

24. $\tan\left(\text{Tan}^{-1} \dfrac{2}{3}\right)$

25. $\cos(\text{Arccos}(-1))$

26. $\sin\left(\text{Arcsin}\left(-\dfrac{\sqrt{3}}{2}\right)\right)$

27. $\text{Arccos}\left(\cos \dfrac{3\pi}{4}\right)$

28. $\text{Arcsec}(\sec \pi)$

29. $\text{Tan}^{-1}\left(\tan \dfrac{\pi}{4}\right)$

30. $\text{Cos}^{-1}(\cos 0)$

31. $\sin\left(\text{Arccos} \dfrac{3}{4}\right)$

32. $\cos(\text{Arctan } 3)$

33. $\cos(\text{Csc}^{-1}(-2))$

34. $\sec\left(2 \text{ Sin}^{-1}\left(\frac{-1}{3}\right)\right)$

35. $\tan\left(\text{Arcsin } \frac{3}{5} + \text{Arccos } \frac{5}{7}\right)$

Write each of the following as an expression in u.

36. $\sin(\text{Tan}^{-1} u)$

37. $\cos\left(\text{Arctan } \frac{u}{\sqrt{1 - u^2}}\right)$

38. $\tan\left(\text{Arcsec } \frac{\sqrt{u^2 + 1}}{u}\right)$

Graph each of the following, and give the domain and range.

39. $y = \text{Sin}^{-1} x$

40. $y = \text{Cos}^{-1} x$

41. $y = \text{Arccot } x$

Find all solutions for the following equations in the interval $0 \leq x < 2\pi$.

42. $\sin^2 x = 1$

43. $2 \tan x - 1 = 0$

44. $3 \sin^2 x - 5 \sin x + 2 = 0$

45. $\tan x = \cot x$

46. $\sec^4 2x = 4$

47. $\tan^2 2x - 1 = 0$

48. $\sec \frac{x}{2} = \cos \frac{x}{2}$

49. $\cos 2x + \cos x = 0$

50. $4 \sin x \cos x = \sqrt{3}$

Find all solutions for the following equations in the interval $0° \leq \theta < 360°$. Round solutions to the nearest tenth of a degree.

51. $\sin^2 \theta + 3 \sin \theta + 2 = 0$

52. $2 \tan^2 \theta = \tan \theta + 1$

53. $\sin 2\theta = \cos 2\theta + 1$

54. $2 \sin 2\theta = 1$

55. $3 \cos^2 \theta + 2 \cos \theta - 1 = 0$

56. $5 \cot^2 \theta - \cot \theta - 2 = 0$

57. $\sin 2\theta + \sin 4\theta = 0$

58. $\cos \theta - \cos 2\theta = 2 \cos \theta$

Solve each equation for x.

59. $4y = 2 \sin x$

60. $y = 3 \cos \frac{x}{2}$

61. $2y = \tan(3x + 2)$

62. $5y = 4 \sin x - 3$

63. $\frac{4}{3} \text{ Arctan } \frac{x}{2} = \pi$

64. $\text{Arccos } x = \text{Arcsin } \frac{2}{7}$

65. $\text{Arccos } x + \text{Arctan } 1 = \frac{11\pi}{12}$

66. $\text{Arccot } x = \text{Arcsin}\left(\frac{-\sqrt{2}}{2}\right) + \frac{3\pi}{4}$

67. Recall Snell's law from Exercises 85–88 of Section 2.3:

$$\frac{c_1}{c_2} = \frac{\sin \theta_1}{\sin \theta_2},$$

where c_1 is the speed of light in one medium, c_2 is the speed of light in a second medium, and θ_1 and θ_2 are the angles shown in the figure on the following page. Suppose that a light is shining up through water into the air as in the figure on the following page. As θ_1 increases, θ_2 approaches 90°, at which point no light will emerge from the water. Assume the ratio c_1/c_2 in this case is .752.

(a) For what value of θ_1 does $\theta_2 = 90°$? This value of θ_1 is called the *criticial angle* for water.

(b) What happens when θ_1 is greater than the critical angle?

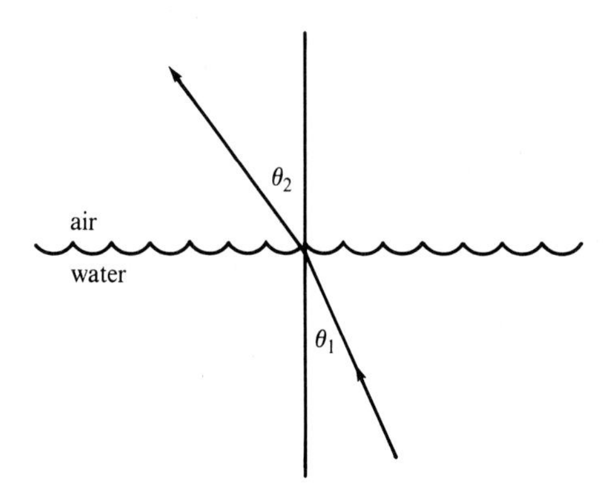

68. In the exercises for Section 4.1 we found the equation

$$y = \frac{1}{3} \sin \frac{4\pi t}{3},$$

where t is time (in seconds) and y is the angle formed by a rhythmically moving arm.

(a) Solve the equation for t.

(b) At what time(s) does the arm form an angle of .3 radians?

7 Triangles and Vectors

Trigonometry is more than three thousand years old. The ancient Egyptians, Babylonians, and Greeks developed trigonometry to find the lengths of the sides of triangles and the measures of their angles. Every triangle has three sides and three angles. In this chapter we show that if any three of the six measures of a triangle (provided at least one measure is a side) are known, then the other three measures can be found. This process is called solving a triangle. In the latter part of the chapter this knowledge is used to solve problems involving vectors.

7.1 Oblique Triangles and the Law of Sines

In the next few sections the methods of solving right triangles are generalized to include all triangles. A triangle that is not a right triangle is called an **oblique triangle.** The measures of the three sides and the three angles of a triangle can be found if at least one side and any other two measures are known. There are four possible cases.

Solving Oblique Triangles

1. One side and two angles are known.
2. Two sides and one angle not included between the two sides are known. This case may lead to more than one triangle.
3. Two sides and the angle included between the two sides are known.
4. Three sides are known.

The first two cases require the *law of sines,* which is discussed in this section and the next. The last two cases require the *law of cosines,* which is discussed in section 7.3.

To derive the law of sines, start with an oblique triangle, such as the acute triangle in Figure 7.1(a) or the obtuse triangle in Figure 7.1(b). (Recall: These terms were defined in section 1.3.) The following discussion applies to both triangles. First, construct the perpendicular from B to side AC. Let h be the length of this perpendicular. Then c is the hypotenuse of right triangle ADB, and a is the hypotenuse of right triangle BDC. By results from chapter 2,

$$\text{in triangle } ADB, \qquad \sin A = \frac{h}{c} \quad \text{or} \quad h = c \sin A,$$

$$\text{in triangle } BDC, \qquad \sin C = \frac{h}{a} \quad \text{or} \quad h = a \sin C.$$

Since $h = c \sin A$ and $h = a \sin C$,

$$a \sin C = c \sin A,$$

or, upon dividing both sides by $\sin A \sin C$,

$$\frac{a}{\sin A} = \frac{c}{\sin C}.$$

In a similar way, by constructing the perpendiculars from other vertices, it can be shown that

$$\frac{a}{\sin A} = \frac{b}{\sin B} \quad \text{and} \quad \frac{b}{\sin B} = \frac{c}{\sin C}.$$

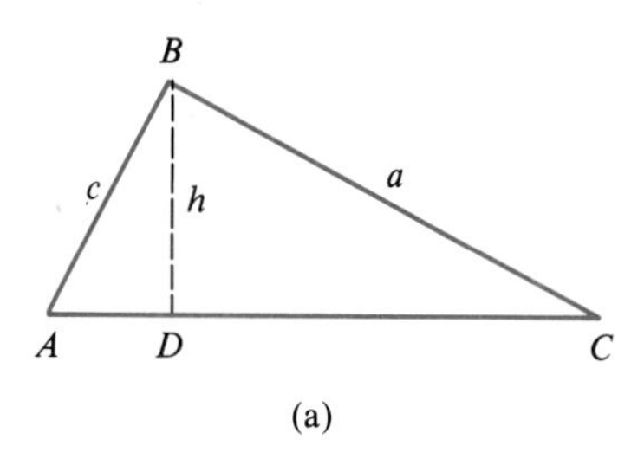

(a)

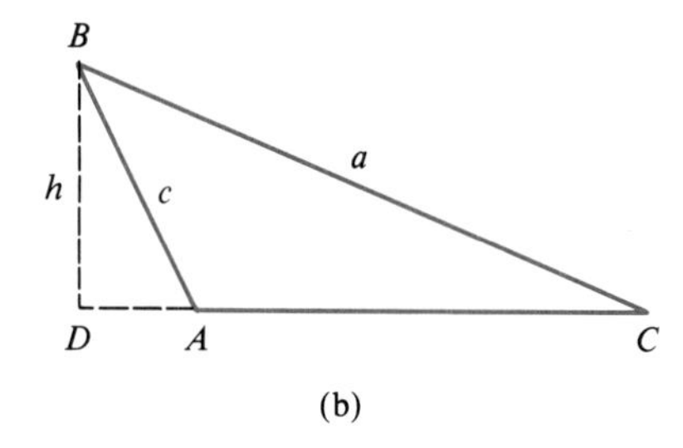

(b)

Figure 7.1

This discussion proves the following theorem.

Law of Sines

In any triangle ABC, with sides a, b, and c,

$$\frac{a}{\sin A} = \frac{b}{\sin B}, \quad \frac{a}{\sin A} = \frac{c}{\sin C}, \quad \text{and} \quad \frac{b}{\sin B} = \frac{c}{\sin C}.$$

The three formulas of the law of sines can be written in a more compact form as

$$\frac{a}{\sin A} = \frac{b}{\sin B} = \frac{c}{\sin C} \quad \text{or} \quad \frac{\sin A}{a} = \frac{\sin B}{b} = \frac{\sin C}{c}.$$

In some cases, this second form is easier to use.

If two angles and one side of a triangle are known, the law of sines can be used to solve the triangle.

Example 1 Solve triangle ABC if $A = 32.0°$, $B = 81.8°$, and $a = 42.9$ cm. See Figure 7.2.

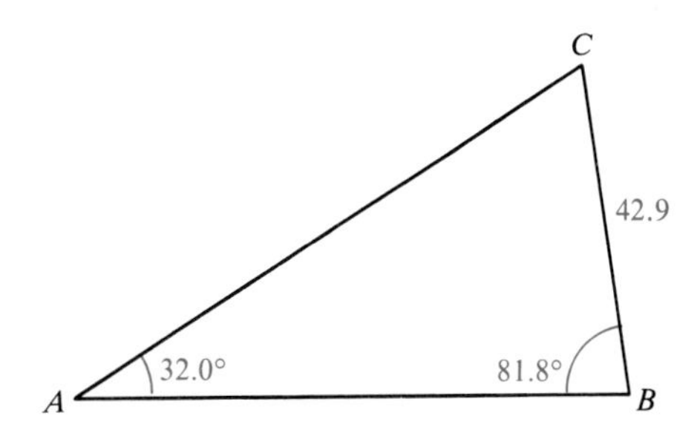

Figure 7.2

Start by drawing a triangle, roughly to scale, and labeling the given parts as in Figure 7.2. Since the values of A, B, and a are known, use the part of the law of sines that involves these variables.

$$\frac{a}{\sin A} = \frac{b}{\sin B}$$

Substituting the known values gives

$$\frac{42.9}{\sin 32.0°} = \frac{b}{\sin 81.8°}.$$

Multiply both sides of the equation by sin 81.8°.

$$b = \frac{42.9 \sin 81.8°}{\sin 32.0°}$$

When using a calculator to find b, keep intermediate answers in the calculator until the final result is found. Then round to the proper number of significant digits. In this case, find sin 81.8°, and then multiply that number by 42.9. Keep the result in the calculator while you find sin 32.0°, and then divide. This final answer should be rounded to 3 significant figures.

$$b = 80.1 \text{ cm}$$

Find C from the fact that the sum of the angles of any triangle is 180°.

$$A + B + C = 180°$$
$$C = 180° - A - B$$
$$C = 180° - \mathbf{32.0°} - \mathbf{81.8°} = 66.2°$$

Now use the law of sines again to find c. (Why should the Pythagorean theorem not be used?)

$$\frac{a}{\sin A} = \frac{c}{\sin C}$$
$$\frac{\mathbf{42.9}}{\sin \mathbf{32.0°}} = \frac{c}{\sin \mathbf{66.2°}}$$
$$c = \frac{42.9 \sin 66.2°}{\sin 32.0°}$$
$$c = 74.1 \text{ cm}$$

Example 2 Shawn Johnson wishes to measure the distance across the Big Muddy River. See Figure 7.3. She finds that $C = 112° \ 53'$, $A = 31° \ 06'$, and $b = 347.6$ ft. Find the required distance.

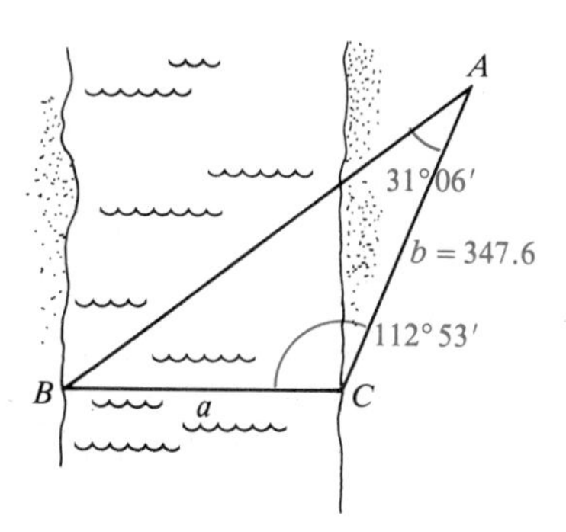

Figure 7.3

To use the law of sines, one side and the angle opposite it must be known. Since the only side whose length is given is b, angle B must be found before the law of sines can be used.

$$B = 180° - A - C$$
$$= 180° - \mathbf{31° \ 06'} - \mathbf{112° \ 53'} = 36° \ 01'$$

Now the required distance a can be found. Use the portion of the law of sines involving A, B, and b.

$$\frac{a}{\sin A} = \frac{b}{\sin B}$$

Substitute the known values.

$$\frac{a}{\sin \mathbf{31^\circ\ 06'}} = \frac{\mathbf{347.6}}{\sin \mathbf{36^\circ\ 01'}}$$

$$a = \frac{347.6 \sin 31^\circ\ 06'}{\sin 36^\circ\ 01'}$$

$$a = 305.3 \text{ ft}$$

Example 3 Two tracking stations are on an east-west line 110 miles apart. A weather balloon is located on a bearing of N 42° E from the western station and a bearing of N 15° E from the eastern station. How far is the balloon from the western station?

Figure 7.4 shows the two stations at points A and B and the weather balloon at point C. Angle $BAC = 90^\circ - 42^\circ = 48^\circ$, the obtuse angle at B equals $90^\circ + 15^\circ = 105^\circ$, and the third angle, C, equals $180^\circ - 105^\circ - 48^\circ = 27^\circ$. Using the law of sines to find side b gives

$$\frac{b}{\sin 105^\circ} = \frac{110}{\sin 27^\circ}$$

$$b = 230,$$

or 230 miles (to two significant digits).

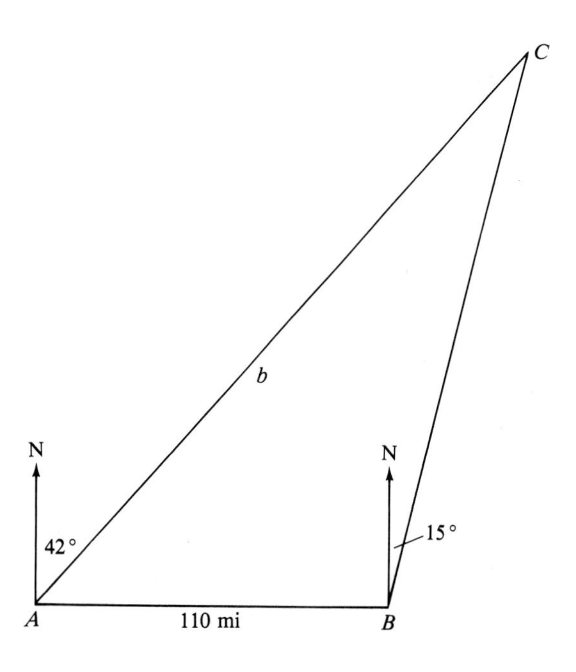

Figure 7.4

Area The method used to derive the law of sines can also be used to derive a useful formula for the area of a triangle. A familiar formula for the area of a triangle is $K = (1/2)bh$, where K represents the area, b the base, and h the height. This formula cannot always be used, since in practice h is often unknown. To find a more useful formula, refer to acute triangle ABC in Figure 7.5(a) or obtuse triangle ABC in Figure 7.5(b).

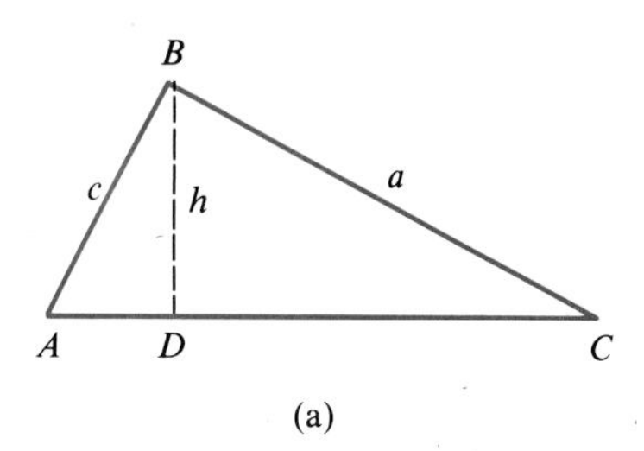

(a)

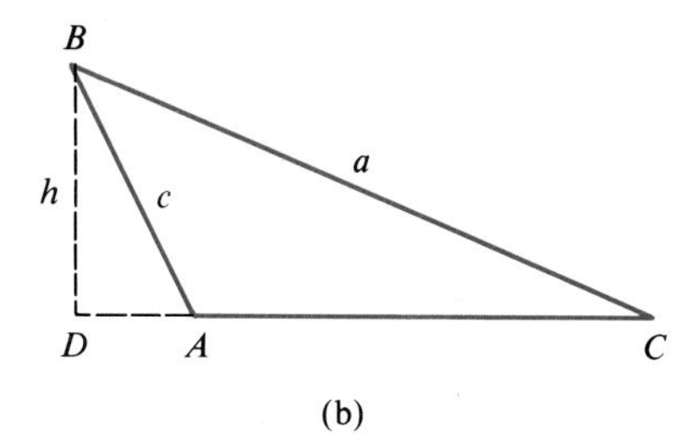

(b)

Figure 7.5

A perpendicular has been drawn from B to the base of the triangle. This perpendicular forms two right triangles. Using triangle ABD,

$$\sin A = \frac{h}{c},$$

or

$$h = c \sin A.$$

Substituting into the formula $K = \frac{1}{2} bh$,

$$K = \frac{1}{2} b(c \sin A)$$

$$\mathbf{K = \frac{1}{2} bc \sin A.}$$

Any other pair of sides and the angle between them could have been used, as stated in the next theorem.

Area of a Triangle

> The area of a triangle is given by half the product of the lengths of two sides and the sine of the angle between the two sides.

Example 4 Find the area of triangle MNP if $m = 29.7$ m, $n = 53.9$ m, and $P = 28° 40'$.
By the last result, the area of the triangle is

$$\frac{1}{2}(29.7)(53.9)\sin 28° 40' = 384 \text{ m}^2.$$

Example 5 Find the area of triangle ABC if $A = 24° 40'$, $b = 27.3$ cm, and $C = 52° 40'$.
Before we can use the formula given above, we must use the law of sines to find either a or c. Let us find a.

$$\frac{a}{\sin 24° 40'} = \frac{27.3}{\sin 102° 40'}$$

(Where did we get $102° 40'$?) Solve for a to verify that $a = 11.7$ cm. Now find the area, K.

$$K = \frac{1}{2}ab\sin C = \frac{1}{2}(11.7)(27.3)\sin 52° 40' = 127$$

The area of triangle ABC is 127 cm^2.

7.1 Exercises

Solve each of the following triangles. Use interpolation or a calculator as necessary. See Example 1.

1. $A = 37°$, $B = 48°$, $c = 18$ m
2. $B = 52°$, $C = 29°$, $a = 43$ cm
3. $A = 46° 30'$, $B = 52° 50'$, $b = 87.3$ mm
4. $A = 59° 30'$, $B = 48° 20'$, $b = 32.9$ m
5. $A = 27.2°$, $C = 115.5°$, $c = 76.0$ ft
6. $B = 124.1°$, $C = 18.7°$, $c = 94.6$ m
7. $A = 68.41°$, $B = 54.23°$, $a = 12.75$ ft
8. $C = 74.08°$, $B = 69.38°$, $c = 45.38$ m
9. $A = 87.2°$, $b = 75.9$ yd, $C = 74.3°$
10. $B = 38° 40'$, $a = 19.7$ cm, $C = 91° 40'$
11. $B = 20° 50'$, $C = 103° 10'$, $AC = 132$ ft
12. $A = 35.3°$, $B = 52.8°$, $AC = 675$ ft

Use a calculator or interpolation to solve each of the following triangles.

13. $A = 39.70°$, $C = 30.35°$, $b = 39.74$ m
14. $C = 71.83°$, $B = 42.57°$, $a = 2.614$ cm
15. $B = 42.88°$, $C = 102.40°$, $b = 3974$ ft
16. $A = 18.75°$, $B = 51.53°$, $c = 2798$ yd
17. $A = 39° 54'$, $a = 268.7$ m, $B = 42° 32'$
18. $C = 79° 18'$, $c = 39.81$ mm, $A = 32° 57'$

Solve each of the following exercises. Recall that bearing was discussed in Section 2.6. Use a calculator or interpolate as necessary. See Examples 2 and 3.

19. To find the distance AB across a river, a distance $BC = 354$ m is laid off on one side of the river. It is found that $B = 112° 10'$ and $C = 15° 20'$. Find AB.

20. To determine the distance RS across a deep canyon, Joanna lays off a distance $TR = 582$ yd. She then finds that $T = 32° 50'$ and $R = 102° 20'$. Find RS.

21. Radio direction finders are placed at points A and B, which are 3.46 mi apart on an east-west line, with A west of B. From A the bearing of a certain radio transmitter is 47.7°, and from B the bearing is 302.5°. Find the distance of the transmitter from A.

22. A ship is sailing due north. At a certain point the bearing of a lighthouse 12.5 km distant is N 38.8° E. Later on, the captain notices that the bearing of the lighthouse has become S 44.2° E. How far did the ship travel between the two observations of the lighthouse?

23. A folding chair is to have a seat 12.0 in deep with angles as shown in the figure. How far down from the seat should the crossing legs be joined? (Find x in the figure.)

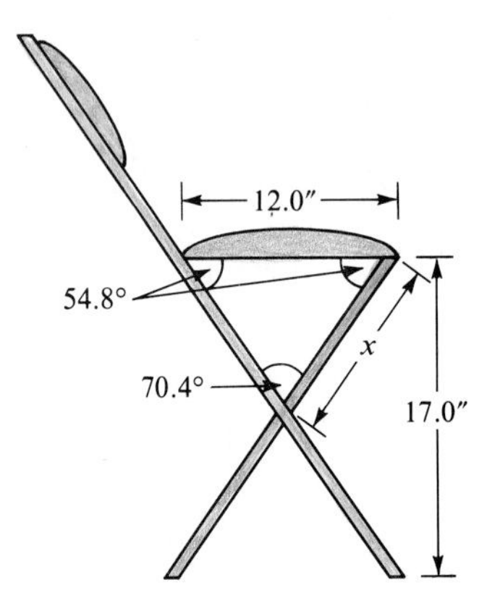

24. Mark notices that the bearing of a tree on the opposite bank of a river flowing north is 115.45°. Lisa is on the same bank as Mark, but 428.3 m away. She notices that the bearing of the tree is 45.47°. The two banks are parallel. What is the distance across the river?

25. Three gears are arranged as shown in the figure. Find angle θ.

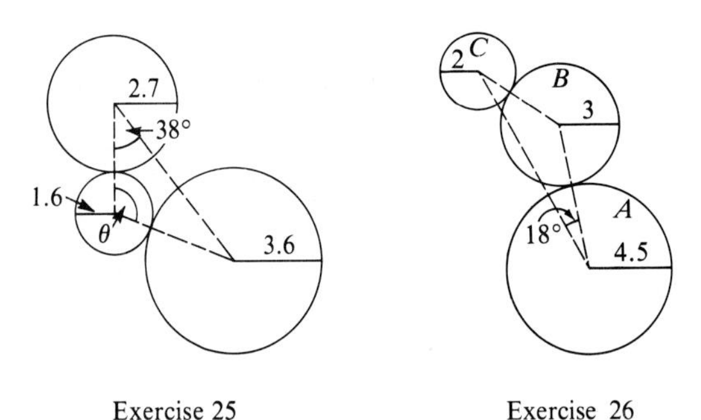

Exercise 25 Exercise 26

26. Three atoms with atomic radii of 2, 3, and 4.5 are arranged as in the figure. Find the distance between the centers of atoms A and C.

Find the area of each of the following triangles. Use a calculator or interpolate as necessary. See Examples 4 and 5.

27. $A = 42.5°$, $b = 13.6$ m, $c = 10.1$ m

28. $C = 72.2°$, $b = 43.8$ ft, $a = 35.1$ ft

29. $B = 124.5°$, $a = 30.4$ cm, $c = 28.4$ cm

30. $C = 142.7°$, $a = 21.9$ km, $b = 24.6$ km

31. $A = 56.80°$, $b = 32.67$ in, $c = 52.89$ in

32. $A = 34.97°$, $b = 35.29$ m, $c = 28.67$ m

33. $A = 24° 25'$, $B = 56° 20'$, $c = 78.40$ cm

34. $B = 48° 30'$, $C = 74° 20'$, $a = 462$ km

35. A painter is going to apply a special coating to a triangular metal plate on a new building. Two sides measure 16.1 m and 15.2 m. She knows that the angle between these sides is 125°. How many square meters should she plan to cover with the coating?

36. A real estate agent wants to find the area of a triangular lot. A surveyor takes measurements and finds that two sides are 52.1 m and 21.3 m, and the angle between them is 42.2°. What is the area of the lot?

37. In any triangle having sides a, b, and c, it must be true that $a + b > c$. Use this fact and the law of sines to show that $\sin A + \sin B > \sin (A + B)$ for any two angles A and B of a triangle.

38. Show that the area of a triangle having sides a, b, and c and corresponding angles A, B, and C is given by

$$\frac{a^2 \sin B \sin C}{2 \sin A}.$$

7.2 The Ambiguous Case of the Law of Sines

The law of sines can be used with any two sides of a triangle and the angles opposite these sides. Also, if two angles and one side are known, the triangle can be solved. What about the case in which two sides and an angle opposite one of them is known? As the next examples show, such a triangle might not even exist, or there might be two triangles satisfying the given conditions.

Example 1 Solve triangle ABC if $C = 55° 40'$, $c = 8.94$ m, and $b = 25.1$ m.

Let us first look for angle B. The work is easier if the unknown is in the numerator, so start with

$$\frac{\sin B}{b} = \frac{\sin C}{c}.$$

Substitute the given values.

$$\frac{\sin B}{25.1} = \frac{\sin 55° 40'}{8.94}$$

$$\sin B = \frac{25.1 \sin 55° 40'}{8.94}$$

$$\sin B = 2.3184$$

Sin B is greater than 1. This is impossible, since $-1 \le \sin B \le 1$, for any angle B. If you try to find the angle whose sine is 2.3184 with a calculator, you get ERROR! This tells you that triangle ABC does not exist. See Figure 7.6. ✤

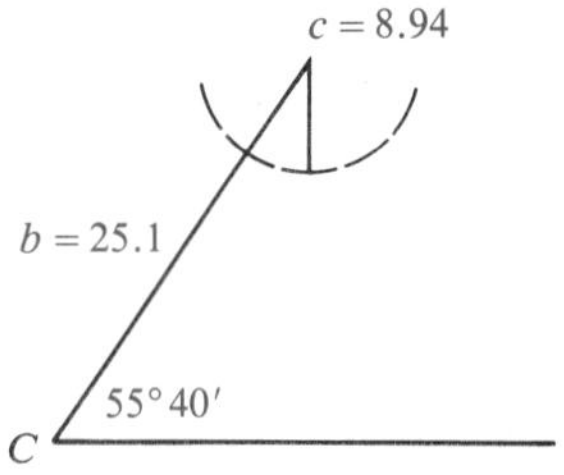

Figure 7.6

When any two angles and the length of a side of a triangle are given, the law of sines can be applied directly to solve the triangle. However, if only one angle and two sides are given, the triangle may not exist, as in Example 1, or there may be more than one triangle satisfying the given conditions. For example, suppose

Number of possible triangles	*Sketch*	*Condition necessary for case to hold*
0	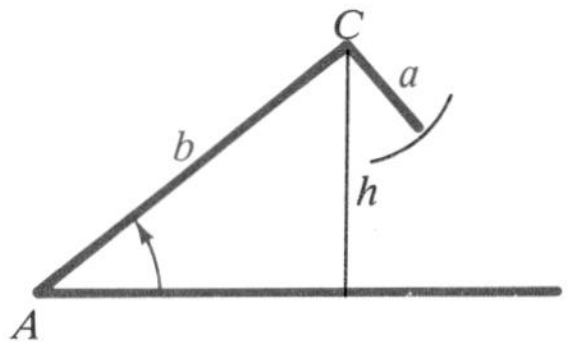	$a < h$ $(h = b \sin A)$
1	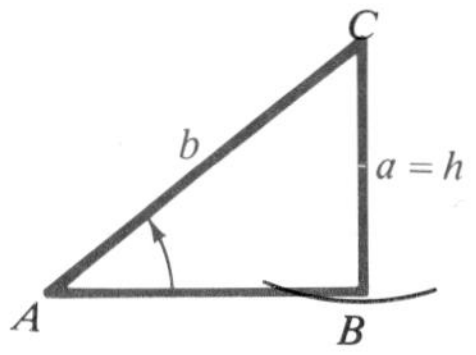	$a = h$
1	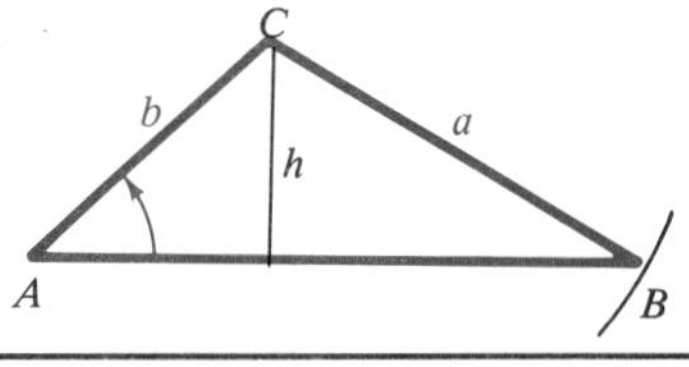	$a > b$
2	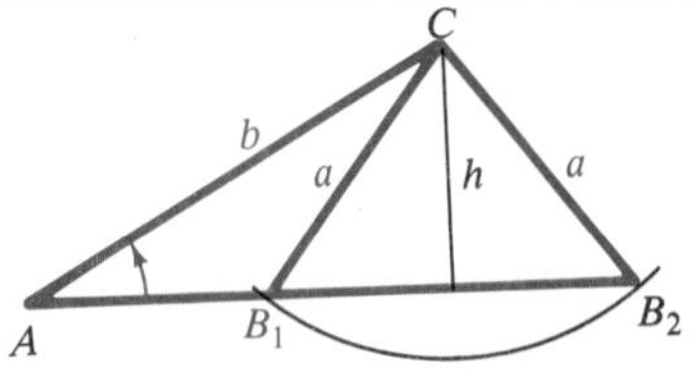	$b > a > h$

that the measure of acute angle A of triangle ABC, the length of side a, and the length of side b are given. To show this information, draw angle A having a terminal side of length b. Now draw a side of length a opposite angle A. The chart on page 278 shows that there might be more than one possible outcome. This situation is called the **ambiguous case of the law of sines.**

If angle A is obtuse, there are two possible outcomes, as shown in the next chart.

Number of possible triangles	*Sketch*	*Condition necessary for case to hold*
0	C, a, b, A	$a \leq b$
1	C, a, b, A, B	$a > b$

It is possible to derive formulas that show which of the various cases exist for a particular set of numerical data. However, this work is unnecessary when using the law of sines. For example, as in Example 1, if the law of sines is used and $\sin B$ is greater than 1, there is no triangle at all. The case in which the result is two different triangles is illustrated in the next example.

Example 2 Solve triangle ABC if $A = 55° 20'$, $a = 22.8$, and $b = 24.9$.

To begin, use the law of sines to find angle B.

$$\frac{a}{\sin A} = \frac{b}{\sin B}$$

$$\frac{22.8}{\sin 55° 20'} = \frac{24.9}{\sin B}$$

$$\sin B = \frac{24.9 \sin 55° 20'}{22.8}$$

$$\sin B = .8992$$

Since $a < b$ and $\sin B < 1$, there are two triangles that satisfy the given information, as shown in Figure 7.7.

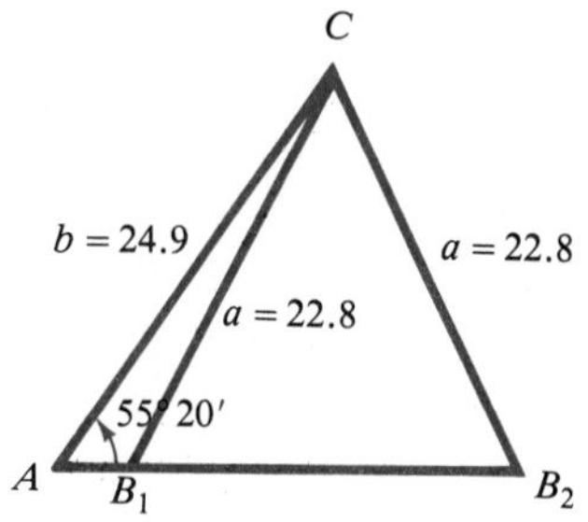

Figure 7.7

Since $\sin B = .8982$, one value of B is

$$B = 64^\circ\ 00'.$$

From the identity $\sin(180^\circ - B) = \sin B$, another possible value of B is

$$B = 180^\circ - 64^\circ\ 00' = 116^\circ\ 00'.$$

To keep track of these two different values of B, let

$$B_1 = 116^\circ\ 00' \qquad \text{and} \qquad B_2 = 64^\circ\ 00'.$$

Now separately solve triangles AB_1C_1 and AB_2C_2 shown in Figure 7.8.

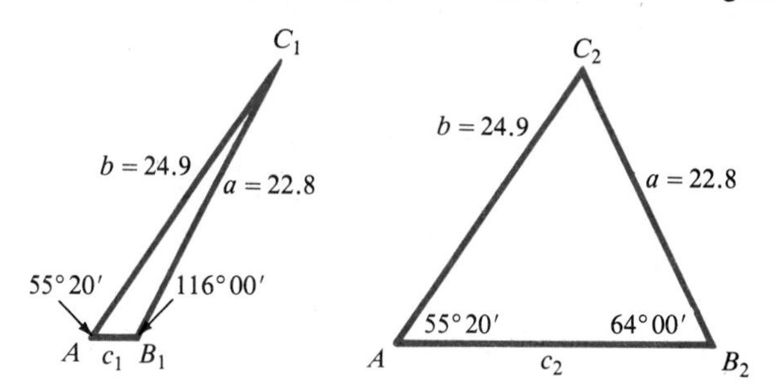

Figure 7.8

Since B_1 is the larger of the two values of B, find C_1 next.

$$C_1 = 180^\circ - A - B_1 = 8^\circ\ 40'.$$

Now, use the law of sines to find c_1.

$$\frac{a}{\sin A} = \frac{c_1}{\sin C_1}$$

$$\frac{22.8}{\sin 55^\circ\ 20'} = \frac{c_1}{\sin 8^\circ\ 40'}$$

$$c_1 = \frac{22.8 \sin 8^\circ\ 40'}{\sin 55^\circ\ 20'}$$

$$c_1 = 4.18$$

To solve triangle AB_2C_2, first find C_2.

$$C_2 = 180° - A - B_2 = 60° \, 40'$$

By the law of sines,

$$\frac{22.8}{\sin 55° \, 20'} = \frac{c_2}{\sin 60° \, 40'}$$

$$c_2 = \frac{22.8 \sin 60° \, 40'}{\sin 55° \, 20'}$$

$$c_2 = 24.2.$$ ✤

The final example illustrates the case where there is one triangle.

Example 3 Solve triangle ABC given $A = 43.5°$, $a = 10.7$, and $b = 7.2$.
To find angle B use the law of sines.

$$\frac{\sin B}{7.2} = \frac{\sin 43.5°}{10.7}$$

$$\sin B = \frac{7.2 \sin 43.5°}{10.7} = .4632$$

$$B = 27.6°$$

If there were two triangles, the angle at B would measure $180° - 27.6° = 152.4°$. Since the sum of this angle and angle A is 195.9°, which is greater than 180°, there can be only one triangle. Then angle $C = 180° - 27.6° - 43.5° = 108.9°$, and side c can be found with the law of sines.

$$\frac{c}{\sin 108.9°} = \frac{10.7}{\sin 43.5°}$$

$$c = \frac{10.7 \sin 108.9°}{\sin 43.5°}$$

$$c = 14.7$$ ✤

7.2 Exercises

Sketch all possible triangles satisfying each of the following conditions. Do not solve the triangles.

1. $A = 35°$, $b = 25$, $a = 37$

2. $A = 56°$, $b = 45$, $a = 52$

3. $B = 35°$, $b = 15$, $a = 20$

4. $C = 68°$, $b = 102$, $c = 95$

5. $A = 128°$, $a = 18$, $c = 12$

6. $C = 114°$, $a = 39$, $c = 20$

Find the missing angles in each of the following triangles that exist. Use a calculator or interpolate as necessary. See Example 1.

7. $A = 29.7°$, $b = 41.5$ ft, $a = 27.2$ ft

8. $B = 48.2°$, $a = 890$ cm, $b = 697$ cm

9. $C = 41° \, 20'$, $b = 25.9$ m, $c = 38.4$ m

10. $B = 48° \, 50'$, $a = 3850$ in, $b = 4730$ in

11. $B = 74.3°$, $a = 859$ m, $b = 783$ m

12. $C = 82.2°$, $a = 10.9$ km, $c = 7.62$ km

13. $A = 142.13°$, $b = 5.432$ ft, $a = 7.297$ ft

14. $B = 113.72°$, $a = 189.6$ yd, $b = 243.8$ yd

15. $C = 129° 18'$, $a = 372.9$ cm, $c = 416.7$ cm

16. $A = 132° 07'$, $b = 7.481$ mi, $a = 8.219$ mi

Solve each of the following triangles that exist. Use a calculator or interpolate as necessary. See Examples 1–3.

17. $A = 42.5°$, $a = 15.6$ ft, $b = 8.14$ ft

18. $C = 52.3°$, $a = 32.5$ yd, $c = 59.8$ yd

19. $B = 72.2°$, $b = 78.3$ m, $c = 145$ m

20. $C = 68.5°$, $c = 258$ cm, $b = 386$ cm

21. $A = 38° 40'$, $a = 9.72$ km, $b = 11.8$ km

22. $C = 29° 50'$, $a = 8.61$ m, $c = 5.21$ m

23. $B = 32° 50'$, $a = 7540$ cm, $b = 5180$ cm

24. $C = 22° 50'$, $b = 159$ mm, $c = 132$ mm

25. $A = 96.80°$, $b = 3.589$ ft, $a = 5.818$ ft

26. $C = 88.70°$, $b = 56.87$ yd, $c = 112.4$ yd

27. $B = 39.68°$, $a = 29.81$ m, $b = 23.76$ m

28. $A = 51.20°$, $c = 7986$ cm, $a = 7208$ cm

29. A surveyor reported the following data about a piece of property: "The property is triangular in shape, with dimensions as shown in the figure." Use the law of sines to see whether such a piece of property could exist.

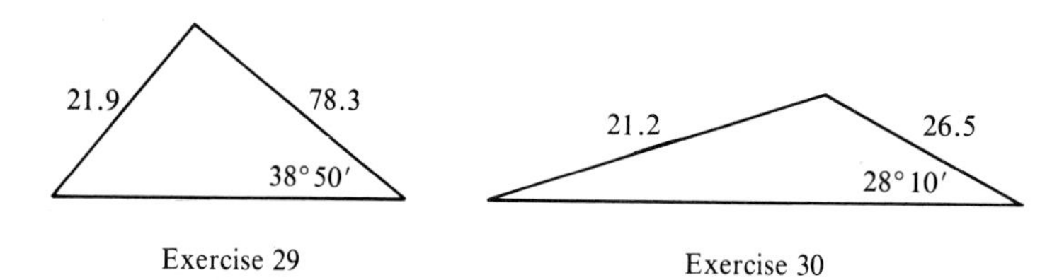

Exercise 29 Exercise 30

30. The surveyor tries again: "A second triangular piece of property has dimensions as shown." This time it turns out that the surveyor did not consider every possible case. Use the law of sines to show why.

Use the law of sines to prove that each of the following statements is true for any triangle ABC, with corresponding sides a, b, and c.

31. $\dfrac{a + b}{b} = \dfrac{\sin A + \sin B}{\sin B}$

32. $\dfrac{a - b}{a + b} = \dfrac{\sin A - \sin B}{\sin A + \sin B}$

33. $\dfrac{a + b}{c} = \dfrac{\cos \frac{1}{2}(A - B)}{\sin \frac{1}{2} C}$

34. $\dfrac{a - b}{c} = \dfrac{\sin \frac{1}{2}(A - B)}{\cos \frac{1}{2} C}$

7.3 The Law of Cosines

If two sides and the angle between the two sides are given, the law of sines cannot be used to solve the triangle. Also, if all three of the sides of a triangle are given, the law of sines cannot be used to find the unknown angles. Both of these cases require the law of cosines.

To derive this law, let ABC be any oblique triangle. Choose a coordinate system so that vertex B is at the origin and side BC is along the positive x-axis. See Figure 7.9.

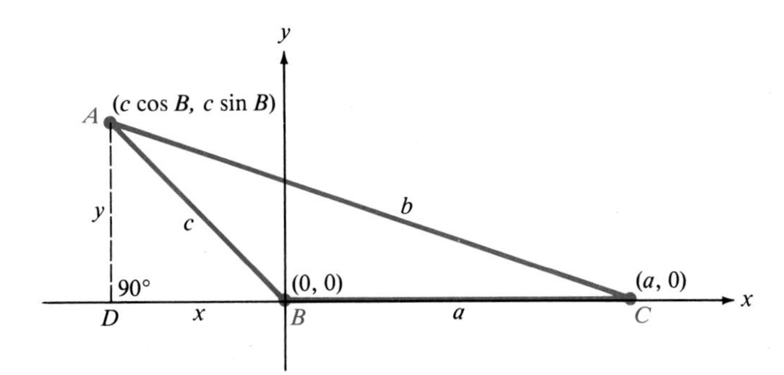

Figure 7.9

Let (x, y) be the coordinates of vertex A of the triangle. Verify that for angle B, whether obtuse or acute,

$$\sin B = \frac{y}{c} \quad \text{and} \quad \cos B = \frac{x}{c}.$$

(Here we assume that x is negative if B is obtuse.) From these results

$$y = c \sin B \quad \text{and} \quad x = c \cos B,$$

so that the coordinates of point A become

$$(c \cos B,\ c \sin B).$$

Point C has coordinates $(a, 0)$, and AC has length b. By the distance formula,

$$b = \sqrt{(c \cos B - a)^2 + (c \sin B)^2}.$$

Squaring both sides and simplifying gives

$$\begin{aligned} b^2 &= (c \cos B - a)^2 + (c \sin B)^2 \\ &= c^2 \cos^2 B - 2ac \cos B + a^2 + c^2 \sin^2 B \\ &= a^2 + c^2 (\cos^2 B + \sin^2 B) - 2ac \cos B \\ &= a^2 + c^2 (1) - 2ac \cos B \\ &= a^2 + c^2 - 2ac \cos B. \end{aligned}$$

This result is one form of the law of cosines. In the work above, we could just as easily have placed A or C at the origin. This would have given the same result, but with the variables rearranged. These various forms of the law of cosines are summarized in the following theorem.

Law of Cosines

In any triangle ABC, with sides a, b, and c,

$$\begin{aligned} a^2 &= b^2 + c^2 - 2bc \cos A \\ b^2 &= a^2 + c^2 - 2ac \cos B \\ c^2 &= a^2 + b^2 - 2ab \cos C. \end{aligned}$$

Example 1 Solve triangle ABC if $A = 42.3°$, $b = 12.9$, and $c = 15.4$. See Figure 7.10.

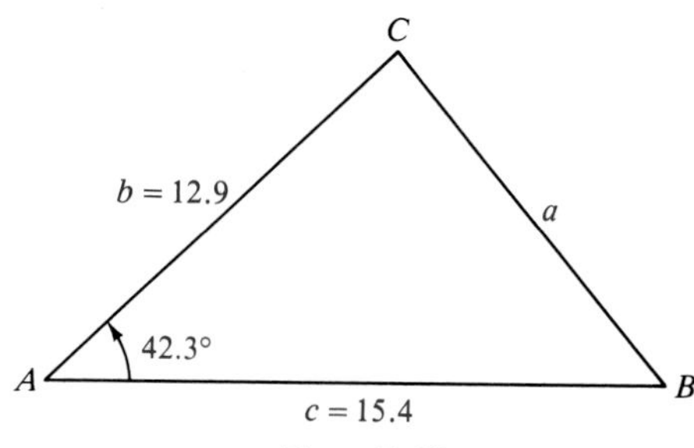

Figure 7.10

Find a with the law of cosines.

$$a^2 = b^2 + c^2 - 2bc \cos A$$
$$a^2 = (\mathbf{12.9})^2 + (\mathbf{15.4})^2 - 2(\mathbf{12.9})(\mathbf{15.4}) \cos \mathbf{42.3°}$$
$$a^2 = 166.41 + 237.16 - (397.32)(.7396)$$
$$a^2 = 403.57 - 293.86$$
$$a^2 = 109.71$$
$$a = 10.5$$

Now that a, b, c, and A are known, the law of sines can be used to find either angle B or angle C. This is the ambiguous case of the law of sines: since $a < b$ and $a < c$, there may be two triangles. If there is an obtuse angle in the triangle, it will be the larger of B and C. Since we cannot tell from the sine of the angle whether it is acute or obtuse, it is a good idea to find the smaller angle (which will be acute) first. We know that $B < C$ because $b < c$, so use the law of sines to find B.

$$\frac{\sin \mathbf{42.3°}}{\mathbf{10.5}} = \frac{\sin B}{\mathbf{12.9}}$$
$$\sin B = \frac{12.9 \sin 42.3°}{10.5}$$
$$\sin B = .8268$$
$$B = 55.8°$$

Now find C.

$$C = 180° - A - B = 81.9°$$ ✦

One way to avoid the problem of the ambiguous case of the law of sines in this example would be to use the law of cosines again, solving for angle B or C.

Example 2 Solve triangle ABC if $C = 132° 40'$, $b = 259$, and $a = 423$.

Here the law of cosines is used in the form $c^2 = a^2 + b^2 - 2ab \cos C$. Inserting the given data gives

$$c^2 = a^2 + b^2 - 2ab \cos C$$
$$c^2 = (423)^2 + (259)^2 - 2(423)(259) \cos 132° 40'.$$

The value of cos 132° 40′ can be found with a calculator or related angles. By either method,

$$\cos 132° 40' = -\cos 47° 20' = -.6777.$$

Now continue finding c.

$$c^2 = (423)^2 + (259)^2 - 2(423)(259)(-.6777)$$
$$c^2 = 178{,}929 + 67{,}081 + 148{,}494$$
$$c^2 = 394{,}504$$
$$c = 628$$

The law of sines can be used to complete the solution. Check that $A = 29° 40'$ and $B = 17° 40'$. ✥

Example 3 Solve triangle ABC if $a = 9.47$, $b = 15.9$, and $c = 21.1$.

Again the law of cosines must be used. It is a good idea to find the largest angle first in case it is obtuse. Since c has the greatest length, angle C will be the largest angle. Start with

$$c^2 = a^2 + b^2 - 2ab \cos C,$$

or

$$\cos C = \frac{a^2 + b^2 - c^2}{2ab}.$$

Inserting the given values leads to

$$\cos C = \frac{(9.47)^2 + (15.9)^2 - (21.1)^2}{2(9.47)(15.9)}$$
$$= \frac{-102.7191}{301.146}$$
$$= -.341094,$$

and

$$C = 109.9°.$$

(Angle C is obtuse since $\cos C$ is negative.) Use the law of sines to find B. Verify that $B = 45.1°$. Since $A = 180° - B - C$,

$$A = 25.0°. ✥$$

As shown in this section and the previous ones, four possible cases can occur in solving an oblique triangle. These cases are summarized in the chart below. The first two cases require the law of sines; the second two require the law of cosines. In all four cases, it is assumed that the given information actually produces a triangle.

Case	*Abbreviation*	*Example*
One side and two angles are known.	SAA	a, B, A known, find b $\quad b = \dfrac{a \sin B}{\sin A}$
Two sides and one angle (not included between the two sides) are known.	SSA	b, c, B known, find C $\quad \sin C = \dfrac{c \sin B}{b}$ (Watch for the ambiguous case; there may be two triangles)
Three sides are known.	SSS	a, b, c, known, find A $\quad \cos A = \dfrac{b^2 + c^2 - a^2}{2bc}$
Two sides and the angle included between the two sides are known.	SAS	a, B, c known, find b $\quad b^2 = a^2 + c^2 - 2ac \cos B$

Area The law of cosines can be used to find a formula for the area of a triangle when only the lengths of the three sides of the triangle are known. This formula is given as the next theorem. For a proof see Exercises 47–52.

Heron's Area Formula

If a triangle has sides of lengths a, b, and c, and if the **semiperimeter** is

$$s = \frac{1}{2}(a + b + c),$$

then the area of the triangle is

$$K = \sqrt{s(s - a)(s - b)(s - c)}.$$

Example 4 Find the area of the triangle having sides of lengths $a = 29.7$ ft, $b = 42.3$ ft, and $c = 38.4$ ft.

To use Heron's area formula, first find s.

$$s = \frac{1}{2}(a + b + c)$$

$$s = \frac{1}{2}(29.7 + 42.3 + 38.4)$$

$$= 55.2$$

The area is

$$K = \sqrt{s(s - a)(s - b)(s - c)}$$

$$K = \sqrt{55.2(55.2 - 29.7)(55.2 - 42.3)(55.2 - 38.4)}$$

$$= \sqrt{55.2(25.5)(12.9)(16.8)}$$

$$= 552 \text{ ft}^2.$$ ✣

7.3 Exercises

Solve each of the following triangles. Use a calculator or interpolate as necessary. See Examples 1 and 2.

1. $C = 28.3°$, $b = 5.71$ in, $a = 4.21$ in

2. $A = 41.4°$, $b = 2.78$ yd, $c = 3.92$ yd

3. $C = 45.6°$, $b = 8.94$ m, $a = 7.23$ m

4. $A = 67.3°$, $b = 37.9$ km, $c = 40.8$ km

5. $A = 80° 40'$, $b = 143$ cm, $c = 89.6$ cm

6. $C = 72° 40'$, $a = 327$ ft, $b = 251$ ft

7. $B = 74.80°$, $a = 8.919$ in, $c = 6.427$ in

8. $C = 59.70°$, $a = 3.725$ mi, $b = 4.698$ mi

9. $A = 112.8°$, $b = 6.28$ m, $c = 12.2$ m

10. $B = 168.2°$, $a = 15.1$ cm, $c = 19.2$ cm

11. $C = 24° 49'$, $a = 251.3$ m, $b = 318.7$ m

12. $B = 52° 28'$, $a = 7598$ in, $c = 6973$ in

Find all the angles in each of the following triangles. Round answers to the nearest ten minutes. See Example 3.

13. $a = 3.0$ ft, $b = 5.0$ ft, $c = 6.0$ ft

14. $a = 4.0$ ft, $b = 5.0$ ft, $c = 8.0$ ft

15. $a = 9.3$ cm, $b = 5.7$ cm, $c = 8.2$ cm

16. $a = 28$ ft, $b = 47$ ft, $c = 58$ ft

17. $a = 42.9$ m, $b = 37.6$ m, $c = 62.7$ m

18. $a = 189$ yd, $b = 214$ yd, $c = 325$ yd

19. $AB = 1240$ ft, $AC = 876$ ft, $BC = 918$ ft

20. $AB = 298$ m, $AC = 421$ m, $BC = 324$ m

Use a calculator or interpolation to find all the angles in each of the following triangles. Round answers to the nearest minute.

21. $a = 12.54$ in, $b = 16.83$ in, $c = 21.62$ in

22. $a = 250.8$ ft, $b = 212.7$ ft, $c = 324.1$ ft

23. $a = 7.095$ m, $b = 5.613$ m, $c = 11.53$ m

24. $a = 15{,}250$ m, $b = 17{,}890$ m, $c = 27{,}840$ m

Solve each of the following problems. Use the laws in this chapter as necessary.

25. Points A and B are on opposite sides of Lake Yankee. From a third point, C, the angle between the lines of sight to A and B is 46.3°. If AC is 350 m long and BC is 286 m long, find AB.

26. The sides of a parallelogram are 4.0 cm and 6.0 cm. One angle is 58° while another is 122°. Find the lengths of the diagonals of the parallelogram.

27. Airports A and B are 450 km apart, on an east-west line. Tom flies in a northeast direction from A to airport C. From C he flies 359 km on a bearing of 128° 40′ to B. How far is C from A?

28. Two ships leave a harbor together, traveling on courses that have an angle of 135° 40′ between them. If they each travel 402 mi, how far apart are they?

29. The plans for a mountain cabin show the dimensions given in the figure on the right. Find x.

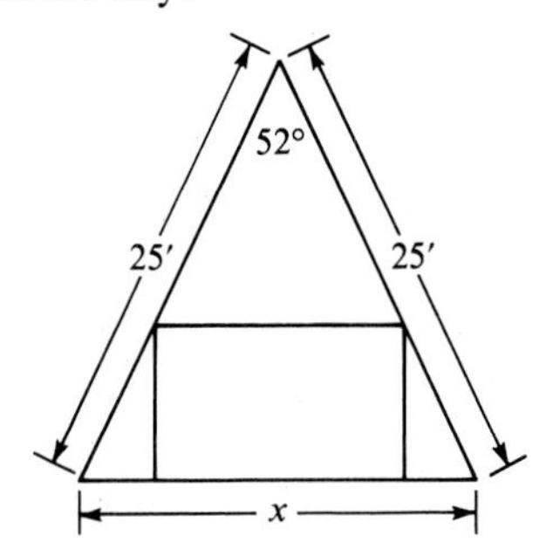

30. A hill slopes at an angle of 12.47° with the horizontal. From the base of the hill, the angle of elevation of a 459.0 ft tower at the top of the hill is 35.98°. How much rope would be required to reach from the top of the tower to the bottom of the hill?

31. A crane with a counterweight is shown in the figure below. Find the distance between points A and B.

32. A weight is supported by cables attached to both ends of a balance beam, as shown in the figure below. What angles are formed between the beam and the cables?

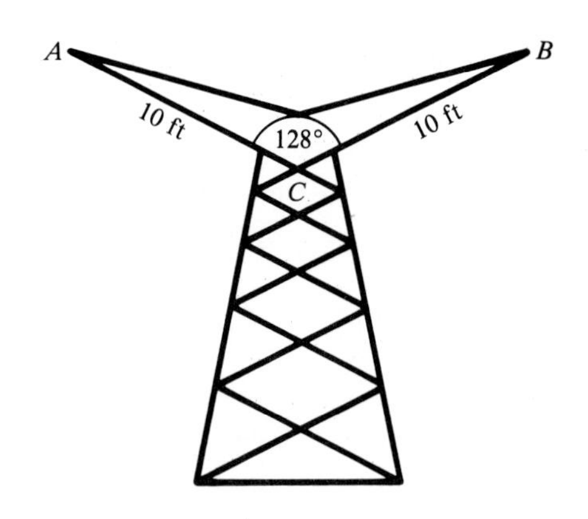

Exercise 31

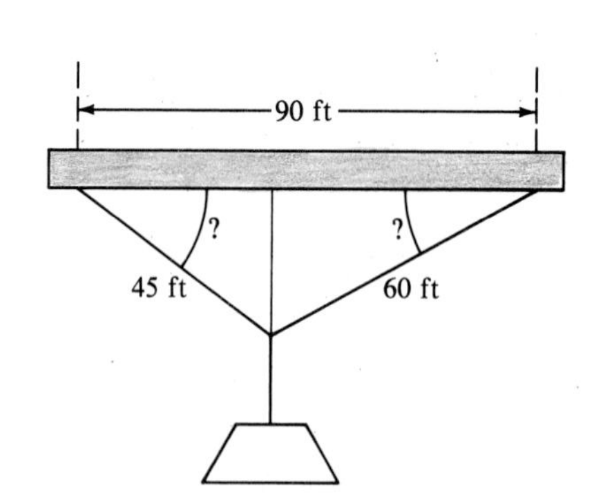

Exercise 32

To help predict eruptions from the volcano Mauna Loa on the island of Hawaii, scientists keep track of the volcano's movement by using a "super triangle" with vertices on the three volcanoes shown on the map below. (For example, in a recent year, Mauna Loa moved 6 inches, a result of increasing internal pressure.) The data in the following exercises have been rounded.

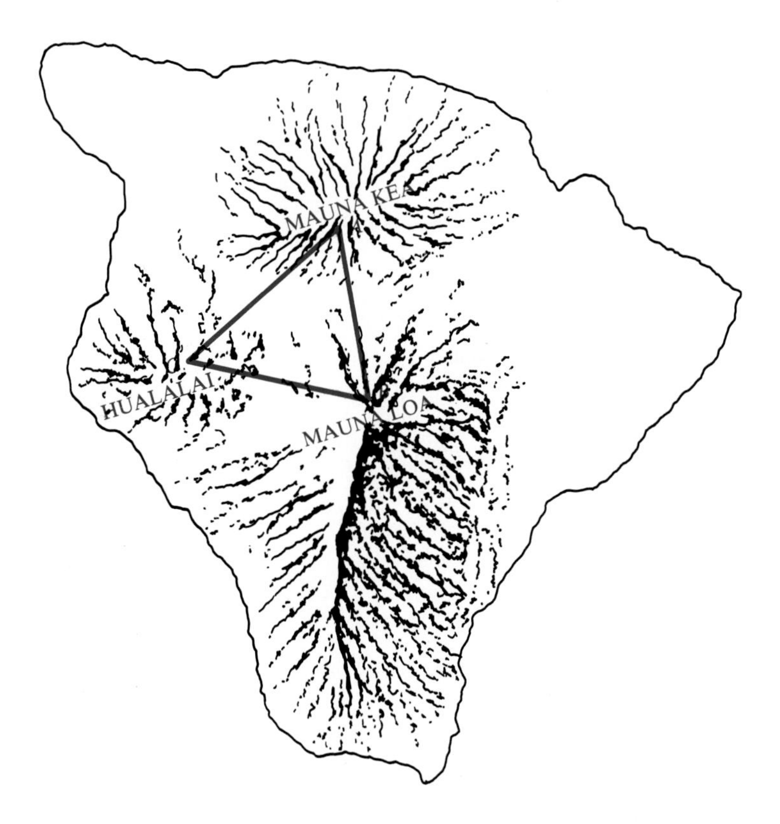

33. $AB = 22.47928$ mi, $AC = 28.14276$ mi, $A = 58.56989°$; find BC

34. $AB = 22.47928$ mi, $BC = 25.24983$ mi, $A = 58.56989°$; find B

*The diagram on the following page is an engineering drawing used in the construction of Michigan's Mackinac Straits Bridge.**

35. Find the angles of the triangle formed by Mackinac West Base, Green Island, and St. Ignace West Base.

36. Find the angles of the triangle formed by A_2, St. Ignace West Base, and St. Ignace East Base.

---

*Reproduced with permission from "Mackinac Bridge," by G. Edwin Pidcock, *Civil Engineering,* May 1956, p. 43.

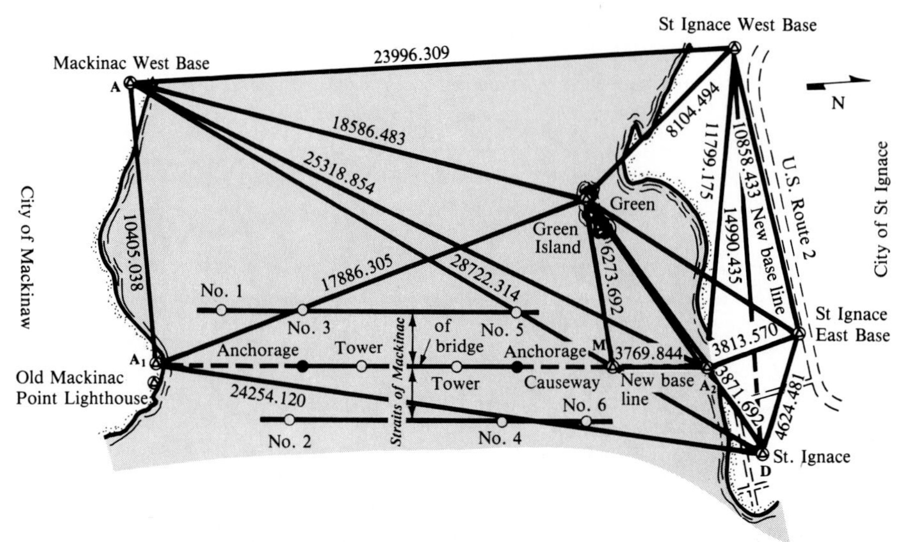

Find the area of each of the following triangles. See Example 4.

37. $a = 12$ m, $b = 16$ m, $c = 25$ m

38. $a = 22$ in, $b = 45$ in, $c = 31$ in

39. $a = 154$ cm, $b = 179$ cm, $c = 183$ cm

40. $a = 25.4$ yd, $b = 38.2$ yd, $c = 19.8$ yd

41. $a = 76.3$ ft, $b = 109$ ft, $c = 98.8$ ft

42. $a = 15.89$ in, $b = 21.74$ in, $c = 10.92$ in

43. $a = 74.14$ ft, $b = 89.99$ ft, $c = 51.82$ ft

44. $a = 1.096$ km, $b = 1.142$ km, $c = 1.253$ km

Solve each of the following problems.

45. A painter needs to cover a triangular region 75 m by 68 m by 85 m. A can of paint covers 75 sq m of area. How many cans (to the next higher number of cans) will be needed?

46. How many cans of paint would be needed in Exercise 45 if the region were 8.2 m by 9.4 m by 3.8 m?

Use the fact that $\cos A = (b^2 + c^2 - a^2)/(2bc)$ *to show that each of the following is true.*

47. $1 + \cos A = \dfrac{(b + c + a)(b + c - a)}{2bc}$

48. $1 - \cos A = \dfrac{(a - b + c)(a + b - c)}{2bc}$

49. $\cos \dfrac{A}{2} = \sqrt{\dfrac{s(s - a)}{bc}}$ $\left(\textit{Hint: } \cos \dfrac{A}{2} = \sqrt{\dfrac{1 + \cos A}{2}}\right)$

50. $\sin \dfrac{A}{2} = \sqrt{\dfrac{(s-b)(s-c)}{bc}}$ $\left(\textit{Hint: } \sin \dfrac{A}{2} = \sqrt{\dfrac{1-\cos A}{2}}\right)$

51. The area of a triangle having sides b and c and angle A is given by $\frac{1}{2}\, bc \sin A$. Show that this result can be written as

$$\sqrt{\frac{1}{2}\, bc\,(1 + \cos A) \cdot \frac{1}{2}\, bc\,(1 - \cos A)}.$$

52. Use the results of Exercises 47–51 to prove Heron's area formula.

7.4 Vectors

The measures of all six parts of a triangle can be found, given at least one side and any two other measures. This section and the next show applications of this work to *vectors*. In this section the basic ideas of vectors are presented, and in the next section the law of sines and the law of cosines are applied to vector problems.

Many quantities in mathematics involve magnitudes, such as 45 lb or 60 mph. These quantities are called **scalars.** Other quantities, called **vector quantities,** involve both magnitude and direction. Typical vector quantities are velocity, acceleration, and force.

A vector quantity is often represented with a directed line segment, which is called a **vector.** The length of the vector represents the magnitude of the vector quantity. The direction of the vector, indicated with an arrowhead, represents the direction of the quantity. For example, the vector in Figure 7.11 represents a force of 10 lb applied at an angle of 30° from the horizontal.

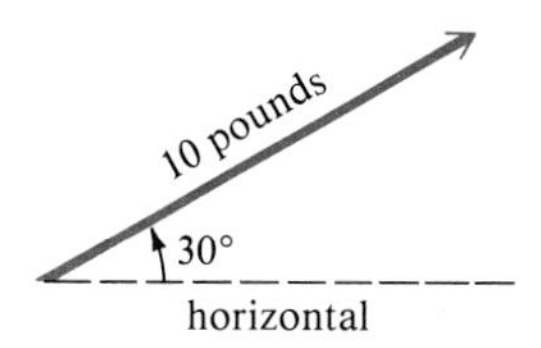

Figure 7.11

The symbol for a vector is often printed in boldface type. When writing vectors by hand, it is customary to use an arrow over the letter or letters. Thus **OP** and $\overrightarrow{OP}$ both represent vector OP. Vectors may be named with either one lowercase or uppercase letter, or two uppercase letters. When two letters are used, the first indicates the *initial point* and the second indicates the *terminal point* of the vector. Knowing these points gives the direction of the vector. For example, vectors **OP** and **PO** in Figure 7.12 are not the same vector. They have the same magnitude, but they have opposite directions. The magnitude of vector **OP** is written |**OP**|.

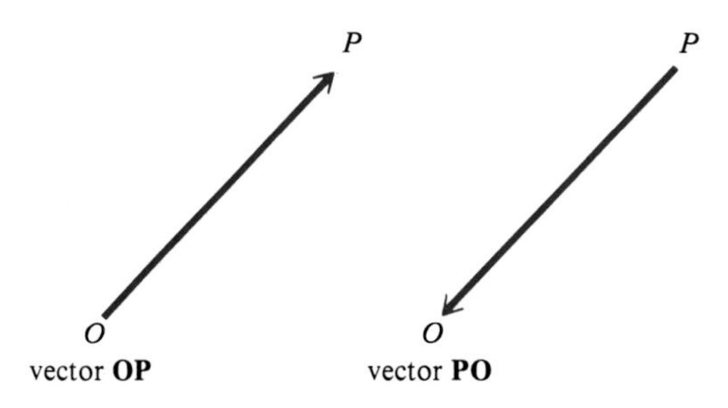

Figure 7.12

Two vectors are *equal* if and only if they both have the same direction and the same magnitude. In Figure 7.13, vectors **A** and **B** are equal, as are vectors **C** and **D.**

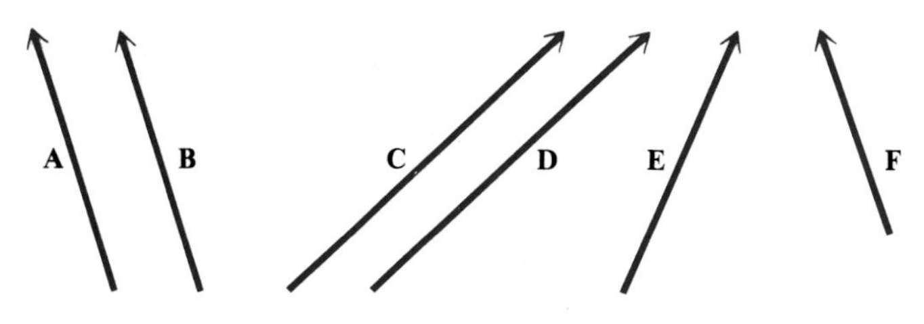

Figure 7.13

As Figure 7.13 shows, equal vectors need not coincide, but they must be parallel. Vectors **A** and **E** are unequal because they do not have the same direction, while **A** ≠ **F** because they have different magnitudes, as indicated by their different lengths.

To find the *sum* of two vectors **A** and **B,** written **A** + **B,** place the initial point of vector **B** at the terminal point of vector **A,** as shown in Figure 7.14. The vector with the same initial point as **A** and the same terminal point as **B** is the sum **A** + **B.** The sum of two vectors is also a vector.

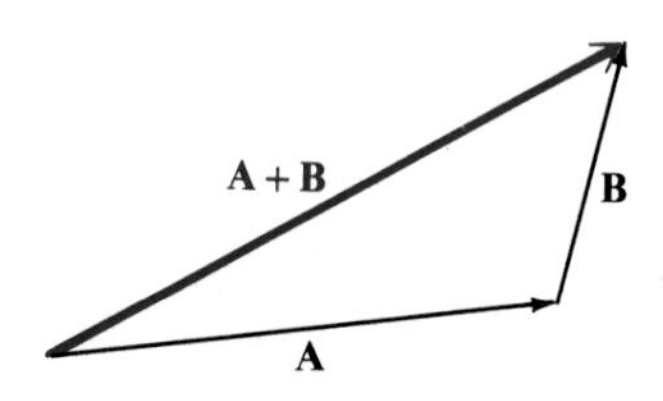

Figure 7.14

Another way to find the sum of two vectors is to use the **parallelogram rule.** Place vectors **A** and **B** so that their initial points coincide. Then complete a parallelogram that has **A** and **B** as two adjacent sides. The diagonal of the parallelogram with the same initial point as **A** and **B** is the same vector sum **A** + **B** found by the definition. See Figure 7.15.

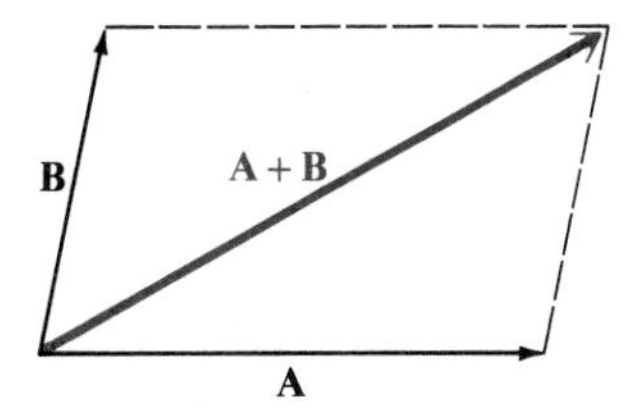

Figure 7.15

The vector sum **A** + **B** is the **resultant** of vectors **A** and **B.** Each of the vectors **A** and **B** is a **component** of vector **A** + **B.** In many practical applications, such as surveying, it is necessary to break a vector into its **vertical** and **horizontal components.** These components are two vectors, one vertical and one horizontal, whose resultant is the original vector. As shown in Figure 7.16, vector **OR** is the vertical component and vector **OS** is the horizontal component of **OP.**

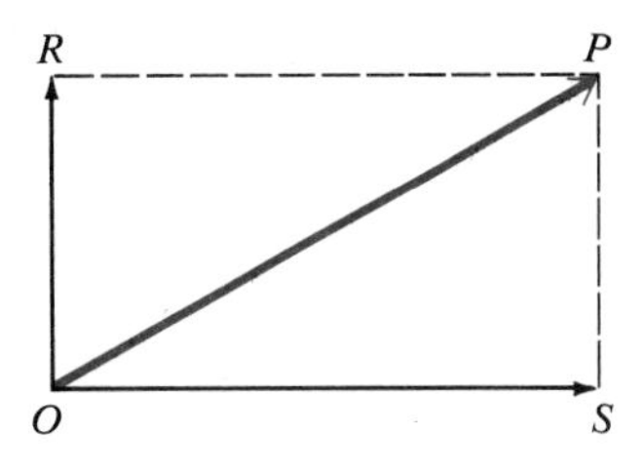

Figure 7.16

For every vector **v** there is a vector −**v** with the same magnitude as **v** but opposite direction. Vector −**v** is the **opposite** of **v.** See Figure 7.17. The sum of **v** and −**v** has magnitude 0 and is a **zero vector.** As with real numbers, to *subtract* vector **B** from vector **A,** find the vector sum **A** + (−**B**)**.** See Figure 7.18.

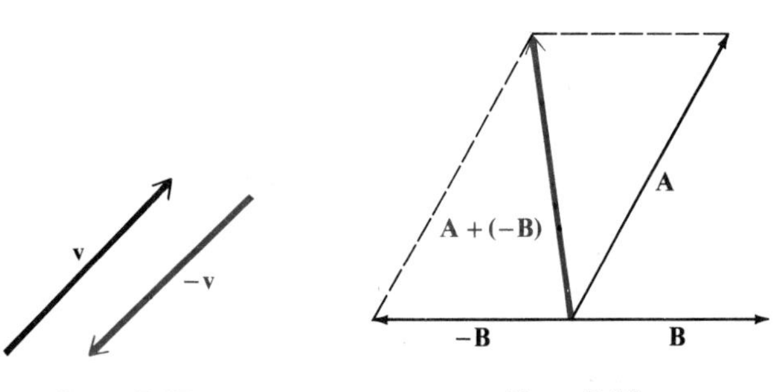

Figure 7.17 **Figure 7.18**

The **scalar product** of a real number (or scalar) k and a vector **u** is the vector k**u,** which has magnitude $|k|$ times the magnitude of **u.** As shown in Figure 7.19, k**u** has the same direction as **u** if $k > 0$, and the opposite direction if $k < 0$.

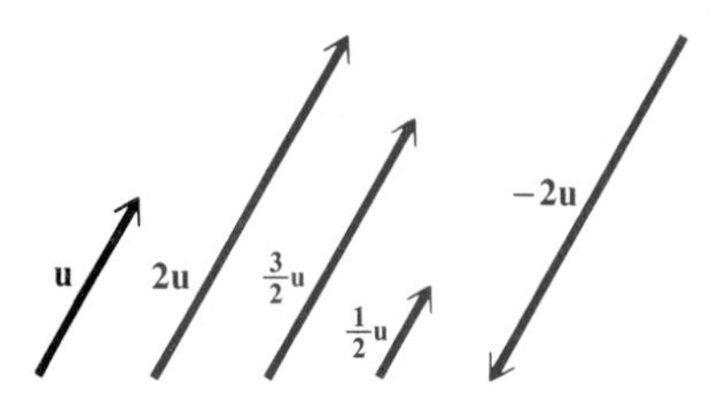

Figure 7.19

Example 1 Vector **w** has magnitude 25.0 and is inclined at an angle of 40° from the horizontal. Find the magnitudes of the horizontal and vertical components of the vector.

In Figure 7.20, the vertical component is labeled **v** and the horizontal component is labeled **u.** Vectors **u, v,** and **w** form a right triangle. In this right triangle,

$$\sin 40° = \frac{|\mathbf{v}|}{|\mathbf{w}|} = \frac{|\mathbf{v}|}{25.0},$$

and

$$|\mathbf{v}| = 25.0 \sin 40° = 25.0(.6428) = 16.1.$$

In the same way,

$$\cos 40° = \frac{|\mathbf{u}|}{25.0},$$

with

$$|\mathbf{u}| = 25.0 \cos 40° = 19.2.$$ ✣

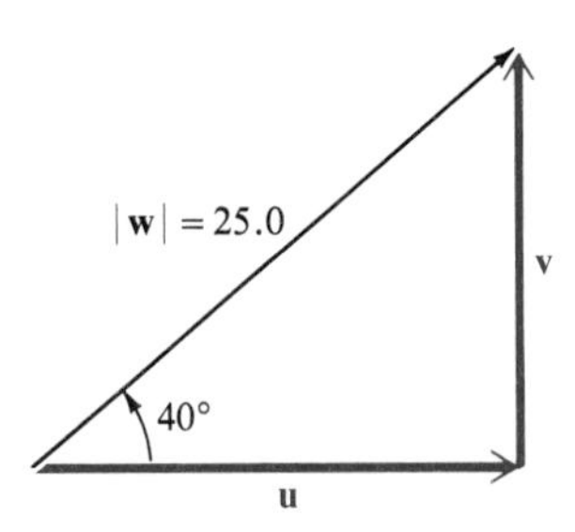

Figure 7.20

Example 2 Two forces of 15 newtons and 22 newtons (a newton is a unit of force used in physics) act at a point in the plane. If the angle between the forces is 100°, find the magnitude of the resultant force.

As shown in Figure 7.21, a parallelogram that has the forces as adjacent sides can be formed. The angles of the parallelogram adjacent to angle P each measure 80°, since adjacent angles of a parallelogram are supplementary. Opposite sides of the parallelogram are equal in length. The resultant force divides the parallelogram into two triangles. (However, the resultant does not necessarily divide angle P into two *equal* angles.) Use the law of cosines to get

$$|\mathbf{v}|^2 = 15^2 + 22^2 - 2(15)(22)\cos 80°$$
$$|\mathbf{v}|^2 = 225 + 484 - 115$$
$$|\mathbf{v}|^2 = 594,$$

and

$$|\mathbf{v}| = 24.$$

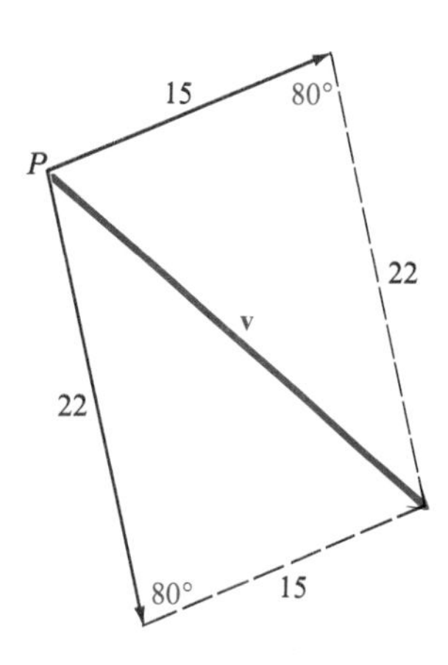

Figure 7.21

7.4 Exercises

Exercises 1–4 refer to the following vectors.

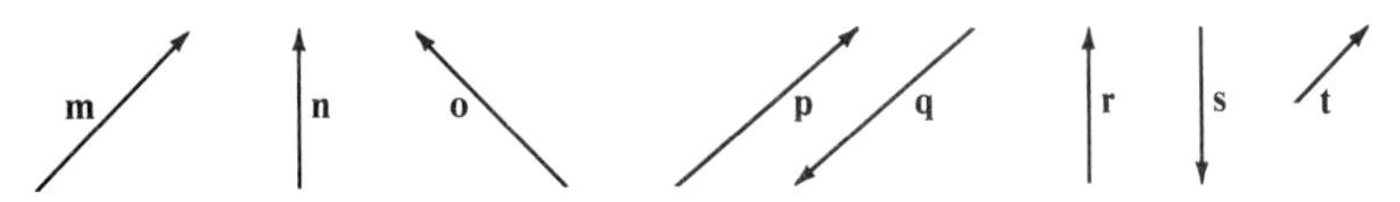

1. Name all pairs of vectors that appear to be equal.
2. Name all pairs of vectors that are opposites.
3. Name all pairs of vectors where the first is a scalar multiple of the other, with the scalar positive.
4. Name all pairs of vectors where the first is a scalar multiple of the other, with the scalar negative.

Exercises 5–22 refer to the vectors pictured here.

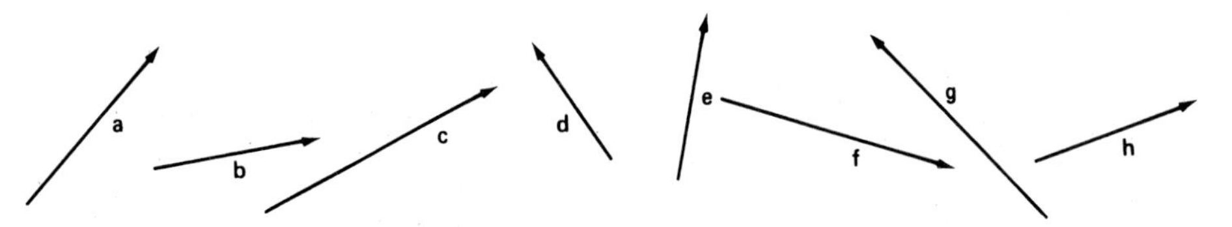

Draw a sketch to represent each of the following vectors. For example, find **a** + **e** *by placing* **a** *and* **e** *so that their initial points coincide. Then use the parallelogram rule to find the resultant, shown in the figure.*

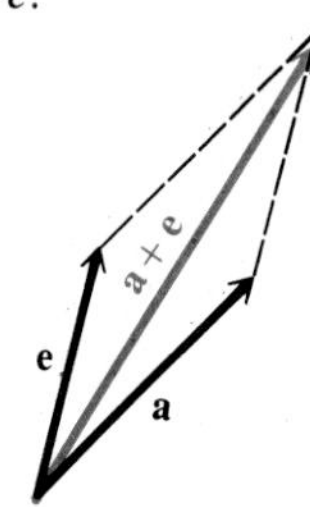

5. −**b**
6. −**g**
7. 3**a**
8. 2**h**
9. **a** + **c**
10. **a** + **b**
11. **h** + **g**
12. **e** + **f**
13. **a** + **h**
14. **b** + **d**
15. **h** + **d**
16. **a** + **f**
17. **a** − **c**
18. **d** − **e**
19. **a** + (**b** + **c**)
20. (**a** + **b**) + **c**
21. **h** + (**e** + **g**)
22. (**h** + **e**) + **g**

23. From the results of Exercises 19–22, do you think that vector addition is associative?

For each pair of vectors **u** *and* **w** *with angle* θ *between them, sketch the resultant.*

24. $|\mathbf{u}| = 18$, $|\mathbf{w}| = 23$, $\theta = 45°$
25. $|\mathbf{u}| = 12$, $|\mathbf{w}| = 20$, $\theta = 27°$
26. $|\mathbf{u}| = 8$, $|\mathbf{w}| = 12$, $\theta = 20°$
27. $|\mathbf{u}| = 20$, $|\mathbf{w}| = 30$, $\theta = 30°$
28. $|\mathbf{u}| = 27$, $|\mathbf{w}| = 50$, $\theta = 12°$
29. $|\mathbf{u}| = 50$, $|\mathbf{w}| = 70$, $\theta = 40°$

For each of the following, vector **v** *has the given magnitude and direction. See the figure. Find the magnitude of the horizontal and vertical components of* **v**. *See Example 1.*

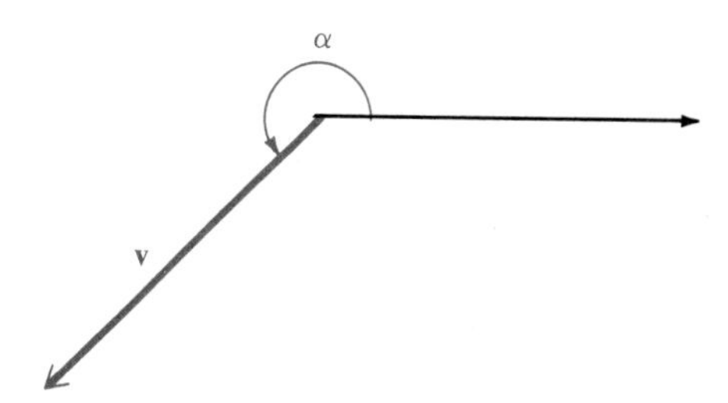

30. $\alpha = 20°$, $|\mathbf{v}| = 50$
31. $\alpha = 38°$, $|\mathbf{v}| = 12$
32. $\alpha = 70°$, $|\mathbf{v}| = 150$
33. $\alpha = 50°$, $|\mathbf{v}| = 26$
34. $\alpha = 35° 50'$, $|\mathbf{v}| = 47.8$
35. $\alpha = 27° 30'$, $|\mathbf{v}| = 15.4$

36. $\alpha = 59° 40'$, $|\mathbf{v}| = 78.9$ **37.** $\alpha = 128.5°$, $|\mathbf{v}| = 198$ **38.** $\alpha = 146.3°$, $|\mathbf{v}| = 238$

39. $\alpha = 251° 10'$, $|\mathbf{v}| = 69.1$ **40.** $\alpha = 302° 40'$, $|\mathbf{v}| = 7890$

In each of the following, two forces act at a point in the plane. The angle between the two forces is given. Find the magnitude of the resultant force. See Example 2.

41. Forces of 250 and 450 newtons, forming an angle of 85°

42. Forces of 19 and 32 newtons, forming an angle of 118°

43. Forces of 17.9 and 25.8 lb, forming an angle of 105° 30′

44. Forces of 75.6 and 98.2 lb, forming an angle of 82° 50′

45. Forces of 116 and 139 lb, forming an angle of 140° 50′

46. Forces of 37.8 and 53.7 lb, forming an angle of 68° 30′

Solve each of the following problems.

47. One rope pulls a barge directly east with a force of 100 newtons. Another rope pulls the barge to the northeast with a force of 200 newtons. Find the resultant force acting on the barge and its direction relative to the first rope.

48. Paula and Steve are pulling their daughter Jessie on a sled. Steve pulls with a force of 18 lb at an angle of 10°. Paula pulls with a force of 12 lb at an angle of 15°. What is the weight of Jessie and the sled? See the figure. (*Hint:* Find the resultant force.)

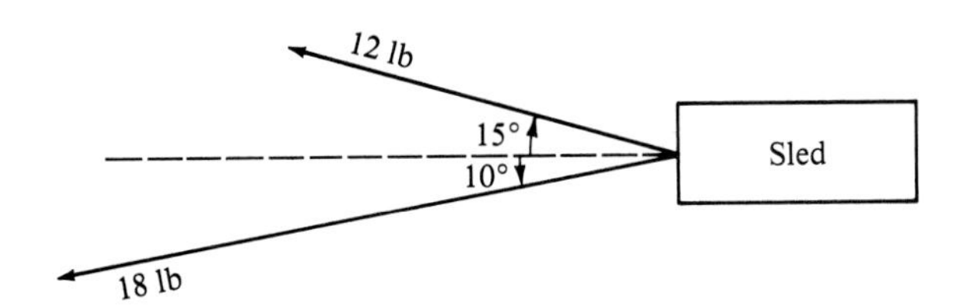

7.5 Applications of Vectors

The previous section covered methods for finding the resultant of two vectors. In many applications it is necessary to find a vector that will counterbalance the resultant. This opposite vector is called the *equilibrant:* the **equilibrant** of vector **u** is the vector **−u.**

Example 1 Find the magnitude of the equilibrant of forces of 48 newtons and 60 newtons acting on a point A, if the angle between the forces is 50°. Then find the angle between the equilibrant and the 48-newton force.

In Figure 7.22 on page 298, the equilibrant is $-\mathbf{v}$. The magnitude of $\mathbf{v}$, and hence of $-\mathbf{v}$, is found by using triangle ABC and the law of cosines:

$$\begin{aligned}|\mathbf{v}|^2 &= \mathbf{48}^2 + \mathbf{60}^2 - 2(\mathbf{48})(\mathbf{60}) \cos \mathbf{130}° \\ &= 2304 + 3600 - 5760(-.6428) \\ &= 9606.5,\end{aligned}$$

and

$$|\mathbf{v}| = 98,$$

to two significant digits.

The required angle, labeled α in Figure 7.22, can be found by subtracting angle CAB from 180°. Use the law of sines to find angle CAB.

$$\frac{98}{\sin 130°} = \frac{60}{\sin CAB}$$

$$\sin CAB = .4690$$

From a calculator or table,

$$\text{angle } CAB = 28°.$$

Finally,

$$\alpha = 180° - 28°$$
$$= 152°.$$

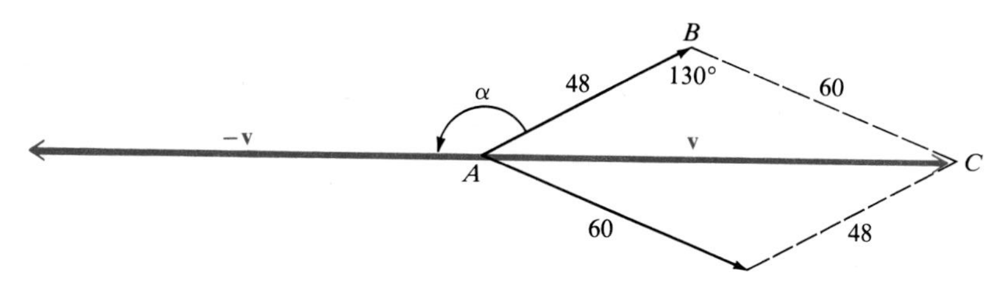

Figure 7.22

Example 2 Find the force required to pull a 50-lb weight up a ramp inclined at 20° to the horizontal.

In Figure 7.23, the vertical 50-lb force represents the force of gravity. The component **BC** represents the force with which the body pushes against the

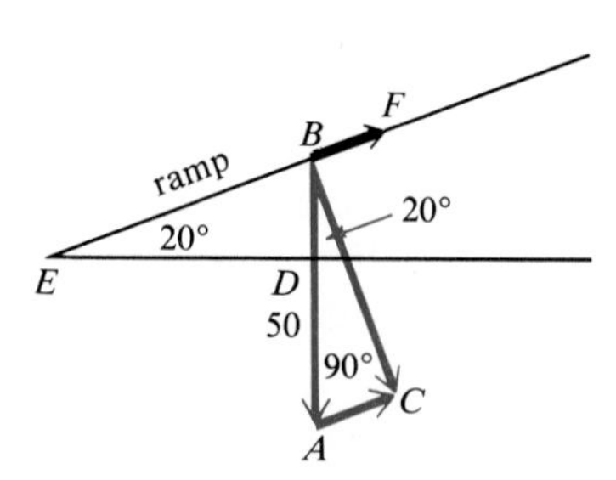

Figure 7.23

ramp, while the component **BF** represents a force that would pull the body up the ramp. Since vectors **BF** and **AC** are equal, $|\mathbf{AC}|$ gives the required force.

Vectors **BF** and **AC** are parallel, so angle *EBD* equals angle *A*. Since angle *BDE* and angle *C* are right angles, triangles *CBA* and *DEB* have two corresponding angles equal and so are similar triangles. Therefore, angle *ABC* equals angle *E*, which is 20°. From right triangle *ABC*,

$$\sin 20^\circ = \frac{|\mathbf{AC}|}{50}$$
$$|\mathbf{AC}| = 50 \sin 20^\circ$$
$$|\mathbf{AC}| = 17.$$

To the nearest pound, a 17-lb force will be required to pull the weight up the ramp. ✥

Problems involving bearing (defined in Section 2.6) can also be worked with vectors, as shown in the next example.

Example 3 A ship leaves port on a bearing of 28° and travels 8.2 mi. The ship then turns due east and travels 4.3 mi. How far is the ship from port? What is its bearing from port?

In Figure 7.24, vectors **PA** and **AE** represent the ship's path. The magnitude and bearing of the resultant **PE** can be found as follows. Triangle *PNA* is a right triangle, so angle $NAP = 90^\circ - 28^\circ = 62^\circ$. Then angle $PAE = 180^\circ - 62^\circ = 118^\circ$. Use the law of cosines to find $|\mathbf{PE}|$, the magnitude of vector **PE.**

$$|\mathbf{PE}|^2 = 8.2^2 + 4.3^2 - 2(8.2)(4.3)\cos 118^\circ$$
$$|\mathbf{PE}|^2 = 67.24 + 18.49 - 70.52(-.4695)$$
$$|\mathbf{PE}|^2 = 118.84$$

Therefore, $$|\mathbf{PE}| = 10.9,$$

or 11 mi, rounded to two significant digits.

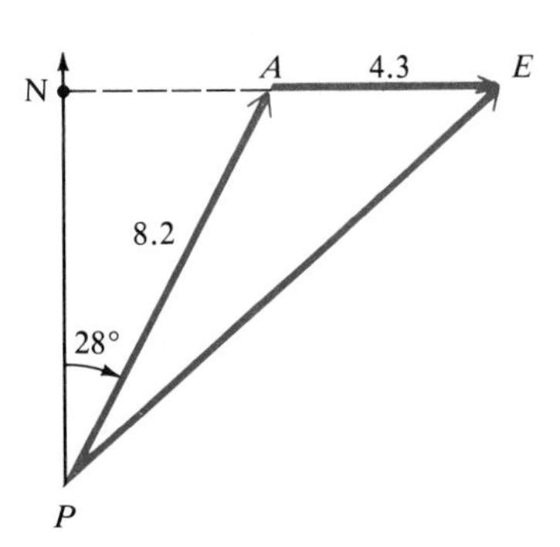

Figure 7.24

To find the bearing of the ship from port, first find angle APE. Use the law of sines, along with the value of $|\mathbf{PE}|$ before rounding.

$$\frac{\sin APE}{4.3} = \frac{\sin 118°}{10.9}$$

$$\sin APE = \frac{4.3 \sin 118°}{10.9}$$

$$\text{angle } APE = 20.4°$$

After rounding, angle APE is 20°, and the ship is 11 mi from port on a bearing of $28° + 20° = 48°$. ✥

In air navigation, the **airspeed** of a plane is its speed relative to the air, while the **groundspeed** is its speed relative to the ground. Because of wind, these two speeds are usually different. The groundspeed of the plane is represented by the vector sum of the airspeed and windspeed vectors. See Figure 7.25.

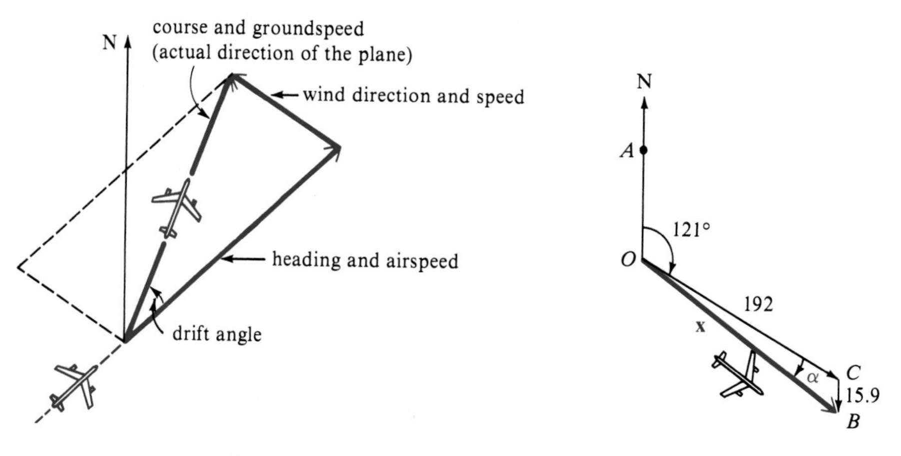

Figure 7.25

Figure 7.26

Example 4 A plane with an airspeed of 192 mph is headed on a bearing of 121°. A north wind is blowing (from north to south) at 15.9 mph. Find the groundspeed and the actual bearing of the plane.

In Figure 7.26 the groundspeed is represented by $|\mathbf{x}|$. We must find angle α to find the bearing, which will be $121° + \alpha$. From Figure 7.26, angle BCO equals angle AOC, which equals 121°. Find $|\mathbf{x}|$ by the law of cosines.

$$|\mathbf{x}|^2 = 192^2 + (15.9)^2 - 2(192)(15.9) \cos 121°$$

$$|\mathbf{x}|^2 = 36{,}864 + 252.81 - 6105.6(-.5150)$$

$$|\mathbf{x}|^2 = 40{,}261$$

Therefore, $$|\mathbf{x}| = 200.7,$$

or 201 mph. Now find α by using the law of sines. As before, use the value of $|\mathbf{x}|$ before rounding.

$$\frac{\sin \alpha}{15.9} = \frac{\sin 121°}{200.7}$$
$$\sin \alpha = .0679$$
$$\alpha = 3.89°$$

After rounding, α is 3.9°. The groundspeed is about 201 mph, on a bearing of 124.9°, or 125° to three significant digits. ✥

7.5 Exercises

Solve each of the following problems. See Examples 1–4.

1. Two forces of 692 newtons and 423 newtons act at a point. The resultant force is 786 newtons. Find the angle between the forces.
2. Two forces of 128 lb and 253 lb act at a point. The equilibrant is 320 lb. Find the angle between the forces.
3. A force of 25 lb is required to push an 80-lb lawn mower up a hill. What angle does the hill make with the horizontal?
4. Find the force required to keep a 3000-lb car parked on a hill that makes an angle of 15° with the horizontal.
5. To build the pyramids in Egypt, it is believed that giant causeways were built to transport the building materials to the site. One such causeway is said to have been 3000 ft long, with a slope of about 2.3°. How much force would be required to pull a 60-ton monolith along this causeway?
6. A force of 500 lb is required to pull a boat up a ramp inclined at 18° with the horizontal. How much does the boat weigh?
7. Two tugboats are pulling a disabled speedboat into port with forces of 1240 lb and 1480 lb. The angle between these forces is 28.2°. Find the direction and magnitude of the equilibrant.
8. Two people are carrying a box. One person exerts a force of 150 lb at an angle of 62.4° with the horizontal. The other person exerts a force of 114 lb at an angle of 54.9°. Find the weight of the box.
9. A crate is supported by two ropes. One rope makes an angle of 46° 20′ with the horizontal and has a tension of 89.6 lb on it. The other rope is horizontal. Find the weight of the crate and the tension in the horizontal rope.
10. Three forces acting at a point are in equilibrium. The forces are 980 lb, 760 lb, and 1220 lb. Find the angles between the directions of the forces. (*Hint:* Arrange the forces to form the sides of a triangle.)
11. A force of 176 lb makes an angle of 78° 50′ with a second force. The resultant of the two forces makes an angle of 41° 10′ with the first force. Find the magnitude of the second force and of the resultant.

12. A force of 28.7 lb makes an angle of 42° 10′ with a second force. The resultant of the two forces makes an angle of 32° 40′ with the first force. Find the magnitude of the second force and of the resultant.

13. A plane files 650 mph on a bearing of 175.3°. A 25 mph wind, from a direction of 266.6°, blows against the plane. Find the resulting bearing of the plane.

14. A pilot wants to fly on a bearing of 74.9°. By flying due east, he finds that a 42 mph wind, blowing from the south, puts him on course. Find the airspeed and the groundspeed.

15. Starting at point *A*, a ship sails 18.5 km on a bearing of 189°, then turns and sails 47.8 km on a bearing of 317°. Find the distance of the ship from point *A*.

16. Two towns 21 mi apart are separated by a dense forest. (See the figure.) To travel from town *A* to town *B*, a person must go 17 mi on a bearing of 325°, then turn and continue for 9 mi to reach town *B*. Find the bearing of *B* from *A*.

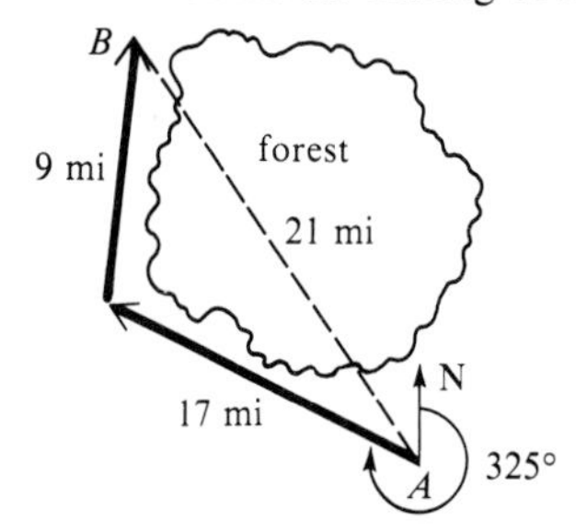

17. An airline route from San Francisco to Honolulu is on a bearing of 233°. A jet flying at 450 mph on that bearing flies into a wind blowing at 39 mph from a direction of 114°. Find the resulting bearing and groundspeed of the plane.

18. A pilot is flying at 168 mph. She wants her flight path to be on a bearing of 57° 40′. A wind is blowing from the south at 27.1 mph. Find the bearing the pilot should fly, and find the plane's groundspeed.

19. What bearing and airspeed are required for a plane to fly 400 mi due north in 2.5 hr if the wind is blowing from a direction of 328° at 11 mph?

20. A plane is headed due south with an airspeed of 192 mph. A wind from a direction of 78° is blowing at 23 mph. Find the groundspeed and resulting bearing of the plane.

21. An airplane is headed on a bearing of 174° at an airspeed of 240 km per hr. A 30 km per hr wind is blowing from a direction of 245°. Find the groundspeed and resulting bearing of the plane.

22. A ship sailing the North Atlantic has been warned to change course to avoid a group of icebergs. The captain turns and sails on a bearing of 62° for a while, then changes course again to a bearing of 115° until the ship reaches its original course. (See the figure.) How much farther did the ship have to travel to avoid the icebergs?

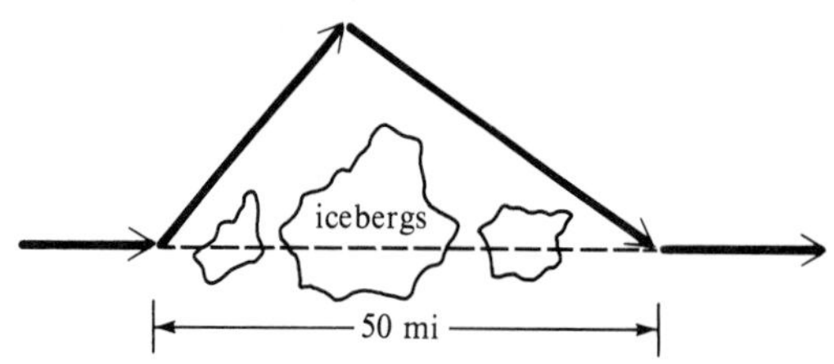

23. The aircraft carrier *Tallahassee* is traveling at sea on a steady course with a bearing of 30° at 32 mph. Patrol planes on the carrier have enough fuel for 2.6 hr of flight when traveling at a speed of 520 mph. One of the pilots takes off on a bearing of 338° and then turns and heads in a straight line, so as to be able to catch the carrier and land on the deck at the exact instant that his fuel runs out. If the pilot left at 2 P.M., at what time did he turn to head for the carrier?

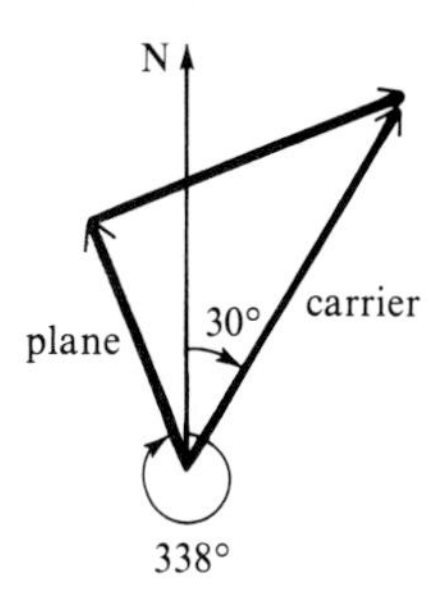

24. A car going around a banked curve is subject to the forces shown in the figure. If the radius of the curve is 100 ft, what value of θ to the nearest degree would allow an automobile to travel around the curve at a speed of 40 ft per sec without depending on friction?

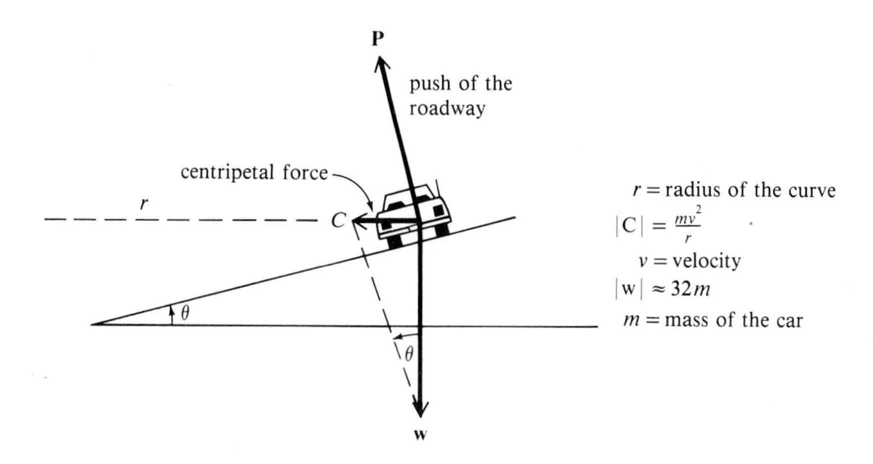

Chapter 7 *Key Concepts*

Law of Sines In any triangle ABC with sides a, b, and c,

$$\frac{a}{\sin A} = \frac{b}{\sin B} \qquad \frac{a}{\sin A} = \frac{c}{\sin C} \qquad \frac{b}{\sin B} = \frac{c}{\sin C}.$$

Law of Cosines In any triangle ABC with sides a, b, and c,

$$a^2 = b^2 + c^2 - 2bc \cos A$$
$$b^2 = a^2 + c^2 - 2ac \cos B$$
$$c^2 = a^2 + b^2 - 2ab \cos C.$$

Area of a Triangle The area of a triangle is given by half the product of the lengths of two sides and the sine of the angle between the two sides.

Heron's Area Formula If a triangle has sides of lengths a, b, and c, and if the semiperimeter is

$$s = \frac{1}{2}(a + b + c),$$

then the area of the triangle is

$$K = \sqrt{s(s - a)(s - b)(s - c)}.$$

Sum of Two Vectors

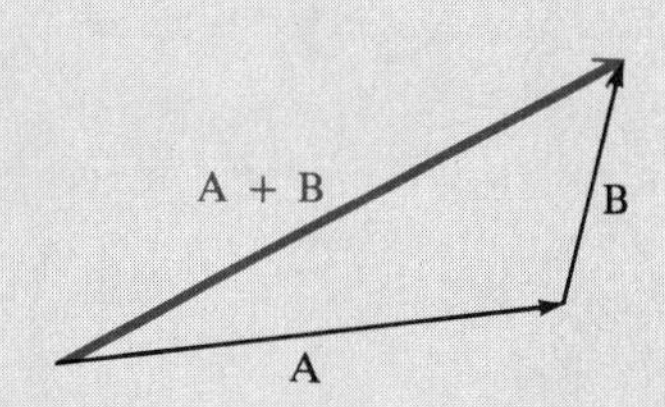

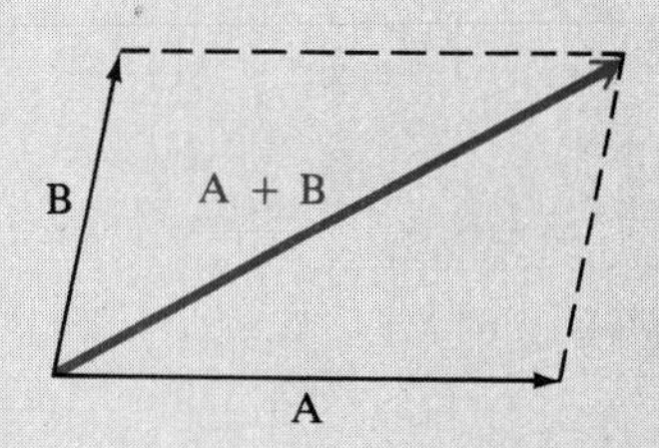

Opposite Vectors

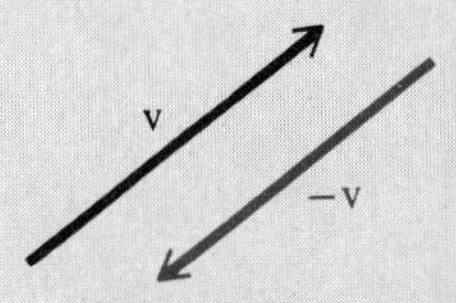

Chapter 7 *Review Exercises*

Find the indicated parts of each of the following triangles.

1. $C = 74° 10'$, $c = 96.3$ m, $B = 39° 30'$; find b

2. $A = 129° 40'$, $a = 127$ ft, $b = 69.8$ ft; find B

3. $C = 51° 20'$, $c = 68.3$ m, $b = 58.2$ m; find B

4. $a = 165$ m, $A = 100.2°$, $B = 25.0°$; find b

5. $B = 39° 50'$, $b = 268$ m, $a = 340$ m; find A

6. $C = 79° 20'$, $c = 97.4$ mm, $a = 75.3$ mm; find A

7. Solve the triangle having $A = 25° 10'$, $a = 6.92$ yd, $b = 4.82$ yd.

8. Solve the triangle having $A = 61.7°$, $a = 78.9$ m, $b = 86.4$ m.

Find the indicated parts of each of the following triangles.

9. $a = 86.14$ in, $b = 253.2$ in, $c = 241.9$ in; find A

10. $B = 120.7°$, $a = 127$ ft, $c = 69.8$ ft; find b

11. $A = 51° 20'$, $c = 68.3$ m, $b = 58.2$ m; find a

12. $a = 14.8$ m, $b = 19.7$ m, $c = 31.8$ m; find B

13. $A = 46° 10'$, $b = 184$ cm, $c = 192$ cm; find a

14. $a = 7.5$ ft, $b = 12.0$ ft, $c = 6.9$ ft; find C

Find the area of each of the following triangles.

15. $b = 840.6$ m, $c = 715.9$ m, $A = 149° 18'$

16. $a = 27.6$ cm, $b = 19.8$ cm, $C = 42° 30'$

17. $a = 6.90$ ft, $b = 10.2$ ft, $C = 35° 10'$

18. $a = 94.6$ yd, $b = 123$ yd, $c = 109$ yd

19. $a = .913$ km, $b = .816$ km, $c = .582$ km

20. $a = 43$ m, $b = 32$ m, $c = 51$ m

Solve each of the following problems.

21. Raoul plans to paint a triangular wall in his A-frame cabin. Two sides measure 7 m each, and the third side measures 6 m. How much paint will he need to buy if a can of paint covers 7.5 sq m?

22. A lot has the shape of the quadrilateral in the figure. What is its area?

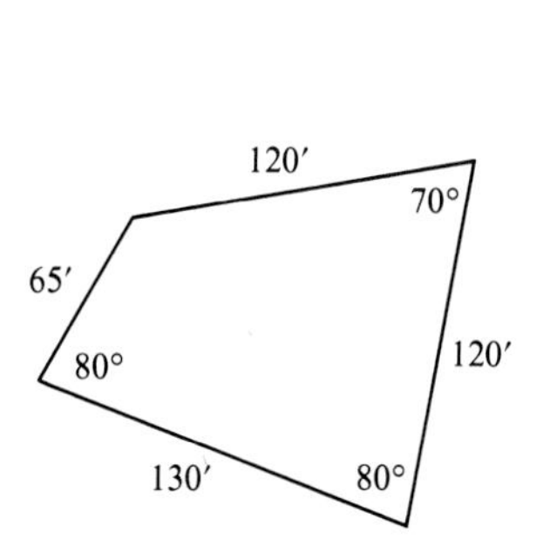

Exercise 22

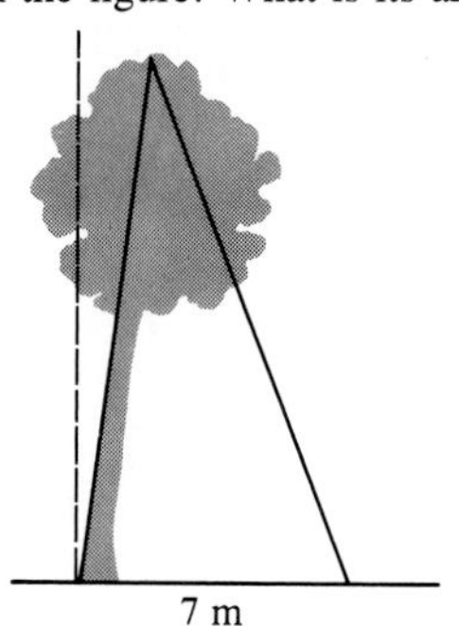

Exercise 23

23. A tree leans at an angle of 8° from the vertical. From a point 7 m from the bottom of the tree, the angle of elevation to the top of the tree is 68°. How tall is the tree?

24. A hill makes an angle of 14.3° with the horizontal. From the base of the hill, the angle of elevation to the top of a tree on top of the hill is 27.2°. The distance along the hill from the base to the tree is 212 ft. Find the height of the tree.

25. A ship is sailing east. At one point, the bearing of a submerged rock is 45° 20′. After sailing 15.2 mi, the bearing of the rock has become 308° 40′. Find the distance of the ship from the rock at the latter point.

26. From an airplane flying over the ocean, the angle of depression to a submarine lying just under the surface is 24° 10′. At the same moment the angle of depression from the airplane to a battleship is 17° 30′. (See the figure.) The distance from the airplane to the battleship is 5120 ft. Find the distance between the battleship and the submarine. (Assume the airplane, submarine, and battleship are in a vertical plane.)

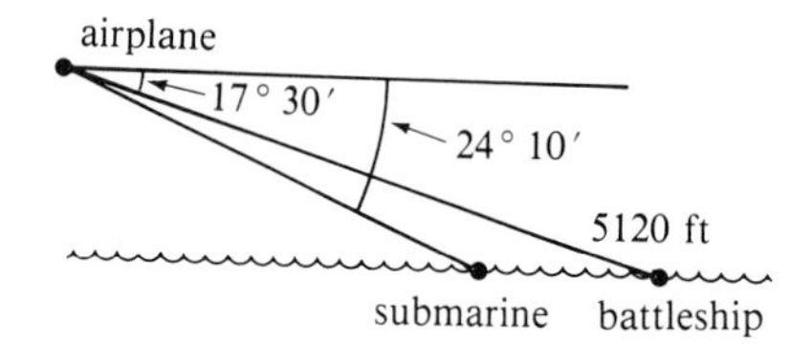

Exercise 26

27. Two boats leave a dock together. Each travels in a straight line. The angle between their courses measures 54° 10′. One boat travels 36.2 km per hr, and the other travels 45.6 km per hr. How far apart will they be after 3 hr?

28. Find the lengths of both diagonals of a parallelogram with adjacent sides of 12 cm and 15 cm if the angle between these sides is 33°.

29. To measure the distance through a mountain for a proposed tunnel, a point C is chosen that can be reached from each end of the tunnel. (See the figure). If $AC = 3800$ m, $BC = 2900$ m, and angle $C = 110°$, find the length of the tunnel.

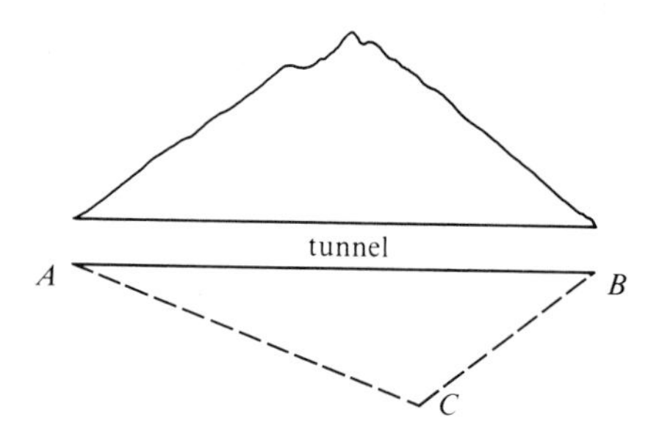

Exercise 29

30. A 35-ft ladder leans against a sloping wall, making an angle of 15° with the wall. If the point at which the ladder touches the wall is 31 ft up from the ground, how far away from the wall is the bottom of the ladder?

In Exercises 31–33 use the vectors pictured here.

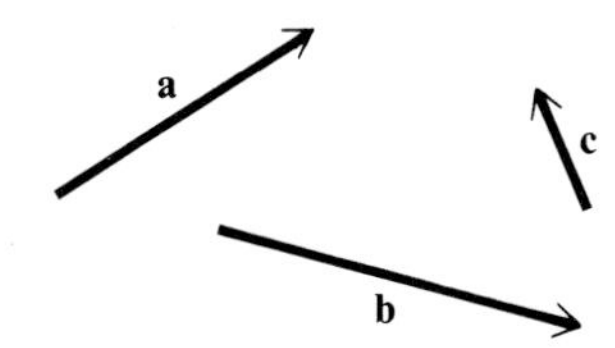

Find each of the following.

31. **a** + **b** **32.** **a** − **b** **33.** **a** + 3**c**

Find the horizontal and vertical components of each of the following vectors, where α is the inclination of the vector from the horizontal.

34. $\alpha = 45°$, magnitude 50 **35.** $\alpha = 75°$, magnitude 69.2

36. $\alpha = 154° 20'$, magnitude 964

Given two forces and the angles between them, find the magnitude of the resultant force.

37. Forces of 15 and 23 lb, forming an angle of 87°

38. Forces of 142 and 215 newtons, forming an angle of 112°

39. Forces of 85.2 and 69.4 newtons, forming an angle of 58° 20′

40. Forces of 475 and 586 lb, forming an angle of 78° 20′

Solve each of the following problems.

41. A 150-lb force acts at a right angle to a 225-lb force. Find the magnitude of the equilibrant and the angle it makes with the 150-lb force.

42. A box is supported above the ground by two ropes. One makes an angle of 52° 40′ with the horizontal. The tension in the rope is 89.6 lb. The second rope makes an angle of 82° 30′ with the first rope, and has a tension of 61.7 lb. Find the weight of the box. (*Hint:* Add the y-components of each tension vector.)

43. A 186-lb force just keeps a 2800-lb car from rolling down a hill. What angle does the hill make with the horizontal?

44. A plane has an airspeed of 520 mph. The pilot wishes to fly on a bearing of 310°. A wind of 37 mph is blowing from a bearing of 212°. What direction should the pilot fly, and what will be her actual speed?

45. A boat travels 15 km per hr in still water. The boat is traveling across a large river, on a bearing of 130°. The current in the river, coming from the west, has a speed of 7 km per hr. Find the resulting speed of the boat and its resulting direction of travel.

46. A long-distance swimmer starts out swimming a steady 3.2 mph due north. A 5.1 mph current is flowing on a bearing of 12°. What is the swimmer's resulting bearing and speed?

8 Complex Numbers

So far this text has dealt only with real numbers. The set of real numbers, however, does not include enough numbers for our needs. For example, there is no real number solution of the equation $x^2 + 1 = 0$. This chapter discusses a set of numbers having the real numbers as a subset, the set of **complex numbers.**

8.1 Operations on Complex Numbers

Complex numbers have the form $a + bi$, where a and b are real numbers and i is the new number defined by

Definition of i

$$i = \sqrt{-1} \quad \text{or} \quad i^2 = -1.$$

Each real number is a complex number, since a real number a may be thought of as the complex number $a + 0i$. A complex number of the form $0 + bi$, where b is nonzero, is an **imaginary number.** Both the set of real numbers and the set of imaginary numbers are subsets of the set of complex numbers. See Figure 8.1. A complex number written in the form $a + bi$ or $a + ib$ is in **standard form.** The form $a + ib$ is used for symbols such as $i\sqrt{5}$, since $\sqrt{5}i$ could be too easily mistaken for $\sqrt{5i}$.

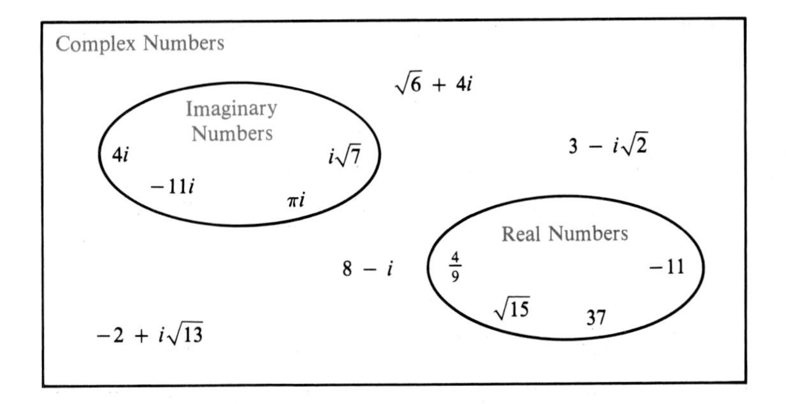

Figure 8.1

Example 1 The list below shows several complex numbers, along with the standard form of each.

Number	*Standard form*
$6i$	$0 + 6i$
-9	$-9 + 0i$
0	$0 + 0i$
$9 - i$	$9 - i$
$i - 1$	$-1 + i$

The symbol $\sqrt{-a}$ is defined as follows.

Definition of $\sqrt{-a}$

For positive real numbers a, $\sqrt{-a} = i\sqrt{a}$.

For example, $\sqrt{-16} = i\sqrt{16} = 4i$, and $\sqrt{-75} = i\sqrt{75} = 5i\sqrt{3}$.

Products or quotients with square roots of negative numbers may be simplified by using the fact that $\sqrt{-a} = i\sqrt{a}$, for positive numbers a. The next example shows how to do this.

Example 2 **(a)** $\sqrt{-7} \cdot \sqrt{-7} = i\sqrt{7} \cdot i\sqrt{7}$
$= i^2 \cdot (\sqrt{7})^2$
$= (-1) \cdot 7 = -7$

(b) $\sqrt{-6} \cdot \sqrt{-10} = i\sqrt{6} \cdot i\sqrt{10} = i^2 \cdot \sqrt{6 \cdot 10} = -1 \cdot 2\sqrt{15} = -2\sqrt{15}$

(c) $\dfrac{\sqrt{-20}}{\sqrt{-2}} = \dfrac{i\sqrt{20}}{i\sqrt{2}} = \sqrt{10}$

(d) $\dfrac{\sqrt{-48}}{\sqrt{24}} = \dfrac{i\sqrt{48}}{\sqrt{24}} = i\sqrt{2}$

When working with negative radicands, it is very important to use the definition $\sqrt{-a} = i\sqrt{a}$ before using any of the other rules for radicals. In particular, the rule $\sqrt{c} \cdot \sqrt{d} = \sqrt{cd}$ is valid only when c and d are not both negative. For example, multiplying $\sqrt{-2} \cdot \sqrt{-32}$ as $\sqrt{64}$ (which is incorrect) gives 8, but

$$\begin{aligned}\sqrt{-2} \cdot \sqrt{-32} &= i\sqrt{2} \cdot i\sqrt{32} \\ &= i^2\sqrt{64} \\ &= (-1)8 \\ &= -8,\end{aligned}$$

which is the correct result.

Complex numbers can be added, subtracted, multiplied, and divided. To *add* two complex numbers, use the following definition.

Addition of Complex Numbers

> For complex numbers $a + bi$ and $c + di$,
> $$(a + bi) + (c + di) = (a + c) + (b + d)i.$$

Example 3

(a) $(3 - 4i) + (-2 + 6i) = [\mathbf{3 + (-2)}] + [\mathbf{-4 + 6}]i = 1 + 2i$

(b) $(-9 + 7i) + (3 - 15i) = -6 - 8i$ ✤

Subtraction with complex numbers is defined as for real numbers.

$$(a + bi) - (c + di) = (a + bi) + (-c - di)$$

Rearranging this result in standard form gives the following definition of subtraction of complex numbers.

Subtraction of Complex Numbers

> For complex numbers $a + bi$ and $c + di$,
> $$(a + bi) - (c + di) = (a - c) + (b - d)i.$$

Example 4 Subtract as indicated.

(a) $(-4 + 3i) - (6 - 7i) = (-4 + 3i) + (-6 + 7i) = -10 + 10i$

(b) $(12 - 5i) - (8 - 3i) = 4 - 2i$ ✤

The *product* of two complex numbers can be found by multiplying as if the numbers were binomials and using the fact that $i^2 = -1$, as follows.

$$\begin{aligned}(a + bi)(c + di) &= ac + adi + bic + bidi \\ &= ac + adi + bci + bdi^2 \\ &= ac + (ad + bc)i + bd(-1) \\ &= (ac - bd) + (ad + bc)i\end{aligned}$$

Thus, the product of the complex numbers $a + bi$ and $c + di$ is defined as follows.

Product of Complex Numbers

For complex numbers $a + bi$ and $c + di$,

$$(a + bi)(c + di) = (ac - bd) + (ad + bc)i.$$

This definition is hard to remember. To find a given product, it is better just to multiply as with binomials. The next example shows this.

Example 5 Find each of the following products.

(a) $(2 - 3i)(3 + 4i) = 2(3) + 2(4i) - 3i(3) - 3i(4i)$
$= 6 + 8i - 9i - 12i^2$
$= 6 - i - 12(-1)$
$= 18 - i$

(b) $(5 - 4i)(7 - 2i) = 5(7) + 5(-2i) - 4i(7) - 4i(-2i)$
$= 35 - 10i - 28i + 8i^2$
$= 35 - 38i + 8(-1)$
$= 27 - 38i$

(c) $(3 - i)(3 + i) = 9 + 3i - 3i - i^2$
$= 9 - (-1) = 10$ ✜

The factors in Example 5(c) are called *conjugates*. The **conjugate** of the complex number $a + bi$ is the complex number $a - bi$. Notice that the product of a pair of conjugates is the difference of squares, so Example 5(c) could have been written as $(3 - i)(3 + i) = 3^2 - i^2 = 9 - (-1) = 10$. The product of conjugates is always a real number.

To find the *quotient* of two complex numbers, use the conjugate of the divisor. The next example shows how to find quotients of complex numbers using the conjugate.

Example 6 **(a)** $\dfrac{3 + 2i}{5 - i}$

Multiply the numerator and denominator by $5 + i$, the conjugate of $5 - i$.

$$\frac{3 + 2i}{5 - i} = \frac{(3 + 2i)(5 + i)}{(5 - i)(5 + i)}$$

$$= \frac{15 + 3i + 10i + 2i^2}{25 - i^2}$$

$$= \frac{13 + 13i}{26} = \frac{1}{2} + \frac{1}{2}i$$

To check this answer, show that

$$(5 - i)\left(\frac{1}{2} + \frac{1}{2}i\right) = 3 + 2i.$$

(b) $\frac{3}{i} = \frac{3(-i)}{i(-i)}$ $\quad$ $-i$ is the conjugate of i

$= \frac{-3i}{-i^2}$

$= \frac{-3i}{1}$ $\quad$ $-i^2 = -(-1) = 1$

$= -3i$ $\quad$ $0 - 3i$ in standard form ✣

Two complex numbers $a + bi$ and $c + di$ are equal if and only if $a = c$ and $b = d$. This fact is used in the next example.

Example 7 Solve $3x - 4yi = (2 + 4i)(3 - 5i)$.

First, find the product on the right side of the equation.

$$\begin{aligned}(2 + 4i)(3 - 5i) &= 6 - 10i + 12i - 20i^2 \\ &= 26 + 2i\end{aligned}$$

Then

$$3x - 4yi = 26 + 2i.$$

Now, by the definition of equality,

$$3x = 26 \quad \text{and} \quad -4y = 2$$

$$x = \frac{26}{3} \quad \text{and} \quad y = -\frac{1}{2}.$$ ✣

8.1 Exercises

Simplify each of the following. See Example 2.

1. $\sqrt{-4}$ **2.** $\sqrt{-49}$ **3.** $\sqrt{-64}$ **4.** $\sqrt{-100}$

5. $\sqrt{-\frac{25}{9}}$ **6.** $\sqrt{-\frac{1}{81}}$ **7.** $\sqrt{-18}$ **8.** $\sqrt{-45}$

9. $\sqrt{-150}$ **10.** $\sqrt{-180}$ **11.** $\sqrt{-27}$ **12.** $\sqrt{-48}$

13. $\sqrt{-80}$ **14.** $\sqrt{-72}$ **15.** $\sqrt{-3} \cdot \sqrt{-3}$ **16.** $\sqrt{-2} \cdot \sqrt{-2}$

17. $\sqrt{-5} \cdot \sqrt{-6}$ **18.** $\sqrt{-27} \cdot \sqrt{-3}$ **19.** $\dfrac{\sqrt{-12}}{\sqrt{-8}}$ **20.** $\dfrac{\sqrt{-15}}{\sqrt{-3}}$

21. $\dfrac{\sqrt{-24}}{\sqrt{72}}$ **22.** $\dfrac{\sqrt{-27}}{\sqrt{9}}$

Perform the following operations and express all results in standard form. See Examples 3–6.

23. $(2 - 5i) + (3 + 2i)$ **24.** $(5 - i) + (3 + 4i)$ **25.** $(-2 + 3i) - (3 + i)$

26. $(4 + 6i) - (-2 - i)$ **27.** $(1 - i) - (5 - 2i)$ **28.** $(-2 + 6i) - (-3 - 8i)$

29. $(2 + i)(3 - 2i)$ **30.** $(-2 + 3i)(4 - 2i)$ **31.** $(2 + 4i)(-1 + 3i)$

32. $(1 + 3i)(2 - 5i)$ **33.** $(5 + 2i)(5 - 3i)$ **34.** $(-3 + 2i)^2$

35. $(2 + i)^2$ **36.** $(\sqrt{6} - i)(\sqrt{6} + i)$ **37.** $(2 - i)(2 + i)$

38. $(5 + 4i)(5 - 4i)$ **39.** $i(3 - 4i)(3 + 4i)$ **40.** $i(2 + 5i)(2 - 5i)$

41. $i(3 - 4i)^2$ **42.** $i(2 + 6i)^2$ **43.** $\dfrac{1 + i}{1 - i}$

44. $\dfrac{2 - i}{2 + i}$ **45.** $\dfrac{4 - 3i}{4 + 3i}$ **46.** $\dfrac{5 + 6i}{5 - 6i}$

47. $\dfrac{4 + i}{6 + 2i}$ **48.** $\dfrac{3 - 2i}{5 + 3i}$ **49.** $\dfrac{5 - 2i}{6 - i}$

50. $\dfrac{3 - 4i}{2 - 5i}$ **51.** $\dfrac{1 - 3i}{1 + i}$ **52.** $\dfrac{-3 + 4i}{2 - i}$

Solve each of the following equations for x and y. See Example 7.

53. $x + yi = 5 + 3i$ **54.** $x + yi = 2 - 4i$

55. $2x + yi = 4 - 3i$ **56.** $x + 3yi = 5 + 2i$

57. $7 - 2yi = 14x - 30i$ **58.** $-5 + yi = x + 6i$

59. $x + yi = (2 + 3i)(4 - 2i)$ **60.** $x + yi = (5 - 7i)(1 + i)$

61. $x + 2i = (3 + yi)(2 - yi)$ **62.** $8 + yi = (x - i)(x + i)$

In work with alternating current, complex numbers are used to describe current, I, voltage, E, and impedance, *Z (the opposition to current). These three quantities are related by the equation $E = IZ$. Thus, if any two of these quantities are known, the third can be found. In each of the following problems, solve the equation $E = IZ$ for the missing variable.*

63. $I = 8 + 6i$, $Z = 6 + 3i$ **64.** $I = 10 + 6i$, $Z = 8 + 5i$

65. $I = 7 + 5i$, $E = 28 + 54i$ **66.** $E = 35 + 55i$, $Z = 6 + 4i$

Find all complex numbers $a + bi$ such that the square $(a + bi)^2$ is

67. real; **68.** imaginary.

69. Show that $\dfrac{\sqrt{2}}{2} + \dfrac{\sqrt{2}}{2}i$ is a square root of i. **70.** Show that $-\dfrac{\sqrt{3}}{2} + \dfrac{1}{2}i$ is a cube root of i.

8.2 Trigonometric Form of Complex Numbers

To graph a complex number such as $2 - 3i$, the familiar coordinate system must be modified. One way to do this is by calling the horizontal axis the **real axis** and the vertical axis the **imaginary axis.** Then complex numbers can be graphed, as shown in Figure 8.2 for the complex number $2 - 3i$.

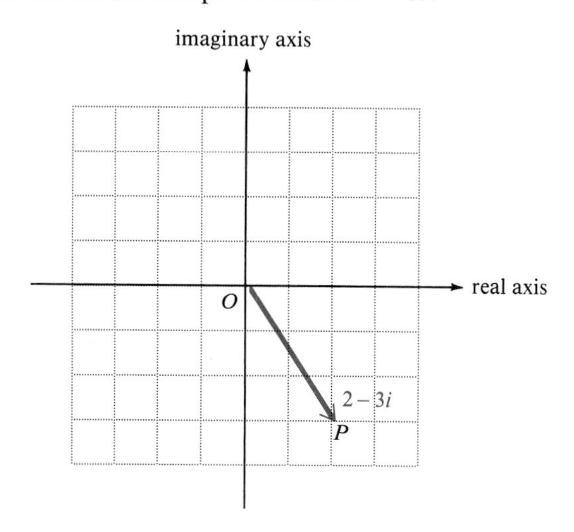

Figure 8.2

Each nonzero complex number graphed in this way determines a unique directed line segment, the segment from the origin to the point representing the complex number. Recall from chapter 7 that such directed line segments (like **OP** of Figure 8.2) are called vectors.

The previous section showed how to find the sum of the two complex numbers $4 + i$ and $1 + 3i$.

$$(4 + i) + (1 + 3i) = 5 + 4i$$

Graphically, the sum of two complex numbers is represented by the vector that is the resultant of the vectors corresponding to the two numbers. The vectors representing the complex numbers $4 + i$ and $1 + 3i$ and the resultant vector that represents their sum, $5 + 4i$, are shown in Figure 8.3.

Example 1 Find the sum of $6 - 2i$ and $-4 - 3i$. Graph both complex numbers and their resultant.

The sum is found by adding the two numbers.

$$(6 - 2i) + (-4 - 3i) = 2 - 5i$$

The graphs are shown in Figure 8.4. ✤

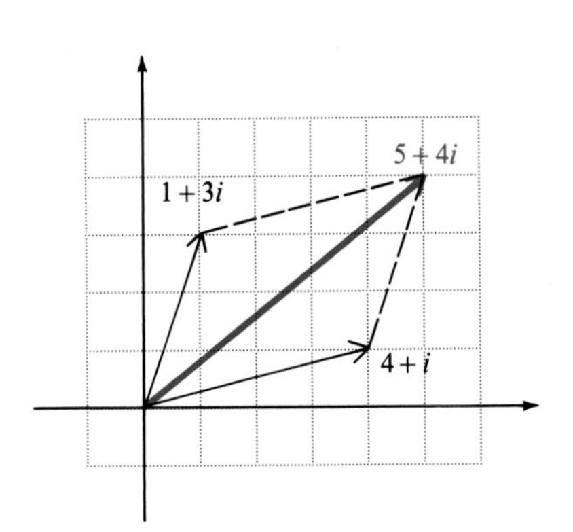

Figure 8.3

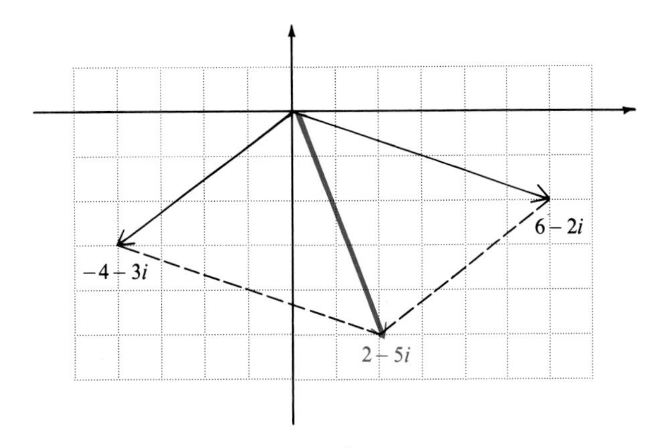

Figure 8.4

Figure 8.5, shows the complex number $x + yi$ that corresponds to a vector **OP** with direction θ and magnitude r. The following relationships between r, θ, x, and y can be verified from Figure 8.5.

$$x = r \cos \theta \qquad r = \sqrt{x^2 + y^2}$$
$$y = r \sin \theta \qquad \tan \theta = \frac{y}{x}$$

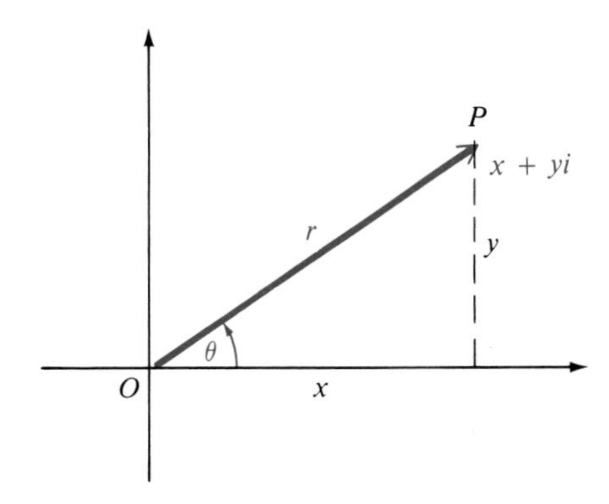

Figure 8.5

Substituting $x = r \cos \theta$ and $y = r \sin \theta$ from the results above into $x + yi$ gives

$$\begin{aligned} x + yi &= r \cos \theta + (r \sin \theta)i \\ &= r(\cos \theta + i \sin \theta). \end{aligned}$$

Trigonometric or Polar Form of a Complex Number

The expression

$$r(\cos\theta + i\sin\theta)$$

is called the **trigonometric form** or **polar form** of the complex number $x + yi$.*

The number r is called the **modulus** or **absolute value** of $x + yi$, while θ is the **argument** of $x + yi$.

Example 2 Write the following complex numbers in trigonometric form.

(a) $-\sqrt{3} + i$

First find r and θ. Since $x = -\sqrt{3}$ and $y = 1$,

$$r = \sqrt{x^2 + y^2} = \sqrt{3 + 1} = 2,$$

$$\tan\theta = \frac{y}{x} = \left(\frac{1}{-\sqrt{3}}\right) = \left(-\frac{\sqrt{3}}{3}\right).$$

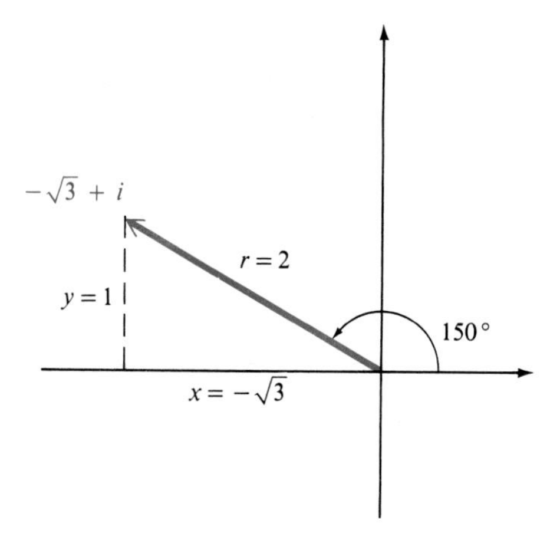

Figure 8.6

As shown in Figure 8.6, the complex number $-\sqrt{3} + i$ is in quadrant II, so that $\theta = 150°$. In trigonometric form,

$$x + yi = r(\cos\theta + i\sin\theta)$$
$$-\sqrt{3} + i = 2(\cos 150° + i\sin 150°).$$

*The expression $\cos\theta + i\sin\theta$ is sometimes abbreviated as cis θ. With this notation $r(\cos\theta + i\sin\theta)$ is written as r cis θ.

(b) $-2 - 2i$

Find r and θ.

$$r = \sqrt{x^2 + y^2} = \sqrt{4 + 4} = 2\sqrt{2}$$

$$\tan\theta = \frac{y}{x} = \left(\frac{-2}{-2}\right) = 1$$

Since $-2 - 2i$ is in quadrant III, $\theta = 225°$. The trigonometric form of $-2 - 2i$ is

$$2\sqrt{2}(\cos 225° + i \sin 225°).$$

Example 3 Express $2(\cos 300° + i \sin 300°)$ in standard form.

Since $\cos 300° = 1/2$ and $\sin 300° = -\sqrt{3}/2$,

$$2(\cos 300° + i \sin 300°) = 2\left(\frac{1}{2} - i\frac{\sqrt{3}}{2}\right) = 1 - i\sqrt{3}.$$

Example 4 **(a)** Write the complex number $6.32(\cos 115° + i \sin 115°)$ in standard form.

Use a calculator or tables to find

$$\cos 115° = -.4226$$

and

$$\sin 115° = .9063,$$

to four decimal places. Then, in standard form,

$$6.32(\cos 115° + i \sin 115°) = 6.32(-.4226 + .9063i)$$
$$= -2.6708 + 5.7278i.$$

(b) Write $5 - 4i$ in trigonometric form.

Here, $r = \sqrt{25 + 16} = \sqrt{41}$, and $\tan\theta = -4/5$. Use a calculator or tables to find $\theta \approx -38.66°$ or $\theta \approx 360° - 38.66° = 321.34°$. Since $5 - 4i$ is in quadrant IV, $\theta \approx 321.34°$. Using these results,

$$5 - 4i = \sqrt{41}(\cos 321.34° + i \sin 321.34°).$$

8.2 Exercises

Graph each of the following complex numbers. See Example 1.

1. $-2 + 3i$ **2.** $-4 + 5i$ **3.** $8 - 5i$ **4.** $6 - 5i$

5. $2 - 2i\sqrt{3}$ **6.** $4\sqrt{2} + 4i\sqrt{2}$ **7.** $-4i$ **8.** $3i$

9. -8 **10.** 2

Find the resultant of each of the following pairs of complex numbers. See Example 1.

11. $4 - 3i,\ -1 + 2i$ **12.** $2 + 3i,\ -4 - i$ **13.** $5 - 6i,\ -2 + 3i$ **14.** $7 - 3i,\ -4 + 3i$

15. $-3,\ 3i$ **16.** $6,\ -2i$ **17.** $2 + 6i,\ -2i$ **18.** $4 - 2i,\ 5$

19. $7 + 6i,\ 3i$ **20.** $-5 - 8i,\ -1$

Write each of the following complex numbers in trigonometric form. See Example 2.

21. $3 - 3i$
22. $-2 + 2i\sqrt{3}$
23. $-3 - 3i\sqrt{3}$
24. $1 + i\sqrt{3}$
25. $\sqrt{3} - i$
26. $4\sqrt{3} + 4i$
27. $-5 - 5i$
28. $-\sqrt{2} + i\sqrt{2}$
29. $2 + 2i$
30. $-\sqrt{3} + i$
31. -4
32. $5i$
33. $-2i$
34. 7

Write the following complex numbers in standard form. See Examples 3 and 4.

35. $2(\cos 45° + i \sin 45°)$
36. $4(\cos 60° + i \sin 60°)$
37. $10(\cos 90° + i \sin 90°)$
38. $8(\cos 270° + i \sin 270°)$
39. $4(\cos 240° + i \sin 240°)$
40. $2(\cos 330° + i \sin 330°)$
41. $(\cos 30° + i \sin 30°)$
42. $3(\cos 150° + i \sin 150°)$
43. $5(\cos 300° + i \sin 300°)$
44. $6(\cos 135° + i \sin 135°)$
45. $\sqrt{2}(\cos 180° + i \sin 180°)$
46. $\sqrt{3}(\cos 315° + i \sin 315°)$

Using a calculator or Table 2, complete the following chart to the nearest ten minutes.

	Standard form	*Trigonometric form*
47.	$2 + 3i$	__________
48.	__________	$(\cos 35° + i \sin 35°)$
49.	__________	$3(\cos 250°\ 10' + i \sin 250°\ 10')$
50.	$-4 + i$	__________
51.	$-1.8794 + .6840i$	__________
52.	__________	$2(\cos 310°\ 20' + i \sin 310°\ 20')$
53.	$3 + 5i$	__________
54.	__________	$(\cos 110°\ 30' + i \sin 110°\ 30')$

In applied work in trigonometry, it is often necessary to find the resultant of more than two vectors graphically. Find the resultant of the indicated vectors in each of the following cases. Give the result in standard form.

55.

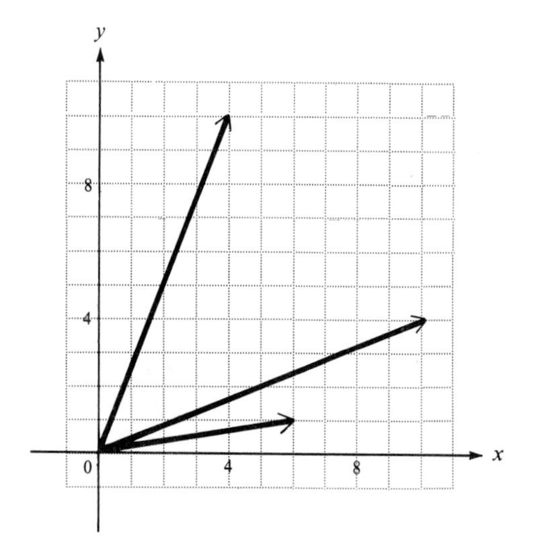

56.

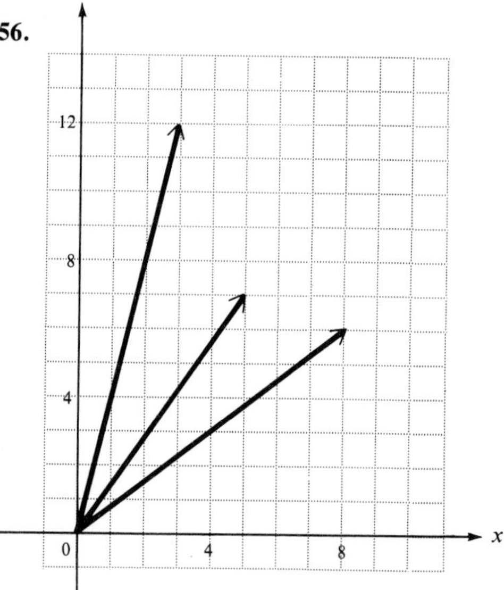

8.3 Product and Quotient Theorems

The product of the two complex numbers $1 + i\sqrt{3}$ and $-2\sqrt{3} + 2i$ can be found by the method shown in section 8.1.

$$\begin{aligned}(1 + i\sqrt{3})(-2\sqrt{3} + 2i) &= -2\sqrt{3} + 2i - 2i(3) + 2i^2\sqrt{3} \\ &= -2\sqrt{3} + 2i - 6i - 2\sqrt{3} \\ &= -4\sqrt{3} - 4i\end{aligned}$$

This same product also can be found by first converting the complex numbers $1 + i\sqrt{3}$ and $-2\sqrt{3} + 2i$ to trigonometric form. Using the method explained in the previous section,

$$1 + i\sqrt{3} = 2(\cos 60° + i \sin 60°)$$

and

$$-2\sqrt{3} + 2i = 4(\cos 150° + i \sin 150°).$$

If the trigonometric forms are now multiplied together and if the trigonometric identities for the cosine and the sine of the sum of two angles are used, the result is

$$\begin{aligned}&[2(\cos 60° + i \sin 60°)][4(\cos 150° + i \sin 150°)] \\ &= 2 \cdot 4(\cos 60° \cdot \cos 150° + i \sin 60° \cdot \cos 150° \\ &\qquad + i \cos 60° \cdot \sin 150° + i^2 \sin 60° \cdot \sin 150°) \\ &= 8[(\cos 60° \cdot \cos 150° - \sin 60° \cdot \sin 150°) \\ &\qquad + i(\sin 60° \cdot \cos 150° + \cos 60° \cdot \sin 150°)] \\ &= 8[\cos (60° + 150°) + i \sin (60° + 150°)] \\ &= 8(\cos 210° + i \sin 210°).\end{aligned}$$

The modulus of the product, 8, is equal to the product of the moduli of the factors, $2 \cdot 4$, while the argument of the product, 210°, is the sum of the arguments of the factors, $60° + 150°$.

Generalizing, the product of the two complex numbers, $r_1(\cos \theta_1 + i \sin \theta_1)$ and $r_2(\cos \theta_2 + i \sin \theta_2)$, is

$$\begin{aligned}&[r_1(\cos \theta_1 + i \sin \theta_1)] \cdot [r_2(\cos \theta_2 + i \sin \theta_2)] \\ &= r_1r_2(\cos \theta_1 \cos \theta_2 + i \sin \theta_1 \cos \theta_2 + i \cos \theta_1 \sin \theta_2 + i^2 \sin \theta_1 \sin \theta_2) \\ &= r_1r_2[(\cos \theta_1 \cos \theta_2 - \sin \theta_1 \sin \theta_2) + i(\sin \theta_1 \cos \theta_2 + \cos \theta_1 \sin \theta_2)] \\ &= r_1r_2[\cos (\theta_1 + \theta_2) + i \sin (\theta_1 + \theta_2)].\end{aligned}$$

This work is summarized in the following *product theorem*.

Product Theorem

If $r_1 (\cos \theta_1 + i \sin \theta_1)$ and $r_2(\cos \theta_2 + i \sin \theta_2)$ are any two complex numbers, then

$$\begin{aligned}&[r_1(\cos \theta_1 + i \sin \theta_1)] \cdot [r_2(\cos \theta_2 + i \sin \theta_2)] \\ &= r_1r_2[\cos (\theta_1 + \theta_2) + i \sin (\theta_1 + \theta_2)].\end{aligned}$$

Example 1 Find the product of $3(\cos 45° + i \sin 45°)$ and $2(\cos 135° + i \sin 135°)$.

Using the product theorem.

$$\begin{aligned}&[3(\cos 45° + i \sin 45°)][2(\cos 135° + i \sin 135°)]\\ &= 3 \cdot 2[\cos (45° + 135°) + i \sin (45° + 135°)]\\ &= 6(\cos 180° + i \sin 180°),\end{aligned}$$

which can be expressed as $6(-1 + i \cdot 0) = 6(-1) = -6$. The two complex numbers in this example are complex factors of -6. ✥

Using the method shown in section 8.1, in standard form the quotient of the complex numbers $1 + i\sqrt{3}$ and $-2\sqrt{3} + 2i$ is

$$\begin{aligned}\frac{1 + i\sqrt{3}}{-2\sqrt{3} + 2i} &= \frac{(1 + i\sqrt{3})(-2\sqrt{3} - 2i)}{(-2\sqrt{3} + 2i)(-2\sqrt{3} - 2i)}\\ &= \frac{-2\sqrt{3} - 2i - 6i - 2i^2\sqrt{3}}{12 - 4i^2}\\ &= \frac{-8i}{16} = -\frac{1}{2}i.\end{aligned}$$

Writing $1 + i\sqrt{3}$, $-2\sqrt{3} + 2i$, and $-\frac{1}{2}i$ in trigonometric form gives

$$\begin{aligned}1 + i\sqrt{3} &= 2(\cos 60° + i \sin 60°)\\ -2\sqrt{3} + 2i &= 4(\cos 150° + i \sin 150°)\\ -\frac{1}{2}i &= \frac{1}{2}[(\cos (-90°) + i \sin (-90°))].\end{aligned}$$

The modulus of the quotient, 1/2, is the quotient of the two moduli, 2 and 4. The argument of the quotient, $-90°$, is the difference of the two arguments, $60° - 150° = -90°$. It would be easier to find the quotient of these two complex numbers in trigonometric form than in standard form. Generalizing from this example leads to another theorem. The proof is similar to the proof of the product theorem, after the numerator and denominator are multiplied by the conjugate of the denominator.

Quotient Theorem

If $r_1(\cos \theta_1 + i \sin \theta_1)$ and $r_2(\cos \theta_2 + i \sin \theta_2)$ are complex numbers, where $r_2(\cos \theta_2 + i \sin \theta_2) \neq 0$, then

$$\frac{r_1(\cos \theta_1 + i \sin \theta_1)}{r_2(\cos \theta_2 + i \sin \theta_2)} = \frac{r_1}{r_2}[\cos (\theta_1 - \theta_2) + i \sin (\theta_1 - \theta_2)].$$

Example 2 Find the quotient of the complex numbers $10(\cos (-60°) + i \sin (-60°))$ and $5(\cos 150° + i \sin 150°)$. Write the result in standard form.

By the quotient theorem,

$$\frac{10(\cos(-60°) + i \sin(-60°))}{5(\cos 150° + i \sin 150°)}$$

$$= \frac{10}{5}[(\cos(-60° - 150°) + i \sin(-60° - 150°)]$$

$$= \frac{10}{5}[\cos(-210°) + i \sin(-210°)]$$

$$= 2(\cos 150° + i \sin 150°).$$

Since $\cos 150° = -\sqrt{3}/2$ and $\sin 150° = 1/2$,

$$2(\cos 150° + i \sin 150°) = 2\left(\frac{-\sqrt{3}}{2} + i \cdot \frac{1}{2}\right)$$

$$= -\sqrt{3} + i.$$

The quotient in standard form is $-\sqrt{3} + i$. ✣

Example 3 Use a calculator to find the following.

(a) $[9.3(\cos 125.2° + i \sin 125.2°)][2.7(\cos 49.8° + i \sin 49.8°)]$

By the product theorem,

$$[9.3(\cos 125.2° + i \sin 125.2°)][2.7(\cos 49.8° + i \sin 49.8°)]$$

$$= (9.3)(2.7)[\cos(125.2° + 49.8°) + i \sin(125.2° + 49.8)°]$$

$$= 25.11(\cos 175° + i \sin 175°)$$

$$= 25.11(-.9962 + .0872i)$$

$$= -25.0146 + 2.1896i.$$

(b) $\dfrac{10.42\left(\cos \dfrac{3\pi}{4} + i \sin \dfrac{3\pi}{4}\right)}{9.03\left(\cos \dfrac{\pi}{5} + i \sin \dfrac{\pi}{5}\right)}$

Use the quotient theorem.

$$\frac{10.42\left(\cos \frac{3\pi}{4} + i \sin \frac{3\pi}{4}\right)}{9.03\left(\cos \frac{\pi}{5} + i \sin \frac{\pi}{5}\right)} = \frac{10.42}{9.03}\left[\cos\left(\frac{3\pi}{4} - \frac{\pi}{5}\right) + i \sin\left(\frac{3\pi}{4} - \frac{\pi}{5}\right)\right]$$

$$= 1.15\left(\cos \frac{11\pi}{20} + i \sin \frac{11\pi}{20}\right)$$

$$= 1.15(\cos 1.7279 + i \sin 1.7279)$$

$$= 1.15(-.1565 + .9877i)$$

$$= -.1800 + 1.1359i$$ ✣

8.3 Exercises

Find each of the following products. Write each product in standard form. See Example 1.

1. $[3(\cos 60° + i \sin 60°)][2(\cos 90° + i \sin 90°)]$

2. $[4(\cos 30° + i \sin 30°)][5(\cos 120° + i \sin 120°)]$

3. $[2(\cos 45° + i \sin 45°)][2(\cos 225° + i \sin 225°)]$

4. $[8(\cos 300° + i \sin 300°)][5(\cos 120° + i \sin 120°)]$

5. $[4(\cos 60° + i \sin 60°)][6(\cos 330° + i \sin 330°)]$

6. $[8(\cos 210° + i \sin 210°)][2(\cos 330° + i \sin 330°)]$

7. $[5(\cos 90° + i \sin 90°)][3(\cos 45° + i \sin 45°)]$

8. $[6(\cos 120° + i \sin 120°)][5(\cos(-30°) + i \sin(-30°))]$

9. $[\sqrt{3}(\cos 45° + i \sin 45°)][\sqrt{3}(\cos 225° + i \sin 225°)]$

10. $[\sqrt{2}(\cos 300° + i \sin 300°)][\sqrt{2}(\cos 270° + i \sin 270°)]$

Find each of the following quotients. Write each answer in standard form. See Example 2.

11. $\dfrac{4(\cos 120° + i \sin 120°)}{2(\cos 150° + i \sin 150°)}$

12. $\dfrac{10(\cos 225° + i \sin 225°)}{5(\cos 45° + i \sin 45°)}$

13. $\dfrac{16(\cos 300° + i \sin 300°)}{8(\cos 60° + i \sin 60°)}$

14. $\dfrac{24(\cos 150° + i \sin 150°)}{2(\cos 30° + i \sin 30°)}$

15. $\dfrac{3(\cos 305° + i \sin 305°)}{9(\cos 65° + i \sin 65°)}$

16. $\dfrac{12(\cos 293° + i \sin 293°)}{6(\cos 23° + i \sin 23°)}$

17. $\dfrac{8}{\sqrt{3} + i}$

18. $\dfrac{2i}{-1 - i\sqrt{3}}$

19. $\dfrac{-i}{1 + i}$

20. $\dfrac{1}{2 - 2i}$

21. $\dfrac{2\sqrt{6} - 2i\sqrt{2}}{\sqrt{2} - i\sqrt{6}}$

22. $\dfrac{4 + 4i}{2 - 2i}$

Use your calculator to work each of the following problems. Give your answers in standard form. See Example 3.

23. $[3.7(\cos 27° 15' + i \sin 27° 15')][4.1(\cos 53° 42' + i \sin 53° 42')]$

24. $[2.81(\cos 54° 12' + i \sin 54° 12')][5.8(\cos 82° 53' + i \sin 82° 53')]$

25. $\dfrac{45.3(\cos 127° 25' + i \sin 127° 25')}{12.8(\cos 43° 32' + i \sin 43° 32')}$

26. $\dfrac{2.94(\cos 1.5032 + i \sin 1.5032)}{10.5(\cos 4.6528 + i \sin 4.6528)}$

27. $\left[1.86\left(\cos \dfrac{5\pi}{9} + i \sin \dfrac{5\pi}{9}\right)\right]^2$

28. $\left[24.3\left(\cos \dfrac{7\pi}{12} + i \sin \dfrac{7\pi}{12}\right)\right]^2$

29. The alternating current in an electric inductor is

$$I = \frac{E}{Z}$$

amperes, where E is the voltage and $Z = R + X_L i$ is the impedance. If $E = 8(\cos 20° + i \sin 20°)$, $R = 6$, and $X_L = 3$, find the current. Give the answer in standard form.

30. The current I in a circuit with voltage E, resistance R, capacitive reactance X_c, and inductive reactance X_L is

$$I = \frac{E}{R + (X_L - X_c)i}.$$

Find I if $E = 12(\cos 25° + i \sin 25°)$, $R = 3$, $X_L = 4$, and $X_c = 6$. Give the answer in standard form.

8.4 Powers and Roots of Complex Numbers

The product theorem can be used to find the square of a complex number. The square of the complex number $r(\cos \theta + i \sin \theta)$ is

$$\begin{aligned}[r(\cos \theta + i \sin \theta)]^2 &= [r(\cos \theta + i \sin \theta)][r(\cos \theta + i \sin \theta)] \\ &= r \cdot r[\cos (\theta + \theta) + i \sin (\theta + \theta] \\ &= r^2(\cos 2\theta + i \sin 2\theta).\end{aligned}$$

In the same way,

$$[r(\cos \theta + i \sin \theta)]^3 = r^3(\cos 3\theta + i \sin 3\theta).$$

These results suggest the plausibility of the following theorem for positive integer values of n. Although the following theorem is stated and can be proved for all n, we will use it only for positive integer values of n and their reciprocals.

De Moivre's Theorem

If $r(\cos \theta + i \sin \theta)$ is a complex number and if n is any real number, then

$$[r(\cos \theta + i \sin \theta)]^n = r^n(\cos n\theta + i \sin n\theta).$$

Example 1 Find $(1 + i\sqrt{3})^8$.

To use De Moivre's theorem, first convert $1 + i\sqrt{3}$ into the trigonometric form shown below.

$$1 + i\sqrt{3} = 2(\cos 60° + i \sin 60°)$$

Now apply De Moivre's theorem.

$$\begin{aligned}(1 + i\sqrt{3})^8 &= [2(\cos 60° + i \sin 60°)]^8 \\ &= 2^8[\cos (8 \cdot 60°) + i \sin (8 \cdot 60°)] \\ &= 256(\cos 480° + i \sin 480°) \\ &= 256(\cos 120° + i \sin 120°) \\ &= 256\left(-\frac{1}{2} + i\frac{\sqrt{3}}{2}\right) \\ &= -128 + 128i\sqrt{3}\end{aligned}$$

De Moivre's theorem is also used to find the nth roots of complex numbers. The nth root of a complex number is defined as follows.

nth Root

For a positive integer n, the complex number $a + bi$ is an **nth root** of the complex number $x + yi$ if

$$(a + bi)^n = x + yi.$$

To find the cube roots of the complex number $8(\cos 135° + i \sin 135°)$, for example, look for a complex number, say $r(\cos \alpha + i \sin \alpha)$, that will satisfy

$$[r(\cos \alpha + i \sin \alpha)]^3 = 8(\cos 135° + i \sin 135°).$$

By De Moivre's theorem, this equation becomes

$$r^3(\cos 3\alpha + i \sin 3\alpha) = 8(\cos 135° + i \sin 135°).$$

One way to satisfy this equation is to set $r^3 = 8$ and also $\cos 3\alpha + i \sin 3\alpha = \cos 135° + i \sin 135°$. The first of these conditions implies that $r = 2$, and the second implies that

$$\cos 3\alpha = \cos 135° \quad \text{and} \quad \sin 3\alpha = \sin 135°.$$

These equations can be satisfied only if

$$3\alpha = 135° + 360° \cdot k, \quad k \text{ any integer},$$

or

$$\alpha = \frac{135° + 360° \cdot k}{3}, \quad k \text{ any integer}.$$

If $k = 0$,
$$\alpha = \frac{135° + 0°}{3} = 45°.$$

For $k = 1$,
$$\alpha = \frac{135° + 360°}{3} = \frac{495°}{3} = 165°.$$

When $k = 2$,
$$\alpha = \frac{135° + 720°}{3} = \frac{855°}{3} = 285°.$$

In the same way, $\alpha = 405°$ when $k = 3$. But note that $\sin 405° = \sin 45°$ and $\cos 405° = \cos 45°$. If $k = 4$, $\alpha = 525°$ which has the same sine and cosine values as 165°. To continue with larger values of k would just be repeating solutions already found. Therefore, all of the cube roots (three of them) can be found by letting $k = 0$, 1, and 2.

When $k = 0$, the root is

$$2(\cos 45° + i \sin 45°).$$

When $k = 1$, the root is

$$2(\cos 165° + i \sin 165°).$$

When $k = 2$, the root is

$$2(\cos 285° + i \sin 285°).$$

In summary, $2(\cos 45° + i \sin 45°)$, $2(\cos 165° + i \sin 165°)$, and $2(\cos 285° + i \sin 285°)$ are the three cube roots of $8(\cos 135° + i \sin 135°)$.

Notice that the formula for α in the example above can be written in an alternative form as

$$\alpha = \frac{135°}{3} + \frac{360° \cdot k}{3} = 45° + 120° \cdot k,$$

for $k = 0, 1,$ and 2, which is easier to use.

Generalizing the work above leads to the following theorem.

nth Root Theorem

If n is any positive integer and r is a positive real number, then the complex number $r(\cos \theta + i \sin \theta)$ has exactly n distinct nth roots, given by

$$r^{1/n}(\cos \alpha + i \sin \alpha),$$

where

$$\alpha = \frac{\theta + 360° \cdot k}{n} \quad \text{or} \quad \alpha = \frac{\theta}{n} + \frac{360° \cdot k}{n},$$

$k = 0, 1, 2, \ldots, n - 1.$

Example 2 Find all fourth roots of $-8 + 8i\sqrt{3}$.

First write $-8 + 8i\sqrt{3}$ in trigonometric form as

$$-8 + 8i\sqrt{3} = 16(\cos 120° + i \sin 120°).$$

Here $r = 16$ and $\theta = 120°$. The fourth roots of this number have modulus $16^{1/4} = 2$ and arguments given as follows. Using the alternative formula for α,

$$\alpha = \frac{120°}{4} + \frac{360° \cdot k}{4} = 30° + 90° \cdot k.$$

If $k = 0$, $\quad \alpha = 30° + 90° \cdot 0 = 30°.$

If $k = 1$, $\quad \alpha = 30° + 90° \cdot 1 = 120°.$

If $k = 2$, $\quad \alpha = 30° + 90° \cdot 2 = 210°.$

If $k = 3$, $\quad \alpha = 30° + 90° \cdot 3 = 300°.$

Using these angles, the fourth roots are

$$2(\cos 30° + i \sin 30°),$$
$$2(\cos 120° + i \sin 120°),$$
$$2(\cos 210° + i \sin 210°),$$

and

$$2(\cos 300° + i \sin 300°).$$

These four roots can be written in standard form as $\sqrt{3} + i$, $-1 + i\sqrt{3}$, $-\sqrt{3} - i$, and $1 - i\sqrt{3}$. The graphs of these roots are all on a circle that has center at the origin and radius 2, as shown in Figure 8.7. Notice that the roots are equally spaced about the circle 90° apart. ✣

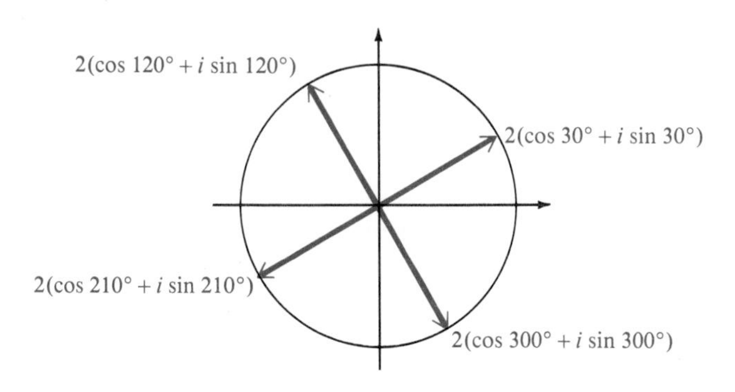

Figure 8.7

Example 3 Find all complex number solutions of $x^5 - 1 = 0$.

Write the equation as

$$x^5 - 1 = 0$$
$$x^5 = 1$$
$$x = \sqrt[5]{1}.$$

While there is only one real number solution, 1, there is a total of five complex number solutions. To find these solutions, first write 1 in trigonometric form as

$$1 = 1 + 0i = 1(\cos 0° + i \sin 0°).$$

The modulus of the fifth roots is $1^{1/5} = 1$, and the arguments are given by

$$0° + 72° \cdot k, \qquad k = 0, 1, 2, 3, \text{ or } 4.$$

By using these arguments, the fifth roots are

$$1(\cos 0° + i \sin 0°),$$
$$1(\cos 72° + i \sin 72°),$$
$$1(\cos 144° + i \sin 144°),$$
$$1(\cos 216° + i \sin 216°),$$

and

$$1(\cos 288° + i \sin 288°).$$

The first of these roots equals 1; the others cannot easily be expressed in standard form. The five fifth roots all lie on a unit circle and are equally spaced around it every 72°, as shown in Figure 8.8. ✣

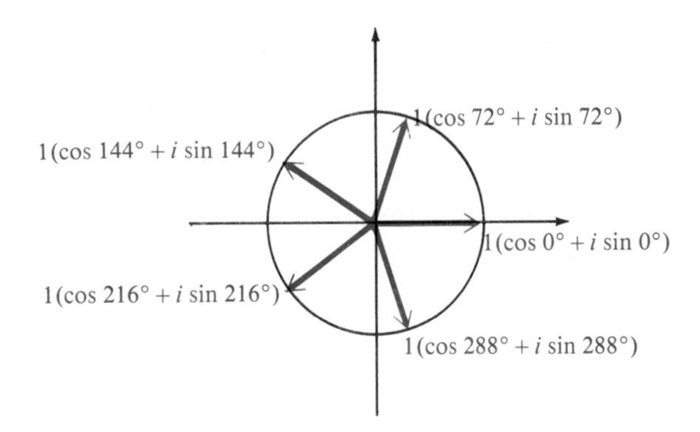

Figure 8.8

8.4 Exercises

Find each of the following powers. Write each answer in standard form. See Example 1.

1. $[3(\cos 30° + i \sin 30°)]^3$
2. $[2(\cos 135° + i \sin 135°)]^4$
3. $(\cos 45° + i \sin 45°)^8$
4. $[2(\cos 120° + i \sin 120°)]^3$
5. $[3(\cos 100° + i \sin 100°)]^3$
6. $[3(\cos 40° + i \sin 40°)]^3$
7. $(\sqrt{3} + i)^5$
8. $(2\sqrt{2} - 2i\sqrt{2})^6$
9. $(2 - 2i\sqrt{3})^4$
10. $\left(\frac{\sqrt{2}}{2} - \frac{\sqrt{2}}{2}i\right)^8$
11. $(-2 - 2i)^5$
12. $(-1 + i)^7$

Find and graph all cube roots of each of the following complex numbers. In Exercises 1–30, leave answers in trigonometric form. See Example 2.

13. $(\cos 0° + i \sin 0°)$
14. $(\cos 90° + i \sin 90°)$
15. $8(\cos 60° + i \sin 60°)$
16. $27(\cos 300° + i \sin 300°)$
17. $-8i$
18. $27i$
19. -64
20. 27
21. $1 + i\sqrt{3}$
22. $2 - 2i\sqrt{3}$
23. $-2\sqrt{3} + 2i$
24. $\sqrt{3} - i$

Find and graph all the following roots of 1.

25. Second
26. Fourth
27. Sixth
28. Eighth

Find and graph all the following roots of i.

29. Second
30. Fourth

Find all solutions of each of the following equations. See Example 3.

31. $x^3 - 1 = 0$
32. $x^3 + 1 = 0$
33. $x^3 + i = 0$
34. $x^4 + i = 0$
35. $x^3 - 8 = 0$
36. $x^3 + 27 = 0$
37. $x^4 + 1 = 0$
38. $x^4 + 16 = 0$
39. $x^4 - i = 0$
40. $x^5 - i = 0$
41. $x^3 - (4 + 4i\sqrt{3}) = 0$
42. $x^4 - (8 + 8i\sqrt{3}) = 0$

Use a calculator to find all solutions of each of the following equations in standard form.

43. $x^3 + 4 - 5i = 0$
44. $x^5 + 2 + 3i = 0$
45. $x^2 + (3.72 + 8.24i) = 0$
46. $x^4 - (5.13 - 4.27i) = 0$

8.5 Polar Equations

Throughout this text we have been using the Cartesian coordinate system to graph equations. Another coordinate system that is particularly useful for graphing many relations is the **polar coordinate system.** The system is based on a point, called the **pole,** and a ray, called the **polar axis.** The polar axis is usually drawn in the direction of the positive x-axis, as shown in Figure 8.9.

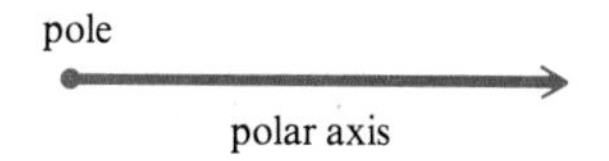

Figure 8.9

In Figure 8.10 the pole has been placed at the origin of a Cartesian coordinate system, so that the polar axis coincides with the positive x-axis. Point P has coordinates (x, y) in the Cartesian coordinate system. Point P can also be located by giving the directed angle θ from the positive x-axis to ray OP and the directed distance r from the pole to point P. The ordered pair (r, θ) gives the **polar coordinates** of point P.

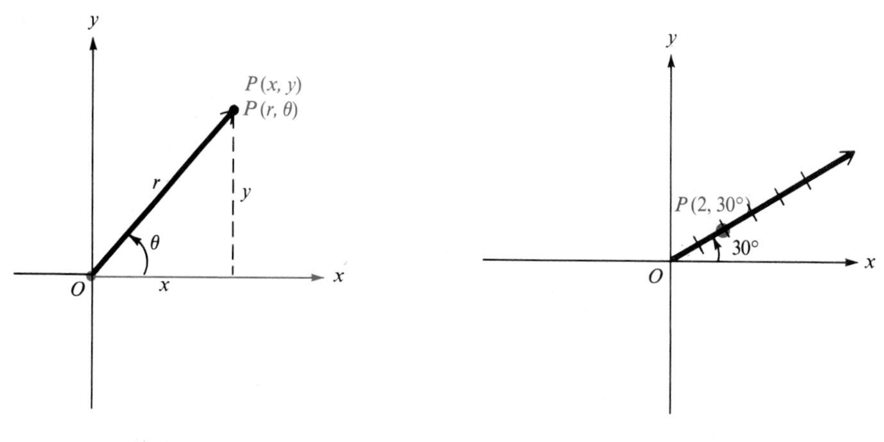

Figure 8.10 **Figure 8.11**

Example 1 Plot each point, given its polar coordinates.

(a) $P(2, 30°)$

In this case, $r = 2$ and $\theta = 30°$, so the point P is located 2 units from the origin in the positive direction on a ray 30° from the polar axis, as shown in Figure 8.11.

(b) $Q(-4, 120°)$

Since r is negative, Q is 4 units in the negative direction from the pole on an extension of the 120° ray. See Figure 8.12.

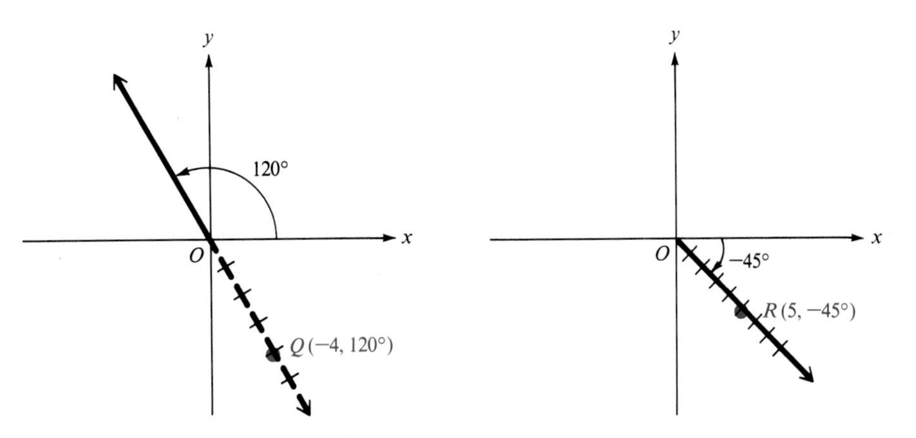

Figure 8.12

Figure 8.13

(c) $R(5, -45°)$

Point R is shown in Figure 8.13. Since θ is negative, the angle is measured in the clockwise direction. ✣

One important difference between Cartesian coordinates and polar coordinates is that while a given point in the plane can have only one pair of Cartesian coordinates, this same point can have an infinite number of pairs of polar coordinates. For example, $(2, 30°)$ locates the same point as $(2, 390°)$ or $(2, -330°)$ or $(-2, 210°)$.

Example 2 Give three other pairs of polar coordinates for the point $P(3, 140°)$.

Three pairs that could be used for the point are $(3, -220°)$, $(-3, 320°)$, and $(-3, -40°)$. See Figure 8.14. ✣

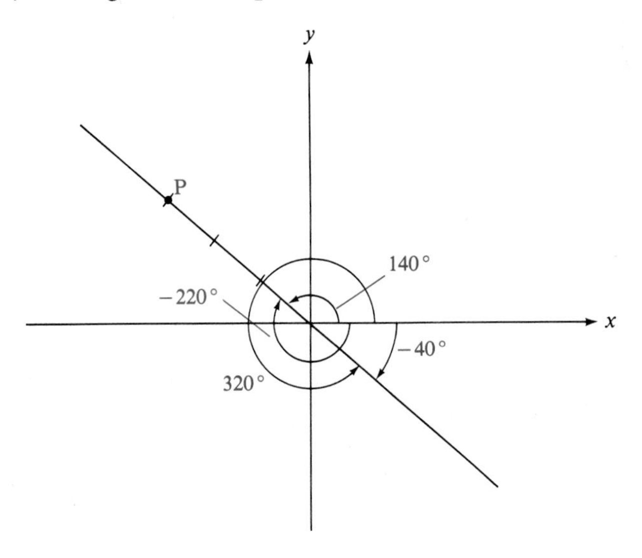

Figure 8.14

An equation like $r = 3 \sin \theta$ where r and θ are the variables, is a **polar equation.** (Equations in x and y are called **rectangular** or **Cartesian equations.**) The simplest equation for many useful curves turns out to be a polar equation.

Graphing a polar equation is much the same as graphing a Cartesian equation. Find some representative ordered pairs, (r, θ), satisfying the equation, and then sketch the graph. For example, to graph $r = 1 + \cos \theta$, first find and graph some ordered pairs (as in the table) and then connect the points in order—from (2, 0°) to (1.9, 30°) to (1.7, 45°) and so on. The graph is shown in Figure 8.15. This curve is called a **cardioid** because of its heart shape.

θ	0°	30°	45°	60°	90°	120°	135°	150°	180°	270°	315°
$\cos \theta$	1	.9	.7	.5	0	−.5	−.7	−.9	−1	0	.7
$r = 1 + \cos \theta$	2	1.9	1.7	1.5	1	.5	.3	.1	0	1	1.7

Once the pattern of values of r becomes clear, it is not necessary to find more ordered pairs. That is why we stopped with the ordered pair (1.7, 315°) in the table above. From the pattern, the pair (1.9, 330°) also would satisfy the relation.

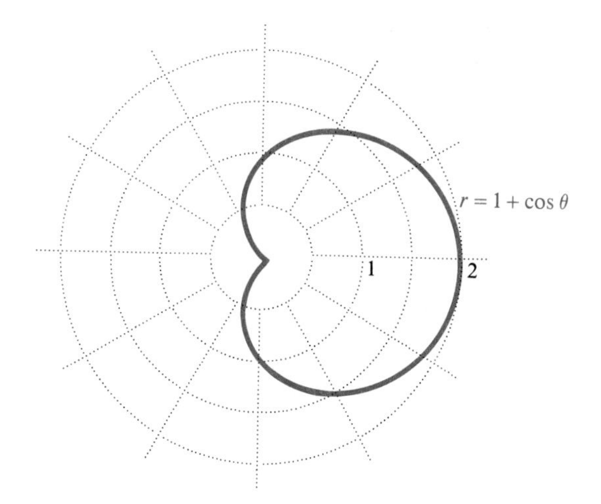

Figure 8.15

Example 3 Graph $r^2 = \cos 2\theta$.

First complete a table of ordered pairs as shown, and then sketch the graph, as in Figure 8.16. The point (−1, 0°), with r negative, may be plotted as (1, 180°). Also, (−.7, 30°) may be plotted as (.7, 210°), and so on. This curve is called a **lemniscate.**

θ	0°	30°	45°	135°	150°	180°
2θ	0°	60°	90°	270°	300°	360°
$\cos 2\theta$	1	.5	0	0	.5	1
$r = \pm\sqrt{\cos 2\theta}$	± 1	$\pm .7$	0	0	$\pm .7$	± 1

Values of θ for $45° < \theta < 135°$ are not included in the table because the corresponding values of $\cos 2\theta$ are negative (quadrants II and III) and so do not have real square roots. Values of θ larger than 180° give 2θ larger than 360°, and would repeat the points already found. ✣

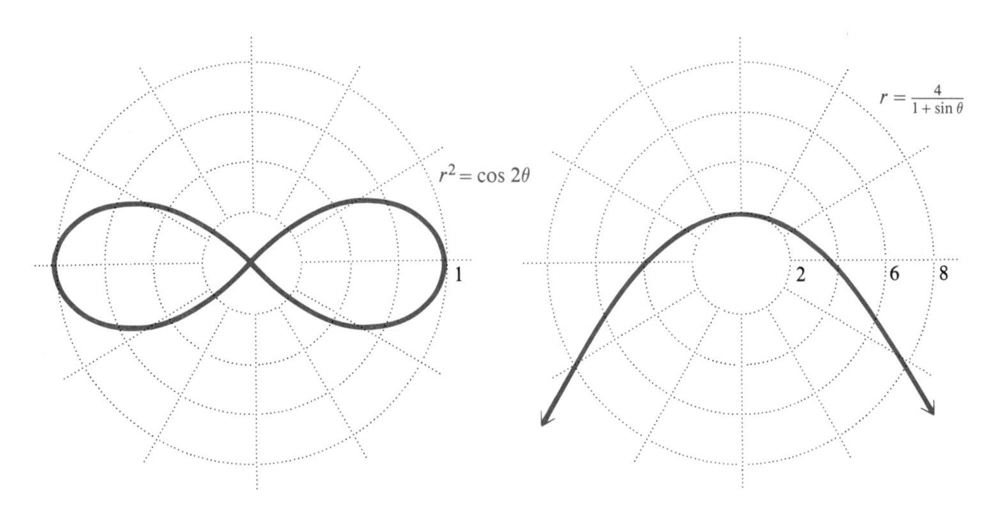

Figure 8.16

Figure 8.17

Example 4 Graph $r = \dfrac{4}{1 + \sin \theta}$.

Again complete a table or ordered pairs, which leads to the graph shown in Figure 8.17.

θ	0°	30°	45°	60°	90°	120°	135°	150°	180°	210°	225°
$\sin \theta$	0	.5	.7	.9	1	.9	.7	.5	0	−.5	−.7
$r = \dfrac{4}{1 + \sin \theta}$	4	2.7	2.3	2.1	2.0	2.1	2.3	2.7	4.0	8.0	13.3

With the points given in the table, the pattern of the graph should start to be clear. If it is not, continue to find additional points. ✣

Example 5 Graph $r = 3 \cos 2\theta$.

Because of the 2θ, the graph requires a large number of points. A few ordered pairs are given below. You should complete the table similarly through the first 360°.

θ	0°	15°	30°	45°	60°	75°	90°
2θ	0°	30°	60°	90°	120°	150°	180°
$\cos 2\theta$	1	.9	.5	0	−.5	−.9	−1
r	3	2.7	1.5	0	−1.5	−2.7	−3

Plotting these points in order gives the graph, called a **four-leaved rose.** Notice in Figure 8.18 how the graph is developed with a continuous curve, beginning with the upper half of the right horizontal leaf and ending with the lower half of that leaf. As the graph is traced, the curve goes through the pole four times.

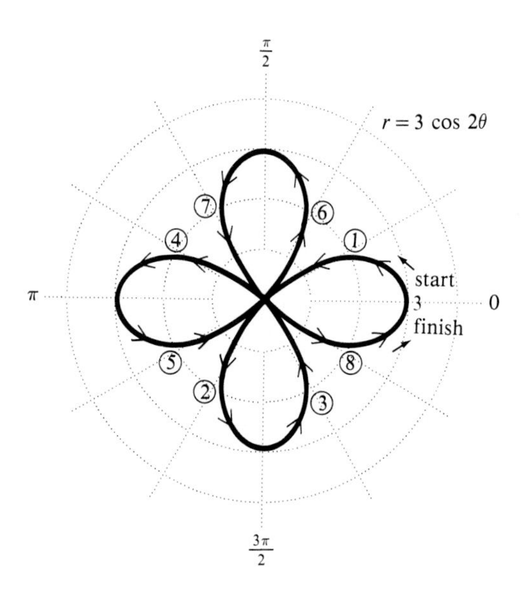

Figure 8.18

As suggested by Example 5, the graphs of $r = \sin n\theta$ and $r = \cos n\theta$ are roses, with n petals if n is odd, and $2n$ petals if n is even.

Example 6 Graph $r = 2\theta$ (θ measured in radians).

Some ordered pairs are shown below. Since $r = 2\theta$, rather than a trigonometric function of θ, it is also necessary to consider negative values of θ. The radian measures have been rounded for simplicity.

θ (degrees)	−180	−90	−45	0	30	60	90	180	270	360
θ (radians)	−3.1	−1.6	−.8	0	.5	1	1.6	3.1	4.7	6.3
$r = 2\theta$	−6.2	−3.2	−1.6	0	1	2	3.2	6.2	9.4	12.6

Figure 8.19 shows this graph, called a **spiral of Archimedes.**

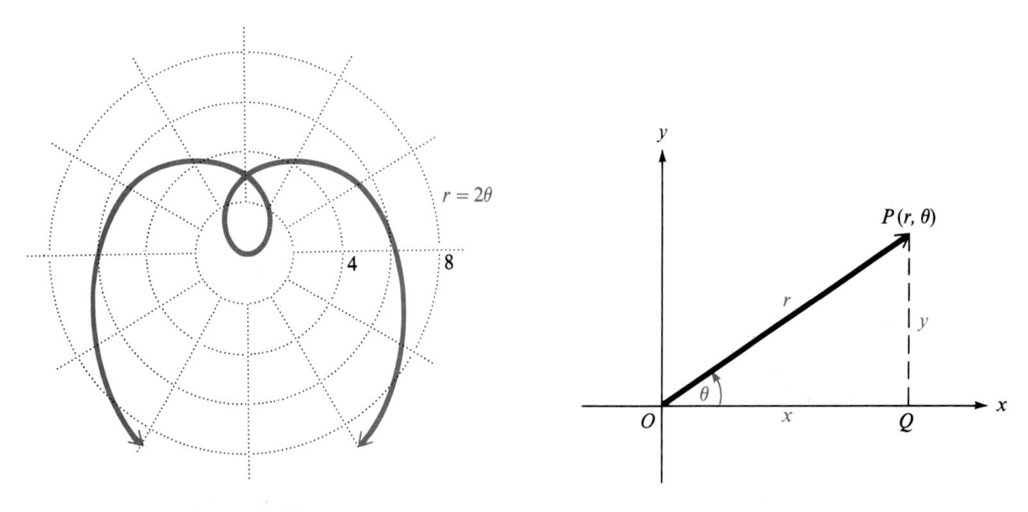

Figure 8.19

Figure 8.20

Sometimes an equation given in polar form is easier to graph in Cartesian form. To convert a polar equation to a Cartesian equation use the following relationships, which were introduced in section 8.2. See triangle *POQ* in Figure 8.20.

Converting Between Polar and Rectangular Coordinates

$$x = r\cos\theta \qquad r = \sqrt{x^2 + y^2}$$

$$y = r\sin\theta \qquad \tan\theta = \frac{y}{x}$$

Example 7 Convert the equation in Example 4,

$$r = \frac{4}{1 + \sin\theta},$$

to Cartesian coordinates.

Multiply both sides of the equation by the denominator on the right, to clear the fraction.

$$r = \frac{4}{1 + \sin\theta}$$

$$r + r\sin\theta = 4$$

Now substitute $\sqrt{x^2 + y^2}$ for r and y for $r \sin\theta$.

$$\sqrt{x^2 + y^2} + y = 4$$

$$\sqrt{x^2 + y^2} = 4 - y$$

Square both sides to eliminate the radical.

$$x^2 + y^2 = (4 - y)^2$$

$$x^2 + y^2 = 16 - 8y + y^2$$

$$x^2 = -8y + 16$$

$$x^2 = -8(y - 2)$$

The final equation represents a parabola and can be graphed using rectangular coordinates. ✤

Example 8 Convert the equation $3x + 2y = 4$ to a polar equation.

Use $x = r\cos\theta$ and $y = r\sin\theta$ to get

$$3x + 2y = 4$$

$$3r\cos\theta + 2r\sin\theta = 4.$$

Now solve for r. First factor out r on the left.

$$\boldsymbol{r}(3\cos\theta + 2\sin\theta) = 4$$

$$r = \frac{4}{3\cos\theta + 2\sin\theta}$$

The polar equation of the line $3x + 2y = 4$ is

$$r = \frac{4}{(3\cos\theta + 2\sin\theta)}.$$ ✤

Example 9 Note the use of polar coordinates in the following example, adapted with permission from *Calculus and Analytic Geometry,* fifth edition, by George Thomas and Ross Finney (Addison-Wesley, 1979).

Karl von Frisch has advanced the following theory about how bees communicate information about newly discovered sources of food. A scout returning to the hive from a flower bed gives away samples of the food and then, if the bed is more than about a hundred yards away, performs a dance to show where the flowers are in relation to the position of the sun. The bee runs straight ahead for a centimeter or so, waggling from side to side, and circles back to the starting place. The bee then repeats the straight run, circling back in the opposite direction. (See Figure 8.21.) The dance continues this way in regular alternation. Exceptionally excited bees have been observed to dance for more than three and a half hours.

If the dance is performed inside, it is performed on the vertical wall of a honeycomb, with gravity substituting for the sun's position. A vertical straight run means that the food is in the direction of the sun. A run 30° to the right of vertical means that the food is 30° to the right of the sun, and so on. Distance (more accurately, the amount of energy required to reach the food) is communicated by the duration of the straight-run portions of the dance. Straight runs lasting three seconds each are typical for distances of about a half mile from the hive. Straight runs that last five seconds each mean about two miles. ✣

The waggle dance of a scout bee

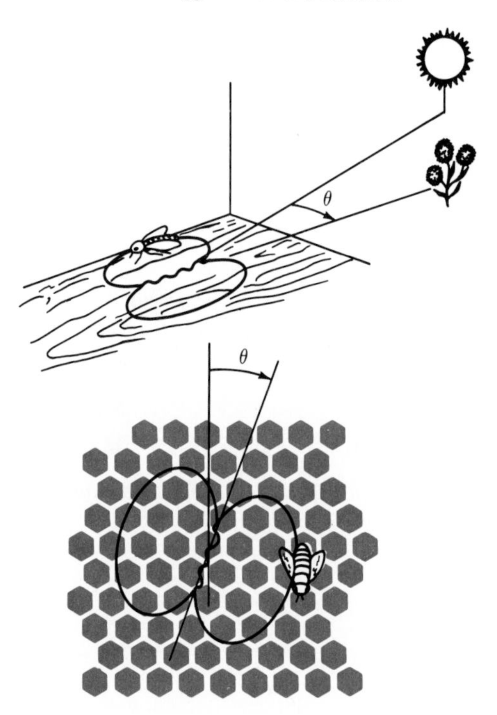

Figure 8.21

8.5 Exercises

Plot each point, given its polar coordinates. See Example 1.

1. (1, 45°) **2.** (3, 120°) **3.** (−2, 135°) **4.** (−4, 27°)

5. (5, −60°) **6.** (2, −45°) **7.** (−3, −210°) **8.** (−1, −120°)

9. (3, 300°) **10.** (4, 270°) **11.** (−5, −420°) **12.** (2, −435°)

Graph each of the following equations for $0° \le \theta < 360°$, unless other domains are specified. See Examples 3–5.

13. $r = 2 + 2 \cos \theta$

14. $r = 2(4 + 3 \cos \theta)$

15. $r = 3 + \cos \theta$ (limaçon)

16. $r = 2 - \cos \theta$ (limaçon)

17. $r = \sin 2\theta$ (four-leaved rose)
(*Hint:* Use $0° \le \theta < 360°$ every 15°.)

18. $r = 3 \cos 5\theta$ (five-leaved rose)

19. $r^2 = 4 \cos 2\theta$ (lemniscate)

20. $r^2 = 4 \sin 2\theta$ (lemniscate)

21. $r = 4(1 - \cos \theta)$ (cardioid)

22. $r = 3(2 - \cos \theta)$ (cardioid)

23. $r = 2 \sin \theta \tan \theta$ (cissoid)

24. $r = \dfrac{\cos 2\theta}{\cos \theta}$

25. $r = \dfrac{3}{2 + \sin \theta}$

26. $r = \sin \theta \cos^2 \theta$

Graph each of the following for $-\pi \le \theta \le \pi$, measuring θ in radians. Exercises 31 and 32 are for students who have studied logarithms. See Example 6.

27. $r = 5\theta$ (spiral of Archimedes)

28. $r = \theta$ (spiral of Archimedes)

29. $r\theta = \pi$ (hyperbolic spiral)

30. $r^2 = \theta$ (parabolic spiral)

31. $\ln r = \theta$ (logarithmic spiral)

32. $\log r = \theta$ (logarithmic spiral)

For each of the following equations, find an equivalent equation in Cartesian coordinates, and sketch the graph. See Example 7.

33. $r = 2 \sin \theta$

34. $r = 2 \cos \theta$

35. $r = \dfrac{2}{1 - \cos \theta}$

36. $r = \dfrac{3}{1 - \sin \theta}$

37. $r + 2 \cos \theta = -2 \sin \theta$

38. $r = \dfrac{3}{4 \cos \theta - \sin \theta}$

39. $r = 2 \sec \theta$

40. $r = -5 \csc \theta$

41. $r(\cos \theta + \sin \theta) = 2$

42. $r(2 \cos \theta + \sin \theta) = 2$

43. $r \sin \theta + 2 = 0$

44. $r \sec \theta = 5$

For each of the following equations, find an equivalent equation in polar coordinates. See Example 8.

45. $x + y = 4$

46. $2x - y = 5$

47. $x^2 + y^2 = 16$

48. $x^2 + y^2 = 9$

49. $y = 2$

50. $x = 4$

Chapter 8 *Key Concepts*

Definition of i $i = \sqrt{-1}$ or $i^2 = -1$

Definition of $\sqrt{-a}$ For positive real numbers a, $\sqrt{-a} = i\sqrt{a}$.

Operations on Complex Numbers For complex numbers $a + bi$ and $c + di$:

Addition of Complex numbers

$$(a + bi) + (c + di) = (a + c) + (b + d)i$$

Subtraction of Complex Numbers

$$(a + bi) - (c + di) = (a - c) + (b - d)i$$

Product of Complex Numbers

$$(a + bi)(c + di) = (ac - bd) + (ad + bc)i$$

Trigonometric Form of Complex Numbers If the complex number $x + yi$ corresponds to the vector with direction angle θ and magnitude r, then

$$x = r \cos \theta \qquad r = \sqrt{x^2 + y^2}$$

$$y = r \sin \theta \qquad \tan \theta = \frac{y}{x},$$

and

$$r(\cos \theta + i \sin \theta)$$

is the trigonometric form (or polar form) of $x + yi$.

nth Root Theorem If n is any positive integer and r is a positive real number, then the complex number $r(\cos \theta + i \sin \theta)$ has exactly n distinct nth roots, given by

$$r^{1/n}(\cos \alpha + i \sin \alpha),$$

where

$$\alpha = \frac{\theta + 360°k}{n} \qquad \text{or} \qquad \alpha = \frac{\theta}{n} + \frac{360°k}{n},$$

$k = 0, 1, 2, \ldots, n - 1$.

Chapter 8 *Key Concepts (continued)*

Operations on Complex Numbers in Trigonometric Form

For any two complex numbers:

$$r_1(\cos \theta_1 + i \sin \theta_1) \text{ and } r_2(\cos \theta_2 + i \sin \theta_2),$$

Product Theorem

$$[r_1(\cos \theta_1 + i \sin \theta_1)] \cdot [r_2(\cos \theta_2 + i \sin \theta_2)]$$
$$= r_1 r_2[(\cos (\theta_1 + \theta_2) + i \sin (\theta_1 + \theta_2)]$$

Quotient Theorem

$$\frac{r_1(\cos \theta_1 + i \sin \theta_1)}{r_2(\cos \theta_2 + i \sin \theta_2)} = \frac{r_1}{r_2} [\cos (\theta_1 - \theta_2) + i \sin (\theta_1 - \theta_2)]$$

De Moivre's Theorem

$$[r(\cos \theta + i \sin \theta)]^n = r^n(\cos n\theta + i \sin n\theta)$$

Chapter 8 *Review Exercises*

Write as a multiple of i.

1. $\sqrt{-9}$
2. $\sqrt{-12}$
3. $\sqrt{-\frac{4}{5}}$
4. $-\sqrt{-\frac{10}{3}}$

Perform the indicated operations. Write answers in standard form.

5. $(1 - i) - (3 + 4i) + 2i$
6. $(2 - 5i) + (9 - 10i) - 3$
7. $(6 - 5i) + (2 + 7i) - (3 - 2i)$
8. $(4 - 2i) - (6 + 5i) - (3 - i)$
9. $(3 + 5i)(8 - i)$
10. $(4 - i)(5 + 2i)$
11. $(2 + 6i)^2$
12. $(6 - 3i)^2$
13. $(1 - i)^3$
14. $(2 + i)^3$
15. $\dfrac{6 + 2i}{3 - i}$
16. $\dfrac{2 - 5i}{1 + i}$
17. $\dfrac{2 + i}{1 - 5i}$
18. $\dfrac{3 + 2i}{i}$
19. $[5(\cos 90° + i \sin 90°)][6(\cos 180° + i \sin 180°)]$
20. $[3(\cos 135° + i \sin 135°)][2(\cos 105° + i \sin 105°)]$
21. $\dfrac{2(\cos 60° + i \sin 60°)}{8(\cos 300° + i \sin 300°)}$
22. $\dfrac{4(\cos 270° + i \sin 270°)}{2(\cos 90° + i \sin 90°)}$
23. $(\sqrt{3} + i)^3$
24. $(2 - 2i)^5$
25. $(\cos 100° + i \sin 100°)^6$
26. $(\cos 20° + i \sin 20°)^3$

Solve each of the following equations for x and y.

27. $x + yi = 2 + 3i$
28. $x + yi = -5 + 6i$
29. $4 - 3yi = 2x + 3i$
30. $3x - 2yi = (i + 5)(2i - 1)$

Find the resultant of each pair of vectors.

31. $7 + 3i$ and $-2 + i$
32. $2 - 4i$ and $5 + i$

Graph each complex number.

33. $5i$
34. $-4 + 2i$
35. $3 - 3i\sqrt{3}$
36. $-5 + i\sqrt{3}$

Complete the chart.

	Standard form	*Trigonometric form*
37.	$-2 + 2i$	____________
38.	____________	$3(\cos 90° + i \sin 90°)$
39.	____________	$2(\cos 225° + i \sin 225°)$
40.	$-4 + 4i\sqrt{3}$	____________
41.	$1 - i$	____________

42. ______________ $4(\cos 240° + i \sin 240°)$

43. $-4i$ ______________

44. ______________ $2(\cos 180° + i \sin 180°)$

45. Find the fifth roots of $-2 + 2i$.

46. Find the cube roots of $1 - i$.

47. Find the tenth roots of 1.

48. Find the fourth roots of $\sqrt{3} + i$.

Solve each equation.

49. $x^3 + 125 = 0$ **50.** $x^4 + 16 = 0$ **51.** $x^4 + 81 = 0$

Graph each polar equation for $0° \leq \theta < 360°$, *unless other domains are given.*

52. $r = 4 \cos \theta$ **53.** $r = -1 + \cos \theta$ **54.** $r = 1 - \cos \theta$

55. $r = 2 \sin 4\theta$ **56.** $r = 3 \cos 3\theta$

57. $r^2 = \sin 2\theta$ **58.** $r = \sin^2 \theta \cos \theta$

59. $r = \theta$ (θ in radians); $-2\pi \leq \theta \leq 2\pi$ **60.** $r = \pi\theta$ (θ in radians); $-2\pi \leq \theta \leq 2\pi$

Find an equivalent equation in rectangular coordinates.

61. $r = \dfrac{3}{1 + \cos \theta}$ **62.** $r = \dfrac{4}{2 \sin \theta - \cos \theta}$ **63.** $r = \sin \theta + \cos \theta$ **64.** $r = 2$

Find an equivalent equation in polar coordinates.

65. $x = -3$ **66.** $y = x$ **67.** $y = x^2$ **68.** $x = y^2$

9

Logarithms

Before small calculators became easily available, computations involving multiplication, division, powers, and roots were performed by using the properties of logarithms. Now many of those problems can be worked simply on a calculator. In fact, calculators are programmed to use logarithms to perform such calculations.

Logarithmic functions are the inverses of exponential functions. This chapter briefly reviews both logarithmic, and exponential functions; the properties of logarithms, exponential and logarithmic equations; and applications involving exponential and logarithmic equations.

9.1 Exponential and Logarithmic Functions

Exponential functions involve exponents. The basic properties of exponents are listed below.

Properties of Exponents

For real numbers a, b, m, and n, where no denominators are zero,

$$a^m \cdot a^n = a^{m+n} \qquad (ab)^m = a^m b^m$$

$$\frac{a^m}{a^n} = a^{m-n} \qquad (a^m)^n = a^{mn}$$

$$a^0 = 1 \qquad (\text{if } a \neq 0).$$

The next two definitions will also be need for work with exponents.

Definitions of a^{-n} and $a^{m/n}$

If a is a nonzero real number, then

$$a^{-n} = \frac{1}{a^n}.$$

For all integers m and n and for all numbers a for which the roots exist,

$$a^{m/n} = \sqrt[n]{a^m} = (\sqrt[n]{a})^m.$$

Example 1 Use the properties and definitions listed above to simplify each of the following.

(a) $2^5 \cdot 2^9 \cdot 2^7 = 2^{5+9+7} = 2^{21}$

(b) $3^{-2} = \frac{1}{3^2} = \frac{1}{9}$

(c) $5^{-4} = \frac{1}{5^4} = \frac{1}{625}$

(d) $\frac{4^9}{4^{-2}} = 4^{9-(-2)} = 4^{11}$

(e) $(7^3 \cdot 3^4)^2 = (7^3)^2(3^4)^2 = 7^6 \cdot 3^8$

(f) $2^0 + (-5)^0 - 3^0 = 1 + 1 - 1 = 1$

(g) $4^{3/2} = (4^{1/2})^3 = (\sqrt{4})^3 = 2^3 = 8$ ✥

An **exponential function** is a function in which the independent variable appears in the exponent, as in $y = 2^x$, where x is the independent variable.

Exponential Function

An exponential function is a function of the form

$$y = a^x,$$

where $a > 0$, and $a \neq 1$.

The equation $y = 2^x$ represents a typical exponential function. To graph this function, first make a table of values of x and y, as follows.

x	-3	-2	-1	0	1	2	3	4
2^x	$\frac{1}{8}$	$\frac{1}{4}$	$\frac{1}{2}$	1	2	4	8	16

Plotting these points and drawing a smooth curve through them produces the graph shown in Figure 9.1. This graph is typical of the graphs of exponential functions of the form $y = a^x$, where $a > 1$. The larger the value of a, the faster the graph rises. As the graph suggests, the domain is the set of all real numbers but the range includes only positive real numbers.

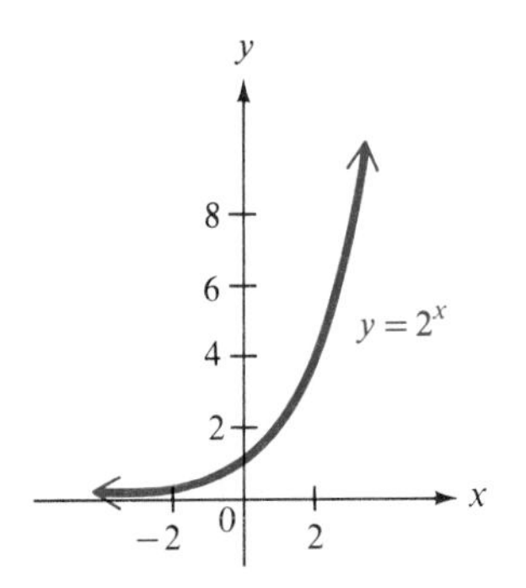

Figure 9.1

Example 2 Graph $y = \left(\frac{1}{2}\right)^x$.

Again, find selected ordered pairs and plot them to get the graph, shown in Figure 9.2. Some ordered pairs are given in the table below. This graph is typical of the graphs of exponential functions of the form $y = a^x$, where $0 < a < 1$. Again, the domain is the set of real numbers and the range is the set of positive real numbers. ✣

x	$\left(\frac{1}{2}\right)^x$
−3	8
−2	4
−1	2
0	1
1	1/2
2	1/4
3	1/8

Figure 9.2

Since neither horizontal nor vertical lines cut the graph of $y = 2^x$ in more than one point, $y = 2^x$ is a one-to-one function and has an inverse. The inverse of $y = 2^x$ is $x = 2^y$. To get the equation of the inverse, x and y were exchanged as in section 6.1.

The expression $x = 2^y$ cannot be solved for y by the methods studied so far. The following definition is used to solve such equations for y.

Definition of Logarithm

For all positive numbers a, with $a \neq 1$,

$$y = \log_a x \quad \textbf{means} \quad x = a^y.$$

Log is an abbreviation for **logarithm.** For example, the logarithmic statement

$$5 = \log_2 32 \quad \text{has the same meaning as} \quad 32 = 2^5$$

(read $5 = \log_2 32$ as "5 is the logarithm of 32 to the base 2"). The number 5 is the exponent on 2 that gives the result 32.

By the definition given here, exponential statements can be turned into logarithmic statements, and logarithmic statements into exponential statements.

Example 3 The chart below lists several pairs of equivalent statements. The same statement is written in both exponential and logarithmic form.

Exponential form	*Logarithmic form*
$3^4 = 81$	$4 = \log_3 81$
$5^3 = 125$	$3 = \log_5 125$
$2^{-4} = \frac{1}{16}$	$-4 = \log_2 \frac{1}{16}$
$5^{-1} = \frac{1}{5}$	$-1 = \log_5 \frac{1}{5}$
$\left(\frac{3}{4}\right)^{-2} = \frac{16}{9}$	$-2 = \log_{3/4} \frac{16}{9}$

By using the definition of logarithm, the logarithmic function with base a is defined as follows.

Logarithmic Function

If $a > 0$, $a \neq 1$, and $x > 0$, then

$$y = \log_a x$$

is the logarithmic function with base a.

As mentioned above, exponential and logarithmic functions are inverses of each other. Since the domain of an exponential function is the set of all real numbers, the range of a logarithmic function is also the set of all real numbers.

In the same way, both the range of an exponential function and the domain of a logarithmic function are the set of all positive real numbers. Because of this domain, only positive numbers have logarithms.

The graph of $y = 2^x$ is shown in Figure 9.3. To get the graph of its inverse, $y = \log_2 x$, reflect $y = 2^x$ about the line $y = x$, as shown in Figure 9.3. The graph is typical of graphs of functions of the form $y = \log_a x$, where $a > 0$.

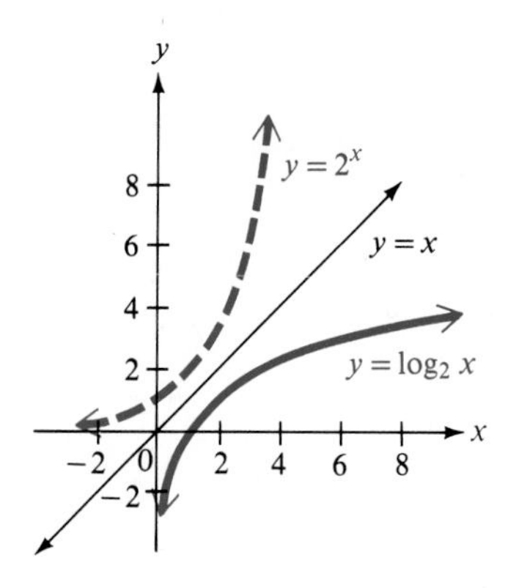

Figure 9.3

Example 4 Graph $y = \log_{1/2} x$.

Some ordered pairs for this function are shown in the table below. Plotting these points and connecting them with a curve leads to the graph in Figure 9.4. This graph is typical of graphs of functions of the form $y = \log_a x$, where $0 < a < 1$. ✜

x	y
8	−3
4	−2
2	−1
1	0
1/2	1
1/4	2

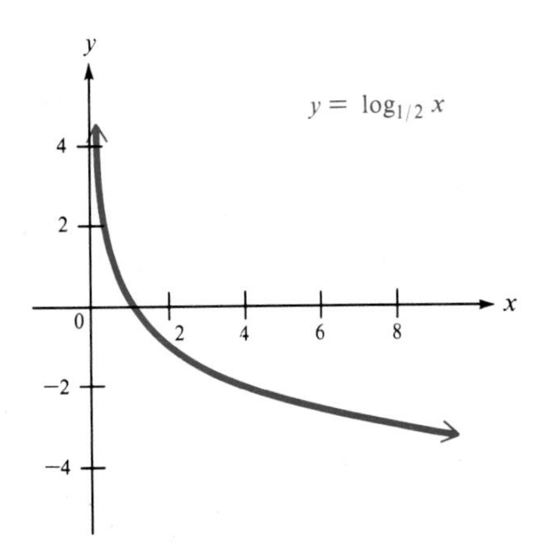

Figure 9.4

The basic properties of logarithms that make them useful in mathematics are summarized below.

Properties of Logarithms

If x and y are positive real numbers, r is any real number, and a is any positive real number, with $a \neq 1$, then

(a) $\log_a xy = \log_a x + \log_a y$ (b) $\log_a \frac{x}{y} = \log_a x - \log_a y$

(c) $\log_a x^r = r \log_a x$ (d) $\log_a a = 1$

(e) $\log_a 1 = 0$.

To prove part (a), let $\log_a x = M$ and $\log_a y = N$. Then by the definition of logarithm,

$$a^M = x \quad \text{and} \quad a^N = y,$$

and
$$(a^M)(a^N) = xy,$$
or
$$a^{M+N} = xy$$
and
$$\log_a xy = M + N.$$

Substituting for M and N gives the desired result,

$$\log_a xy = \log_a x + \log_a y.$$

The proofs of parts (b) and (c) are left for the exercises. Parts (d) and (e) come directly from the definition of logarithm.

Example 5 The properties given above are used to rewrite each of the following logarithmic expressions.

(a) $\log_6 9 \cdot 4 = \log_6 9 + \log_6 4$

(b) $\log_{12} \frac{17}{18} = \log_{12} 17 - \log_{12} 18$

(c) $\log_5 9^8 = 8 \cdot \log_5 9$

(d) $\log_7 \sqrt{19} = \log_7 19^{1/2}$

$$= \frac{1}{2} \cdot \log_7 19 \quad ✣$$

Example 6 Assume that

$$\log_{10} 2 = .3010 \quad \text{and} \quad \log_{10} 3 = .4771.$$

Find the base 10 logarithms of 4, 5, 6, and 12.

Use the properties of logarithms.

$$\textbf{(a)}\ \log_{10} 4 = \log_{10} 2^2 = 2 \cdot \log_{10} 2 = 2(.3010) = .6020$$

$$\textbf{(b)}\ \log_{10} 5 = \log_{10} \frac{10}{2} = \log_{10} 10 - \log_{10} 2 = 1 - .3010 = .6990$$

$$\textbf{(c)}\ \log_{10} 6 = \log_{10} 2 \cdot 3 = \log_{10} 2 + \log_{10} 3 = .3010 + .4771 = .7781$$

$$\textbf{(d)}\ \log_{10} 12 = \log_{10} 4 \cdot 3 = \log_{10} 4 + \log_{10} 3 = .6020 + .4771 = 1.0791$$

9.1 Exercises

Use the properties of exponents to simplify each of the following. Leave answers in exponential form. See Example 1.

1. $3^4 \cdot 3^7$ **2.** $4^5 \cdot 4^8$ **3.** $\frac{7^4}{7^3}$ **4.** $\frac{9^5}{9^2}$

5. $\frac{6^4 \cdot 6^5}{6^6}$ **6.** $\frac{9^{12} \cdot 9^4}{9^{11}}$ **7.** $(8^9)^2$ **8.** $(3^4)^3$

9. $\frac{(2^5)^2 \cdot 2^4}{2^6}$ **10.** $\frac{(3^6)^3 \cdot 3^7}{3^8}$ **11.** 2^{-3} **12.** 4^{-2}

13. $3^{-7} \cdot 3^8$ **14.** $5^{-9} \cdot 5^6$ **15.** $\frac{9^{-8} \cdot 9^2}{(9^2)^{-2}}$ **16.** $\frac{(4^{-3})^2 \cdot 4^5}{(4^3)^{-3}}$

17. $25^{1/2}$ **18.** $8^{2/3}$ **19.** $16^{3/4}$ **20.** $9^{3/2}$

Graph each of the following exponential functions. See Example 2.

21. $y = \left(\frac{1}{3}\right)^x$ **22.** $y = \left(\frac{1}{4}\right)^x$ **23.** $y = 2^{x+1}$ **24.** $y = 3^{x+1}$

25. $y = 2^{-x}$ **26.** $y = 3^{-x}$ **27.** $y = 3^x - 3$ **28.** $y = 2^x - 2$

Write each of the following in logarithmic form. See Example 3.

29. $3^5 = 243$ **30.** $2^7 = 128$ **31.** $10^4 = 10{,}000$ **32.** $8^2 = 64$

33. $6^{-2} = \frac{1}{36}$ **34.** $2^{-3} = \frac{1}{8}$ **35.** $\left(\frac{2}{3}\right)^{-2} = \frac{9}{4}$ **36.** $\left(\frac{3}{8}\right)^{-1} = \frac{8}{3}$

Write each of the following in exponential form. See Example 3.

37. $\log_4 16 = 2$ **38.** $\log_5 25 = 2$ **39.** $\log_{10} 1000 = 3$ **40.** $\log_7 2401 = 4$

41. $\log_{3/4} \frac{9}{16} = 2$ **42.** $\log_{5/8} \frac{125}{512} = 3$ **43.** $\log_4 \frac{1}{16} = -2$ **44.** $\log_5 \frac{1}{125} = -3$

Graph each of the following. See Example 4.

45. $y = \log_4 x$ **46.** $y = \log_{1/4} x$

47. $y = \log_{1/3} x$ **48.** $y = \log_3 x$

49. $y = \log_2 (x - 1)$

50. $y = \log_2 x^2$

51. $y = 1 + \log_3 x$

52. $y = 1 - \log_3 x$

Simplify each of the following using the properties of logarithms. See Example 5.

53. $\log_6 9 + \log_6 5$

54. $\log_2 7 + \log_2 3$

55. $\log_5 12 - \log_5 7$

56. $\log_9 17 - \log_9 23$

57. $\log_2 7 + \log_2 4 - \log_2 5$

58. $\log_7 8 + \log_7 16 - \log_7 3$

59. $4 \cdot \log_3 2$

60. $3 \cdot \log_5 2$

61. $\frac{1}{2} \cdot \log_8 7$

62. $\frac{1}{2} \cdot \log_2 14$

63. $4 \cdot \log_2 3 - 2 \cdot \log_2 5$

64. $3 \cdot \log_4 5 - 4 \cdot \log_4 2$

Use the properties of logarithms to find each of the following. See Example 6. Assume that

$$\log_{10} 2 = .3010 \qquad \log_{10} 7 = .8451$$
$$\log_{10} 3 = .4771 \qquad \log_{10} 11 = 1.0414.$$

65. $\log_{10} 14$

66. $\log_{10} 22$

67. $\log_{10} 28$

68. $\log_{10} 63$

69. $\log_{10} 2^8$

70. $\log_{10} 3^7$

71. $\log_{10} 11^3$

72. $\log_{10} \sqrt{3}$

73. $\log_{10} \sqrt[3]{11}$

74. $\log_{10} \sqrt[4]{3}$

75. Prove property (b) of logarithms. (*Hint:* The proof is very similar to that for property (a).)

76. Prove property (c) of logarithms.

9.2 Common and Natural Logarithms

The logarithms most often used in applications are either base 10 or base e ($e \approx 2.7182818$) logarithms. Since our number system uses base 10, logarithms to base 10 are convenient for numerical calculations, historically the main application of logarithms. Base e logarithms, called natural logarithms, are important in advanced work in mathematics, as well as in many applications.

Common Logarithms Base 10 logarithms are called **common logarithms.** The common logarithm of the number x, or $\log_{10} x$, is abbreviated **log x.** In this section we briefly explain the use of the table of common logarithms (Table 3) at the back of this book. The table gives the common logarithm of numbers between 1 and 10. Any number can be written in scientific notation as a number between 1 and 10 times a power of 10. This plus the properties of logarithms can be used with the table to find the common logarithm of any number.

Example 1 Find log 5.74.

Look down the left column of Table 3 to find the first two digits of 5.74, that is, 5.7. Read across that row until you reach the column that has 4 at the top. You should find the digits .7589. Thus,

$$\log 5.74 = .7589.$$

All the numbers in Table 3 are approximations, but we shall use the equals sign for convenience.

To find the logarithm of a number greater than 10 or less than 1, first write the number in scientific notation and then use the properties of logarithms.

Example 2 Find log 596.

Write the number 596 in scientific notation as

$$596 = 5.96 \times 10^2.$$

Then

$$\begin{aligned} \log 596 &= \log (5.96 \times 10^2) \\ &= \log 5.96 + \log 10^2 && \text{Property (a) of logarithms} \\ &= \log 5.96 + 2. && \log_{10} 10^2 = 2 \end{aligned}$$

From Table 3, $\log 5.96 = .7752$. Thus,

$$\log 596 = .7752 + 2 = 2.7752.$$ ✣

The integer part of this logarithm, 2, is called the **characteristic;** the decimal part, .7752, is called the **mantissa.** The mantissa is always a positive number. The characteristic may be a positive or a negative integer.

Example 3 Find log .000421.

$$\begin{aligned} \log .000421 &= \log (4.21 \times 10^{-4}) \\ &= \log 4.21 + \log 10^{-4} \\ &= \log 4.21 + (-4) \end{aligned}$$

From Table 3,

$$\log .000421 = .6243 + (-4) \quad \text{or} \quad .6243 - 4.$$ ✣

If you use a calculator with a logarithm key to find log .000421, the display will show -3.3757, the algebraic sum of $.6243 + (-4)$.

Example 4 Find the value of N that makes $\log N = 1.6955$.

Since the logarithm is given and the number must be found, reverse the process used above. On a calculator, this is done by pressing the inverse key, and then the logarithm key. To use Table 3, work in the opposite direction from the way the table was used before. First find the mantissa, .6955, in the body of the table. Then look at the left of the row containing .6955 to find the first two digits of the answer. The first two digits are 4.9. Then find the third digit by looking at the top (or bottom) of the column containing .6955. The third digit is 6. Thus, the *digits* of N are 4.96. Since the characteristic is 1,

$$N = 4.96 \times 10^1 = 49.6.$$

The number 49.6 is called the **antilogarithm** of 1.6955. ✣

Example 5 Find the antilogarithm of each of the following.

(a) .8785 − 3

To use a calculator, first write the logarithm as −2.1215 (found by adding .8785 and −3). Then use the inverse and logarithm keys. If Table 3 is used, find the digits corresponding to a mantissa of .8785. Reading from the left column and the top row gives 7.56. The characteristic is −3. Thus, the antilogarithm is

$$7.56 \times 10^{-3} = .00756.$$

(b) 4.774

From Table 3 or a calculator, the antilogarithm is

$$5.99 \times 10^{4} = 59{,}900.$$

Interpolation Table 3 gives the logarithms of numbers with three significant digits. For greater accuracy in finding the logarithms of numbers with more significant digits, interpolation can be used.

Example 6 Use Table 3 to find log 8.874.

First, find two numbers in Table 3 that are closest to 8.874, and then note their logarithms.

$$.010\left[\begin{array}{l} .004\left[\begin{array}{l} \log 8.870 = .9479 \\ \log 8.874 = \ ? \end{array}\right] d \\ \log 8.880 = .9484 \end{array}\right] .0005$$

Approximate d by forming equal ratios with the numbers on each side of this display.

$$\frac{.004}{.010} = \frac{d}{.0005}$$

$$d \approx .0002$$

Add the value of d to the value of log 8.870 to get the answer.

$$\log 8.874 \approx .9479 + .0002 = .9481$$

Interpolation can also be used to find the antilogarithm of a given logarithm.

Example 7 Find the antilogarithm of 3.4042.

Locate the two numbers in the body of the logarithm table that are closest to the mantissa, .4042. Then set up an array as above.

$$10\left[\begin{array}{l} d\left[\begin{array}{l} \log 2530 = 3.4031 \\ \log \ ? \ = 3.4042 \end{array}\right] .0011 \\ \log 2540 = 3.4048 \end{array}\right] .0017$$

Form equal ratios to solve for d.

$$\frac{d}{10} = \frac{.0011}{.0017}$$
$$d \approx 6$$

Finally, the antilogarithm of 3.4042 is

$$2530 + 6 = 2536.$$ ✥

The properties of logarithms can be used to simplify numerical calculations. Logarithms are particularly useful for finding powers and roots since they make it possible to use the simpler operations of multiplication and division.

Example 8 Find $(.0839)^{2/3}$.

Let $R = (.0839)^{2/3}$. Then

$$\begin{aligned} \log R &= \log (.0839)^{2/3} \\ &= \frac{2}{3} \log .0839 \\ &= \frac{2}{3}(.9238 - 2). \end{aligned}$$

To simplify the computation, change the characteristic, -2, to a number that is a multiple of 3. One way to do this is to write the characteristic as $1 - 3$. Adding the positive part of $1 - 3$ to the mantissa gives

$$\log R = \frac{2}{3}(1.9238 - 3).$$

Now, multiply 1.9238 and -3 by 2/3 to get

$$\log R = 1.2825 - 2 = .2825 - 1,$$

from which

$$R = .192 \quad \text{(to three significant digits).}$$ ✥

Natural Logarithms Logarithms to base e are called **natural logarithms,** and are used in many applications and in theoretical work, such as in calculus. The number e is an irrational number, like π. To nine decimal places,

$$e = 2.718281828.$$

The common notation for $\log_e x$ is $\ln x$, where ln is an abbreviation for "natural logarithm." The abbreviation $\ln x$ is read "el-en x" or "natural logarithm of x". The values of natural logarithms can be found from a table of natural logarithms, such as Table 4 in the back of this book, or by using a calculator that has an ln key.

Example 9 Find ln 20.

From Table 4 or a calculator, ln 20 = 2.9957. ✦

Example 10 Find ln 870.

The number 870 is not given in Table 4. However, properties of logarithms can be used to find this logarithm using the table, as shown below.

$$\begin{aligned}\ln 870 &= \ln (8.7)(100)\\ &= \ln 8.7 + \ln 100\\ &= 2.1633 + 4.6052\\ &= 6.7685\end{aligned}$$ ✦

Example 11 Find ln .38.

If Table 4 is used,

$$\begin{aligned}\ln .38 &= \ln (3.8)(.1)\\ &= \ln 3.8 + \ln .1\\ &= 1.3350 + (-2.3026)\\ &= -.9676.\end{aligned}$$ ✦

9.2 Exercises

Find the common logarithm of each of the following numbers. See Examples 1–3.

1. 2.79 **2.** 794 **3.** 98,300 **4.** 59.2

5. 749,000 **6.** .000976 **7.** .00591 **8.** .0105

Find the common antilogarithm of each of the following. See Examples 4 and 5.

9. 2.8585 **10.** 5.7945 **11.** 3.3674 **12.** .7767 − 3

13. .0569 − 1 **14.** .9991 − 2 **15.** .7566 − 3 **16.** .6821

Use interpolation or a calculator to find each of the following. See Example 6.

17. log 4.897 **18.** log 358.7 **19.** log 37,990 **20.** log .8276

21. log .02913 **22.** log .0009744

Use interpolation or a calculator to find the common antilogarithm of each of the following. See Example 7.

23. .4383 **24.** 1.4691 **25.** 3.4207 **26.** .2745 − 1

27. .7060 − 3 **28.** .9678 − 4

Use common logarithms to find each of the following to three significant digits. See Example 8.

29. $(.00432)^2$ **30.** $(21.9)^{3.1}$ **31.** $\sqrt{91.7}$ **32.** $\sqrt[4]{39,900}$

33. $(596)^{2/3}$ **34.** $(1.74)^{4/3}$ **35.** $\dfrac{(7.18)^3(2.41)}{(59.8)^2}$ **36.** $\dfrac{(38.4)^2(5.17)^3}{(98.6)^5}$

Find each of the following natural logarithms. See Examples 9–11.

37. ln 25 **38.** ln 50 **39.** ln 86 **40.** ln 970

41. ln 28,000,000 **42.** ln 37,000 **43.** ln .08 **44.** ln .004

45. ln .58 **46.** ln .47

9.3 Exponential and Logarithmic Equations

Equations with variables as exponents are called **exponential equations.** Equations that contain logarithms are called **logarithmic equations.** This section reviews solving these types of equations. Solving these equations depends on the following two properties.

Properties of Exponential and Logarithmic Equations

1. If a is a positive number, with $a \neq 1$, and x and y are real numbers, then

$$a^x = a^y \textbf{ if and only if } x = y.$$

2. If x, y, and a are positive with $a, \neq 1$, then

$$x = y \textbf{ if and only if } \log_a x = \log_a y.$$

Property 1 is used to solve the equations in the first two examples.

Example 1 Solve $2^x = 8$.

Write both equations as exponential expressions with the same base. Since $8 = 2^3$, the equation $2^x = 8$ becomes

$$2^x = 2^3,$$

from which

$$x = 3,$$

by property 1. ✣

Example 2 Solve the exponential equation $9^x = 27$.

Write 9 as 3^2 and 27 as 3^3. Then, $9^x = 27$ becomes

$$(3^2)^x = 3^3$$
$$3^{2x} = 3^3.$$

By property 1 above,

$$2x = 3$$
$$x = \frac{3}{2}.$$ ✣

Other uses of these properties to solve exponential and logarithmic equations are shown in the following examples.

Example 3 Solve $5^x = 15$.

Here it is not possible to write both sides as powers of the same base. Use property 2 above, and take common logarithms of both sides of the equation.

$$\log 5^x = \log 15$$

$$x \cdot \log 5 = \log 15 \qquad \text{Property (c) of logarithms}$$

$$x = \frac{\log 15}{\log 5}$$

The exact solution is $x = \log 15/\log 5$. A decimal approximation can be found by evaluating log 15 and log 5 in Table 3 or by using a calculator, and then dividing to get

$$x \approx \frac{1.1761}{.6990} \approx 1.683,$$

to the nearest thousandth. ✥

Example 4 Solve $\log_2 (x - 3) + \log_2 x = \log_2 4$.

By property (a) of logarithms,

$$\log_2(x - 3) + \log_2 x = \log_2 x(x - 3),$$

so the equation becomes

$$\log_2 x(x - 3) = \log_2 4.$$

Then, by property 2 above,

$$x(x - 3) = 4$$

$$x^2 - 3x = 4$$

$$x^2 - 3x - 4 = 0$$

$$(x - 4)(x + 1) = 0$$

$$x = 4 \quad \text{or} \quad x = -1.$$

The domain of the common logarithm function does not include negative numbers, so the only solution is

$$x = 4.$$ ✥

Example 5 Solve $\log_y 7 = 2$.

Use the definition of logarithms to convert this equation from logarithmic form to exponential form.

$$y^2 = 7$$

Take the square root of both sides.

$$y = \pm\sqrt{7}$$

Since the base of a logarithm must be positive,

$$y = \sqrt{7}.$$

Example 6 Solve $\log_3 (1 - 2x) = \log_3 x - 1$.

Subtract $\log_3 x$ from both sides to get

$$\log_3(1 - 2x) - \log_3 x = -1.$$

$$\log_3 \frac{1 - 2x}{x} = -1 \quad \text{Property (b) of logarithms}$$

$$3^{-1} = \frac{1 - 2x}{x} \quad \text{Definition of logarithm}$$

$$\frac{1}{3} = \frac{1 - 2x}{x}$$

$$x = 3 - 6x$$

$$7x = 3$$

$$x = \frac{3}{7}$$

Since 3/7 makes both $1 - 2x$ and x positive numbers, the solution is 3/7.

Exponential and logarithmic equations arise in many important applications, particularly in applications of population growth or decay. The next example is a problem of this type.

Example 7 Suppose that the population of a city is given by

$$P = 1{,}000{,}000(2^{.02t}),$$

where t is time measured in months.

(a) Find the population at time $t = 0$.

Let $t = 0$ in the equation and solve for P.

$$\begin{aligned} P &= 1{,}000{,}000(2^{.02t}) \\ &= 1{,}000{,}000 \cdot 2^{(.02)(\mathbf{0})} \\ &= 1{,}000{,}000 \cdot 2^0 \\ &= 1{,}000{,}000(1) = 1{,}000{,}000 \end{aligned}$$

The population is 1,000,000 when t is 0.

(b) How long will it take for the population to reach 1,300,000?
Let $P = 1{,}300{,}000$ and solve for t.

$$P = 1{,}000{,}000(2^{.02t})$$
$$\mathbf{1{,}300{,}000} = 1{,}000{,}000(2^{.02t})$$

Divide both sides by 1,000,000.

$$1.3 = 2^{.02t}$$

Take logarithms on both sides. Then use a property of logarithms on the right side.

$$\log 1.3 = \log 2^{.02t}$$
$$\log 1.3 = .02t \log 2$$

To solve for t, divide both sides of the equation by .02 log 2. Then evaluate the quotient.

$$\frac{\log 1.3}{.02 \log 2} = t$$
$$t = \frac{.1139}{(.02)(.3010)}$$
$$t \approx 18.9$$

The population will reach 1,300,000 in about 18.9 or 19 months.

9.3 Exercises

Solve each of the following equations. Do not use tables or a calculator. See Examples 1 and 2.

1. $5^x = 125$ **2.** $3^p = 81$ **3.** $32^k = 2$ **4.** $64^z = 4$
5. $100^r = 1000$ **6.** $32^m = 16$ **7.** $27^z = 81$ **8.** $49^r = 343$
9. $2^{5x} = 16^{x+2}$ **10.** $3^{6x} = 9^{2x+1}$ **11.** $25^{2k} = 125^{k+1}$ **12.** $36^{2p+1} = 6^{3p-2}$

Solve each of the following equations. Express all answers with three significant digits. See Examples 3–6.

13. $10^p = 3$ **14.** $10^m = 7$ **15.** $10^{r+2} = 15$ **16.** $10^{m-3} = 28$
17. $6^k = 10$ **18.** $12^x = 10$ **19.** $14^z = 28$ **20.** $21^p = 63$
21. $7^a = 32$ **22.** $12^b = 70$ **23.** $15^x = 8^{2x+1}$ **24.** $21^{y+1} = 32^{2y}$
25. $16^{z+3} = 28^{2z-1}$ **26.** $40^{r-3} = 32^{3r+1}$
27. $\log x - \log (x - 14) = \log 8$ **28.** $\log (k - 3) = 1 + \log (k - 21)$
29. $\log z = 1 - \log (3z - 13)$ **30.** $\log x + \log (x + 2) = 0$
31. $\log_x 10 = 3$ **32.** $\log_k 25 = 5$
33. $2 + \log x = 0$ **34.** $-6 + \log 2k = 0$

35. The number of species in a sample is given by

$$S(n) = a \ln\left(1 + \frac{n}{a}\right),$$

where n is the number of individuals in the sample and a is a constant that indicates the diversity of species in the community. If $a = .36$, find $S(n)$ for the following values of n.

(a) 100 **(b)** 200 **(c)** 150 **(d)** 500

(e) Find n if $S(n) = 4$ and $a = .36$.

36. Suppose that the number of rabbits in a colony increases according to the relationship

$$y = y_0 e^{.4t},$$

where t represents time in months and y_0 is the initial population of rabbits.

(a) Find the number of rabbits present at time $t = 4$ if $y_0 = 100$.

(b) How long will it take for the number of rabbits to triple?

37. A city in Ohio finds its residents moving into the countryside. Its population is declining according to the relationship

$$P = P_0 e^{-.04t},$$

where t is time measured in years and P_0 is the population at time $t = 0$.

(a) If $P_0 = 1{,}000{,}000$, find the population at time $t = 1$.

(b) If $P_0 = 1{,}000{,}000$, estimate the time it will take for the population to be reduced to 750,000.

(c) How long will it take for the population to be cut in half?

38. It can be shown that P dollars compounded continuously (every instant) at an annual rate of interest i would amount to

$$A = Pe^{ni}$$

at the end of n years. How much would \$20,000 compounded continuously at 8% amount to for the following number of years?

(a) 1 year **(b)** 5 years

(c) Find the number of years it will take for \$1000 compounded continuously at 7% to double.

39. Suppose that a large cloud of radioactive debris from a nuclear explosion has floated over the Pacific Northwest, contaminating much of the hay supply. Consequently, farmers in the area are concerned that the cows eating this hay will give contaminated milk. (The tolerance level for radioactive iodine in milk is 0.) The percent of the initial amount of radioactive iodine still present in the hay after t days is approximated by

$$y = 100(2.7)^{-.1t},$$

where t is time measured in days.

(a) Some scientists believe that the hay is safe after the percent of radioactive iodine has declined to 10% of the original amount. Find the number of days before the hay can be used.

(b) Other scientists believe that the hay is not safe until the level of radioactive iodine has declined to only 1% of the original level. Find the number of days this would take.

Chapter 9 *Key Concepts*

Exponential Function

An exponential function is a function of the form

$$y = a^x,$$

where $a > 0$ and $a \neq 1$.

Definition of Logarithm

For all positive numbers a, with $a \neq 1$,

$$y = \log_a x \quad \text{means} \quad x = a^y.$$

Logarithmic Function

If $a > 0$, $a \neq 1$, and $x > 0$, then

$$y = \log_a x$$

is the logarithmic function with base a.

Properties of Logarithms

If x and y are positive real numbers, r is any real number, and a is any positive real number, with $a \neq 1$, then

(a) $\log_a xy = \log_a x + \log_a y$

(b) $\log_a \frac{x}{y} = \log_a x - \log_a y$

(c) $\log_a x^r = r \log_a x$

(d) $\log_a a = 1$ (e) $\log_a 1 = 0$

Properties of Exponential and Logarithmic Equations

1. If a is a positive number, with a $\neq$ 1, and x and y are real numbers, then

$$a^x = a^y \text{ if and only if } x = y.$$

2. If x, y, and a are positive, with $a \neq 1$, then

$$x = y \text{ if and only if } \log_a x = \log_a y.$$

Chapter 9 *Review Exercises*

Simplify each expression.

1. $\dfrac{3^4 \cdot 3^5}{3^6}$ **2.** $(4^3)^5$ **3.** $(2^3)^2 \cdot 2^5$

4. $3^{-12} \cdot 3^5$ **5.** $16^{3/4}$ **6.** $\dfrac{2^{1/3}}{2^{5/3}}$

Graph each function.

7. $y = 5^x$ **8.** $y = \left(\dfrac{1}{3}\right)^x$ **9.** $y = 2^{-x}$

10. $y = 3^{x+2}$ **11.** $y = \log_3 x$ **12.** $y = 1 + \log_2 x$

13. $y = \log_2(2 - x)$ **14.** $y = \log_3 2x$

Convert to logarithmic form.

15. $2^9 = 512$ **16.** $10^{-2} = .01$ **17.** $\left(\dfrac{2}{3}\right)^{-1} = \dfrac{3}{2}$ **18.** $4^{3/2} = 8$

Write in exponential form.

19. $\log_3 81 = 4$ **20.** $\log_4 2 = \dfrac{1}{2}$ **21.** $\log_{1/3} 3 = -1$ **22.** $\log_{2/5} \dfrac{8}{125} = 3$

Use the properties of logarithms to simplify each expression.

23. $\log_5 4 + \log_5 3$ **24.** $\log_6 8 + \log_6 4$ **25.** $5 \log_2 x$

26. $\dfrac{1}{2} \log_{10} p$ **27.** $2 \log_3 x + 4 \log_3 y$ **28.** $\dfrac{2}{3} \ln x^3 - \dfrac{1}{4} \ln y$

Find the common logarithm of each number.

29. 2.48 **30.** 63.5 **31.** 1240

32. .00421 **33.** 5,270,000 **34.** .643

Find the common antilogarithm of each number.

35. 3.4314 **36.** 0.7497 **37.** $.2279 - 2$ **38.** $.6493 - 1$

39. Use interpolation or a calculator to find log 16.92.

40. Use interpolation or a calculator to find the common antilogarithm of 1.2354.

Use common logarithms to calculate each of the following to three significant digits.

41. $(2.65)^{1.3}$ **42.** $\sqrt[5]{.00647}$ **43.** $(9.81)^{-2}$ **44.** $\left(\dfrac{7.19}{2.43}\right)^5$

Find each of the following natural logarithms.

45. ln 15 **46.** ln 120 **47.** ln .12 **48.** ln .03

Solve each equation without using tables or a calculator.

49. $3^x = 27$ **50.** $25^p = 625$ **51.** $2^{m-4} = 8^{2m}$ **52.** $49^{-k+3} = 7^{5k}$

Solve each equation. Give answers as decimals rounded to the nearest hundredth.

53. $5^n = 33$

54. $4^{y+2} = 17$

55. $\log x = 2 - \log (x - 3)$

56. $\ln p - \ln (p + 3) = 4$

57. $\log_t 16 = 2$

58. $\ln k = 1$

59. The amount of a radioactive specimen present at time t (measured in days) is

$$A(t) = 5000(3)^{-.02t},$$

where $A(t)$ is measured in grams. Find the *half-life* of the specimen, that is, the time when only half of the specimen remains.

60. The number of fish in a pond is

$$N = 10 \log_3 (x + 1),$$

where x is time in months. How long will it take for the number of fish to reach 30?

Appendix

Table 1 Squares and Square Roots

n	n^2	$\sqrt{n}$	$\sqrt{10n}$	n	n^2	$\sqrt{n}$	$\sqrt{10n}$
1	1	1.000	3.162	51	2601	7.141	22.583
2	4	1.414	4.472	52	2704	7.211	22.804
3	9	1.732	5.477	53	2809	7.280	23.022
4	16	2.000	6.325	54	2916	7.348	23.238
5	25	2.236	7.071	55	3025	7.416	23.452
6	36	2.449	7.746	56	3136	7.483	23.664
7	49	2.646	8.367	57	3249	7.550	23.875
8	64	2.828	8.944	58	3364	7.616	24.083
9	81	3.000	9.487	59	3481	7.681	24.290
10	100	3.162	10.000	60	3600	7.746	24.495
11	121	3.317	10.488	61	3721	7.810	24.698
12	144	3.464	10.954	62	3844	7.874	24.900
13	169	3.606	11.402	63	3969	7.937	25.100
14	196	3.742	11.832	64	4096	8.000	25.298
15	225	3.873	12.247	65	4225	8.062	25.495
16	256	4.000	12.649	66	4356	8.124	25.690
17	289	4.123	13.038	67	4489	8.185	25.884
18	324	4.243	13.416	68	4624	8.246	26.077
19	361	4.359	13.784	69	4761	8.307	26.268
20	400	4.472	14.142	70	4900	8.367	26.458
21	441	4.583	14.491	71	5041	8.426	26.646
22	484	4.690	14.832	72	5184	8.485	26.833
23	529	4.796	15.166	73	5329	8.544	27.019
24	576	4.899	15.492	74	5476	8.602	27.203
25	625	5.000	15.811	75	5625	8.660	27.386
26	676	5.099	16.125	76	5776	8.718	27.568
27	729	5.196	16.432	77	5929	8.775	27.749
28	784	5.292	16.733	78	6084	8.832	27.928
29	841	5.385	17.029	79	6241	8.888	28.107
30	900	5.477	17.321	80	6400	8.944	28.284
31	961	5.568	17.607	81	6561	9.000	28.460
32	1024	5.657	17.889	82	6724	9.055	28.636
33	1089	5.745	18.166	83	6889	9.110	28.810
34	1156	5.831	18.439	84	7056	9.165	28.983
35	1225	5.916	18.708	85	7225	9.220	29.155
36	1296	6.000	18.974	86	7396	9.274	29.326
37	1369	6.083	19.235	87	7569	9.327	29.496
38	1444	6.164	19.494	88	7744	9.381	29.665
39	1521	6.245	19.748	89	7921	9.434	29.833
40	1600	6.325	20.000	90	8100	9.487	30.000
41	1681	6.403	20.248	91	8281	9.539	30.166
42	1764	6.481	20.494	92	8464	9.592	30.332
43	1849	6.557	20.736	93	8649	9.644	30.496
44	1936	6.633	20.976	94	8836	9.695	30.659
45	2025	6.708	21.213	95	9025	9.747	30.822
46	2116	6.782	21.448	96	9216	9.798	30.984
47	2209	6.856	21.679	97	9409	9.849	31.145
48	2304	6.928	21.909	98	9604	9.899	31.305
49	2401	7.000	22.136	99	9801	9.950	31.464
50	2500	7.071	22.361	100	10000	10.000	31.623

Table 2 Trigonometric Functions in Degrees and Radians

θ (degrees)	θ (radians)	$\sin\theta$	$\cos\theta$	$\tan\theta$	$\cot\theta$	$\sec\theta$	$\csc\theta$		
0°00′	.0000	.0000	1.0000	.0000	—	1.000	—	1.5708	**90°00′**
10	.0029	.0029	1.0000	.0029	343.8	1.000	343.8	1.5679	50
20	.0058	.0058	1.0000	.0058	171.9	1.000	171.9	1.5650	40
30	.0087	.0087	1.0000	.0087	114.6	1.000	114.6	1.5621	30
40	.0116	.0116	.9999	.0116	85.94	1.000	85.95	1.5592	20
50	.0145	.0145	.9999	.0145	68.75	1.000	68.76	1.5563	10
1°00′	.0175	.0175	.9998	.0175	57.29	1.000	57.30	1.5533	**89°00′**
10	.0204	.0204	.9998	.0204	49.10	1.000	49.11	1.5504	50
20	.0233	.0233	.9997	.0233	42.96	1.000	42.98	1.5475	40
30	.0262	.0262	.9997	.0262	38.19	1.000	38.20	1.5446	30
40	.0291	.0291	.9996	.0291	34.37	1.000	34.38	1.5417	20
50	.0320	.0320	.9995	.0320	31.24	1.001	31.26	1.5388	10
2°00′	.0349	.0349	.9994	.0349	28.64	1.001	28.65	1.5359	**88°00′**
10	.0378	.0378	.9993	.0378	26.43	1.001	26.45	1.5330	50
20	.0407	.0407	.9992	.0407	24.54	1.001	24.56	1.5301	40
30	.0436	.0436	.9990	.0437	22.90	1.001	22.93	1.5272	30
40	.0465	.0465	.9989	.0466	21.47	1.001	21.49	1.5243	20
50	.0495	.0494	.9988	.0495	20.21	1.001	20.23	1.5213	10
3°00′	.0524	.0523	.9986	.0524	19.08	1.001	19.11	1.5184	**87°00′**
10	.0553	.0552	.9985	.0553	18.07	1.002	18.10	1.5155	50
20	.0582	.0581	.9983	.0582	17.17	1.002	17.20	1.5126	40
30	.0611	.0610	.9981	.0612	16.35	1.002	16.38	1.5097	30
40	.0640	.0640	.9980	.0641	15.60	1.002	15.64	1.5068	20
50	.0669	.0669	.9978	.0670	14.92	1.002	14.96	1.5039	10
4°00′	.0698	.0698	.9976	.0699	14.30	1.002	14.34	1.5010	**86°00′**
10	.0727	.0727	.9974	.0729	13.73	1.003	13.76	1.4981	50
20	.0756	.0756	.9971	.0758	13.20	1.003	13.23	1.4952	40
30	.0785	.0785	.9969	.0787	12.71	1.003	12.75	1.4923	30
40	.0814	.0814	.9967	.0816	12.25	1.003	12.29	1.4893	20
50	.0844	.0843	.9964	.0846	11.83	1.004	11.87	1.4864	10
5°00′	.0873	.0872	.9962	.0875	11.43	1.004	11.47	1.4835	**85°00′**
10	.0902	.0901	.9959	.0904	11.06	1.004	11.10	1.4806	50
20	.0931	.0929	.9957	.0934	10.71	1.004	10.76	1.4777	40
30	.0960	.0958	.9954	.0963	10.39	1.005	10.43	1.4748	30
40	.0989	.0987	.9951	.0992	10.08	1.005	10.13	1.4719	20
50	.1018	.1016	.9948	.1022	9.788	1.005	9.839	1.4690	10
6°00′	.1047	.1045	.9945	.1051	9.514	1.006	9.567	1.4661	**84°00′**
10	.1076	.1074	.9942	.1080	9.255	1.006	9.309	1.4632	50
20	.1105	.1103	.9939	.1110	9.010	1.006	9.065	1.4603	40
30	.1134	.1132	.9936	.1139	8.777	1.006	8.834	1.4573	30
40	.1164	.1161	.9932	.1169	8.556	1.007	8.614	1.4544	20
50	.1193	.1190	.9929	.1198	8.345	1.007	8.405	1.4515	10
		$\cos\theta$	$\sin\theta$	$\cot\theta$	$\tan\theta$	$\csc\theta$	$\sec\theta$	θ (radians)	θ (degrees)

θ (degrees)	θ (radians)	$\sin\theta$	$\cos\theta$	$\tan\theta$	$\cot\theta$	$\sec\theta$	$\csc\theta$		
7°00′	.1222	.1219	.9925	.1228	8.144	1.008	8.206	1.4486	**83°00′**
10	.1251	.1248	.9922	.1257	7.953	1.008	8.016	1.4457	50
20	.1280	.1276	.9918	.1287	7.770	1.008	7.834	1.4428	40
30	.1309	.1305	.9914	.1317	7.596	1.009	7.661	1.4399	30
40	.1338	.1334	.9911	.1346	7.429	1.009	7.496	1.4370	20
50	.1376	.1363	.9907	.1376	7.269	1.009	7.337	1.4341	10
8°00′	.1396	.1392	.9903	.1405	7.115	1.010	7.185	1.4312	**82°00′**
10	.1425	.1421	.9899	.1435	6.968	1.010	7.040	1.4283	50
20	.1454	.1449	.9894	.1465	6.827	1.011	6.900	1.4254	40
30	.1484	.1478	.9890	.1495	6.691	1.011	6.765	1.4224	30
40	.1513	.1507	.9886	.1524	6.561	1.012	6.636	1.4195	20
50	.1542	.1536	.9881	.1554	6.435	1.012	6.512	1.4166	10
9°00′	.1571	.1564	.9877	.1584	6.314	1.012	6.392	1.4137	**81°00′**
10	.1600	.1593	.9872	.1614	6.197	1.013	6.277	1.4108	50
20	.1629	.1622	.9868	.1644	6.084	1.013	6.166	1.4079	40
30	.1658	.1650	.9863	.1673	5.976	1.014	6.059	1.4050	30
40	.1687	.1679	.9858	.1703	5.871	1.014	5.955	1.4021	20
50	.1716	.1708	.9853	.1733	5.769	1.015	5.855	1.3992	10
10°00′	.1745	.1736	.9848	.1763	5.671	1.015	5.759	1.3963	**80°00′**
10	.1774	.1765	.9843	.1793	5.576	1.016	5.665	1.3934	50
20	.1804	.1794	.9838	.1823	5.485	1.016	5.575	1.3904	40
30	.1833	.1822	.9833	.1853	5.396	1.017	5.487	1.3875	30
40	.1862	.1851	.9827	.1883	5.309	1.018	5.403	1.3846	20
50	.1891	.1880	.9822	.1914	5.226	1.018	5.320	1.3817	10
11°00′	.1920	.1908	.9816	.1944	5.145	1.019	5.241	1.3788	**79°00′**
10	.1949	.1937	.9811	.1974	5.066	1.019	5.164	1.3759	50
20	.1978	.1965	.9805	.2004	4.989	1.020	5.089	1.3730	40
30	.2007	.1994	.9799	.2035	4.915	1.020	5.016	1.3701	30
40	.2036	.2022	.9793	.2065	4.843	1.021	4.945	1.3672	20
50	.2065	.2051	.9787	.2095	4.773	1.022	4.876	1.3643	10
12°00′	.2094	.2079	.9781	.2126	4.705	1.022	4.810	1.3614	**78°00′**
10	.2123	.2108	.9775	.2156	4.638	1.023	4.745	1.3584	50
20	.2153	.2136	.9769	.2186	4.574	1.024	4.682	1.3555	40
30	.2182	.2164	.9763	.2217	4.511	1.024	4.620	1.3526	30
40	.2211	.2193	.9757	.2247	4.449	1.025	4.560	1.3497	20
50	.2240	.2221	.9750	.2278	4.390	1.026	4.502	1.3468	10
13°00′	.2269	.2250	.9744	.2309	4.331	1.026	4.445	1.3439	**77°00′**
10	.2298	.2278	.9737	.2339	4.275	1.027	4.390	1.3410	50
20	.2327	.2306	.9730	.2370	4.219	1.028	4.336	1.3381	40
30	.2356	.2334	.9724	.2401	4.165	1.028	4.284	1.3352	30
40	.2385	.2363	.9717	.2432	4.113	1.029	4.232	1.3323	20
50	.2414	.2391	.9710	.2462	4.061	1.030	4.182	1.3294	10
		$\cos\theta$	$\sin\theta$	$\cot\theta$	$\tan\theta$	$\csc\theta$	$\sec\theta$	θ (radians)	θ (degrees)

Table 2 Trigonometric Functions in Degrees and Radians (continued)

θ (degrees)	θ (radians)	sin θ	cos θ	tan θ	cot θ	sec θ	csc θ		
14°00′	.2443	.2419	.9703	.2493	4.011	1.031	4.134	1.3265	**76°00′**
10	.2473	.2447	.9696	.2524	3.962	1.031	4.086	1.3235	50
20	.2502	.2476	.9689	.2555	3.914	1.032	4.039	1.3206	40
30	.2531	.2504	.9681	.2586	3.867	1.033	3.994	1.3177	30
40	.2560	.2532	.9674	.2617	3.821	1.034	3.950	1.3148	20
50	.2589	.2560	.9667	.2648	3.776	1.034	3.906	1.3119	10
15°00′	.2618	.2588	.9659	.2679	3.732	1.035	3.864	1.3090	**75°00′**
10	.2647	.2616	.9652	.2711	3.689	1.036	3.822	1.3061	50
20	.2676	.2644	.9644	.2742	3.647	1.037	3.782	1.3032	40
30	.2705	.2672	.9636	.2773	3.606	1.038	3.742	1.3003	30
40	.2734	.2700	.9628	.2805	3.566	1.039	3.703	1.2974	20
50	.2763	.2728	.9621	.2836	3.526	1.039	3.665	1.2945	10
16°00′	.2793	.2756	.9613	.2867	3.487	1.040	3.628	1.2915	**74°00′**
10	.2822	.2784	.9605	.2899	3.450	1.041	3.592	1.2886	50
20	.2851	.2812	.9596	.2931	3.412	1.042	3.556	1.2857	40
30	.2880	.2840	.9588	.2962	3.376	1.043	3.521	1.2828	30
40	.2909	.2868	.9580	.2994	3.340	1.044	3.487	1.2799	20
50	.2938	.2896	.9572	.3026	3.305	1.045	3.453	1.2770	10
17°00′	.2967	.2924	.9563	.3057	3.271	1.046	3.420	1.2741	**73°00′**
10	.2996	.2952	.9555	.3089	3.237	1.047	3.388	1.2712	50
20	.3025	.2979	.9546	.3121	3.204	1.048	3.356	1.2683	40
30	.3054	.3007	.9537	.3153	3.172	1.049	3.326	1.2654	30
40	.3083	.3035	.9528	.3185	3.140	1.049	3.295	1.2625	20
50	.3113	.3062	.9520	.3217	3.108	1.050	3.265	1.2595	10
18°00′	.3142	.3090	.9511	.3249	3.078	1.051	3.236	1.2566	**72°00′**
10	.3171	.3118	.9502	.3281	3.047	1.052	3.207	1.2537	50
20	.3200	.3145	.9492	.3314	3.018	1.053	3.179	1.2508	40
30	.3229	.3173	.9483	.3346	2.989	1.054	3.152	1.2479	30
40	.3258	.3201	.9474	.3378	2.960	1.056	3.124	1.2450	20
50	.3287	.3228	.9465	.3411	2.932	1.057	3.098	1.2421	10
19°00′	.3316	.3256	.9455	.3443	2.904	1.058	3.072	1.2392	**71°00′**
10	.3345	.3283	.9446	.3476	2.877	1.059	3.046	1.2363	50
20	.3374	.3311	.9436	.3508	2.850	1.060	3.021	1.2334	40
30	.3403	.3338	.9426	.3541	2.824	1.061	2.996	1.2305	30
40	.3432	.3365	.9417	.3574	2.798	1.062	2.971	1.2275	20
50	.3462	.3393	.9407	.3607	2.773	1.063	2.947	1.2246	10
20°00′	.3491	.3420	.9397	.3640	2.747	1.064	2.924	1.2217	**70°00′**
10	.3520	.3448	.9387	.3673	2.723	1.065	2.901	1.2188	50
20	.3549	.3475	.9377	.3706	2.699	1.066	2.878	1.2159	40
30	.3578	.3502	.9367	.3739	2.675	1.068	2.855	1.2130	30
40	.3607	.3529	.9356	.3772	2.651	1.069	2.833	1.2101	20
50	.3636	.3557	.9346	.3805	2.628	1.070	2.182	1.2072	10
		cos θ	sin θ	cot θ	tan θ	csc θ	sec θ	θ (radians)	θ (degrees)

θ (degrees)	θ (radians)	sin θ	cos θ	tan θ	cot θ	sec θ	csc θ		
21°00′	.3665	.3584	.9336	.3839	2.605	1.071	2.790	1.2043	**69°00′**
10	.3694	.3611	.9325	.3872	2.583	1.072	2.769	1.2014	50
20	.3723	.3638	.9315	.3906	2.560	1.074	2.749	1.1985	40
30	.3752	.3665	.9304	.3939	2.539	1.075	2.729	1.1956	30
40	.3782	.3692	.9293	.3973	2.517	1.076	2.709	1.1926	20
50	.3811	.3719	.9283	.4006	2.496	1.077	2.689	1.1897	10
22°00′	.3840	.3746	.9272	.4040	2.475	1.079	2.669	1.1868	**68°00′**
10	.3869	.3773	.9261	.4074	2.455	1.080	2.650	1.1839	50
20	.3898	.3800	.9250	.4108	2.434	1.081	2.632	1.1810	40
30	.3927	.3827	.9239	.4142	2.414	1.082	2.613	1.1781	30
40	.3956	.3854	.9228	.4176	2.394	1.084	2.595	1.1752	20
50	.3985	.3881	.9216	.4210	2.375	1.085	2.577	1.1723	10
23°00′	.4014	.3907	.9205	.4245	2.356	1.086	2.559	1.1694	**67°00′**
10	.4043	.3934	.9194	.4279	2.337	1.088	2.542	1.1665	50
20	.4072	.3961	.9182	.4314	2.318	1.089	2.525	1.1636	40
30	.4102	.3987	.9171	.4348	2.300	1.090	2.508	1.1606	30
40	.4131	.4014	.9159	.4383	2.282	1.092	2.491	1.1577	20
50	.4160	.4041	.9147	.4417	2.264	1.093	2.475	1.1548	10
24°00′	.4189	.4067	.9135	.4452	2.246	1.095	2.459	1.1519	**66°00′**
10	.4218	.4094	.9124	.4487	2.229	1.096	2.443	1.1490	50
20	.4247	.4120	.9112	.4522	2.211	1.097	2.427	1.1461	40
30	.4276	.4147	.9100	.4557	2.194	1.099	2.411	1.1432	30
40	.4305	.4173	.9088	.4592	2.177	1.100	2.396	1.1403	20
50	.4334	.4200	.9075	.4628	2.161	1.102	2.381	1.1374	10
25°00′	.4363	.4226	.9063	.4663	2.145	1.103	2.366	1.1345	**65°00′**
10	.4392	.4253	.9051	.4699	2.128	1.105	2.352	1.1316	50
20	.4422	.4279	.9038	.4734	2.112	1.106	2.337	1.1286	40
30	.4451	.4305	.9026	.4770	2.097	1.108	2.323	1.1257	30
40	.4480	.4331	.9013	.4806	2.081	1.109	2.309	1.1228	20
50	.4509	.4358	.9001	.4841	2.066	1.111	2.295	1.1199	10
26°00′	.4538	.4384	.8988	.4877	2.050	1.113	2.281	1.1170	**64°00′**
10	.4567	.4410	.8975	.4913	2.035	1.114	2.268	1.1141	50
20	.4596	.4436	.8962	.4950	2.020	1.116	2.254	1.1112	40
30	.4625	.4462	.8949	.4986	2.006	1.117	2.241	1.1083	30
40	.4654	.4488	.8936	.5022	1.991	1.119	2.228	1.1054	20
50	.4683	.4514	.8923	.5059	1.977	1.121	2.215	1.1025	10
27°00′	.4712	.4540	.8910	.5095	1.963	1.122	2.203	1.0996	**63°00′**
10	.4741	.4566	.8897	.5132	1.949	1.124	2.190	1.0966	50
20	.4771	.4592	.8884	.5169	1.935	1.126	2.178	1.0937	40
30	.4800	.4617	.8870	.5206	1.921	1.127	2.166	1.0908	30
40	.4829	.4643	.8857	.5243	1.907	1.129	2.154	1.0879	20
50	.4858	.4669	.8843	.5280	1.894	1.131	2.142	1.0850	10
		cos θ	sin θ	cot θ	tan θ	csc θ	sec θ	θ (radians)	θ (degrees)

Table 2 Trigonometric Functions in Degrees and Radians (continued)

θ (degrees)	θ (radians)	sin θ	cos θ	tan θ	cot θ	sec θ	csc θ		
28°00′	.4887	.4695	.8829	.5317	1.881	1.133	2.130	1.0821	**62°00′**
10	.4916	.4720	.8816	.5354	1.868	1.134	2.118	1.0792	50
20	.4945	.4746	.8802	.5392	1.855	1.136	2.107	1.0763	40
30	.4974	.4772	.8788	.5430	1.842	1.138	2.096	1.0734	30
40	.5003	.4797	.8774	.5467	1.829	1.140	2.085	1.0705	20
50	.5032	.4823	.8760	.5505	1.816	1.142	2.074	1.0676	10
29°00′	.5061	.4848	.8746	.5543	1.804	1.143	2.063	1.0647	**61°00′**
10	.5091	.4874	.8732	.5581	1.792	1.145	2.052	1.0617	50
20	.5120	.4899	.8718	.5619	1.780	1.147	2.041	1.0588	40
30	.5149	.4924	.8704	.5658	1.767	1.149	2.031	1.0559	30
40	.5178	.4950	.8689	.5696	1.756	1.151	2.020	1.0530	20
50	.5207	.4975	.8675	.5735	1.744	1.153	2.010	1.0501	10
30°00′	.5236	.5000	.8660	.5774	1.732	1.155	2.000	1.0472	**60°00′**
10	.5265	.5025	.8646	.5812	1.720	1.157	1.990	1.0443	50
20	.5294	.5050	.8631	.5851	1.709	1.159	1.980	1.0414	40
30	.5323	.5075	.8616	.5890	1.698	1.161	1.970	1.0385	30
40	.5352	.5100	.8601	.5930	1.686	1.163	1.961	1.0356	20
50	.5381	.5125	.8587	.5969	1.675	1.165	1.951	1.0327	10
31°00′	.5411	.5150	.8572	.6009	1.664	1.167	1.942	1.0297	**59°00′**
10	.5440	.5175	.8557	.6048	1.653	1.169	1.932	1.0268	50
20	.5469	.5200	.8542	.6088	1.643	1.171	1.923	1.0239	40
30	.5498	.5225	.8526	.6128	1.632	1.173	1.914	1.0210	30
40	.5527	.5250	.8511	.6168	1.621	1.175	1.905	1.0181	20
50	.5556	.5275	.8496	.6208	1.611	1.177	1.896	1.0152	10
32°00′	.5585	.5299	.8480	.6249	1.600	1.179	1.887	1.0123	**58°00′**
10	.5614	.5324	.8465	.6289	1.590	1.181	1.878	1.0094	50
20	.5643	.5348	.8450	.6330	1.580	1.184	1.870	1.0065	40
30	.5672	.5373	.8434	.6371	1.570	1.186	1.861	1.0036	30
40	.5701	.5398	.8418	.6412	1.560	1.188	1.853	1.0007	20
50	.5730	.5422	.8403	.6453	1.550	1.190	1.844	.9977	10
33°00′	.5760	.5446	.8387	.6494	1.540	1.192	1.836	.9948	**57°00′**
10	.5789	.5471	.8371	.6536	1.530	1.195	1.828	.9919	50
20	.5818	.5495	.8355	.6577	1.520	1.197	1.820	.9890	40
30	.5847	.5519	.8339	.6619	1.511	1.199	1.812	.9861	30
40	.5876	.5544	.8323	.6661	1.501	1.202	1.804	.9832	20
50	.5905	.5568	.8307	.6703	1.492	1.204	1.796	.9803	10
34°00′	.5934	.5592	.8290	.6745	1.483	1.206	1.788	.9774	**56°00′**
10	.5963	.5616	.8274	.6787	1.473	1.209	1.781	.9745	50
20	.5992	.5640	.8258	.6830	1.464	1.211	1.773	.9716	40
30	.6021	.5664	.8241	.6873	1.455	1.213	1.766	.9687	30
40	.6050	.5688	.8225	.6916	1.446	1.216	1.758	.9657	20
50	.6080	.5712	.8208	.6959	1.437	1.218	1.751	.9628	10
		cos θ	sin θ	cot θ	tan θ	csc θ	sec θ	θ (radians)	θ (degrees)

θ (degrees)	θ (radians)	sin θ	cos θ	tan θ	cot θ	sec θ	csc θ		
35°00′	.6109	.5736	.8192	.7002	1.428	1.221	1.743	.9599	**55°00′**
10	.6138	.5760	.8175	.7046	1.419	1.223	1.736	.9570	50
20	.6167	.5783	.8158	.7089	1.411	1.226	1.729	.9541	40
30	.6196	.5807	.8141	.7133	1.402	1.228	1.722	.9512	30
40	.6225	.5831	.8124	.7177	1.393	1.231	1.715	.9483	20
50	.6254	.5854	.8107	.7221	1.385	1.233	1.708	.9454	10
36°00′	.6283	.5878	.8090	.7265	1.376	1.236	1.701	.9425	**54°00′**
10	.6312	.5901	.8073	.7310	1.368	1.239	1.695	.9396	50
20	.6341	.5925	.8056	.7355	1.360	1.241	1.688	.9367	40
30	.6370	.5948	.8039	.7400	1.351	1.244	1.681	.9338	30
40	.6400	.5972	.8021	.7445	1.343	1.247	1.675	.9308	20
50	.6429	.5995	.8004	.7490	1.335	1.249	1.668	.9279	10
37°00′	.6458	.6018	.7986	.7536	1.327	1.252	1.662	.9250	**53°00′**
10	.6487	.6041	.7969	.7581	1.319	1.255	1.655	.9221	50
20	.6516	.6065	.7951	.7627	1.311	1.258	1.649	.9192	40
30	.6545	.6088	.7934	.7673	1.303	1.260	1.643	.9163	30
40	.6574	.6111	.7916	.7720	1.295	1.263	1.636	.9134	20
50	.6603	.6134	.7898	.7766	1.288	1.266	1.630	.9105	10
38°00′	.6632	.6157	.7880	.7813	1.280	1.269	1.624	.9076	**52°00′**
10	.6661	.6180	.7862	.7860	1.272	1.272	1.618	.9047	50
20	.6690	.6202	.7844	.7907	1.265	1.275	1.612	.9018	40
30	.6720	.6225	.7826	.7954	1.257	1.278	1.606	.8988	30
40	.6749	.6248	.7808	.8002	1.250	1.281	1.601	.8959	20
50	.6778	.6271	.7790	.8050	1.242	1.284	1.595	.8930	10
39°00′	.6807	.6293	.7771	.8098	1.235	1.287	1.589	.8901	**51°00′**
10	.6836	.6316	.7753	.8146	1.228	1.290	1.583	.8872	50
20	.6865	.6338	.7735	.8195	1.220	1.293	1.578	.8843	40
30	.6894	.6361	.7716	.8243	1.213	1.296	1.572	.8814	30
40	.6923	.6383	.7698	.8292	1.206	1.299	1.567	.8785	20
50	.6952	.6406	.7679	.8342	1.199	1.302	1.561	.8756	10
40°00′	.6981	.6428	.7660	.8391	1.192	1.305	1.556	.8727	**50°00′**
10	.7010	.6450	.7642	.8441	1.185	1.309	1.550	.8698	50
20	.7039	.6472	.7623	.8491	1.178	1.312	1.545	.8668	40
30	.7069	.6494	.7604	.8541	1.171	1.315	1.540	.8639	30
40	.7098	.6517	.7585	.8591	1.164	1.318	1.535	.8610	20
50	.7127	.6539	.7566	.8642	1.157	1.322	1.529	.8581	10
41°00′	.7156	.6561	.7547	.8693	1.150	1.325	1.524	.8552	**49°00′**
10	.7185	.6583	.7528	.8744	1.144	1.328	1.519	.8523	50
20	.7214	.6604	.7509	.8796	1.137	1.332	1.514	.8494	40
30	.7243	.6626	.7490	.8847	1.130	1.335	1.509	.8465	30
40	.7272	.6648	.7470	.8899	1.124	1.339	1.504	.8436	20
50	.7301	.6670	.7451	.8952	1.117	1.342	1.499	.8407	10
		cos θ	sin θ	cot θ	tan θ	csc θ	sec θ	θ (radians)	θ (degrees)

Table 2 Trigonometric Functions in Degrees and Radians (continued)

θ (degrees)	θ (radians)	sin θ	cos θ	tan θ	cot θ	sec θ	csc θ		
42°00′	.7330	.6691	.7431	.9004	1.111	1.346	1.494	.8378	**48°00′**
10	.7359	.6713	.7412	.9057	1.104	1.349	1.490	.8348	50
20	.7389	.6734	.7392	.9110	1.098	1.353	1.485	.8319	40
30	.7418	.6756	.7373	.9163	1.091	1.356	1.480	.8290	30
40	.7447	.6777	.7353	.9217	1.085	1.360	1.476	.8261	20
50	.7476	.6799	.7333	.9271	1.079	1.364	1.471	.8232	10
43°00′	.7505	.6820	.7314	.9325	1.072	1.367	1.466	.8203	**47°00′**
10	.7534	.6841	.7294	.9380	1.066	1.371	1.462	.8174	50
20	.7563	.6862	.7274	.9435	1.060	1.375	1.457	.8145	40
30	.7592	.6884	.7254	.9490	1.054	1.379	1.453	.8116	30
40	.7621	.6905	.7234	.9545	1.048	1.382	1.448	.8087	20
50	.7660	.6926	.7214	.9601	1.042	1.386	1.444	.8058	10
44°00′	.7679	.6947	.7193	.9657	1.036	1.390	1.440	.8029	**46°00′**
10	.7709	.6967	.7173	.9713	1.030	1.394	1.435	.7999	50
20	.7738	.6988	.7153	.9770	1.024	1.398	1.431	.7970	40
30	.7767	.7009	.7133	.9827	1.018	1.402	1.427	.7941	30
40	.7796	.7030	.7112	.9884	1.012	1.406	1.423	.7912	20
50	.7825	.7050	.7092	.9942	1.006	1.410	1.418	.7883	10
45°00′	.7854	.7071	.7071	1.000	1.000	1.414	1.414		
		cos θ	sin θ	cot θ	tan θ	csc θ	sec θ	θ (radians)	θ (degrees)

Table 3 Common Logarithms

n	0	1	2	3	4	5	6	7	8	9
1.0	.0000	.0043	.0086	.0128	.0170	.0212	.0253	.0294	.0334	.0374
1.1	.0414	.0453	.0492	.0531	.0569	.0607	.0645	.0682	.0719	.0755
1.2	.0792	.0828	.0864	.0899	.0934	.0969	.1004	.1038	.1072	.1106
1.3	.1139	.1173	.1206	.1239	.1271	.1303	.1335	.1367	.1399	.1430
1.4	.1461	.1492	.1523	.1553	.1584	.1614	.1644	.1673	.1703	.1732
1.5	.1761	.1790	.1818	.1847	.1875	.1903	.1931	.1959	.1987	.2014
1.6	.2041	.2068	.2095	.2122	.2148	.2175	.2201	.2227	.2253	.2279
1.7	.2304	.2330	.2355	.2380	.2405	.2430	.2455	.2480	.2504	.2529
1.8	.2553	.2577	.2601	.2625	.2648	.2672	.2695	.2718	.2742	.2765
1.9	.2788	.2810	.2833	.2856	.2878	.2900	.2923	.2945	.2967	.2989
2.0	.3010	.3032	.3054	.3075	.3096	.3118	.3139	.3160	.3181	.3201
2.1	.3222	.3243	.3263	.3284	.3304	.3324	.3345	.3365	.3385	.3404
2.2	.3424	.3444	.3464	.3483	.3502	.3522	.3541	.3560	.3579	.3598
2.3	.3617	.3636	.3655	.3674	.3692	.3711	.3729	.3747	.3766	.3784
2.4	.3802	.3820	.3838	.3856	.3874	.3892	.3909	.3927	.3945	.3962
2.5	.3979	.3997	.4014	.4031	.4048	.4065	.4082	.4099	.4116	.4133
2.6	.4150	.4166	.4183	.4200	.4216	.4232	.4249	.4265	.4281	.4298
2.7	.4314	.4330	.4346	.4362	.4378	.4393	.4409	.4425	.4440	.4456
2.8	.4472	.4487	.4502	.4518	.4533	.4548	.4564	.4579	.4594	.4609
2.9	.4624	.4639	.4654	.4669	.4683	.4698	.4713	.4728	.4742	.4757
3.0	.4771	.4786	.4800	.4814	.4829	.4843	.4857	.4871	.4886	.4900
3.1	.4914	.4928	.4942	.4955	.4969	.4983	.4997	.5011	.5024	.5038
3.2	.5051	.5065	.5079	.5092	.5105	.5119	.5132	.5145	.5159	.5172
3.3	.5185	.5198	.5211	.5224	.5237	.5250	.5263	.5276	.5289	.5302
3.4	.5315	.5328	.5340	.5353	.5366	.5378	.5391	.5403	.5416	.5428
3.5	.5441	.5453	.5465	.5478	.5490	.5502	.5514	.5527	.5539	.5551
3.6	.5563	.5575	.5587	.5599	.5611	.5623	.5635	.5647	.5658	.5670
3.7	.5682	.5694	.5705	.5717	.5729	.5740	.5752	.5763	.5775	.5786
3.8	.5798	.5809	.5821	.5832	.5843	.5855	.5866	.5877	.5888	.5899
3.9	.5911	.5922	.5933	.5944	.5955	.5966	.5977	.5988	.5999	.6010
4.0	.6021	.6031	.6042	.6053	.6064	.6075	.6085	.6096	.6107	.6117
4.1	.6128	.6138	.6149	.6160	.6170	.6180	.6191	.6201	.6212	.6222
4.2	.6232	.6243	.6253	.6263	.6274	.6284	.6294	.6304	.6314	.6325
4.3	.6335	.6345	.6355	.6365	.6375	.6385	.6395	.6405	.6415	.6425
4.4	.6435	.6444	.6454	.6464	.6474	.6484	.6493	.6503	.6513	.6522
4.5	.6532	.6542	.6551	.6561	.6571	.6580	.6590	.6599	.6609	.6618
4.6	.6628	.6637	.6646	.6656	.6665	.6675	.6684	.6693	.6702	.6712
4.7	.6721	.6730	.6739	.6749	.6758	.6767	.6776	.6785	.6794	.6803
4.8	.6812	.6821	.6830	.6839	.6848	.6857	.6866	.6875	.6884	.6893
4.9	.6902	.6911	.6920	.6928	.6937	.6946	.6955	.6964	.6972	.6981
5.0	.6990	.6998	.7007	.7016	.7024	.7033	.7042	.7050	.7059	.7067
5.1	.7076	.7084	.7093	.7101	.7110	.7118	.7126	.7135	.7143	.7152
5.2	.7160	.7168	.7177	.7185	.7193	.7202	.7210	.7218	.7226	.7235
5.3	.7243	.7251	.7259	.7267	.7275	.7284	.7292	.7300	.7308	.7316
5.4	.7324	.7332	.7340	.7348	.7356	.7364	.7372	.7380	.7388	.7396
n	0	1	2	3	4	5	6	7	8	9

Table 3 Common Logarithms (continued)

n	0	1	2	3	4	5	6	7	8	9
5.5	.7404	.7412	.7419	.7427	.7435	.7443	.7451	.7459	.7466	.7474
5.6	.7482	.7490	.7497	.7505	.7513	.7520	.7528	.7536	.7543	.7551
5.7	.7559	.7566	.7574	.7582	.7589	.7597	.7604	.7612	.7619	.7627
5.8	.7634	.7642	.7649	.7657	.7664	.7672	.7679	.7686	.7694	.7701
5.9	.7709	.7716	.7723	.7731	.7738	.7745	.7752	.7760	.7767	.7774
6.0	.7782	.7789	.7796	.7803	.7810	.7818	.7825	.7832	.7839	.7846
6.1	.7853	.7860	.7868	.7875	.7882	.7889	.7896	.7903	.7910	.7917
6.2	.7924	.7931	.7938	.7945	.7952	.7959	.7966	.7973	.7980	.7987
6.3	.7993	.8000	.8007	.8014	.8021	.8028	.8035	.8041	.8048	.8055
6.4	.8062	.8069	.8075	.8082	.8089	.8096	.8102	.8109	.8116	.8122
6.5	.8129	.8136	.8142	.8149	.8156	.8162	.8169	.8176	.8182	.8189
6.6	.8195	.8202	.8209	.8215	.8222	.8228	.8235	.8241	.8248	.8254
6.7	.8261	.8267	.8274	.8280	.8287	.8293	.8299	.8306	.8312	.8319
6.8	.8325	.8331	.8338	.8344	.8351	.8357	.8363	.8370	.8376	.8382
6.9	.8388	.8395	.8401	.8407	.8414	.8420	.8426	.8432	.8439	.8445
7.0	.8451	.8457	.8463	.8470	.8476	.8482	.8488	.8494	.8500	.8506
7.1	.8513	.8519	.8525	.8531	.8537	.8543	.8549	.8555	.8561	.8567
7.2	.8573	.8579	.8585	.8591	.8597	.8603	.8609	.8615	.8621	.8627
7.3	.8633	.8639	.8645	.8651	.8657	.8663	.8669	.8675	.8681	.8686
7.4	.8692	.8698	.8704	.8710	.8716	.8722	.8727	.8733	.8739	.8745
7.5	.8751	.8756	.8762	.8768	.8774	.8779	.8785	.8791	.8797	.8802
7.6	.8808	.8814	.8820	.8825	.8831	.8837	.8842	.8848	.8854	.8859
7.7	.8865	.8871	.8876	.8882	.8887	.8893	.8899	.8904	.8910	.8915
7.8	.8921	.8927	.8932	.8938	.8943	.8949	.8954	.8960	.8965	.8971
7.9	.8976	.8982	.8987	.8993	.8998	.9004	.9009	.9015	.9020	.9025
8.0	.9031	.9036	.9042	.9047	.9053	.9058	.9063	.9069	.9074	.9079
8.1	.9085	.9090	.9096	.9101	.9106	.9112	.9117	.9122	.9128	.9133
8.2	.9138	.9143	.9149	.9154	.9159	.9165	.9170	.9175	.9180	.9186
8.3	.9191	.9196	.9201	.9206	.9212	.9217	.9222	.9227	.9232	.9238
8.4	.9243	.9248	.9253	.9258	.9263	.9269	.9274	.9279	.9284	.9289
8.5	.9294	.9299	.9304	.9309	.9315	.9320	.9325	.9330	.9335	.9340
8.6	.9345	.9350	.9355	.9360	.9365	.9370	.9375	.9380	.9385	.9390
8.7	.9395	.9400	.9405	.9410	.9415	.9420	.9425	.9430	.9435	.9440
8.8	.9445	.9450	.9455	.9460	.9465	.9469	.9474	.9479	.9484	.9489
8.9	.9494	.9499	.9504	.9509	.9513	.9518	.9523	.9528	.9533	.9538
9.0	.9542	.9547	.9552	.9557	.9562	.9566	.9571	.9576	.9581	.9586
9.1	.9590	.9595	.9600	.9605	.9609	.9614	.9619	.9624	.9628	.9633
9.2	.9638	.9643	.9647	.9652	.9657	.9661	.9666	.9671	.9675	.9680
9.3	.9685	.9689	.9694	.9699	.9703	.9708	.9713	.9717	.9722	.9727
9.4	.9731	.9736	.9741	.9745	.9750	.9754	.9759	.9763	.9768	.9773
9.5	.9777	.9782	.9786	.9791	.9795	.9800	.9805	.9809	.9814	.9818
9.6	.9823	.9827	.9832	.9836	.9841	.9845	.9850	.9854	.9859	.9863
9.7	.9868	.9872	.9877	.9881	.9886	.9890	.9894	.9899	.9903	.9908
9.8	.9912	.9917	.9921	.9926	.9930	.9934	.9939	.9943	.9948	.9952
9.9	.9956	.9961	.9965	.9969	.9974	.9978	.9983	.9987	.9991	.9996
n	0	1	2	3	4	5	6	7	8	9

Table 4 Natural Logarithms

x	ln x	x	ln x	x	ln x
0.0		4.5	1.5041	9.0	2.1972
0.1	−2.3026	4.6	1.5261	9.1	2.2083
0.2	−1.6094	4.7	1.5476	9.2	2.2192
0.3	−1.2040	4.8	1.5686	9.3	2.2300
0.4	−0.9163	4.9	1.5892	9.4	2.2407
0.5	−0.6931	5.0	1.6094	9.5	2.2513
0.6	−0.5108	5.1	1.6292	9.6	2.2618
0.7	−0.3567	5.2	1.6487	9.7	2.2721
0.8	−0.2231	5.3	1.6677	9.8	2.2824
0.9	−0.1054	5.4	1.6864	9.9	2.2925
1.0	0.0000	5.5	1.7047	10	2.3026
1.1	0.0953	5.6	1.7228	11	2.3979
1.2	0.1823	5.7	1.7405	12	2.4849
1.3	0.2624	5.8	1.7579	13	2.5649
1.4	0.3365	5.9	1.7750	14	2.6391
1.5	0.4055	6.0	1.7918	15	2.7081
1.6	0.4700	6.1	1.8083	16	2.7726
1.7	0.5306	6.2	1.8245	17	2.8332
1.8	0.5878	6.3	1.8405	18	2.8904
1.9	0.6419	6.4	1.8563	19	2.9444
2.0	0.6931	6.5	1.8718	20	2.9957
2.1	0.7419	6.6	1.8871	25	3.2189
2.2	0.7885	6.7	1.9021	30	3.4012
2.3	0.8329	6.8	1.9169	35	3.5553
2.4	0.8755	6.9	1.9315	40	3.6889
2.5	0.9163	7.0	1.9459	45	3.8067
2.6	0.9555	7.1	1.9601	50	3.9120
2.7	0.9933	7.2	1.9741	55	4.0073
2.8	1.0296	7.3	1.9879	60	4.0943
2.9	1.0647	7.4	2.0015	65	4.1744
3.0	1.0986	7.5	2.0149	70	4.2485
3.1	1.1314	7.6	2.0281	75	4.3175
3.2	1.1632	7.7	2.0412	80	4.3820
3.3	1.1939	7.8	2.0541	85	4.4427
3.4	1.2238	7.9	2.0669	90	4.4998
3.5	1.2528	8.0	2.0794	95	4.5539
3.6	1.2809	8.1	2.0919	100	4.6052
3.7	1.3083	8.2	2.1041		
3.8	1.3350	8.3	2.1163		
3.9	1.3610	8.4	2.1281		
4.0	1.3863	8.5	2.1401		
4.1	1.4110	8.6	2.1518		
4.2	1.4351	8.7	2.1633		
4.3	1.4586	8.8	2.1748		
4.4	1.4816	8.9	2.1861		

Answers to Selected Exercises

CHAPTER 1

Section 1.1 (page 7)

1. 8 **3.** -5 **5.** -15 **7.** -13 **9.** 10 **11.** -5.39579 **13.** II **15.** III **17.** None **19.** None **21.** II **23.** IV **25.** $3\sqrt{2}$ **27.** $5\sqrt{2}$ **29.** $\sqrt{34}$ **31.** $\sqrt{29}$ **33.** 4 **35.** $\sqrt{133}$ **37.** 8.1480 **39.** Yes **41.** No **43.** No **45.** Yes **47.** No **49.** 5, -1 **51.** $9 + \sqrt{119}$, $9 - \sqrt{119}$ **53.** $x^2 + y^2 = 25$

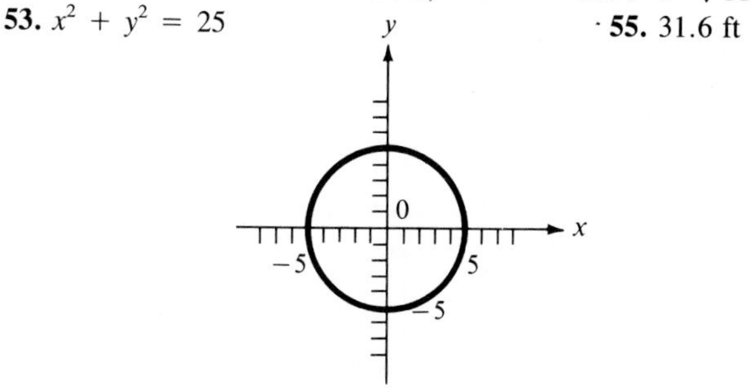

55. 31.6 ft **57.** 231.3 m **59.** 6 **61.** 0 **63.** -24 **65.** 7.994 **67.** $-2a^2 + 4a + 6$ **69.** $-2a^2 + 8$ **71.** $(-2, 11)$; $(-1, 8)$; $(0, 5)$; $(1, 2)$; $(2, -1)$; $(3, -4)$

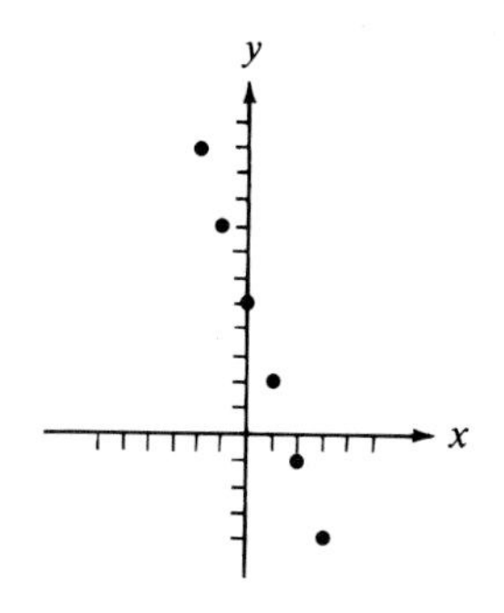

73. $(-2, -8)$; $(-1, -3)$; $(0, 0)$; $(1, 1)$; $(2, 0)$; $(3, -3)$

75. Domain: all reals; range: all reals; function **77.** Domain: all reals; range: $y \geq 4$; function **79.** Domain: all reals; range: $y \leq 4$; function **81.** Domain: $x \geq 0$; range: all reals; not a function **83.** Domain: $x \geq -4$; range: $y \geq 0$; function **85.** Domain: all reals; range: $y \geq 1$; function **87.** Domain: $-5 \leq x \leq 4$; range: $-2 \leq y \leq 6$; function **89.** Domain: all reals; range: $y \leq 12$; function **91.** Domain: $-4 \leq x \leq 4$; range: $-3 \leq y \leq 3$; not a function **93.** $x \neq 0$ **95.** $x \neq 7/3$, $x \neq -1/2$ **97.** All reals

Section 1.2 (page 18)

1. 320° **3.** 235° **5.** 90° **7.** 179° **9.** 130° **11.** 94.5937°

Answers other than the ones we give are possible in Exercises 13–27.

13. 435°; −285°; quadrant I

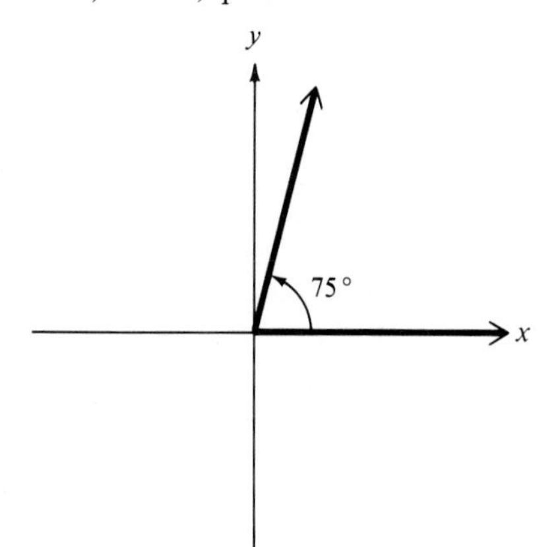

15. 482°; −238°; quadrant II

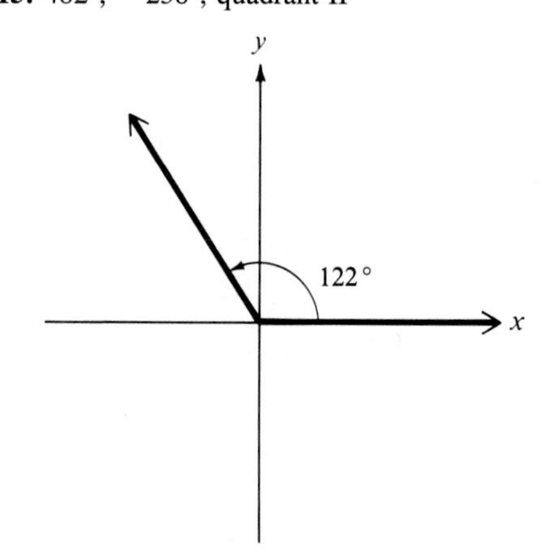

17. 594°; −126°; quadrant III

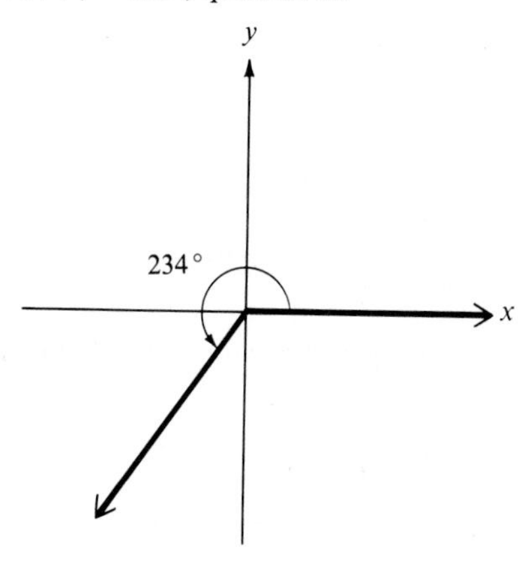

19. 660°; −60°; quadrant IV

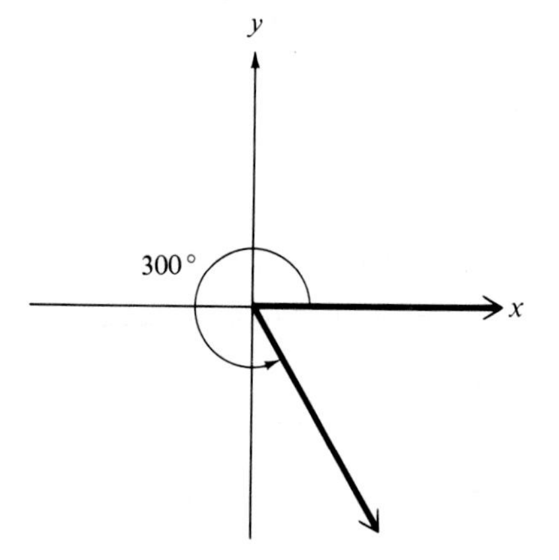

21. 78°; −282°; quadrant I

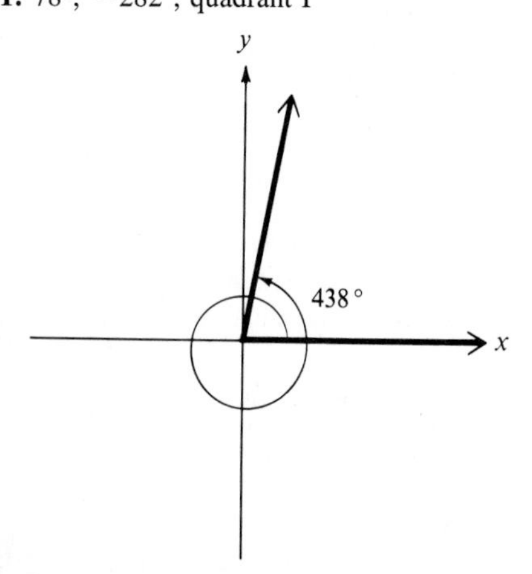

23. 152°; −208°; quadrant II

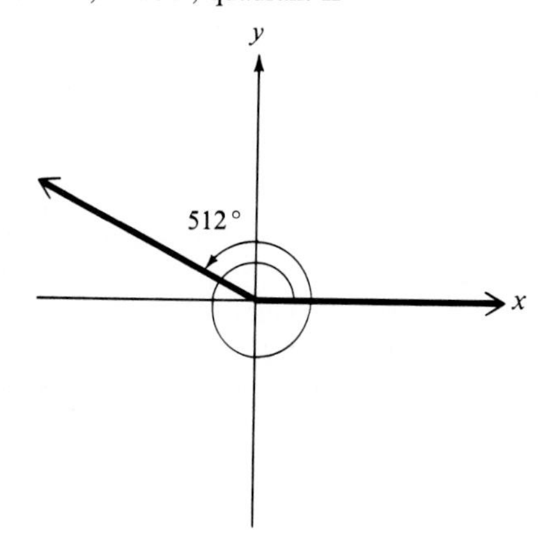

25. 308°; −412°; quadrant IV

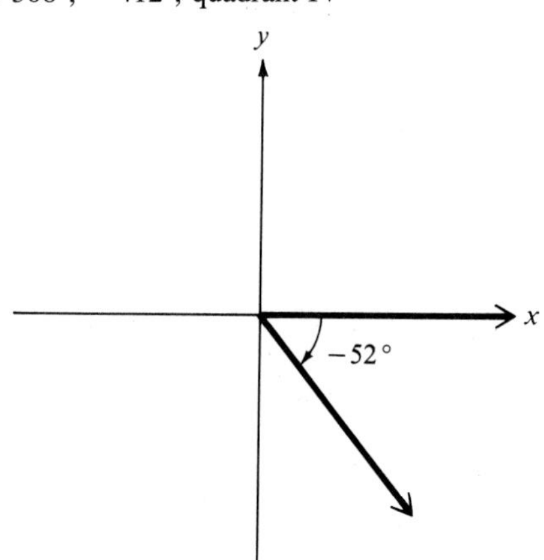

27. 201°; −519°; quadrant III

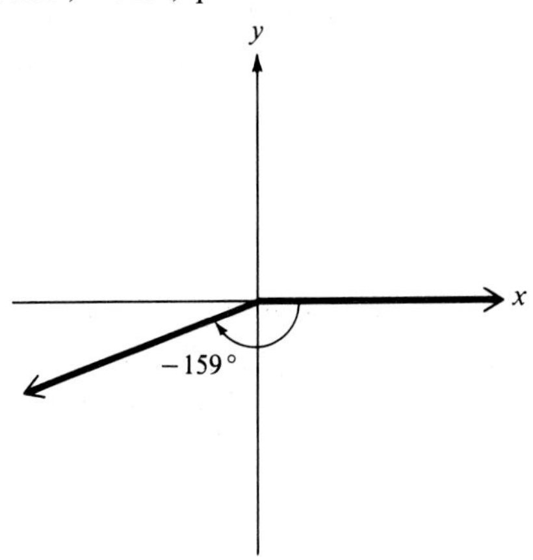

Your answers may differ a little from ours in Exercises 29–33.

29. 225° or −135°

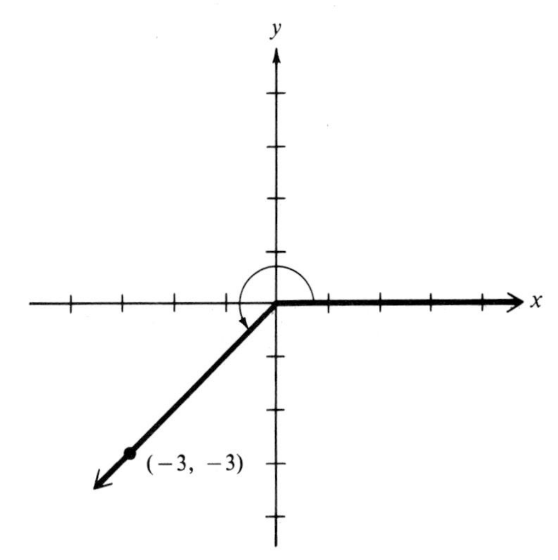

31. 239° or −121°

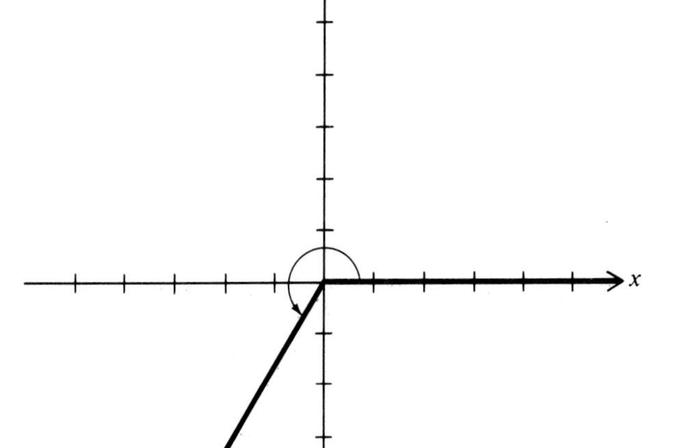

33. 120° or −240°

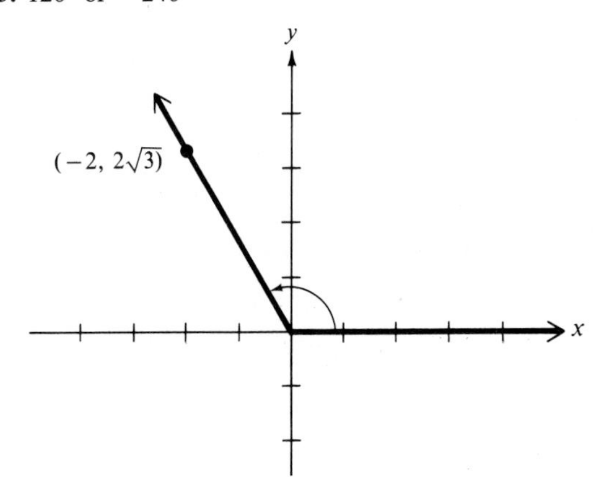

35. 70°; 110° **37.** 55°; 35° **39.** 100°; 80° **41.** 158° 47′ **43.** 112° 42′ **45.** 38° 32′ **47.** 27° 17′ **49.** 53° 41′ 13″ **51.** 59° 17′ 23″ **53.** 21° 48′ **55.** 2° 29′ **57.** 24° 48′ 56″ **59.** 20.9° **61.** 91.598° **63.** 274.316° **65.** 31° 25′ 47″ **67.** 89° 54′ 01″ **69.** 178° 35′ 58″ **71.** 1800° **73.** 12.5 rotations per hour **75.** $95° + n \cdot 360°$

Section 1.3 (page 25)

1. Acute; scalene **3.** Acute; equilateral **5.** Right; scalene **7.** Right; isosceles **9.** Obtuse; scalene **11.** Acute; isosceles **13.** A and P; C and R; B and Q; AC and PR; CB and RQ; AB and PQ **15.** M and Q; N and R; P and S; PN and SR; MN and RQ; MP and SQ **17.** H and F; K and E; HGK and FGE; HK and EF; GK and GE; HG and GF **19.** Similar **21.** Not similar **23.** Similar **25.** $P = 78°$; $M = 46°$; $A = N = 56°$ **27.** $T = 74°$; $Y = 28°$; $Z = W = 78°$ **29.** $T = 20°$; $V = 64°$; $R = U = 96°$ **31.** $a = 5$; $b = 3$ **33.** $a = 6$; $b = 7\ 1/2$ **35.** $x = 6$ **37.** 500 m, 700 m **39.** 30 m **41.** 110 **43.** 111.1 **45.** Missing side in first quadrilateral is 40 cm; missing sides in second quadrilateral are 27 cm and 36 cm **47.** $x = 10$; $y = 5$ **49.** $x = 20$; $y = 10$

Section 1.4 (page 33)

1. −3 **3.** −3 **5.** 5 **7.** 1 **9.** −1 **11.** 3 **13.** 1

In Exercises 15–43 we give, in order, sine, cosine, tangent, cotangent, secant, and cosecant. **15.** 4/5; −3/5; −4/3; −3/4; −5/3; 5/4 **17.** −12/13; 5/13; −12/5; −5/12; 13/5; −13/12 **19.** 4/5; 3/5; 4/3; 3/4; 5/3; 5/4 **21.** 24/25; −7/25; −24/7; −7/24; −25/7; 25/24 **23.** 1; 0; undefined; 0; undefined; 1 **25.** 0; 1; 0; undefined; 1; undefined **27.** $\sqrt{3}/2$; 1/2; $\sqrt{3}$; $\sqrt{3}/3$; 2; $2\sqrt{3}/3$ **29.** −1/2; $\sqrt{3}/2$; $-\sqrt{3}/3$; $-\sqrt{3}$; $2\sqrt{3}/3$; −2 **31.** $-\sqrt{2}/2$; $\sqrt{2}/2$; −1; −1; $\sqrt{2}$; $-\sqrt{2}$ **33.** −2/3; $\sqrt{5}/3$; $-2\sqrt{5}/5$; $-\sqrt{5}/2$; $3\sqrt{5}/5$; −3/2 **35.** $\sqrt{3}/4$; $-\sqrt{13}/4$; $-\sqrt{39}/13$; $-\sqrt{39}/3$; $-4\sqrt{13}/13$; $4\sqrt{3}/3$ **37.** $-\sqrt{10}/5$; $\sqrt{15}/5$; $-\sqrt{6}/3$; $-\sqrt{6}/2$; $\sqrt{15}/3$; $-\sqrt{10}/2$ **39.** −.34727; .93777; −.37031; −2.7004; 1.0664; −2.8796 **41.** −.5638; −.8259; .6826; 1:465; −1.211; −1.774 **43.** −.633; .774; −.818; −1.22; 1.29; −1.58 **45.** Both are positive **47.** Both are negative **49.** Positive **51.** Negative **53.** Positive **55.** Positive **57.** Negative **59.** Negative

Section 1.5 (page 39)

1. 1/3 **3.** −5 **5.** $2\sqrt{2}$ **7.** $-3\sqrt{5}/5$ **9.** .700692 **11.** 2.27789 **13.** 1/2 **15.** $\sqrt{3}$ **17.** −100 **19.** 5 **21.** 3 **23.** 4 **25.** II **27.** III **29.** IV **31.** II or IV **33.** I or II **35.** I or III

In Exercises 37–47 we give, in order, sine and cosecant, cosine and secant, tangent and cotangent.

37. +; +; + **39.** −; −; + **41.** −; +; − **43.** +; +; + **45.** −; +; − **47.** −; −; + **49.** Impossible **51.** Possible **53.** Impossible **55.** Possible **57.** Possible **59.** Impossible **61.** Possible **63.** Impossible **65.** Impossible **67.** $-\sqrt{5}/3$ **69.** $-\sqrt{5}/2$ **71.** −4/3 **73.** −.405092

In Exercises 75–87 we give, in order, sine, cosine, tangent, cotangent, secant, and cosecant. **75.** −4/5, −3/5; 4/3; 3/4; −5/3; −5/4 **77.** 7/25; −24/25; −7/24; −24/7; −25/24; 25/7 **79.** 1/2; $-\sqrt{3}/2$; $-\sqrt{3}/3$; $-\sqrt{3}$; $-2\sqrt{3}/3$; 2 **81.** $8\sqrt{67}/67$; $\sqrt{201}/67$; $8\sqrt{3}/3$; $\sqrt{3}/8$; $\sqrt{201}/3$; $\sqrt{67}/8$ **83.** $3\sqrt{13}/13$; $2\sqrt{13}/13$; 3/2; 2/3; $\sqrt{13}/2$; $\sqrt{13}/3$ **85.** .164215; −.986425; −.166475; −6.00691; −1.01376; 6.08958 **87.** a; $\sqrt{1 - a^2}$; $a\sqrt{1 - a^2}/(1 - a^2)$; $\sqrt{1 - a^2}/a$; $\sqrt{1 - a^2}/(1 - a^2)$; $1/a$

Chapter 1 Review Exercises (page 43)

1. −20 **3.** 5 **5.** $2\sqrt{11}$ **7.** 2 **9.** 3.207559 **11.** Domain: all reals; range: all reals; function **13.** Domain: all reals; range: $y \leq 3$; function **15.** Domain: $x \geq -1$; range: all reals; not a function **17.** Domain: all reals; range: $y \geq 1$; function **19.** 309° **21.** 72° **23.** 1280°

25. 47.420° **27.** 74° 17′ 54″ **29.** 183° 05′ 50′ **31.** $V = 41°$; $Z = 32°$; $Y = U = 107°$ **33.** $N = 12°$; $R = 82°$; $M = 86°$ **35.** $m = 45$; $n = 60$
37. $r = 108/7 \approx 15.43$ **39.** 5/2, 5 **41.** $-\sqrt{2}/2$; $-\sqrt{2}/2$; 1 **43.** 0; -1; 0
In Exercises 45, 47, 49, 59, and 61 we give, in order, sine, cosine, tangent, cotangent, secant, and cosecant.
45. 15/17; $-8/17$; $-15/8$; $-8/15$; $-17/8$; 17/15 **47.** $-5\sqrt{26}/26$; $\sqrt{26}/26$; -5; $-1/5$; $\sqrt{26}$; $-\sqrt{26}/5$
49. $-1/2$; $\sqrt{3}/2$; $-\sqrt{3}/3$; $-\sqrt{3}$; $2\sqrt{3}/3$; -2 **51.** -6 **53.** -3 **55.** Possible
57. Possible **59.** $\sqrt{3}/5$; $-\sqrt{22}/5$; $-\sqrt{66}/22$; $-\sqrt{66}/3$; $-5\sqrt{22}/22$; $5\sqrt{3}/3$ **61.** $-2\sqrt{5}/5$; $-\sqrt{5}/5$; 2; 1/2; $-\sqrt{5}$; $-\sqrt{5}/2$ **63.** $-\sqrt{21}/5$

CHAPTER 2

Section 2.1 (page 50)

In Exercises 1–9 we give, in order, sine, cosine, tangent, cotangent, secant, and cosecant.
1. 3/5; 4/5; 3/4; 4/3; 5/4; 5/3 **3.** 21/29; 20/29; 21/20; 20/21; 29/20; 29/21 **5.** n/p; m/p; n/m; m/n; p/m; p/n **7.** .7593; .6508; 1.1667; .8571; 1.5365; 1.3170 **9.** .8717; .4901; 1.7785; .5623; 2.0403; 1.1472 **11.** cot 40° **13.** sec 43° **15.** sin 37° 11′ **17.** cot 64° 17′
19. cos $(90° - \gamma)$ **21.** csc $(90° - 2A)$ **23.** sin $(70° - \alpha)$ **25.** 30° **27.** 20°
29. 12° **31.** 8° **33.** 70° **35.** True **37.** True **39.** False **41.** True
43. True

Section 2.2 (page 58)

In Exercises 1–19 we give, in order, sine, cosine, tangent, cotangent, secant, and cosecant. **1.** $\sqrt{3}/2$; $-1/2$; $-\sqrt{3}$; $-\sqrt{3}/3$; -2; $2\sqrt{3}/3$ **3.** 1/2; $-\sqrt{3}/2$; $-\sqrt{3}/3$; $-\sqrt{3}$; $-2\sqrt{3}/3$; 2 **5.** $-\sqrt{3}/2$; $-1/2$; $\sqrt{3}$; $\sqrt{3}/3$; -2; $-2\sqrt{3}/3$ **7.** $-1/2$; $\sqrt{3}/2$; $-\sqrt{3}/3$; $-\sqrt{3}$; $2\sqrt{3}/3$; -2 **9.** $\sqrt{3}/2$; 1/2; $\sqrt{3}$; $\sqrt{3}/3$; 2; $2\sqrt{3}/3$ **11.** 1/2; $-\sqrt{3}/2$; $-\sqrt{3}/3$; $-\sqrt{3}$; $-2\sqrt{3}/3$; 2 **13.** 1/2; $\sqrt{3}/2$; $\sqrt{3}/3$; $\sqrt{3}$; $2\sqrt{3}/3$; 2 **15.** $\sqrt{3}/2$; 1/2; $\sqrt{3}$; $\sqrt{3}/3$; 2; $2\sqrt{3}/3$ **17.** $-1/2$; $\sqrt{3}/2$; $-\sqrt{3}/3$; $-\sqrt{3}$; $2\sqrt{3}/3$; -2
19. $\sqrt{3}/2$; 1/2; $\sqrt{3}$; $\sqrt{3}/3$; 2; $2\sqrt{3}/3$ **21.** $\sqrt{3}/3$; $\sqrt{3}$ **23.** $\sqrt{3}/2$; $\sqrt{3}/3$; $2\sqrt{3}/3$ **25.** -1; -1
27. $-\sqrt{3}/2$; $-2\sqrt{3}/3$ **29.** 1 **31.** 23/4 **33.** $1/2 + \sqrt{3}$ **35.** $-29/12$ **37.** $-\sqrt{3}/3$
39. False **41.** True **43.** False **45.** True **47.** True **49.** 30°; 150° **51.** 60°; 240° **53.** 120°; 240° **55.** 240°; 300° **57.** 135°; 315° **59.** 90° 270° **61.** 0°; 180°
63. .70710678 **65.** $a = 12$; $b = 12\sqrt{3}$; $d = 12\sqrt{3}$; $c = 12\sqrt{6}$ **67.** $m = 7\sqrt{3}/3$; $a = 14\sqrt{3}/3$; $n = 14\sqrt{3}/3$; $q = 14\sqrt{6}/3$ **69.** $s^2\sqrt{3}/4$

Section 2.3 (page 69)

1. 225° **3.** 123° **5.** 98° 20′ **7.** 18° 30′ **9.** 109° 30′ **11.** 161° 50′ **13.** 82°
15. 32° **17.** 74° 30′ **19.** 29° 10′ **21.** 1° 21′ **23.** 65° 13′ **25.** 35° 46′
27. 69° 50′ **29.** 34° 30′ **31.** 39° 12′ **33.** .6248 **35.** 1.137 **37.** .8526
39. .3121 **41.** $-.7954$ **43.** .0116 **45.** $-.4120$ **47.** $-.3228$ **49.** $-.9969$
51. 1.142 **53.** -3.179 **55.** $-.3217$ **57.** 58° 00′ **59.** 30° 30′ **61.** 46° 10′
63. 81° 10′ **65.** .86288 **67.** -6.04837 **69.** $-.20092$ **71.** $-.35831$ **73.** .27256
75. $-.72307$ **77.** $-.26676$ **79.** -1.89263 **81.** -1 **83.** 1
85. 2×10^8 m per sec **87.** 19° **89.** 48.7°

Section 2.4 (page 74)

1. 4.5 to 5.5 lb **3.** 9.55 to 9.65 tons **5.** 8.945 to 8.955 m **7.** 19.65 to 19.75 liters
9. 253.7405 to 253.7415 m **11.** .015 to .025 ft **13.** 28,999.5 to 29,000.5 ft **15.** 3
17. 4 **19.** 4 **21.** 2 **23.** 5 **25.** 3 **27.** 769; 7.7×10^2 **29.** 12.5; 13
31. 1.50×10^2; 1.5×10^2 **33.** 9.00; 9.0 **35.** 7.13; 7.1 **37.** 11.6; 12 **41.** False
43. True **45.** False **47.** False **49.** False **51.** True **53.** 76.42 **55.** 1.91
57. 28,300 or 2.83×10^4 **59.** .10 **61.** 8.6423 **63.** -4.39 **65.** .5738 **67.** 369
69. 4.86×10^7 **71.** 1.29×10^{-4} **73.** -6.04×10^{11}

Section 2.5 (page 80)

1. $B = 53° 40'$; $a = 571$ m; $b = 777$ m **3.** $M = 38.8°$; $n = 154$ m; $p = 198$ m **5.** $A = 47.9108°$; $c = 84.816$ cm; $a = 62.942$ cm **7.** $B = 62°$; $C = 90°$; $a = 8.17$ ft; $b = 15.4$ ft **9.** $A = 17°$; $C = 90°$; $a = 39.1$ in; $c = 134$ in **11.** $c = 85.9$ yd; $A = 62° 50'$; $B = 27° 10'$; $C = 90°$ **13.** $b = 42.3$ cm; $A = 24° 10'$; $B = 65° 50'$; $C = 90°$ **15.** $B = 36° 36'$; $C = 90°$; $a = 310.8$ ft; $b = 230.8$ ft **17.** $A = 50° 51'$; $C = 90°$; $a = .4832$ m; $b = .3934$ m **19.** $A = 71° 36'$; $B = 18° 24'$; $a = 7.413$ m **21.** $A = 47.568°$; $b = 143.97$ m; $c = 213.38$ m **23.** $B = 32.791°$; $a = 156.77$ cm; $b = 101.00$ cm **25.** $a = 115.072$ m; $A = 33.4901°$; $B = 56.5099°$ **27.** 26.6 m **29.** 52° 30′ **31.** 19.46 ft **33.** 45.7 m **35.** 28.94 in **37.** 11.1 ft **39.** 35° 50′ **41.** 40,600 ft **43.** 26° 20′ **45.** 54° 40′ **47.** 6.993792×10^9 mi

Section 2.6 (page 88)

1. 31° 20′ **3.** 3.3 ft **5.** 13.4 ft **7.** 84.7 ft **9.** 59.8 m **11.** 446 **13.** 114 ft **15.** 5.18 m **17.** 70 m per sec **19.** 320 mi **21.** 7.49 mi **23.** 156 mi **25.** 120 mi

Appendix (page 94)

1. .5704 **3.** 1.162 **5.** 2.171 **7.** .9052 **9.** .9686 **11.** .3404 **13.** .1951 **15.** 1.478 **17.** 1.562 **19.** .8121 **21.** 35° 44′ **23.** 25° 53′ **25.** 21° 48′ **27.** 23° 08′ **29.** 73° 27′

Chapter 2 Review Exercises (page 97)

In Exercises 1–7 we give, in order, sine, cosine, tangent, cotangent, secant, and cosecant.

1. 60/61; 11/61; 60/11; 11/60; 61/11; 61/60 **3.** $\sqrt{3}/2$; $-1/2$; $-\sqrt{3}$; $-\sqrt{3}/3$; -2; $2\sqrt{3}/3$ **5.** $-\sqrt{3}/2$; $1/2$; $-\sqrt{3}$; $-\sqrt{3}/3$; 2; $-2\sqrt{3}/3$ **7.** $1/2$; $-\sqrt{3}/2$; $-\sqrt{3}/3$; $-\sqrt{3}$; $-2\sqrt{3}/3$; 2 **9.** 10° **11.** 7° **13.** True **15.** True **17.** 1 **19.** $-5/3$ **21.** False **23.** True **25.** True **27.** .9537 **29.** $-.7159$ **31.** 1.558 **33.** 1.93621 **35.** .999849 **37.** 37° 40′ **39.** 38.94° **41.** 55° 40′ **43.** 12° 44′ **45.** 63° 00′ **47.** 5 **49.** 3 **51.** 976 **53.** 44.0 **55.** $B = 31° 30'$; $a = 638$; $b = 391$ **57.** $B = 50.28°$; $a = 32.38$ m; $c = 50.66$ m **59.** 82.6 ft **61.** 18.75 cm **63.** 1200 m **65.** 140 mi

CHAPTER 3

Section 3.1 (page 106)

1. $\pi/3$ **3.** $\pi/2$ **5.** $5\pi/6$ **7.** $7\pi/6$ **9.** $5\pi/3$ **11.** $5\pi/2$ **13.** $\pi/9$ **15.** $7\pi/9$ **17.** .68 **19.** .742 **21.** 2.43 **23.** 1.122 **25.** .9847 **27.** .832391 **29.** $-.518537$ **31.** -3.66119 **33.** 60° **35.** 315° **37.** 330° **39.** $-30°$ **41.** 288° **43.** 132° **45.** 63° **47.** 66° **49.** 114.5916°; 114° 35′ **51.** 99.6947°; 99° 42′ **53.** 5.2254°; 5° 14′ **55.** 564.228°; 564° 14′ **57.** $-198.925°$; $-198° 55'$ **59.** $-240.531°$; $-240° 32'$ **61.** $\sqrt{3}/2$ **63.** 1 **65.** $2\sqrt{3}/3$ **67.** 1 **69.** $-\sqrt{3}$ **71.** 1/2 **73.** -1 **75.** $-\sqrt{3}/2$ **77.** 1 **79.** 0 **81.** $-\sqrt{3}$ **83.** 1/2 **85.** We begin the answers with the blank next to 30°, and then proceed counterclockwise from there: $\pi/6$; 45°; $\pi/3$; $\pi/2$; 120°; 135°; $5\pi/6$; π; $7\pi/6$; $5\pi/4$; 240°; 300°; $7\pi/4$; $11\pi/6$ **87.** 4π, $2\pi/3$ **89.** $\pi/2$ **91.** π **93.** $2\pi/3$

Section 3.2 (page 113)

1. 25.1 in **3.** 25.8 cm **5.** 318 m **7.** 5.05 m **9.** 53.429 m **11.** 1600 km **13.** 1200 km **15.** 3500 km **17.** 5900 km **19.** 7300 km **21.** **(a)** 11.6 in **(b)** 37° 05′ **23.** 38° 30′ **25.** 146 in **27.** 21 m **29.** .20 km **31.** 2100 mi **33.** 850 ft **35.** 42 m^2 **37.** 1120 m^2 **39.** 1300 cm^2 **41.** 114 cm^2 **43.** 359 m^2 **45.** 760.34 m^2 **47.** **(a)** 11.25° or 11° 15′; $\pi/16$ **(b)** 480 ft **(c)** 15 ft **(d)** 570 ft^2 **51.** 282.488 cm^2 **53.** 164.964 mm^2 **55.** $V = r^2\theta h/2$ (θ in radians)

Section 3.3 (page 123)

Answers may differ in the last digit depending on whether a table or a calculator was used. **1.** .4877 **3.** .5317 **5.** .7314 **7.** .8004 **9.** .9981 **11.** 1.017 **13.** .9636 **15.** .4245 **17.** 1.213 **19.** $-.4436$ **21.** $-.7547$ **23.** $-.9967$ **25.** -3.867 **27.** $-.3090$ **29.** .3173 **31.** 2.282 **33.** 14.33 **35.** 4.636 **37.** .2095 **39.** 1.4428 **41.** 1.0152 **43.** .9541 **45.** 1.4748 **47.** 1.0181 **49.** .5405 **51.** .3093 **53.** .9320 **55.** .2675 **57.** .9093 **59.** .9160 **61.** III **63.** IV **65.** **(a)** 30° **(b)** 60° **(c)** 75° **(d)** 86° **(f)** 60° **69.** .8417 **71.** .1801 **73.** .0100

Section 3.4 (page 128)

1. $5\pi/4$ radians **3.** $\pi/25$ radians/sec **5.** 9 min **7.** 10.768 radians **9.** $72\pi/5$ cm/sec **11.** 6 radians/sec **13.** 9.29755 cm/sec **15.** 18π cm **17.** 12 sec **19.** $3\pi/32$ radians/sec **21.** 24.8647 cm **23.** $\pi/6$ radians/hr **25.** $\pi/30$ radians/sec **27.** $7\pi/30$ cm/min **29.** 168π m/min **31.** 1500π m/min **33.** About 29 sec **35.** **(a)** 2π radians/day; $\pi/12$ radians/hr **(b)** 0 **(c)** $12{,}800\pi$ km/day or 533π km/hr **(d)** 9050π km/day or 377π km/hr **37.** .24 radians/sec **39.** .303 m

Chapter 3 Review Exercises (page 133)

1. $\pi/4$ **3.** $4\pi/9$ **5.** $11\pi/6$ **7.** $17\pi/3$ **9.** 225° **11.** 480° **13.** $-110°$ **15.** 168° **17.** $\sqrt{3}$ **19.** $-1/2$ **21.** $-\sqrt{3}$ **23.** 2 **25.** 35.8 cm **27.** 7.683 cm **29.** 273 m^2 **31.** 4500 km **33.** 1.0999 **35.** .7269 **37.** .8660 **39.** .9703 **41.** 1.675 **43.** $-.5048$ **45.** .3898 **47.** .5148 **49.** 1.105 **51.** 15/32 sec **53.** $\pi/20$ rad/sec **55.** 285.3 cm

CHAPTER 4

Section 4.1 (page 145)

1. 2

3. 2/3

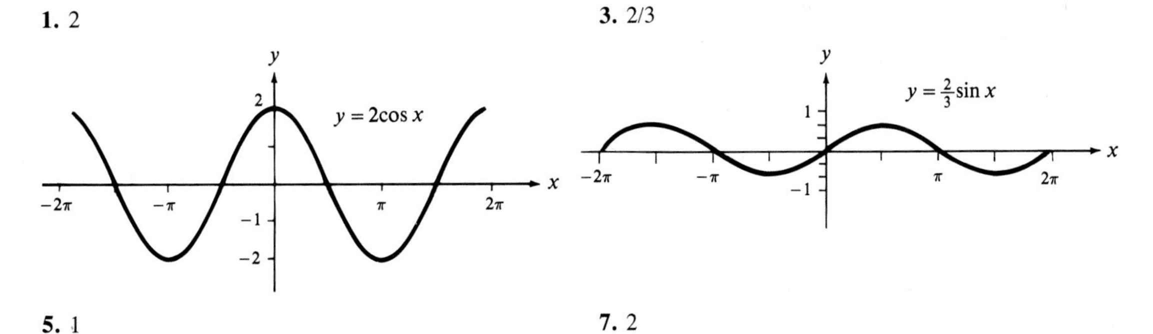

5. 1

7. 2

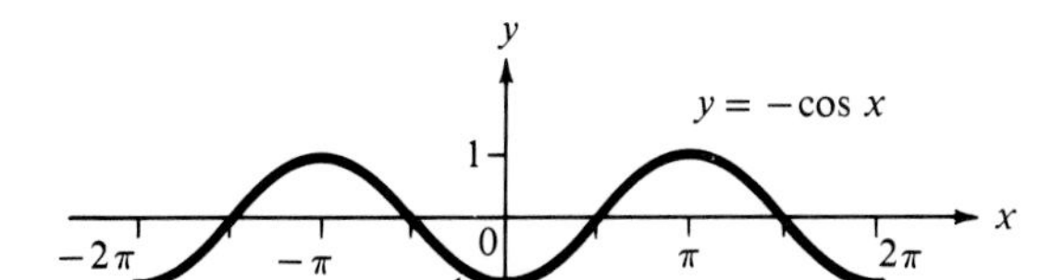

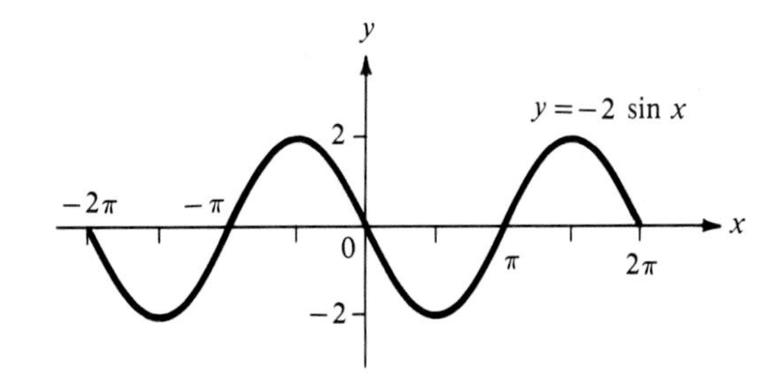

9. 4π; 1; none

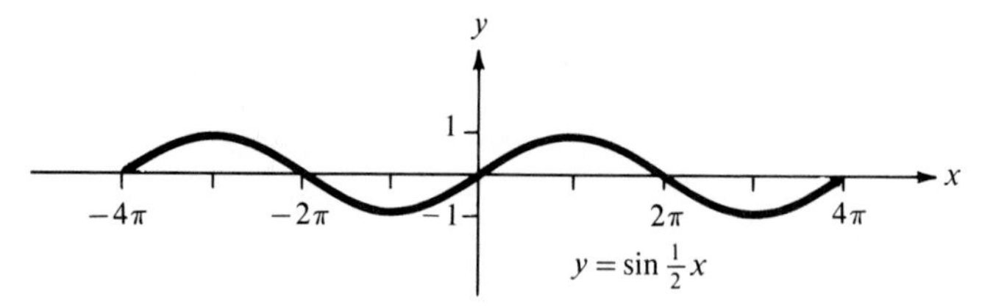

11. 6π; 1; none

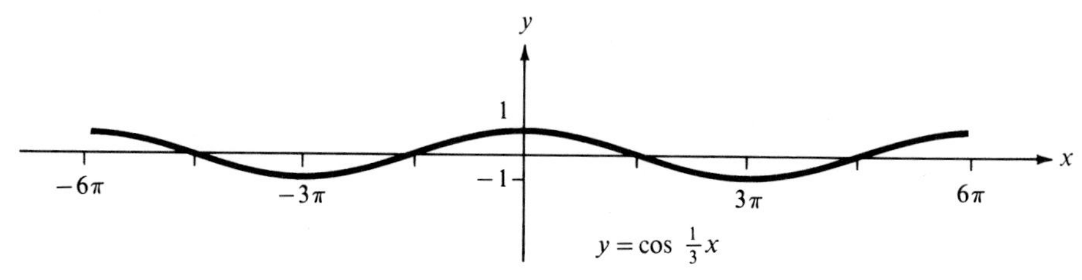

13. $2\pi/3$; 1; none

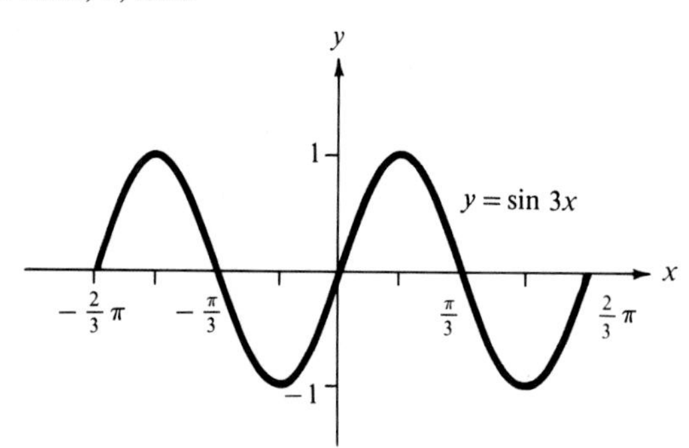

15. π; 1; none

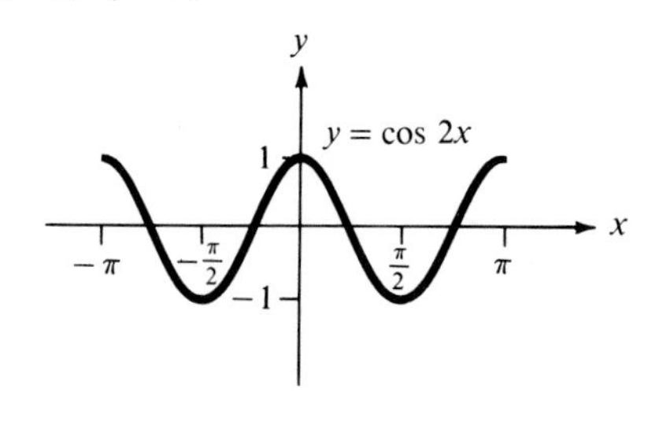

17. $\pi/2$; 1; none

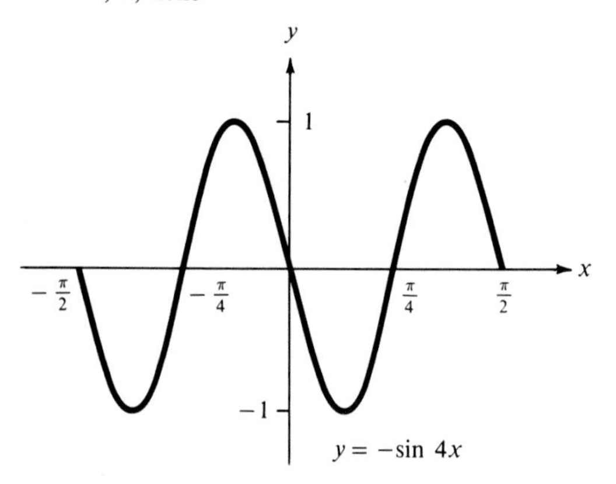

19. 8π; 2; none

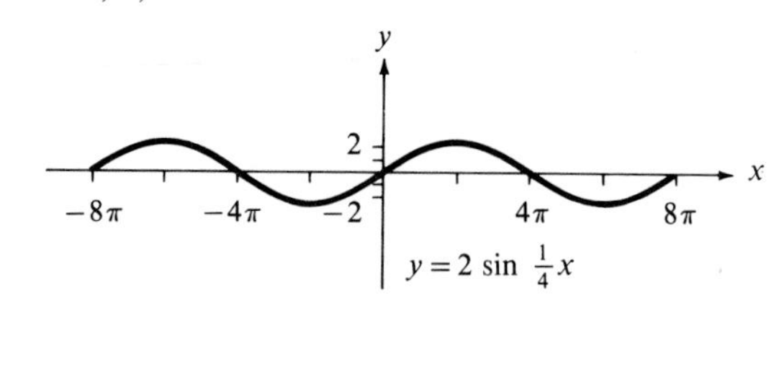

21. $2\pi/3$; 2; none

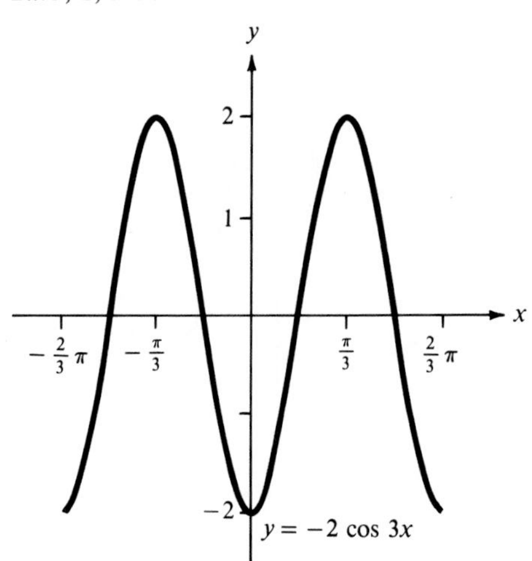

23. 2π; 2; -3

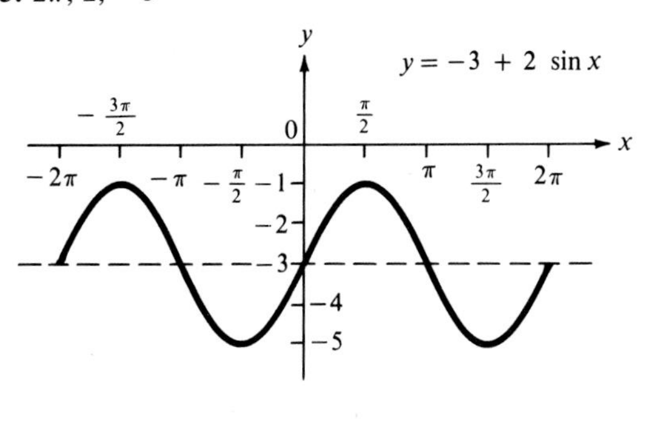

25. $8\pi/3$; 2/3; 1

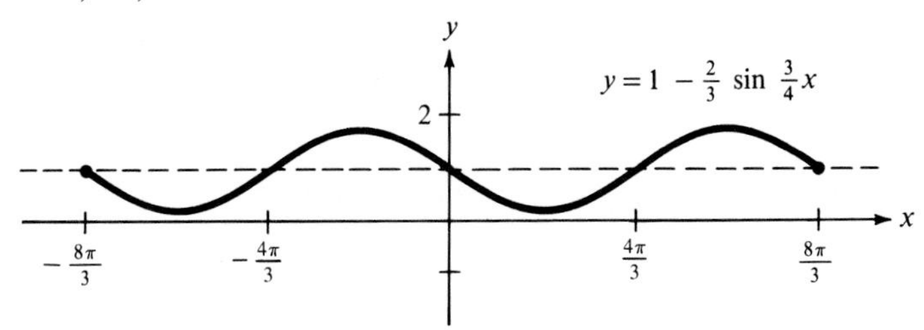

27. 2π; 1; 2

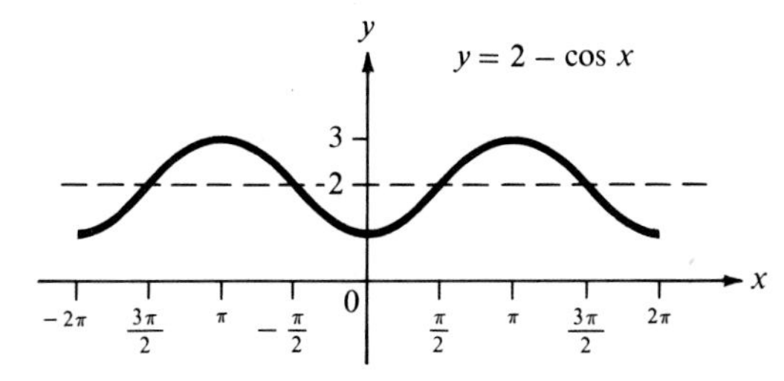

29. 4π; 2;1

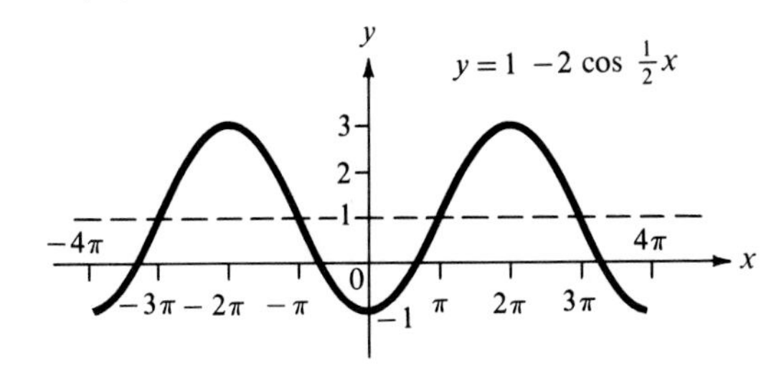

31. $2\pi/3$; $1/2$; -2

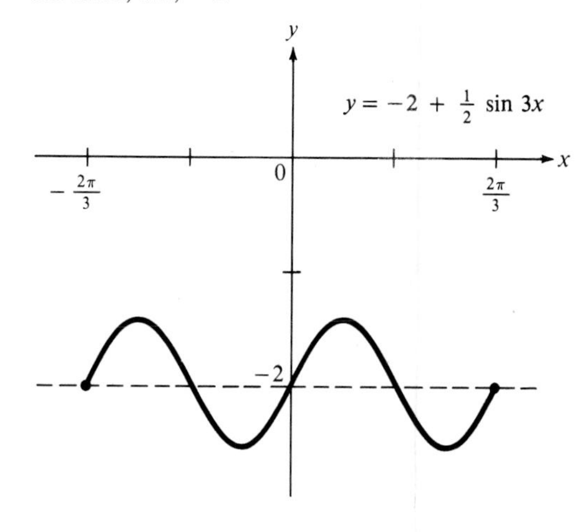

33. 2; 1; none

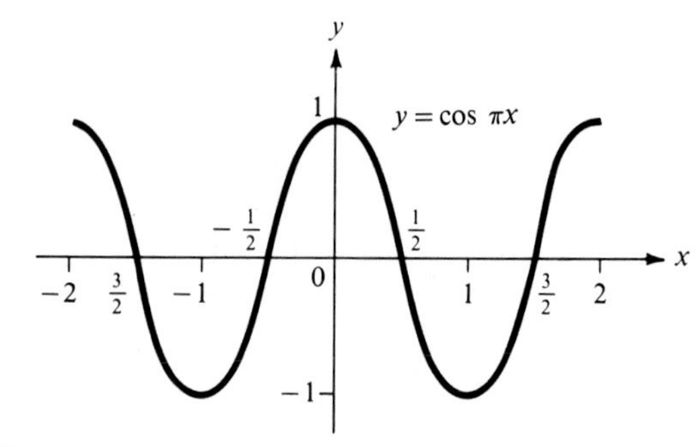

35.

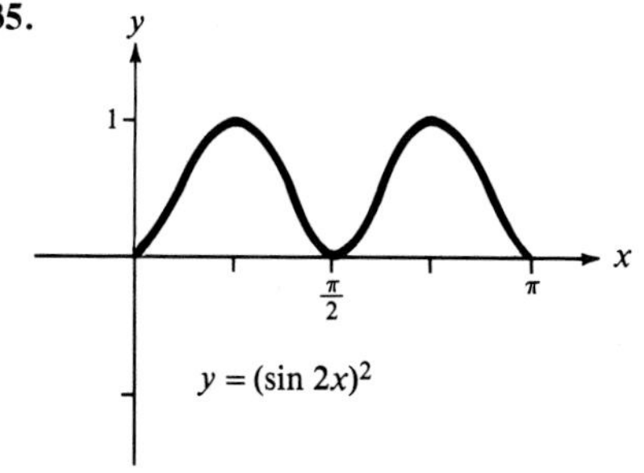

37.

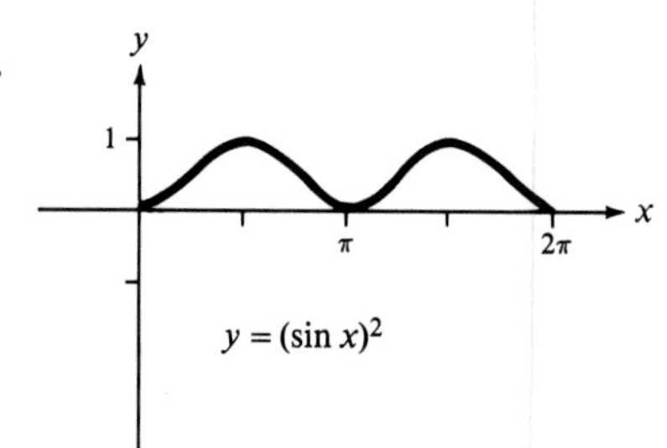

39. **(a)** 20 **(b)** 75 **41.** **(a)** About 7/4 hr **(b)** 1 yr

43. 1; $4\pi/3$

45. **(a)** 5; 1/60 **(b)** 60 **(c)** 5; 1.545; −4.045; −4.045; 1.545

(d)

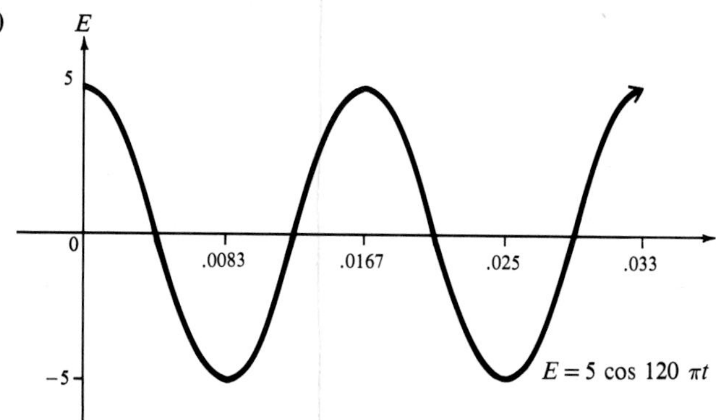

Section 4.2 (page 155)

1. 2; 2π; none; π to the right **3.** 4; 4π; none; π to the left **5.** 3; π; none; $\pi/4$ to the right

7. 1; $2\pi/3$; up 2; $\pi/15$ to the right

9.

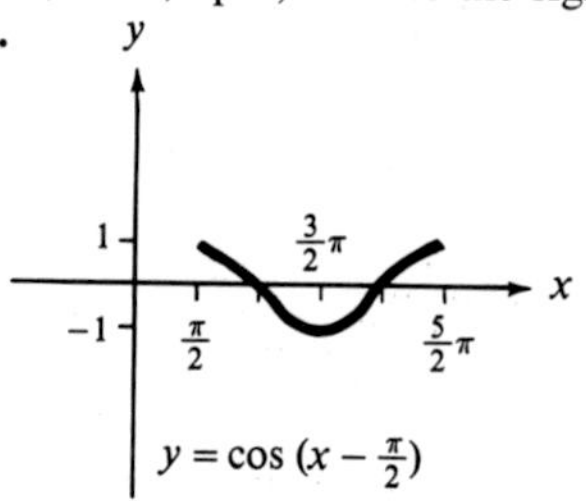

11.

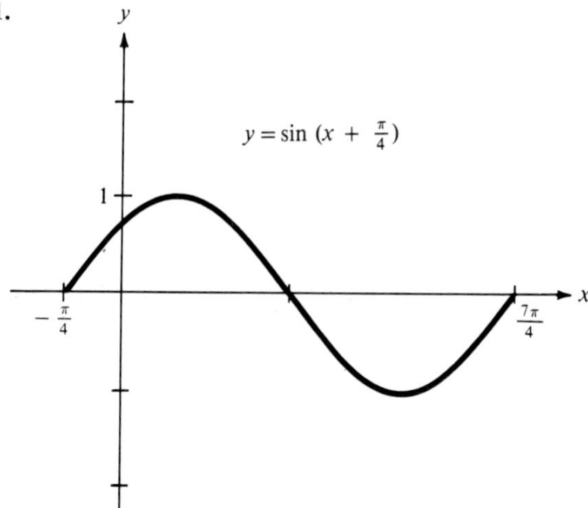

13.

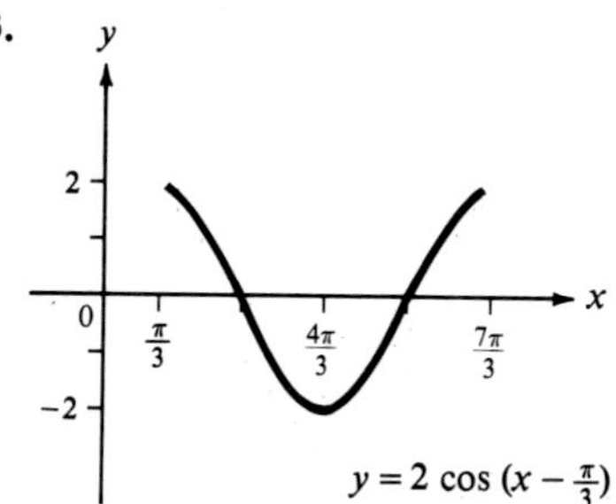

15.

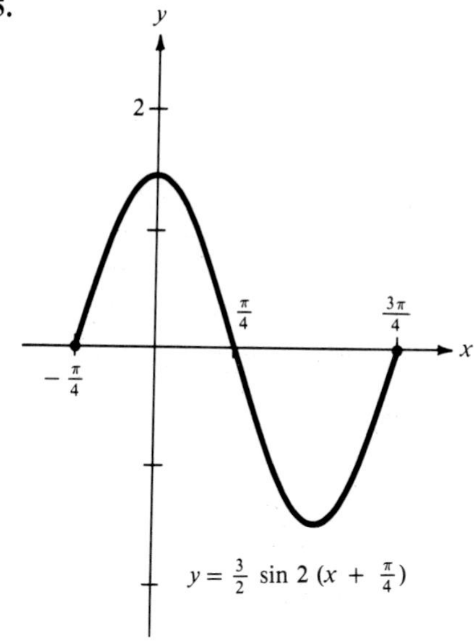

17.

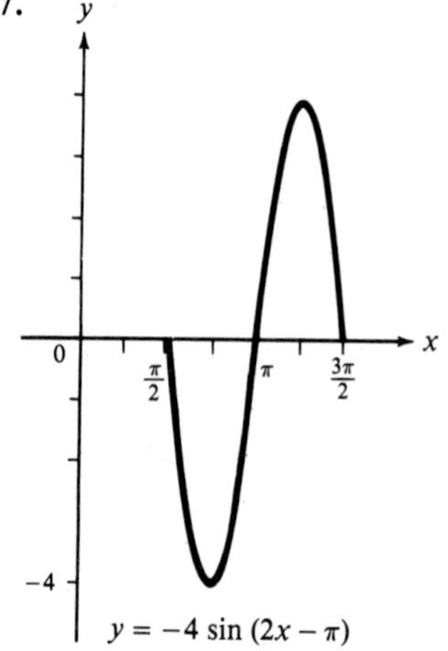

19.

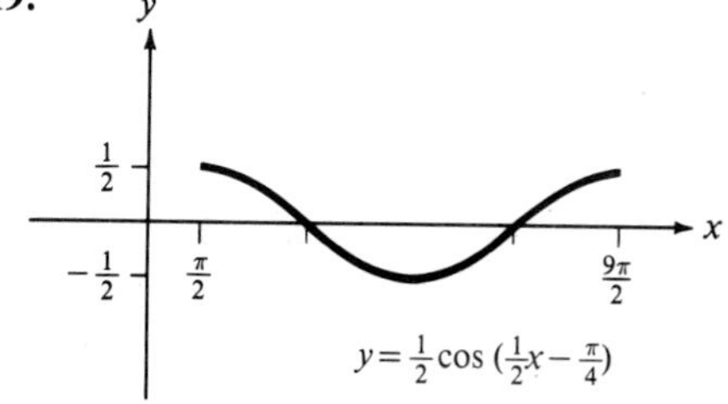

21.

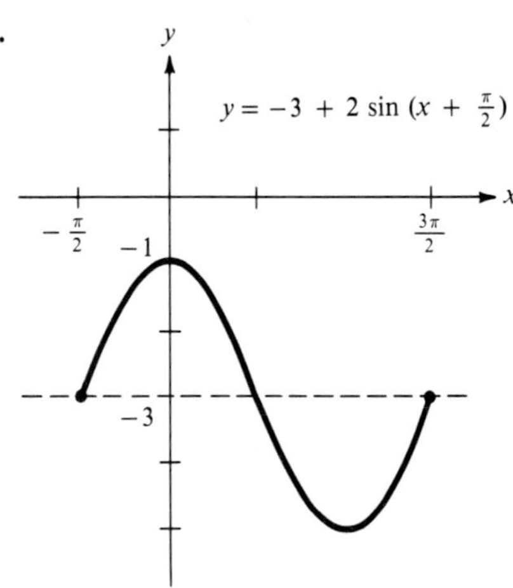

23.

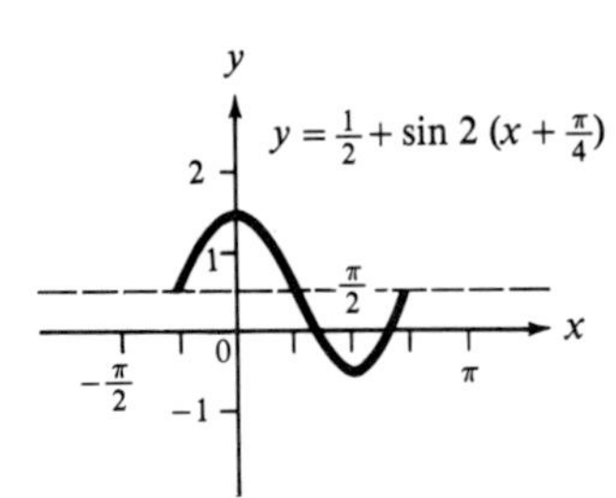

25.

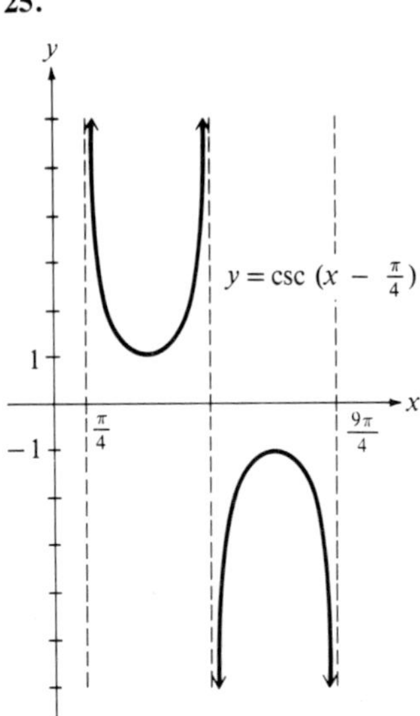

27.

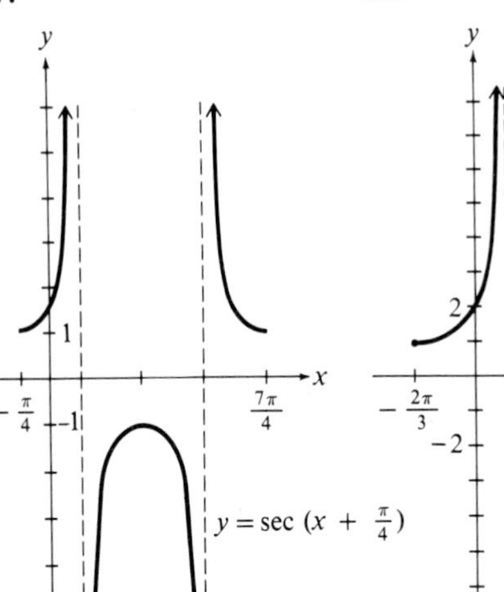

29.

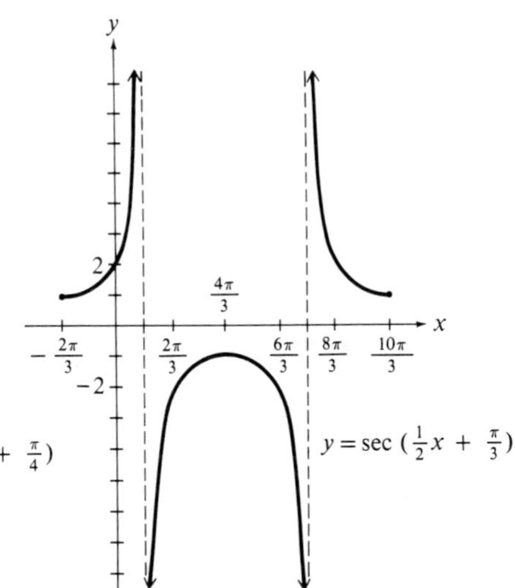

31.

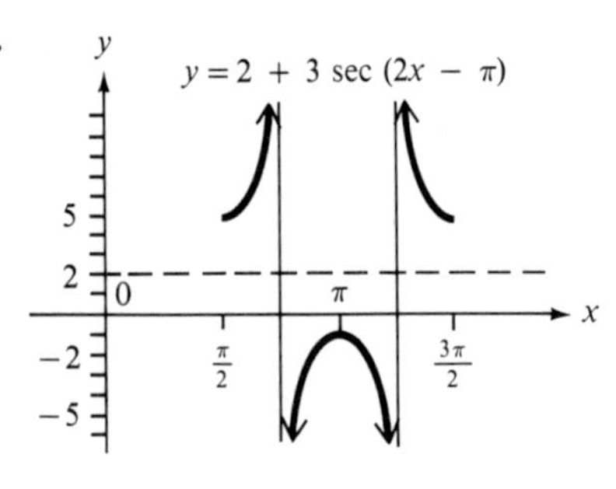

33.

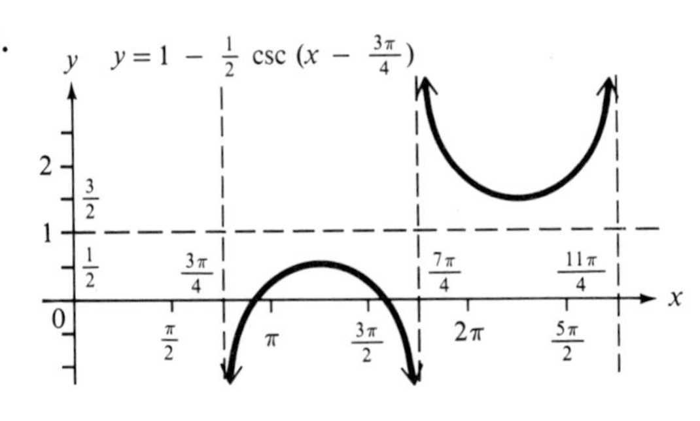

35.

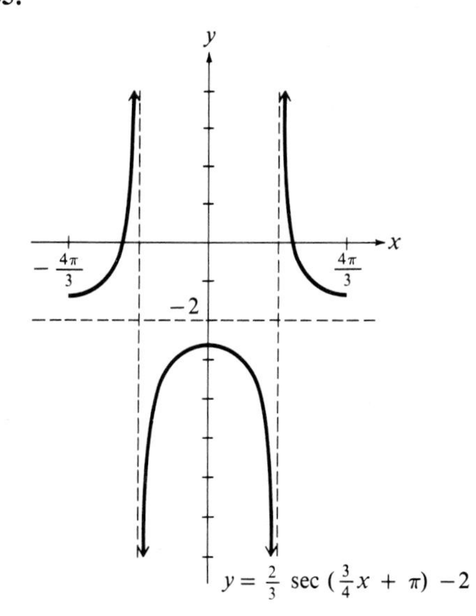

Section 4.3 (page 161)

1.

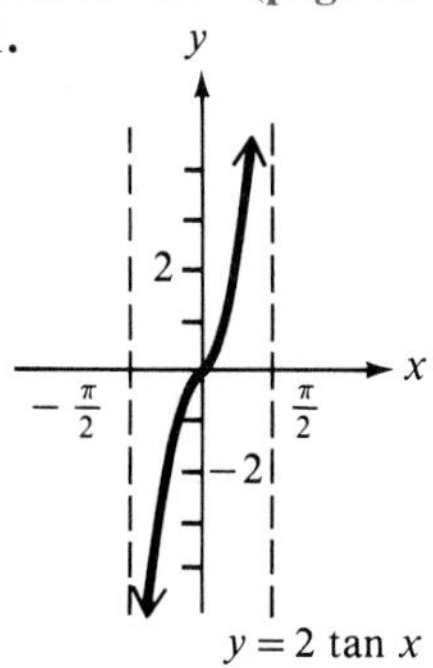

3.

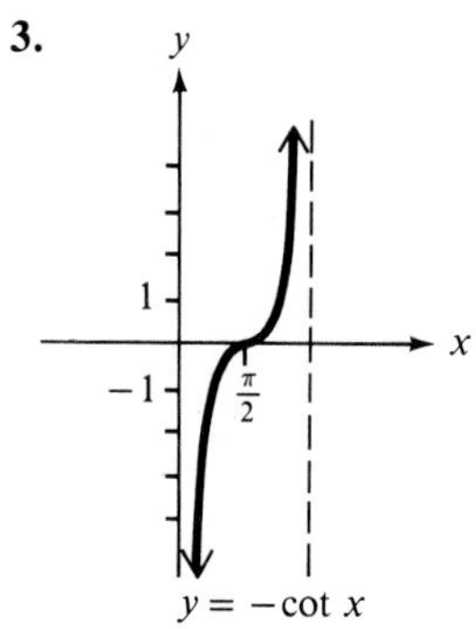

5.

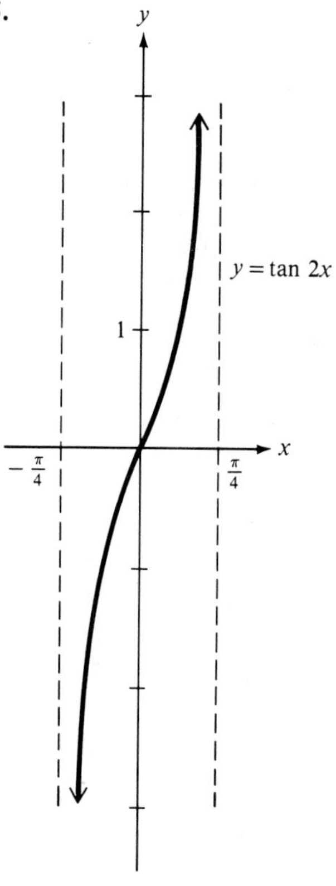

7.

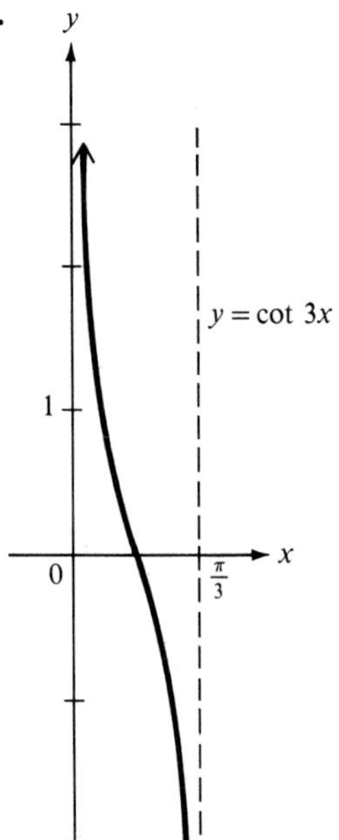

9. π

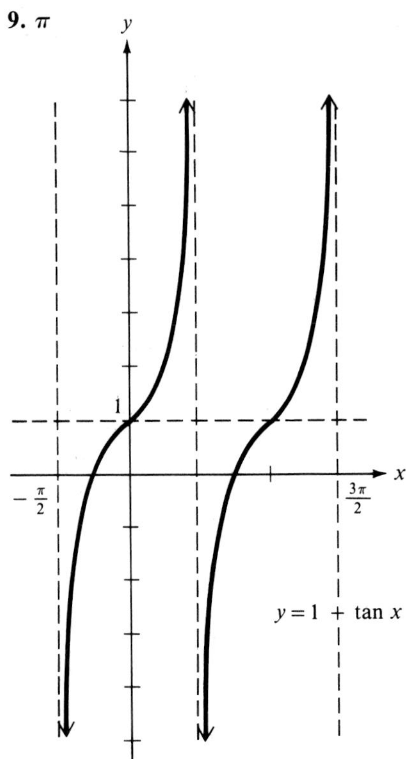

11. π

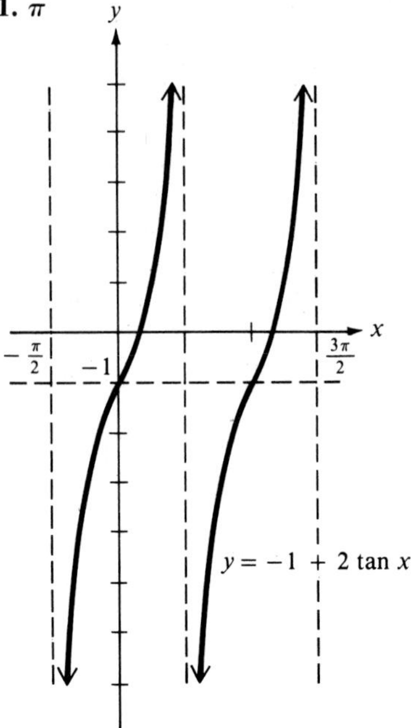

13. π

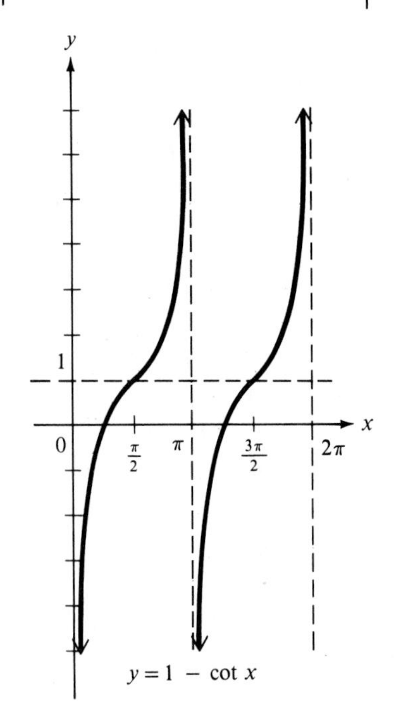

15. $\pi/2$

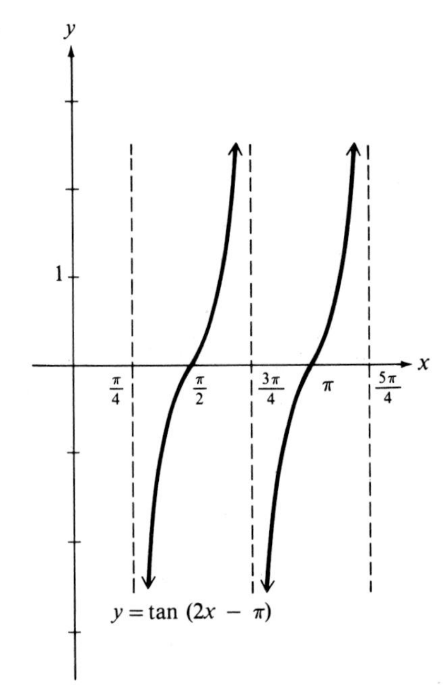

17. $\pi/3$

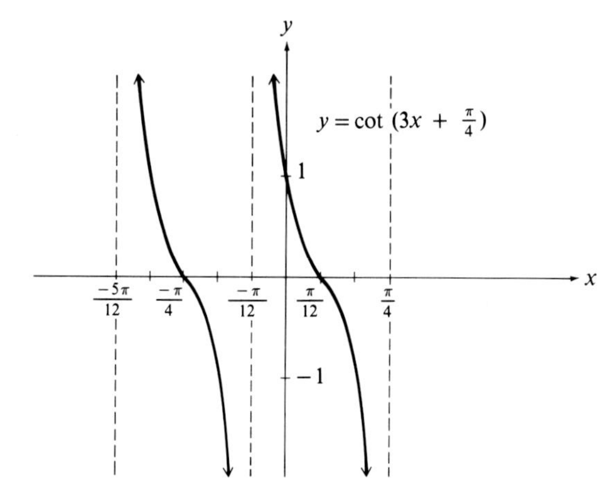

19. $\pi/2$

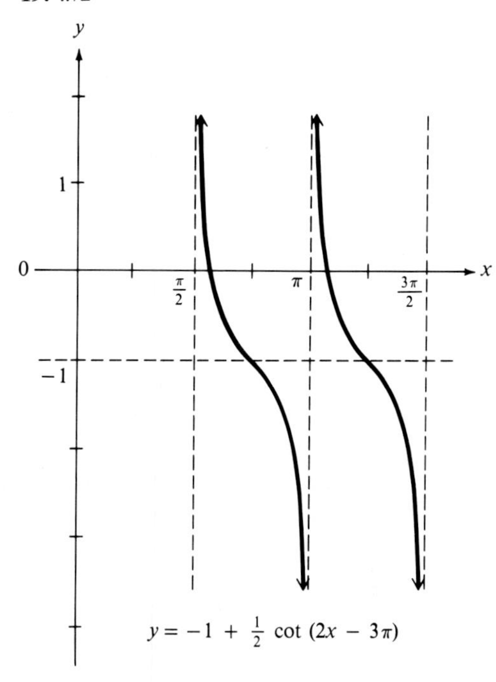

21. $4\pi/3$

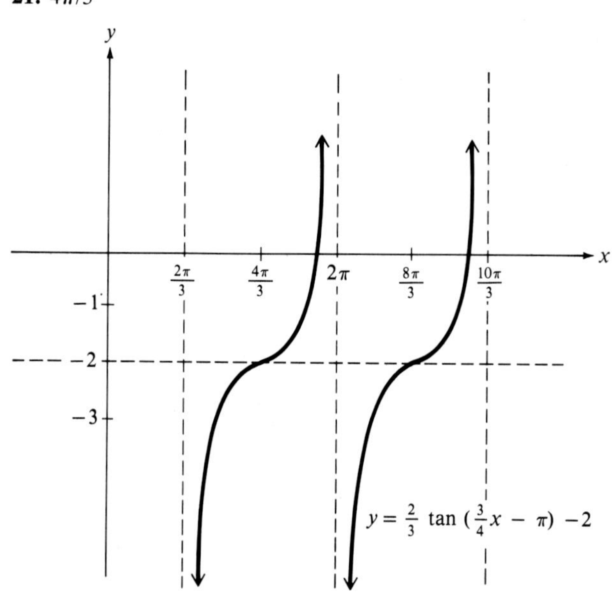

23. **(a)** 0 **(b)** −2.9 m **(c)** −12.3 m **(d)** 12.3 m **(e)** It leads to $\tan \pi/2$, which does not exist. **(f)** All t except $.25\,k$, where k is an integer

25. **(b)**

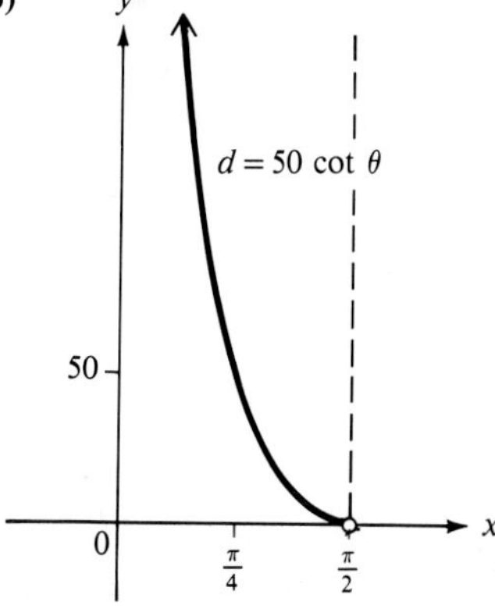

Section 4.4 (page 166)

1.

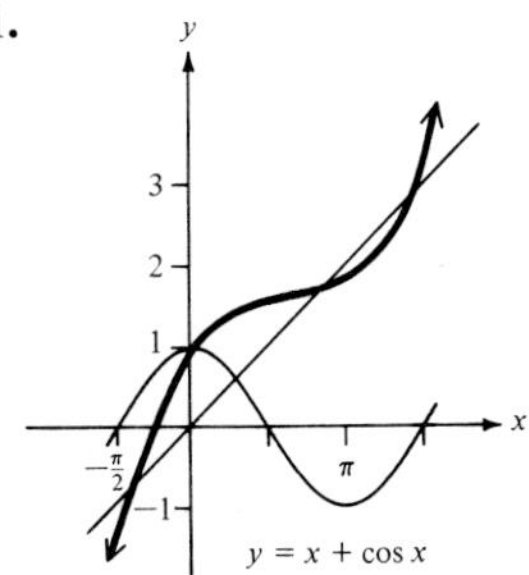

3.

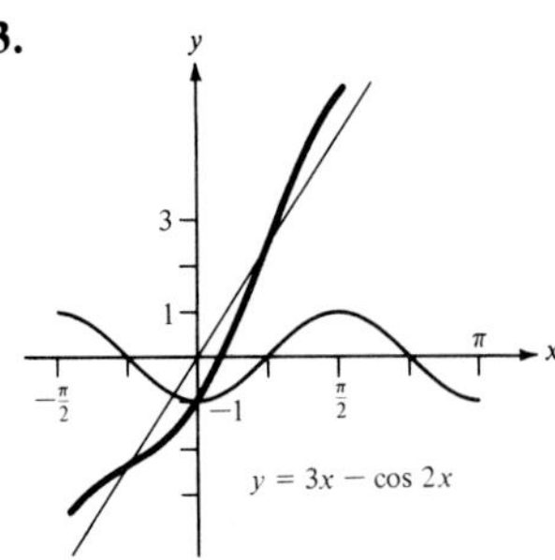

5.

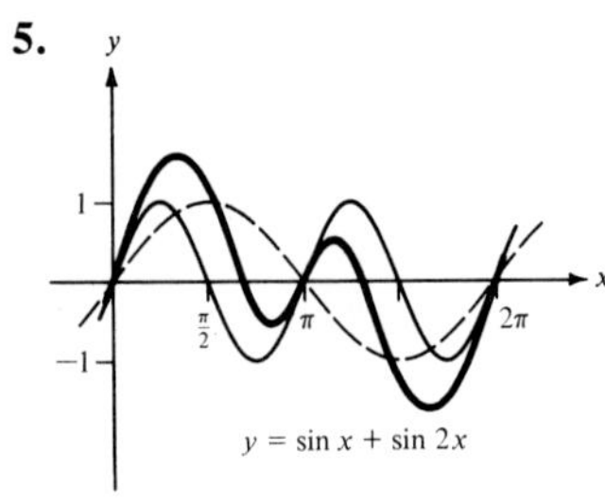

7.

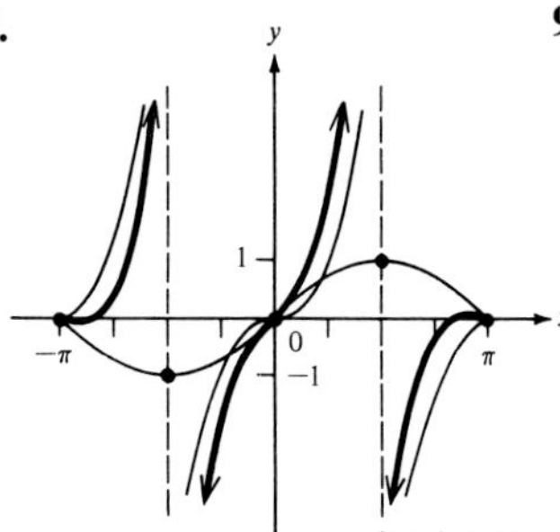

9.

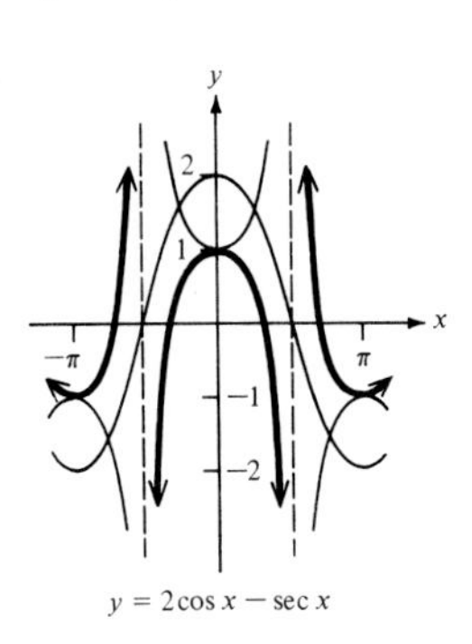

11.

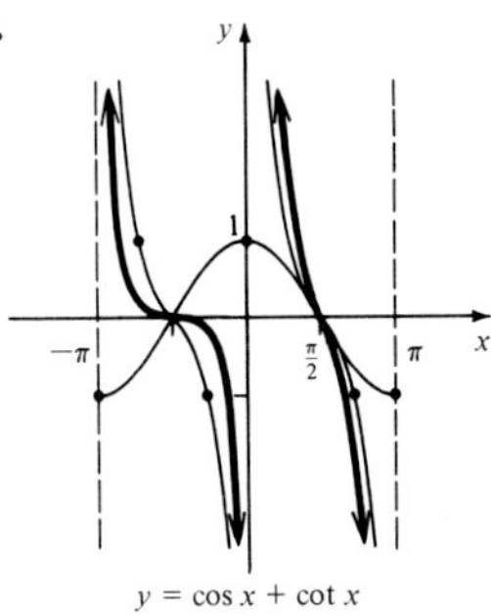

13.

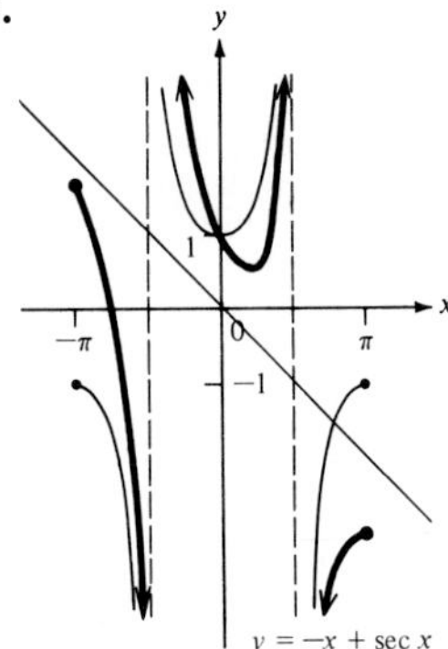

15.

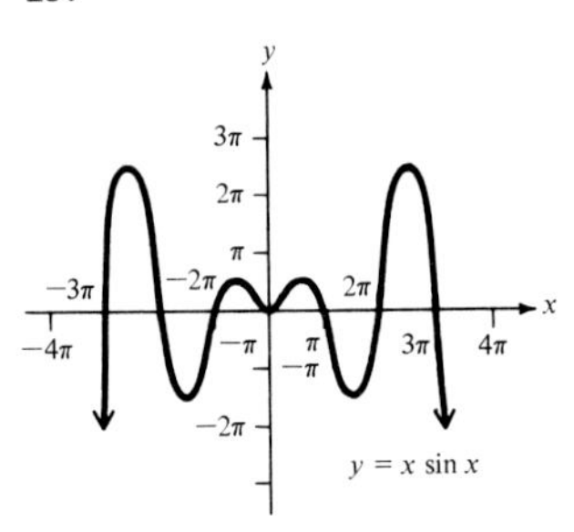

17.

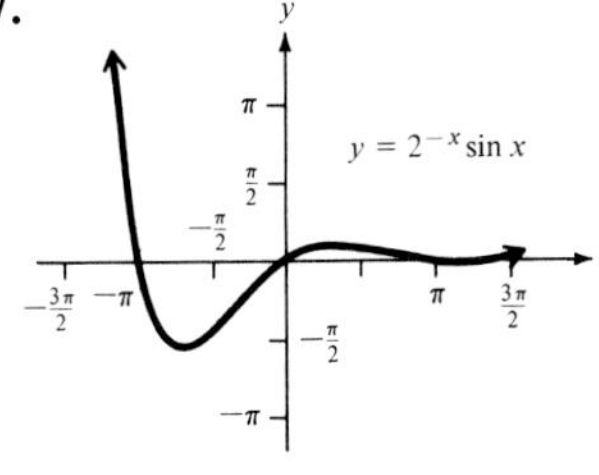

19.

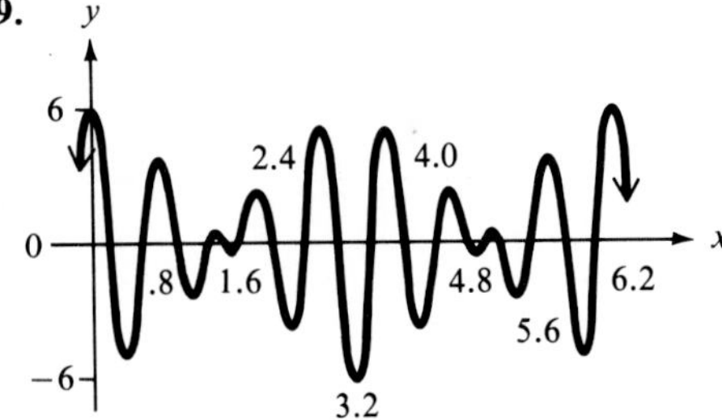

21.

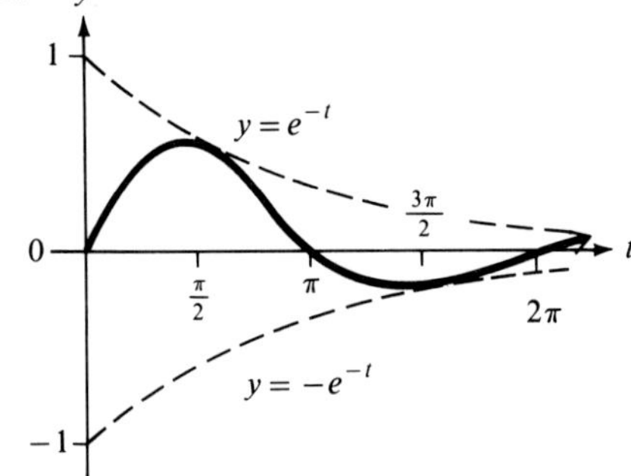

Section 4.5 (page 172)

1. (a) $y = \sin 2t$; 1; π; $1/\pi$ **(b)** $y = \sin 3t$; 1; $2\pi/3$; $3/(2\pi)$ **(c)** $y = \sin 4t$; 1; $\pi/2$; $2/\pi$ **3.** $y = 2\sin(t + \pi/3)$; 2; 2π; $1/(2\pi)$ **5.** $8/\pi^2$ ft **7. (a)** period, $2\pi\sqrt{m/k}$; frequency, $(1/(2\pi))\sqrt{k/m}$ **(b)** $1/\pi^2$

Chapter 4 Review Exercises (page 175)

1. 2; 2π; none; none **3.** 1/2; $2\pi/3$; none; none **5.** 2; 8π; 1; none **7.** 3; 2π; none; $\pi/2$ to the left **9.** None; π; none; $\pi/8$ to the right **11.** 1/3; $2\pi/3$; none; $\pi/9$ to the right

13.

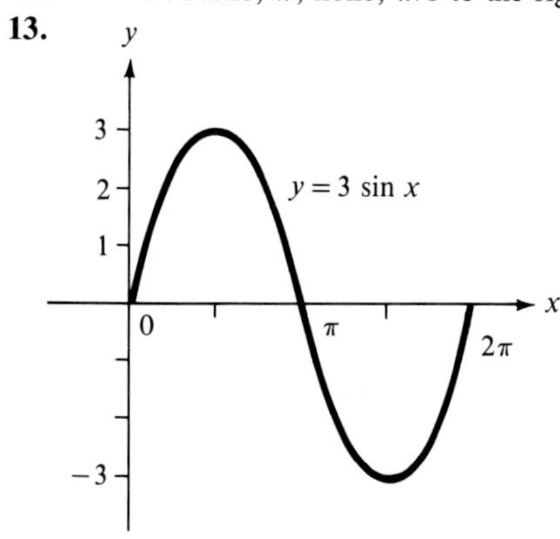

15.

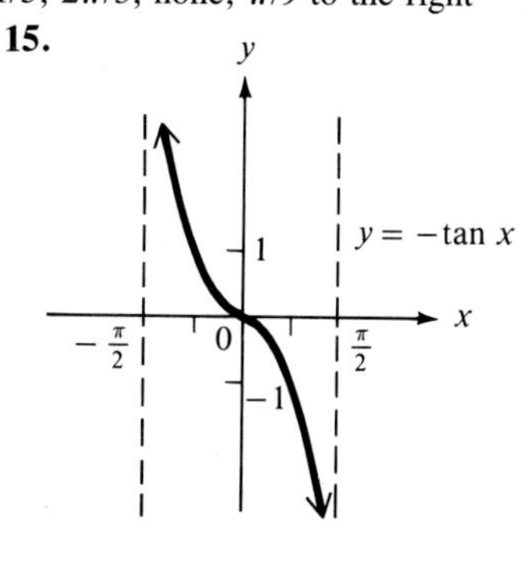

17.

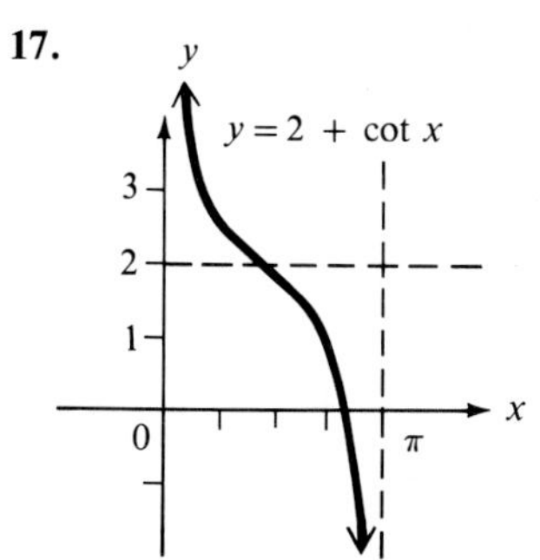

19.

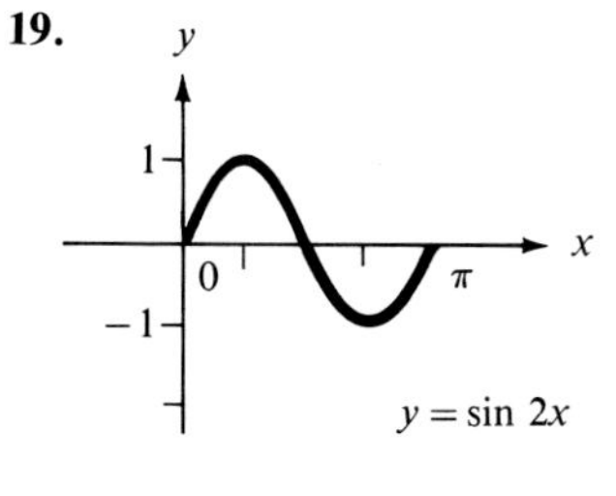

21.

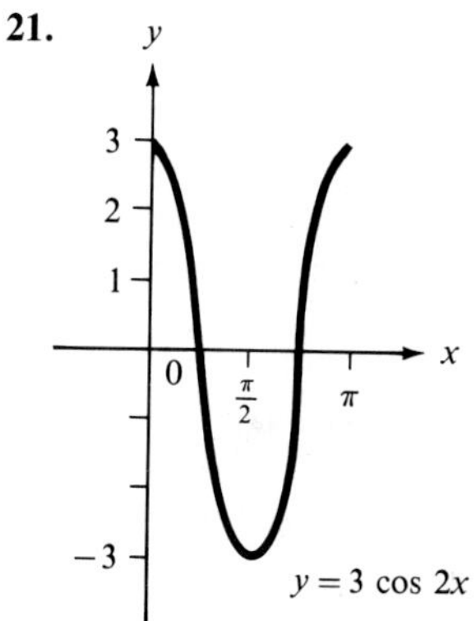

23.

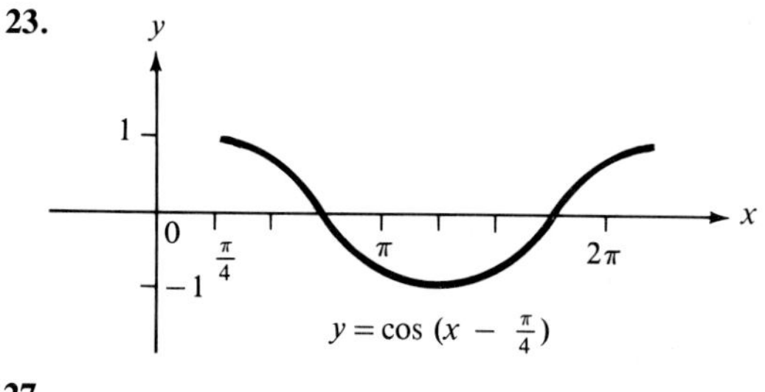

25.

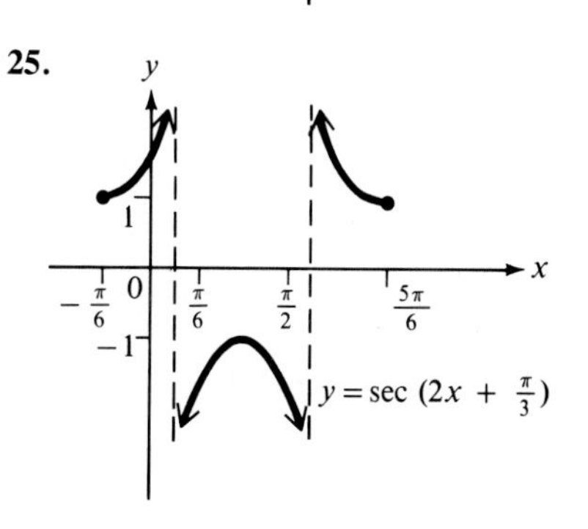

27.

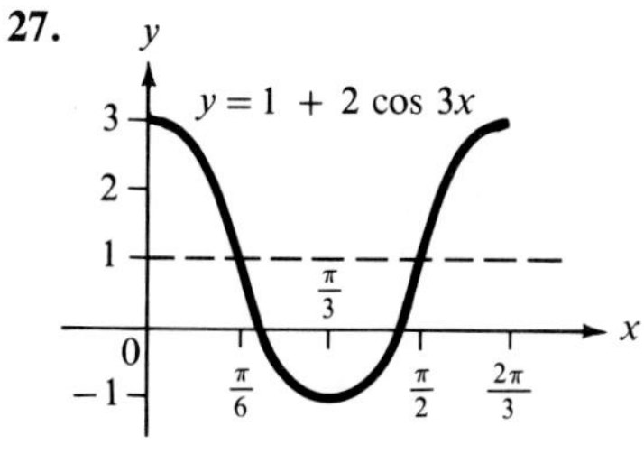

29.

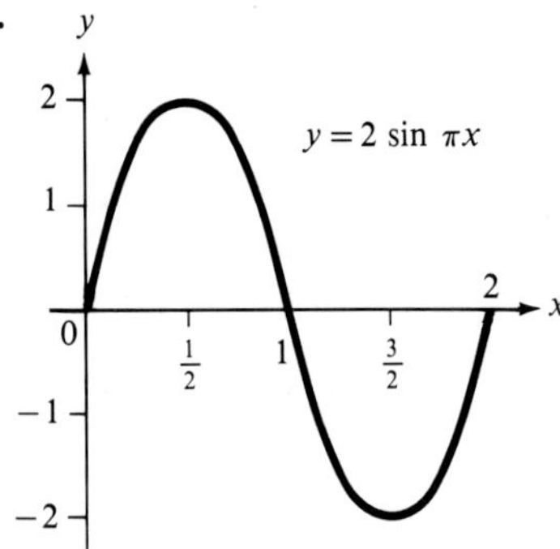

31.

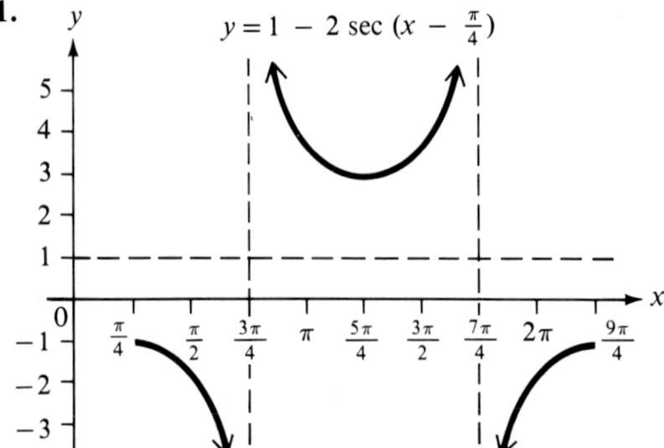

33.

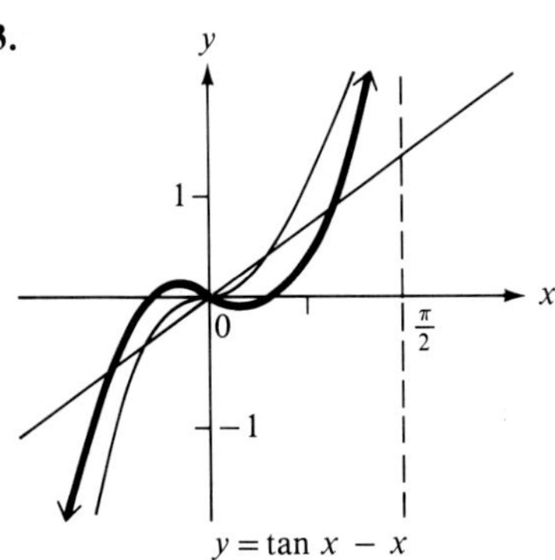

35.

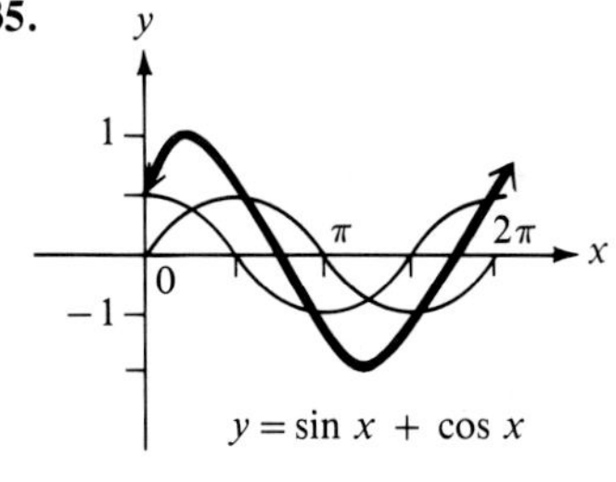

37. **(a)** About 20 years **(b)** From about 10,000 to about 150,000

CHAPTER 5

Section 5.1 (page 183)

1. $\sqrt{7}/4$ **3.** $-2\sqrt{5}/5$ **5.** $\sqrt{21}/2$ **7.** $\cos\theta = -\sqrt{5}/3$; $\tan\theta = -2\sqrt{5}/5$; $\cot\theta = -\sqrt{5}/2$; $\sec\theta = -3\sqrt{5}/5$; $\csc\theta = 3/2$ **9.** $\sin\theta = -\sqrt{17}/17$; $\cos\theta = 4\sqrt{17}/17$; $\cot\theta = -4$; $\sec\theta = \sqrt{17}/4$; $\csc\theta = -\sqrt{17}$ **11.** $\sin\theta = 2\sqrt{2}/3$; $\cos\theta = -1/3$; $\tan\theta = -2\sqrt{2}$; $\cot\theta = -\sqrt{2}/4$; $\csc\theta = 3\sqrt{2}/4$ **13.** $\sin\theta = 3/5$; $\cos\theta = 4/5$; $\tan\theta = 3/4$; $\sec\theta = 5/4$; $\csc\theta = 5/3$ **15.** $\sin\theta = -\sqrt{7}/4$; $\cos\theta = 3/4$; $\tan\theta = -\sqrt{7}/3$; $\cot\theta = -3\sqrt{7}/7$; $\csc\theta = -4\sqrt{7}/7$ **17.** (b) **19.** (e) **21.** (a) **23.** (a) **25.** (d) **27.** 1 **29.** $-\sin\alpha$ **31.** 0 **33.** $(1 + \sin\theta)/\cos\theta$ **35.** 1 **37.** -1 **39.** $\sin^2\theta/\cos^4\theta$ **41.** 1 **43.** $(\cos^2\alpha + 1)/(\sin^2\alpha\cos^2\alpha)$ **45.** $(\sin^2 s - \cos^2 s)/\sin^4 s$ **47.** $\pm\sqrt{1 + \cot^2\theta}/(1 + \cot^2\theta)$; $\pm\sqrt{\sec^2\theta - 1}/\sec\theta$ **49.** $\pm\sin\theta\sqrt{1 - \sin^2\theta}/(1 - \sin^2\theta)$; $\pm\sqrt{1 - \cos^2\theta}/\cos\theta$; $\pm\sqrt{\sec^2\theta - 1}$; $\pm\sqrt{\csc^2\theta - 1}/(\csc^2\theta - 1)$ **51.** $\pm\sqrt{1 - \sin^2\theta}/(1 - \sin^2\theta)$; $\pm\sqrt{\tan^2\theta + 1}$; $\pm\sqrt{1 + \cot^2\theta}/\cot\theta$; $\pm\csc\theta\sqrt{\csc^2\theta - 1}/(\csc^2\theta - 1)$ **53.** $\sin\theta = \pm\sqrt{2x + 1}/(x + 1)$ **61.** $4\sec\theta$; $3x/\sqrt{16 + 9x^2}$; $4/\sqrt{16 + 9x^2}$ **63.** $\sin^3\theta$; $\sqrt{1 - x^2}$; $\sqrt{1 - x^2}/x$ **65.** $(\tan^2\theta\sec\theta)/16$; $4x/\sqrt{1 + 16x^2}$; $1/\sqrt{1 + 16x^2}$ **67.** $(25\sqrt{6} - 60)/12$; $-(25\sqrt{6} + 60)/12$

Section 5.2 (page 191)

1. $1/(\sin\theta\cos\theta)$ **3.** $1 + \cos s$ **5.** 1 **7.** 1 **9.** $2 + 2\sin t$ **11.** $-2\cos x/[(1 + \cos x)(1 - \cos x)]$ or $-2\cos x/\sin^2 x$ **13.** $(\sin\gamma + 1)(\sin\gamma - 1)$ **15.** $4\sin x$

17. $(2 \sin x + 1)(\sin x + 1)$ **19.** $(4 \sec x - 1)(\sec x + 1)$ **21.** $(\cos^2 x + 1)^2$ **23.** $(\sin x - \cos x)(1 + \sin x \cos x)$ **25.** $\sin \theta$ **27.** 1 **29.** $\tan^2 \beta$ **31.** $\tan^2 x$ **33.** $\sec^2 x$ **75.** Identity **77.** Not an identity **79.** Not an identity **81.** Not an identity

Section 5.3 **(page 198)**
1. cot 3° **3.** $\sin 5\pi/12$ **5.** sec 104° 24′ **7.** $\cos(-\pi/8)$ **9.** csc (−56° 42′) **11.** tan (−86.9814°) **13.** tan **15.** cos **17.** csc **19.** True **21.** False **23.** True **25.** True **27.** 15° **29.** (140/3)° **31.** 20° **33.** $(\sqrt{6} - \sqrt{2})/4$ **35.** $(\sqrt{2} - \sqrt{6})/4$ **37.** $(\sqrt{2} - \sqrt{6})/4$ **39.** 0 **41.** $\sqrt{2}/2$ **43.** 0 **45.** 0 **47.** $(\sqrt{3}\cos\theta - \sin\theta)/2$ **49.** $(\cos\theta - \sqrt{3}\sin\theta)/2$ **51.** $-\sin x$ **53.** $(4 - 6\sqrt{6})/25$; $(4 + 6\sqrt{6})/25$ **55.** 16/65; −56/65 **57.** −77/85; −13/85 **59.** $(2\sqrt{638} - \sqrt{30})/56$; $(2\sqrt{638} + \sqrt{30})/56$ **71.** **(a)** $\cos\theta$ **(b)** $\sin\theta$ **(c)** $-\cos\theta$ **(d)** $-\sin\theta$ **(e)** $\cos\theta$ **(f)** $-\sin\theta$ **(g)** $-\cos\theta$ **(h)** $\sin\theta$ **75.** −.88331 **77.** .39441

Section 5.4 **(page 204)**
1. $(\sqrt{6} - \sqrt{2})/4$ **3.** $2 - \sqrt{3}$ **5.** $-(\sqrt{6} + \sqrt{2})/4$ **7.** $(\sqrt{6} + \sqrt{2})/4$ **9.** $\sqrt{2}/2$ **11.** −1 **13.** 0 **15.** 1 **17.** 0 **19.** 0 **21.** $\sqrt{2}(\sin\theta + \cos\theta)/2$ **23.** $(\sqrt{3}\tan\theta + 1)/(\sqrt{3} - \tan\theta)$ **25.** $(1 + \tan s)/(1 - \tan s)$ **27.** $\sin\theta$ **29.** $\tan\theta$ **31.** $-\sin\theta$ **33.** 63/65; 33/65; 63/16; 33/56 **35.** $(4\sqrt{2} + \sqrt{5})/9$; $(4\sqrt{2} - \sqrt{5})/9$; $(-8\sqrt{5} - 5\sqrt{2})/(20 - 2\sqrt{10})$ or $(4\sqrt{2} + \sqrt{5})/(2 - 2\sqrt{10})$; $(-8\sqrt{5} + 5\sqrt{2})/(20 + 2\sqrt{10})$ or $(-4\sqrt{2} + \sqrt{5})/(2\sqrt{10} + 2)$ **37.** 77/85; 13/85; −77/36; 13/84 **39.** −33/65; −63/65; 33/56; 63/16 **41.** 1; −161/289; undefined; −161/240 **43.** $-(3\sqrt{22} + \sqrt{21})/20$; $(-3\sqrt{22} + \sqrt{21})/20$; $-(66\sqrt{7} + 7\sqrt{66})/(154 - 3\sqrt{462})$; $(-66\sqrt{7} + 7\sqrt{66})/(154 + 3\sqrt{462})$ **59.** .357072 **61.** .382273 **63.** −2.09121 **67.** $\sin A \cos B \cos C + \cos A \sin B \cos C + \cos A \cos B \sin C - \sin A \sin B \sin C$ **71.** 18.4°

Section 5.5 **(page 210)**
1. $\sqrt{3}/2$ **3.** $\sqrt{3}/3$ **5.** $\sqrt{3}/2$ **7.** $\sqrt{2}/2$ **9.** $-\sqrt{2}/2$ **11.** $\sqrt{2}/4$ **13.** (1/2) tan 102° **15.** (1/4) cos 94.2° **17.** $-\cos 4\pi/5$ **19.** $\sin 10x$ **21.** $\cos 4\alpha$ **23.** $\tan 2x/9$ **25.** 1 **27.** −1/2 **29.** −1/2 **31.** $\sqrt{3}$ **33.** $\sqrt{3}$ **35.** 0 **37.** $\cos\theta = 2\sqrt{5}/5$; $\sin\theta = \sqrt{5}/5$; $\tan\theta = 1/2$; $\sec\theta = \sqrt{5}/2$; $\csc\theta = \sqrt{5}$; $\cot\theta = 2$ **39.** $\cos x = -\sqrt{42}/12$; $\sin x = \sqrt{102}/12$; $\tan x = -\sqrt{119}/7$; $\sec x = -2\sqrt{42}/7$; $\csc x = 2\sqrt{102}/17$; $\cot x = -\sqrt{119}/17$ **41.** $\cos 2\theta = 17/25$; $\sin 2\theta = -4\sqrt{21}/25$; $\tan 2\theta = -4\sqrt{21}/17$; $\sec 2\theta = 25/17$; $\csc 2\theta = -25\sqrt{21}/84$; $\cot 2\theta = -17\sqrt{21}/84$ **43.** $\tan 2x = -4/3$; $\sec 2x = -5/3$; $\cos 2x = -3/5$; $\cot 2x = -3/4$; $\sin 2x = 4/5$; $\csc 2x = 5/4$ **45.** $\sin 2\alpha = -4\sqrt{55}/49$; $\cos 2\alpha = 39/49$; $\tan 2\alpha = -4\sqrt{55}/39$; $\cot 2\alpha = -39\sqrt{55}/220$; $\sec 2\alpha = 49/39$; $\csc 2\alpha = -49\sqrt{55}/220$ **47.** $\tan^2 2x = 4\tan^2 x/(1 - 2\tan^2 x + \tan^4 x)$ **49.** $\cos 3x = 4\cos^3 x - 3\cos x$ **51.** $\tan 3x = (3\tan x - \tan^3 x)/(1 - 3\tan^2 x)$ **53.** $\tan 4x = 4(\tan x - \tan^3 x)/(1 - 6\tan^2 x + \tan^4 x)$ **79.** −.843580 **81.** −1.57091 **83.** 1 **85.** $\csc^2(3r)$ **87.** 980.799 cm per sec^2

Section 5.6 **(page 216)**
1. sin 20° **3.** tan 73.5° **5.** tan 29.87° **7.** $\cos 9x$ **9.** $\tan 4\theta$ **11.** $\cos x/8$ **13.** − **15.** + **17.** $\sin 22.5° = (\sqrt{2 - \sqrt{2}})/2$; $\cos 22.5° = (\sqrt{2 + \sqrt{2}})/2$; $\tan 22.5° = \sqrt{3 - 2\sqrt{2}}$ or $\sqrt{2} - 1$ **19.** $\sin 195° = -(\sqrt{2 - \sqrt{3}})/2$; $\cos 195° = -(\sqrt{2 + \sqrt{3}})/2$; $\tan 195° = \sqrt{7 - 4\sqrt{3}}$ or $2 - \sqrt{3}$ **21.** $\sin(-\pi/8) = -(\sqrt{2 - \sqrt{2}})/2$; $\cos(-\pi/8) = (\sqrt{2 + \sqrt{2}})/2$; $\tan(-\pi/8) = -\sqrt{3 - 2\sqrt{2}}$ or $1 - \sqrt{2}$ **23.** $\sin 5\pi/2 = 1$; $\cos 5\pi/2 = 0$; $\tan 5\pi/2$ does not exist **25.** $\sqrt{10}/4$ **27.** 3 **29.** $\sqrt{50 - 10\sqrt{5}}/10$ **31.** $-\sqrt{7}$ **33.** $\sqrt{5}/5$ **35.** $-\sqrt{42}/12$ **51.** 84° **53.** 60° **55.** 3.9 **57.** .892230 **59.** −1.97579

Section 5.7 (page 222)

1. $(1/2)(\sin 70° - \sin 20°)$ **3.** $(3/2)(\cos 8x + \cos 2x)$ **5.** $(1/2)[\cos 2\theta - \cos(-4\theta)] = (1/2)(\cos 2\theta - \cos 4\theta)$ **7.** $-4[\cos 9y + \cos(-y)] = -4(\cos 9y + \cos y)$ **9.** $2 \cos 45° \sin 15°$ **11.** $2 \cos 95° \cos(-53°) = 2 \cos 95° \cos 53°$ **13.** $2 \cos(15\beta/2) \sin(9\beta/2)$ **15.** $-6 \cos(7x/2) \sin(-3x/2) = 6 \cos(7x/2) \sin(3x/2)$

Section 5.8 (page 225)

1. $\sqrt{2} \sin(x + 135°)$ **3.** $13 \sin(\theta + 293°)$ **5.** $17 \sin(x + 152°)$ **7.** $25 \sin(\theta + 254°)$ **9.** $5 \sin(x + 53°)$ **11.** $3 \sin(x + 42°)$ **13.** $-\sin\theta + \sqrt{3}\cos\theta$ **15.** $\sin\theta - \cos\theta$ **17.** $\cos\theta$

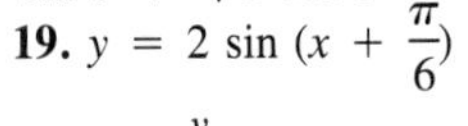
19. $y = 2 \sin(x + \frac{\pi}{6})$

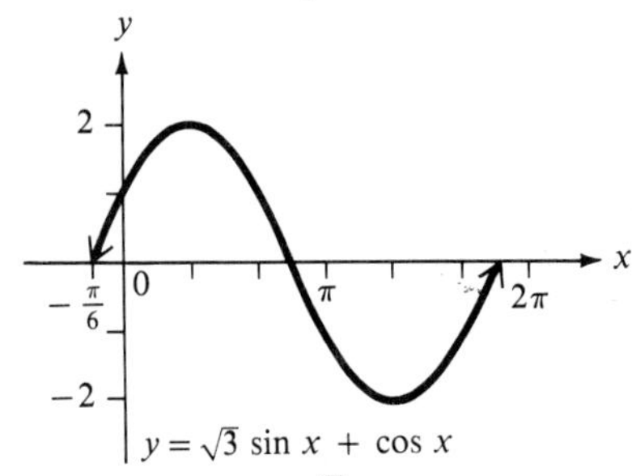

21. $y = \sqrt{2} \sin(x + \frac{3\pi}{4})$

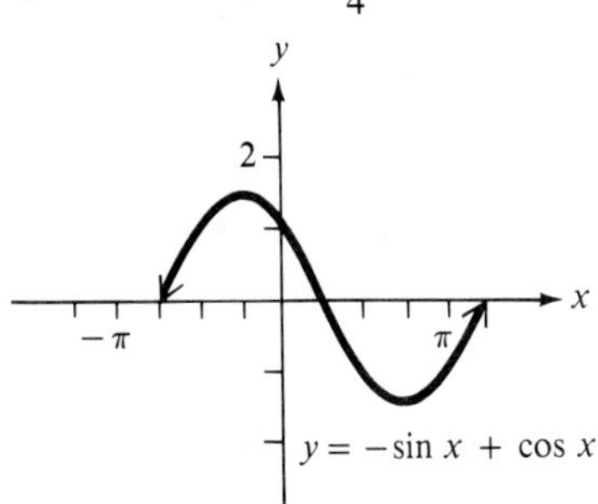

23. $y = 2 \sin(2x + \frac{\pi}{6})$

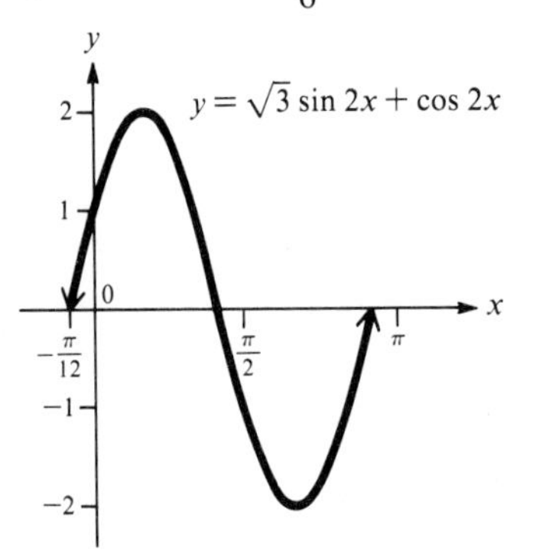

Chapter 5 Review Exercises (page 229)

1. $\sin x = -4/5$; $\tan x = -4/3$; $\sec x = 5/3$; $\csc x = -5/4$; $\cot x = -3/4$ **3.** $\sin(\pi/8) = \sqrt{2 - \sqrt{2}}/2$; $\cos(\pi/8) = \sqrt{2 + \sqrt{2}}/2$; $\tan(\pi/8) = \sqrt{3 - 2\sqrt{2}}$ **5.** (e) **7.** (j) **9.** (i) **11.** (h) **13.** (g) **15.** (a) **17.** (f) **19.** (e) **21.** 1 **23.** $1/\cos^2\theta$ or $\sec^2\theta$ **25.** $1/(\sin^2\theta \cos^2\theta)$ **27.** $(4 + 3\sqrt{15})/20$; $(4\sqrt{15} + 3)/20$; $(192 + 25\sqrt{15})/231$ **29.** $(4 - 9\sqrt{11})/50$; $(12\sqrt{11} - 3)/50$; $(\sqrt{11} - 16)/21$ **31.** $\sin\theta = \sqrt{8 - 2\sqrt{2}}/4$; $\cos\theta = -\sqrt{8 + 2\sqrt{2}}/4$ **33.** $\sin 2x = -3/5$ or $-.6$; $\cos 2x = -4/5$ or $-.8$ **35.** 1/2 **37.** $(\sqrt{5} - 1)/2$ **73.** $(\sqrt{6} - \sqrt{2})/4$

CHAPTER 6

Section 6.1 (page 238)

1. One-to-one **3.** Not one-to-one **5.** One-to-one **7.** Not one-to-one **9.** One-to-one **11.** One-to-one

13.

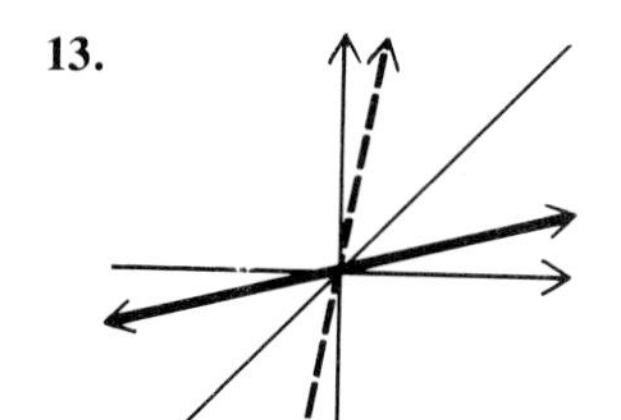

15. Not one-to-one

17.

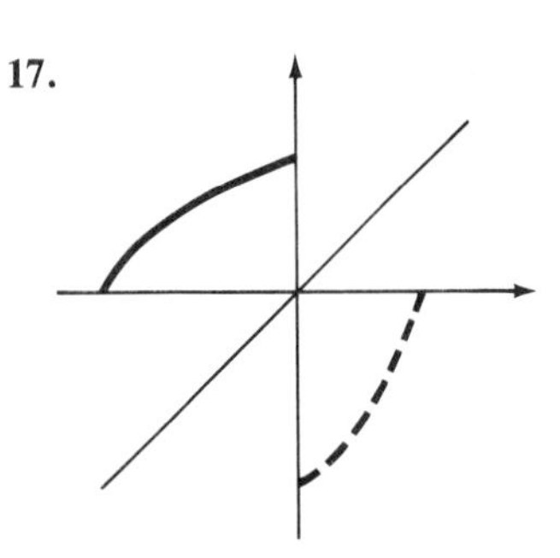

19.

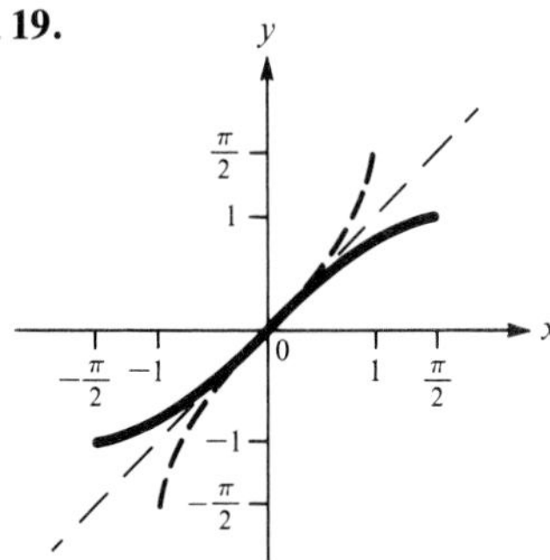

21.

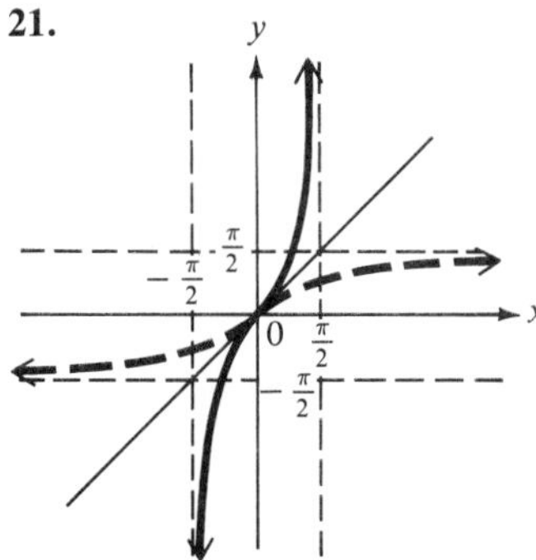

23. $y = x - 3$

25. $y = (2x + 1)/3$ **27.** Not one-to-one **29.** $y = (x + 1)^3$ **31.** Not one-to-one **33.** Yes
35. No **37.** Yes

Section 6.2 (page 246)

1. $-\pi/6$ **3.** $\pi/4$ **5.** 0 **7.** $\pi/3$ **9.** $5\pi/4$ **11.** $5\pi/6$ **13.** $-7.666°$
15. 113.5° **17.** 30.99° **19.** .8378 from table; .8379 from calculator **21.** 2.3155
23. 1.1897 from table; 1.1898 from calculator **25.** $\sqrt{7}/3$ **27.** $\sqrt{5}/5$ **29.** $-\sqrt{5}/2$
31. $\sqrt{34}/3$ **33.** 1/2 **35.** -1 **37.** 2 **39.** $\pi/4$ **41.** $\pi/3$ **43.** 120/169
45. $-7/25$ **47.** $4\sqrt{6}/25$ **49.** $-24/7$ **51.** $(\sqrt{10} - 3\sqrt{30})/20$ **53.** $-16/65$
55. .894427 **57.** .123430 **59.** $\sqrt{1 - u^2}$ **61.** $\sqrt{u^2 + 1}/u$ **63.** $\sqrt{1 - u^2}/u$
65. $\sqrt{u^2 - 4}/u$ **67.** $u/\sqrt{2}$ or $\sqrt{2}u/2$ **69.** $2/\sqrt{4 - u^2}$ or $2\sqrt{4 - u^2}/(4 - u^2)$
71. Real numbers; $0 < y < \pi$ **73.** $x \leq -1$ or $x \geq 1$; $0 \leq y < \pi/2$ or $\pi/2 < y \leq \pi$

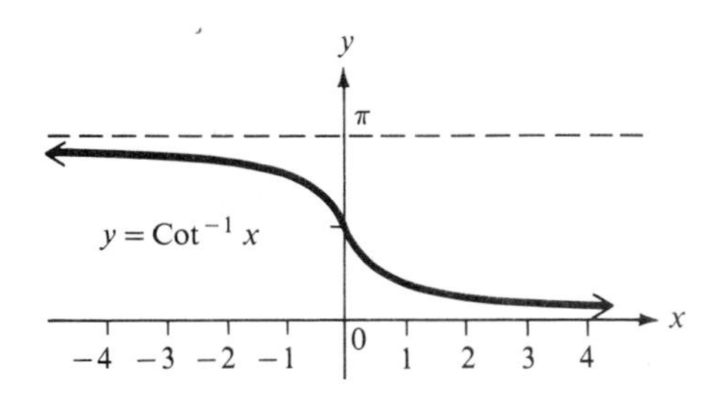

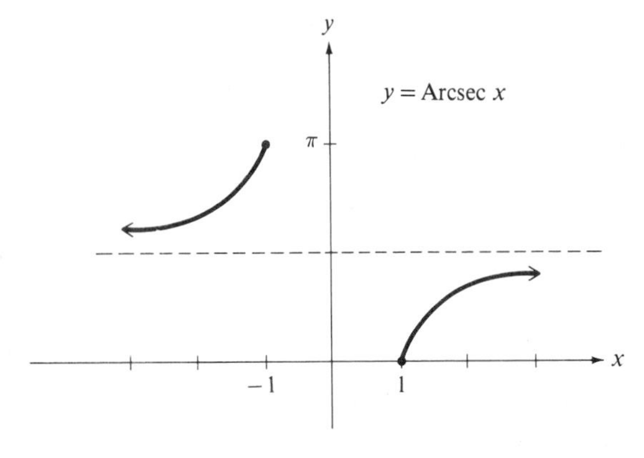

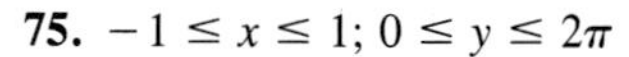
75. $-1 \le x \le 1;\ 0 \le y \le 2\pi$

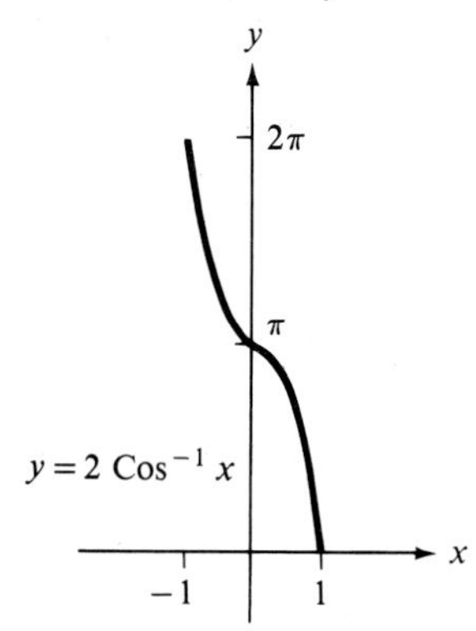

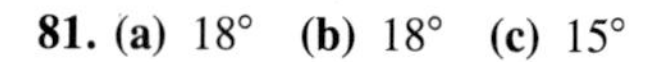
81. **(a)** 18° **(b)** 18° **(c)** 15°

Section 6.3 (page 253)

1. $3\pi/4, 7\pi/4$ **3.** $\pi/6, 5\pi/6$ **5.** $\pi/3, 2\pi/3\ 4\pi/3, 5\pi/3$ **7.** $\pi/6, 7\pi/6, 4\pi/3, 5\pi/3$ **9.** $\pi/4, 5\pi/4, 2\pi/3, 5\pi/3$ **11.** π **13.** $7\pi/6, 3\pi/2, 11\pi/6$ **15.** $\pi/4, 3\pi/4, 5\pi/4, 7\pi/4$ **17.** 90°, 210°, 330° **19.** 45°, 135°, 225°, 315° **21.** 0°, 180° **23.** 0°, 60°, 120°, 180°, 240°, 300° **25.** 90°, 270° **27.** 33° 30′, 326° 30′ **29.** 0°, 90°, 180°, 270° **31.** 30°, 150°, 210°, 330° **33.** 0°, 90° **35.** 53° 40′, 126° 20′, 188° 00′, 352° 00′ **37.** 106° 20′, 149° 40′, 286° 20′, 329° 40′ **39.** No solution **41.** 57° 40′, 159° 10′ **43.** **(a)** 1/4 sec **(b)** 1/6 sec **(c)** .21 sec **45.** 14°

Section 6.4 (page 256)

1. $\pi/12; 11\pi/12; 13\pi/12; 23\pi/12$ **3.** $\pi/2, 7\pi/6, 11\pi/6$ **5.** $\pi/18, 7\pi/18, 13\pi/18, 19\pi/18, 25\pi/18, 31\pi/18$ **7.** $3\pi/8, 5\pi/8, 11\pi/8, 13\pi/8$ **9.** $\pi/2, 3\pi/2$ **11.** $0, \pi/4, \pi/2, 3\pi/4, \pi, 5\pi/4, 3\pi/2, 7\pi/4$ **13.** $\pi/8, 3\pi/8, 5\pi/8, 7\pi/8, 9\pi/8, 11\pi/8, 13\pi/8, 15\pi/8$ **15.** $\pi/2$ **17.** $\pi/3, \pi, 5\pi/3$ **19.** 15°, 45°, 135°, 165°, 255°, 285° **21.** 0° **23.** 120°, 240° **25.** 30°, 150°, 270° **27.** 0°, 30°, 150°, 180° **29.** 60°, 300° **31.** 11° 50′, 78° 10′, 191° 50′, 258° 10′ **33.** 30°, 90°, 150°, 210°, 270°, 330° **35.** 0°, 120°, 240° **37.** 110° 20′ **39.** $\pi/12, 5\pi/12, \pi/2, 13\pi/12, 17\pi/12, 3\pi/2$ **41.** $0, \pi/4, 3\pi/4, \pi, 5\pi/4, 7\pi/4$ **43.** $\pi/6, \pi/2, 5\pi/6, 3\pi/2$ **45.** **(a)** 91.3 days after March 21, on June 20 **(b)** 273.8 days after March 21, on December 19 **(c)** 228.7 days after March 21, on November 4, and again after 318.8 days, on February 2 **47.** .001 sec **49.** .004 sec

Section 6.5 (page 262)

1. $x = \text{Cos}^{-1}(y/5)$ **3.** $x = (1/3)\text{ Cot}^{-1} 2y$ **5.** $x = (1/2)\text{ Arctan }(y/3)$ **7.** $x = 4\text{ Arccos } y/6$ **9.** $x = (1/5)\text{ Cos}^{-1}(-y/2)$ **11.** $x = -3 + \text{Cos}^{-1} y$ **13.** $x = \text{Sin}^{-1}(y + 2)$ **15.** $x = \text{Arcsin }[(y + 4)/2]$ **17.** $2\sqrt{2}$ **19.** $\pi - 3$ **21.** 3/5 **23.** 4/5 **25.** 0 **27.** 1/2 **29.** $-1/2$ **31.** 0 **33.** $t = (50/\pi)\ [\text{Arccos }(d - 550)/450]$ **35.** **(a)** $t = [1/(2\pi f)][\text{Arcsin } e/E_{max}]$ **(b)** .00068 **37.** **(a)** $x = \text{Sin } u,\ -\pi/2 \le u \le \pi/2$ **(b)**

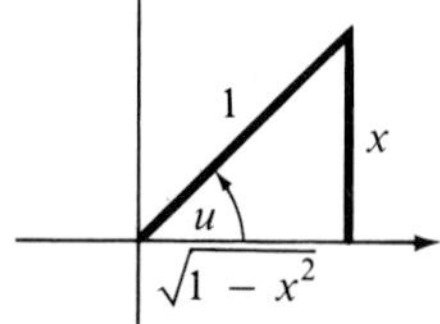

(c) $\tan u = x/\sqrt{1 - x^2}$ or $x\sqrt{1 - x^2}/(1 - x^2)$ **(d)** $u = \text{Arctan }(x/\sqrt{1 - x^2})$ or $\text{Arctan }[x\sqrt{1 - x^2}/(1 - x^2)]$

Chapter 6 Review Exercises (page 266)

1. Not one-to-one **3.** Not one-to-one **5.** Not one-to-one **7.** One-to-one **9.** $y = (x - 3)/12$ **11.** $y = (2/x) + 9$ or $y = (2 + 9x)/x$ **13.** Yes **15.** No **17.** $\pi/4$ **19.** $-\pi/3$ **21.** $-41° 20'$ **23.** $1/2$ **25.** -1 **27.** $3\pi/4$ **29.** $\pi/4$ **31.** $\sqrt{7}/4$ **33.** $\sqrt{3}/2$ **35.** $(15 + 8\sqrt{6})/(20 - 6\sqrt{6})$ or $(294 + 125\sqrt{6})/92$ **37.** $\sqrt{1 - u^2}$ **39.** $-1 \le x \le 1$, $-\pi/2 \le y \le \pi/2$ **41.** $-\infty < x < \infty$, $0 < y < \pi$

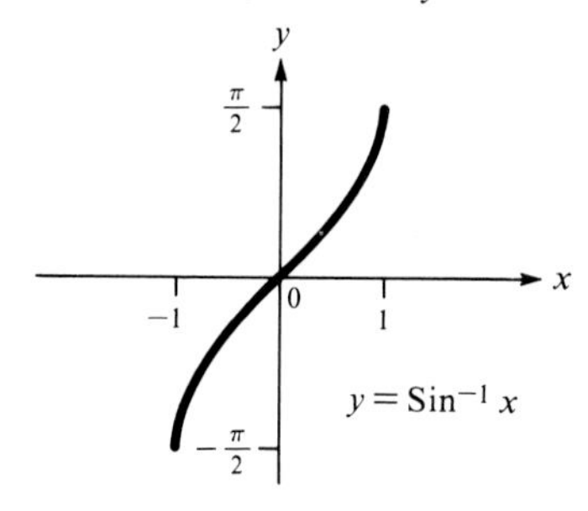

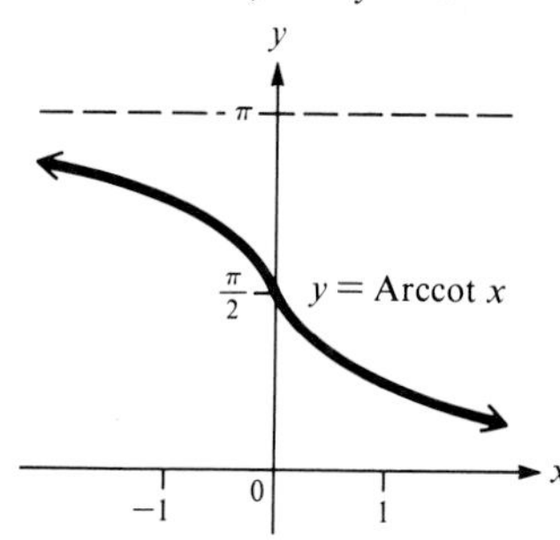

43. .4636, 3.6052 **45.** $\pi/4$, $3\pi/4$, $5\pi/4$, $7\pi/4$ **47.** $\pi/8$, $3\pi/8$, $5\pi/8$, $7\pi/8$, $9\pi/8$, $11\pi/8$, $13\pi/8$, $15\pi/8$ **49.** $\pi/3$, π, $5\pi/3$ **51.** 270° **53.** 45°, 90°, 225°, 270° **55.** 70.5°, 180°, 289.5° **57.** 0°, 60°, 90°, 120°, 180°, 240°, 270°, 300° **59.** $x = \text{Sin}^{-1} 2y$ **61.** $x = (1/3)(-2 + \text{Tan}^{-1} 2y)$ **63.** $x = -2$ **65.** $x = -1/2$ **67. (a)** 48.8° **(b)** The light beam is completely under water.

CHAPTER 7

Section 7.1 (page 275)

1. $C = 95°$, $b = 13$ m, $a = 11$ m **3.** $C = 80° 40'$, $a = 79.5$ mm, $c = 108$ mm **5.** $B = 37.3°$, $a = 38.5$ ft, $b = 51.0$ ft **7.** $C = 57.36°$, $b = 11.13$ ft, $c = 11.55$ ft **9.** $B = 18.5°$, $a = 239$ yd, $c = 230$ yd **11.** $A = 56° 00'$, $c = 361$ ft, $a = 308$ ft **13.** $B = 110.0°$, $a = 27.01$ m, $c = 21.36$ m **15.** $A = 34.72°$, $a = 3326$ ft, $c = 5704$ ft **17.** $C = 97° 34'$, $b = 283.2$ m, $c = 415.2$ m **19.** 118 m **21.** 1.93 mi **23.** 10.4 in **25.** 111° **27.** 46.4 m^2 **29.** 356 cm^2 **31.** 722.9 in^2 **33.** 1071 cm^2 **35.** 100

Section 7.2 (page 281)

1.

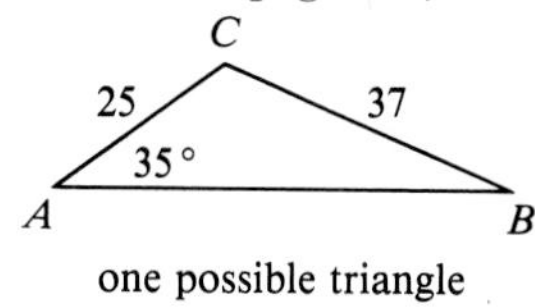

one possible triangle

3.

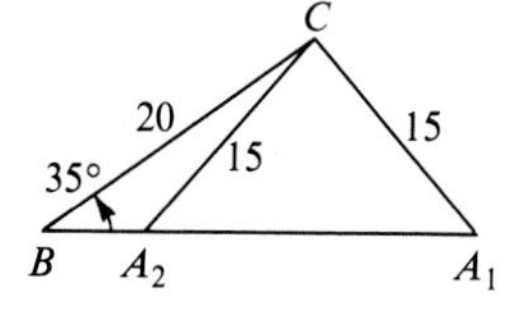

two triangles: A_1BC and A_2BC

5. B 18 12 128° A C

one possible triangle

7. $B_1 = 49.1°$, $C_1 = 101.2°$, $B_2 = 130.9°$, $C_2 = 19.4°$ **9.** $B = 26° 30'$, $A = 112° 10'$ **11.** No such triangle **13.** $B = 27.19°$, $C = 10.68°$ **15.** $A = 43° 50'$, $B = 6° 52'$ **17.** $B = 20.6°$, $C = 116.9°$, $c = 20.6$ ft **19.** No such triangle **21.** $B_1 = 49° 20'$, $C_1 = 92° 00'$, $c_1 = 15.5$ km, $B_2 = 130° 40'$, $C_2 = 10° 40'$, $c_2 = 2.88$ km **23.** $A_1 = 52° 10'$, $C_1 = 95° 00'$, $c_1 = 9520$ cm, $A_2 = 127° 50'$, $C_2 = 19° 20'$, $c_2 = 3160$ cm **25.** $B = 37.77°$, $C = 45.43°$, $c = 4.174$ ft **27.** $A_1 = 53.23°$, $C_1 = 87.09°$, $c_1 = 37.16$ m, $A_2 = 126.77°$, $C_2 = 13.55°$, $c_2 = 8.719$ m **29.** Does not exist

Section 7.3 (page 287)

1. $c = 2.83$ in, $A = 44.9°$, $B = 106.8°$ **3.** $c = 6.46$ m, $A = 53.1°$, $B = 81.3°$ **5.** $a = 156$ cm, $B = 64° 50'$, $C = 34° 30'$ **7.** $b = 9.529$ in, $A = 64.59°$, $C = 40.61°$ **9.** $a = 15.7$ m, $B = 21.6°$, $C = 45.6°$ **11.** $c = 139.0$ m, $A = 49° 20'$, $B = 105° 51'$ **13.** $A = 30°$, $B = 56°$, $C = 94°$ **15.** $A = 81° 50'$, $B = 37° 20'$, $C = 60° 50'$ **17.** $A = 42° 00'$, $B = 35° 50'$, $C = 102° 10'$ **19.** $A = 47° 40'$, $B = 44° 50'$, $C = 87° 20'$ **21.** $A = 35° 22'$, $B = 50° 58'$, $C = 93° 40'$ **23.** $A = 28° 10'$, $B = 21° 56'$, $C = 129° 54'$ **25.** 257 m **27.** 281 km **29.** 22 ft **31.** 18 ft **33.** 25.24983 mi **35.** The angle at Mackinac West Base is 16.42821°; the angle at Green Island is 123.13624°; the angle at St. Ignace West Base is 40.43555°. **37.** 78 m^2 **39.** 12,600 cm^2 **41.** 3650 ft^2 **43.** 1921 ft^2 **45.** 33 cans

Section 7.4 (page 295)

1. m and p; n and r **3.** m and p equal $2t$, or t is one half m or p; also, $m = 1p$ and $n = 1r$

5. −**b**

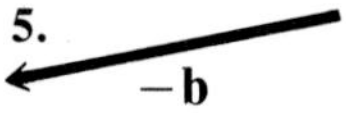

7. 3**a**

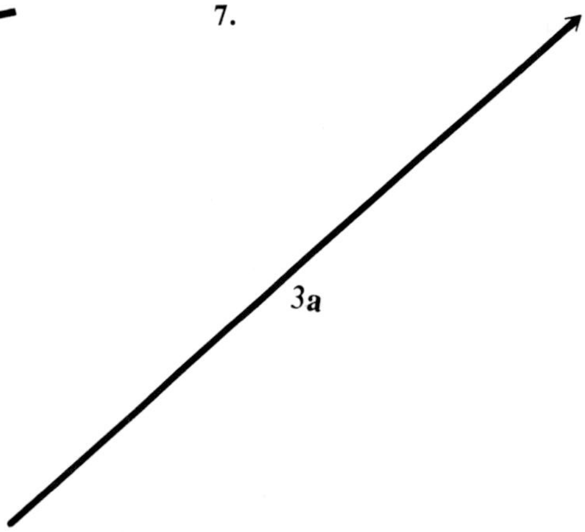

9. **a**, **c**, **a** + **c**

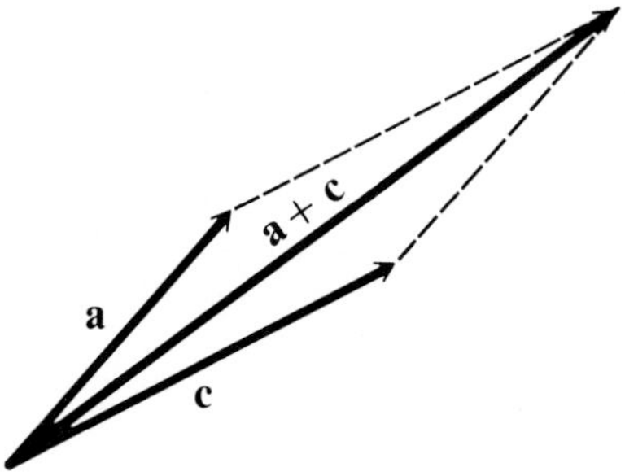

11. **g**, **h**, **h** + **g**

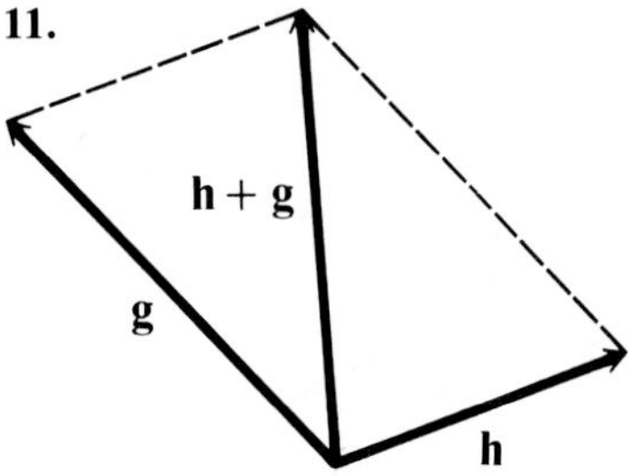

13. **a**, **h**, **a** + **h**

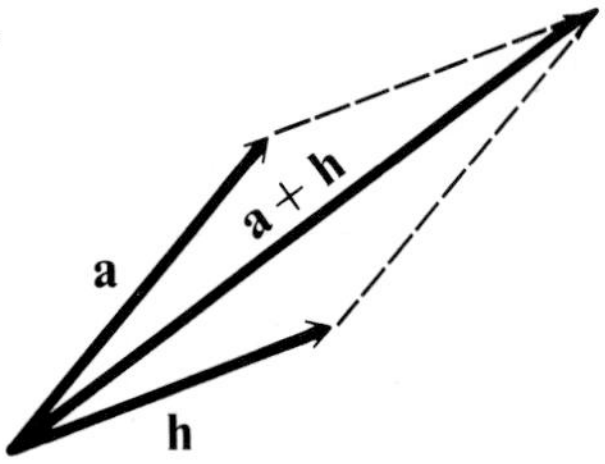

15. **d**, **h**, **h** + **d**

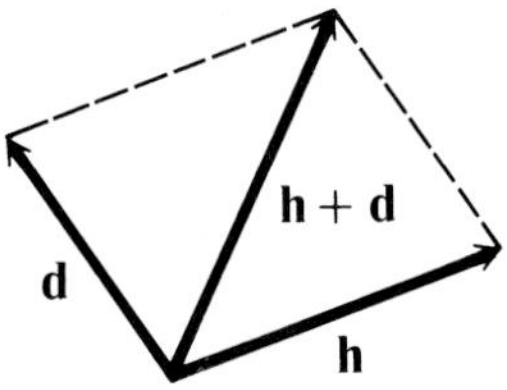

17. **a**, −**c**, **a** − **c**

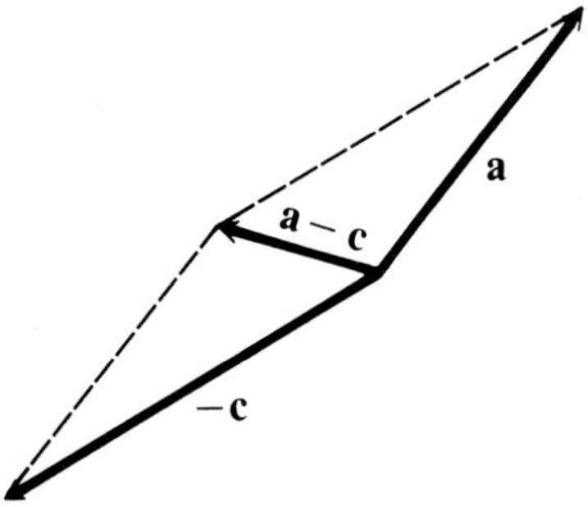

19. **a**, **b**, **c**, **b** + **c**, **a** + (**b** + **c**)

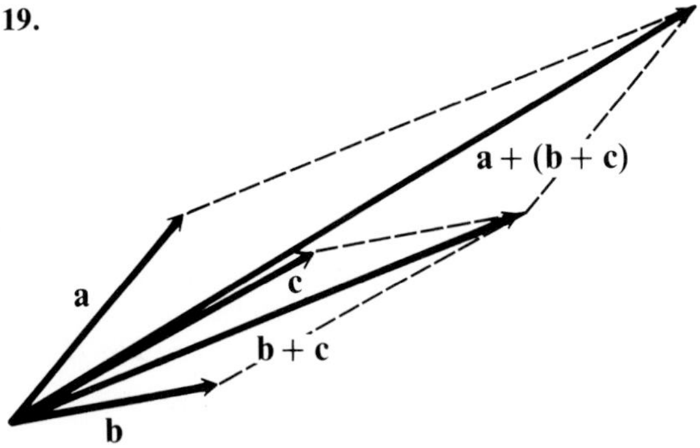

21.

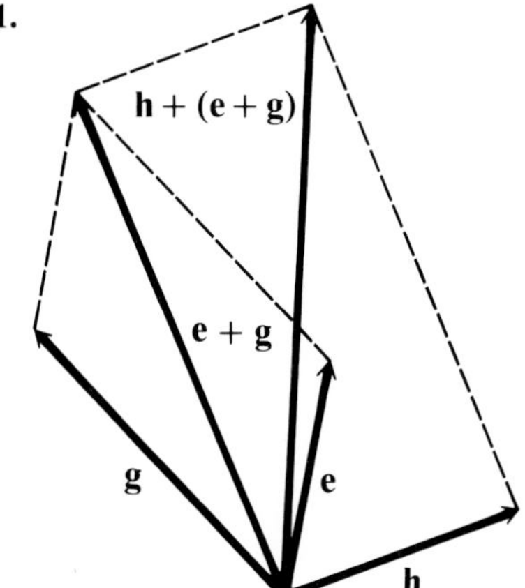

23. Yes

25.

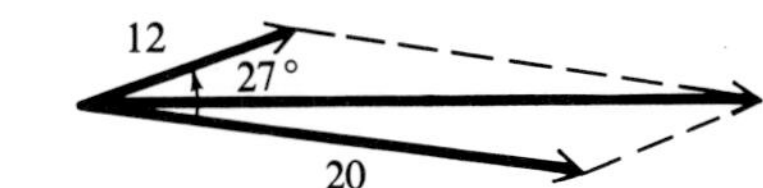

27.

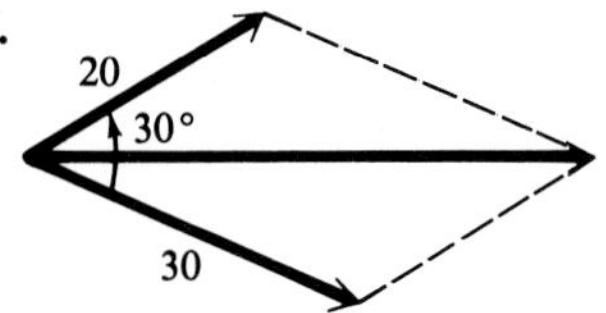

29.

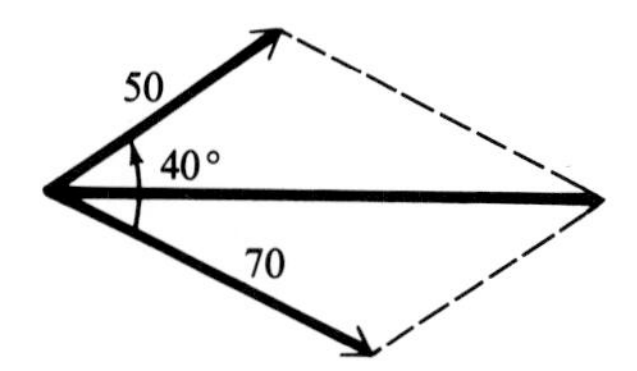

31. 9.5, 7.4 **33.** 17, 20 **35.** 13.7, 7.11 **37.** 123, 155 **39.** 22.3, 65.4 **41.** 530 newtons **43.** 27.2 lb **45.** 88.2 lb **47.** 280 newtons, 30.4°

Section 7.5 (page 301)

1. 94° 00′ **3.** 18° **5.** 2.4 tons **7.** 2640 lb at an angle of 167.2° with the 1480-lb force **9.** Weight 64.8 lb, tension 61.9 lb **11.** 190, 283 lb, respectively **13.** 173.1° **15.** 39.2 km **17.** 237°, 470 mph **19.** 358° 00′, 170 mph **21.** 230 kph, 167° **23.** Turn at 3:21 P.M. on a bearing of 152 °

Chapter 7 Review Exercises (page 305)

1. 63.7 m **3.** 41° 40′ **5.** 54° 20′ or 125° 40′ **7.** $B = 17° 10'$, $C = 137° 40'$, $c = 11.0$ yd **9.** 19.87° or 19° 52′ **11.** 55.5 m **13.** 148 cm **15.** 153,600 m^2 **17.** 20.3 ft^2 **19.** .234 km^2 **21.** About 2.5 cans (better buy 3). **23.** 13 m **25.** 10.8 mi **27.** 115 km **29.** 5500 m **31.**

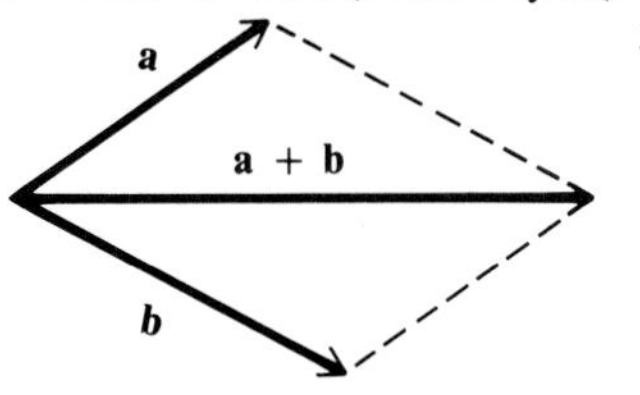

33.

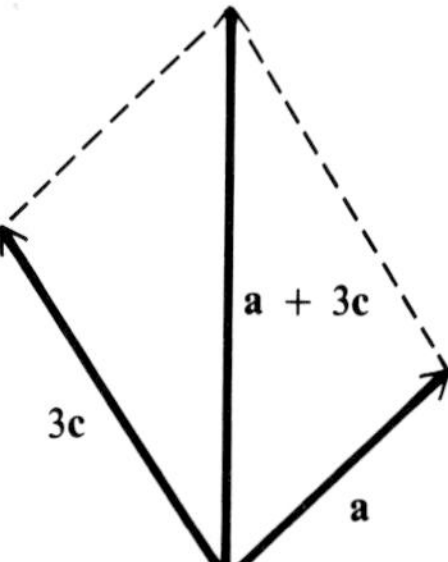

35. Horizontal 17.9, vertical 66.8 **37.** 28 lb **39.** 135 newtons **41.** 270, 123° 40′ **43.** 3° 50′ **45.** 21 kph, bearing 118°

CHAPTER 8

Section 8.1 (page 312)

1. $2i$ **3.** $8i$ **5.** $5i/3$ **7.** $3i\sqrt{2}$ **9.** $5i\sqrt{6}$ **11.** $3i\sqrt{3}$ **13.** $4i\sqrt{5}$ **15.** -3
17. $-\sqrt{30}$ **19.** $\sqrt{6}/2$ **21.** $i\sqrt{3}/3$ **23.** $5 - 3i$ **25.** $-5 + 2i$ **27.** $-4 + i$
29. $8 - i$ **31.** $-14 + 2i$ **33.** $31 - 5i$ **35.** $3 + 4i$ **37.** 5 **39.** $25i$
41. $24 - 7i$ **43.** i **45.** $7/25 - 24i/25$ **47.** $13/20 - i/20$ **49.** $32/37 - 7i/37$
51. $-1 - 2i$ **53.** $5, 3$ **55.** $2, -3$ **57.** $1/2, 15$ **59.** $14, 8$ **61.** $10, -2$
63. $E = 30 + 60i$ **65.** $Z = 233/37 + (119/37)i$ **67.** $a = 0$ or $b = 0$

Section 8.2 (page 317)

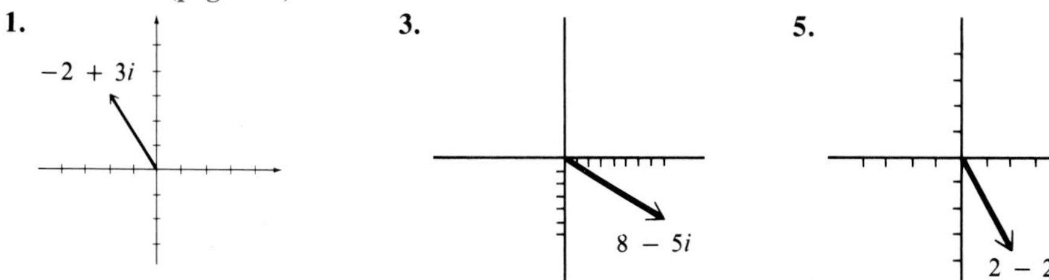

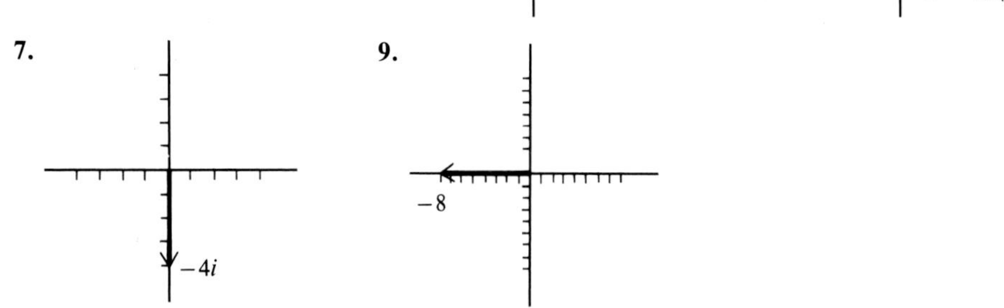

11. $3 - i$ **13.** $3 - 3i$ **15.** $-3 + 3i$ **17.** $2 + 4i$ **19.** $7 + 9i$
21. $3\sqrt{2}(\cos 315° + i \sin 315°)$ **23.** $6(\cos 240° + i \sin 240°)$ **25.** $2(\cos 330° + i \sin 330°)$
27. $5\sqrt{2}(\cos 225° + i \sin 225°)$ **29.** $2\sqrt{2}(\cos 45° + i \sin 45°)$ **31.** $4(\cos 180° + i \sin 180°)$
33. $2(\cos 270° + i \sin 270°)$ **35.** $\sqrt{2} + i\sqrt{2}$ **37.** $10i$ **39.** $-2 - 2i\sqrt{3}$
41. $(\sqrt{3}/2) + (1/2)i$ **43.** $(5/2) - (5\sqrt{3}/2)i$ **45.** $-\sqrt{2}$ **47.** $\sqrt{13}(\cos 56° 20' + i \sin 56° 20')$
49. $-1.0179 - 2.8221i$ **51.** $2(\cos 160° 00' + i \sin 160° 00')$
53. $\sqrt{34}(\cos 59° 00' + i \sin 59° 00')$ **55.** $20 + 15i$

Section 8.3 (page 322)

1. $-3\sqrt{3} + 3i$ **3.** $-4i$ **5.** $12\sqrt{3} + 12i$ **7.** $(-15\sqrt{2}/2) + (15\sqrt{2}/2)i$ **9.** $-3i$
11. $\sqrt{3} - i$ **13.** $-1 - i\sqrt{3}$ **15.** $-1/6 - i\sqrt{3}/6$ **17.** $2\sqrt{3} - 2i$ **19.** $-1/2 - i/2$
21. $\sqrt{3} + i$ **23.** $2.39 + 15.0i$ **25.** $.378 + 3.52i$ **27.** $5520 + 9550i$ **29.** $1.2 - .14i$

Section 8.4 (page 327)

1. $27i$ **3.** 1 **5.** $(27/2) - (27\sqrt{3}/2)i$ **7.** $-16\sqrt{3} + 16i$ **9.** $-128 + 128i\sqrt{3}$
11. $128 + 128i$

13. (cos 0° + *i* sin 0°),
(cos 120° + *i* sin 120°),
(cos 240° + *i* sin 240°)

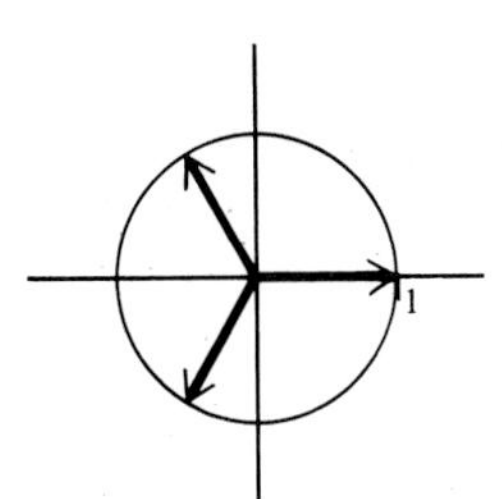

15. 2(cos 20° + *i* sin 20°),
2(cos 140° + *i* sin 140°),
2(cos 260° + *i* sin 260°)

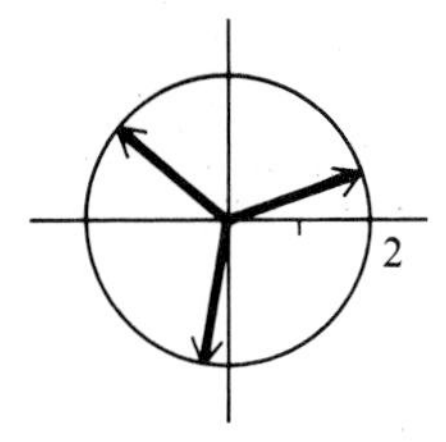

17. 2(cos 90° + *i* sin 90°),
2(cos 210° + *i* sin 210°),
2(cos 330° + *i* sin 330°)

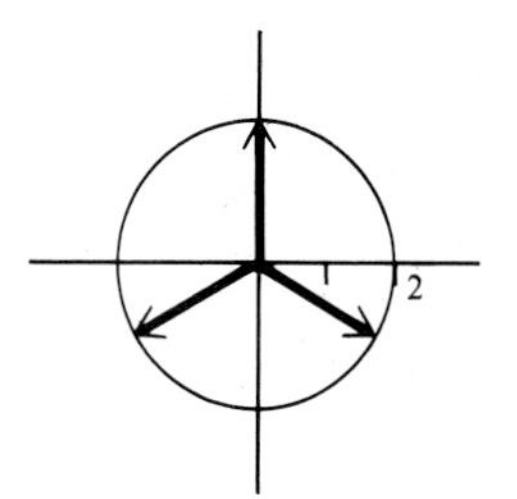

19. 4(cos 60° + *i* sin 60°),
4(cos 180° + *i* sin 180°),
4(cos 300° + *i* sin 300°)

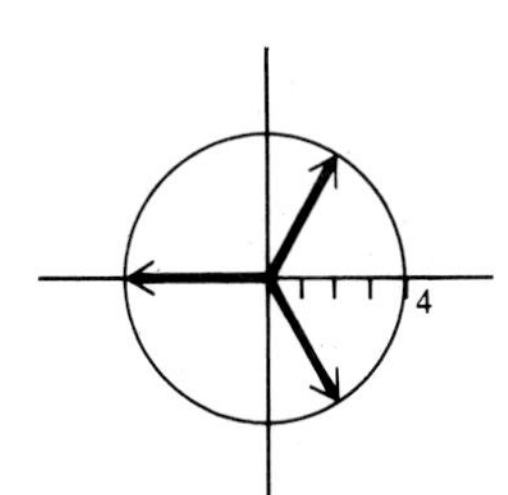

21. $\sqrt[3]{2}$(cos 20° + *i* sin 20°),
$\sqrt[3]{2}$(cos 140° + *i* sin 140°),
$\sqrt[3]{2}$(cos 260° + *i* sin 260°)

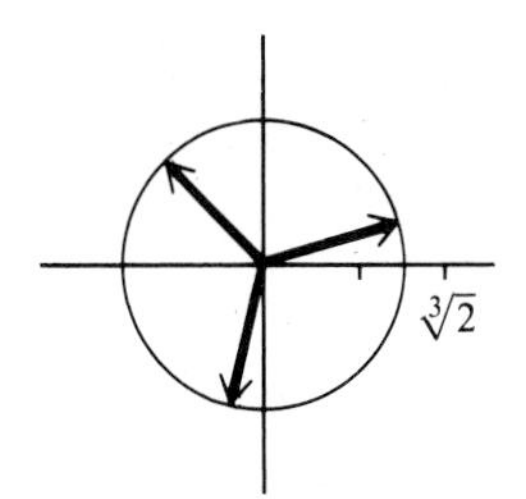

23. $\sqrt[3]{4}$(cos 50° + *i* sin 50°),
$\sqrt[3]{4}$(cos 170° + *i* sin 170°),
$\sqrt[3]{4}$(cos 290° + *i* sin 290°)

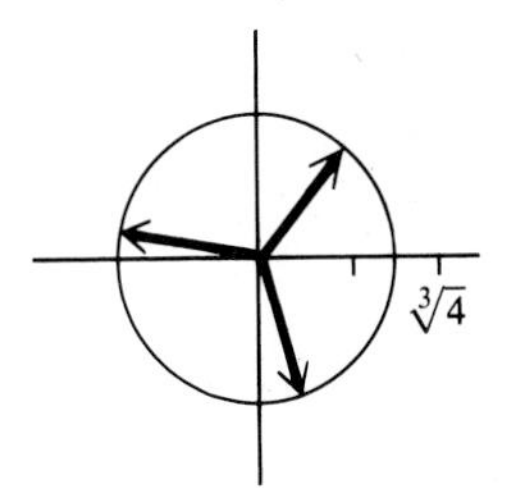

25. (cos 0° + *i* sin 0°),
(cos 180° + *i* sin 180°)

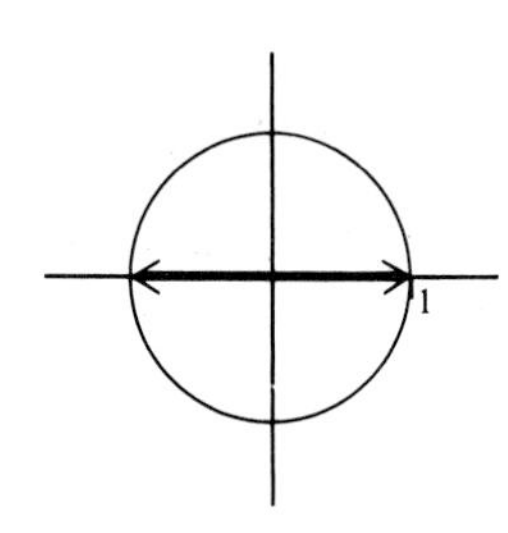

27. (cos 0° + *i* sin 0°),
(cos 60° + *i* sin 60°),
(cos 120° + *i* sin 120°),
(cos 180° + *i* sin 180°),
(cos 240° + *i* sin 240°),
(cos 300° + *i* sin 300°)

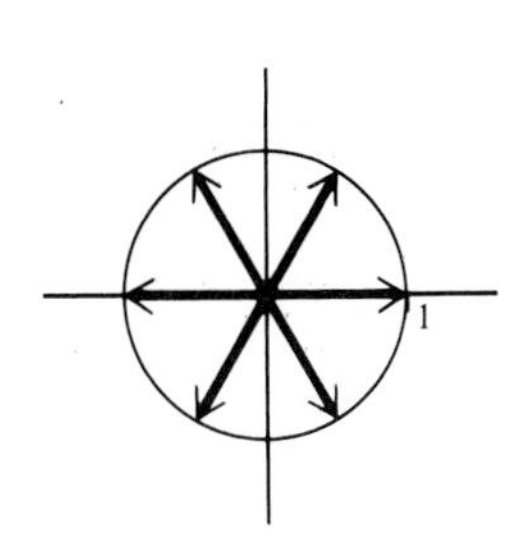

29. (cos 45° + *i* sin 45°),
(cos 225° + *i* sin 225°)

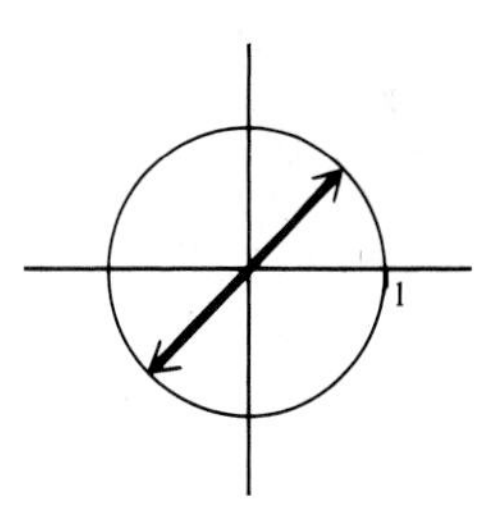

31. (cos 0° + *i* sin 0°), (cos 120° + *i* sin 120°), (cos 240° + *i* sin 240°) **33.** (cos 90° + *i* sin 90°), (cos 210° + *i* sin 210°), (cos 330° + *i* sin 330°) **35.** 2(cos 0° + *i* sin 0°), 2(cos 120° + *i* sin 120°), 2(cos 240° + *i* sin 240°) **37.** (cos 45° + *i* sin 45°), (cos 135° + *i* sin 135°), (cos 225° + *i* sin 225°), (cos 315° + *i* sin 315°) **39.** (cos 22 1/2° + *i* sin 22 1/2°), (cos 112 1/2° + *i* sin 112 1/2°), (cos 202 1/2° + *i* sin 202 1/2°), (cos 292 1/2° + *i* sin 292 1/2°) **41.** 2(cos 20° + *i* sin 20°), 2(cos 140° + *i* sin 140°), 2(cos 260° + *i* sin 260°) **43.** 1.3606 + 1.2637*i*, −1.7747 + .5464*i*, .4141 − 1.8102*i*
45. 1.6309 − 2.5259*i*, − 1.6309 + 2.5259*i*

Section 8.5 (page 335)

1–11.

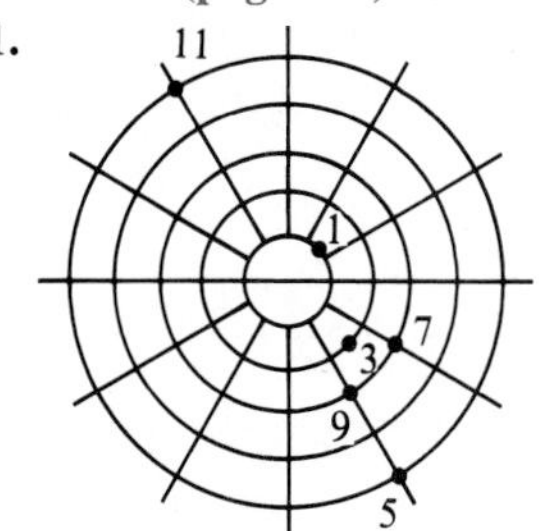

13.

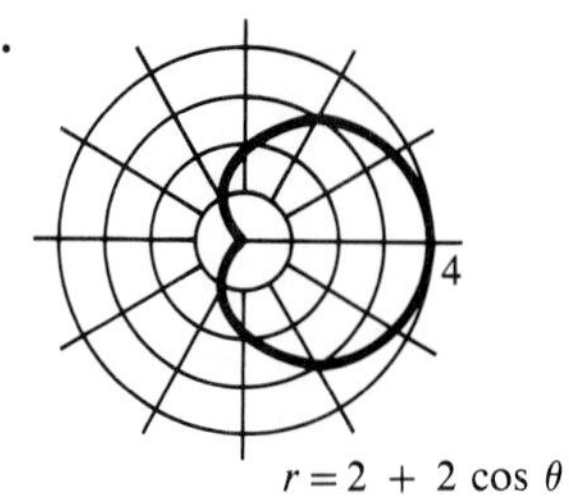

$r = 2 + 2\cos\theta$

15.

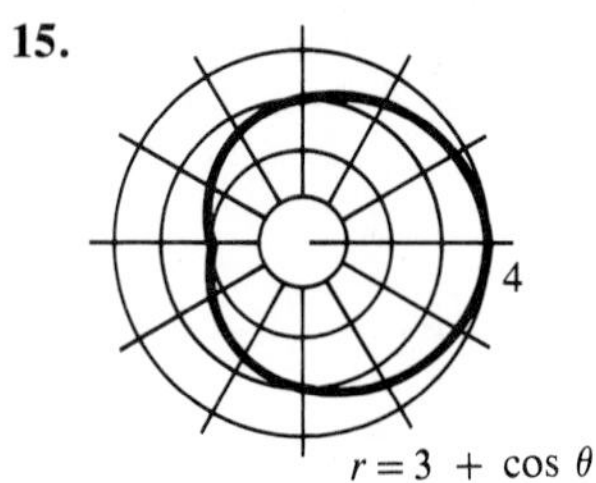

$r = 3 + \cos\theta$

17.

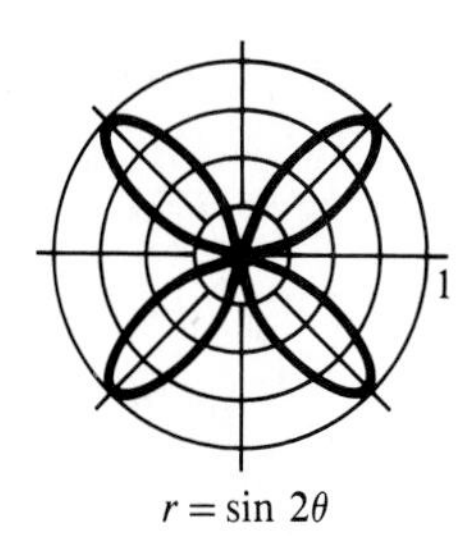

$r = \sin 2\theta$

19.

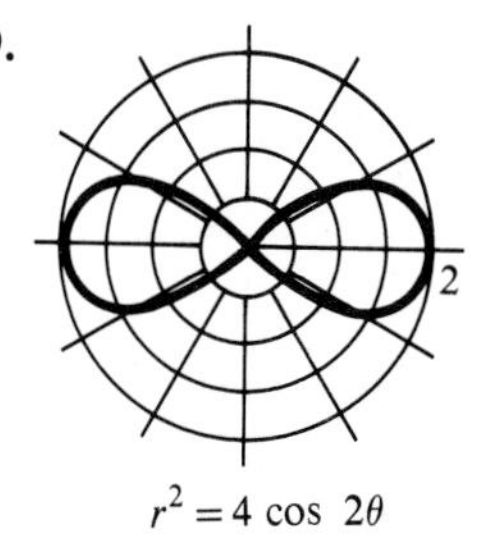

$r^2 = 4\cos 2\theta$

21.

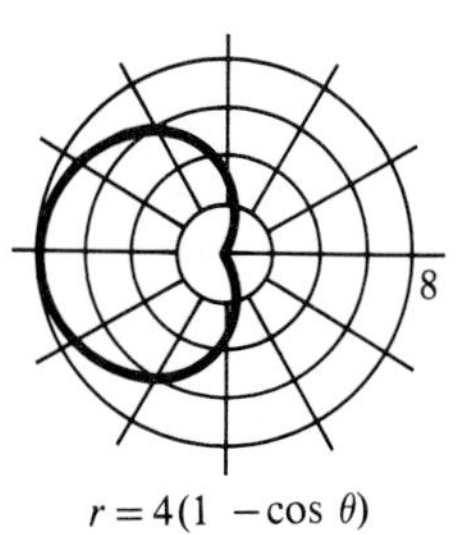

$r = 4(1 - \cos\theta)$

23.

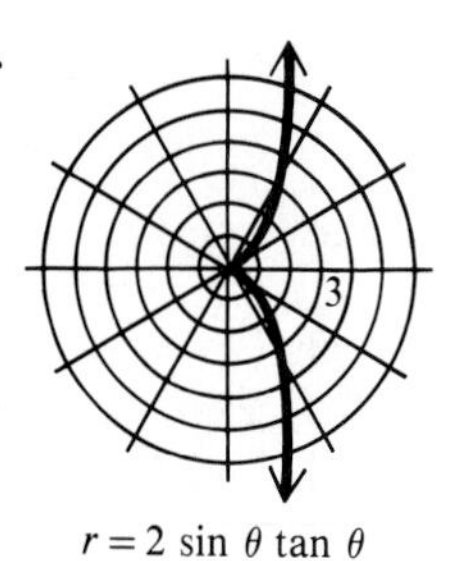

$r = 2\sin\theta\tan\theta$

25.

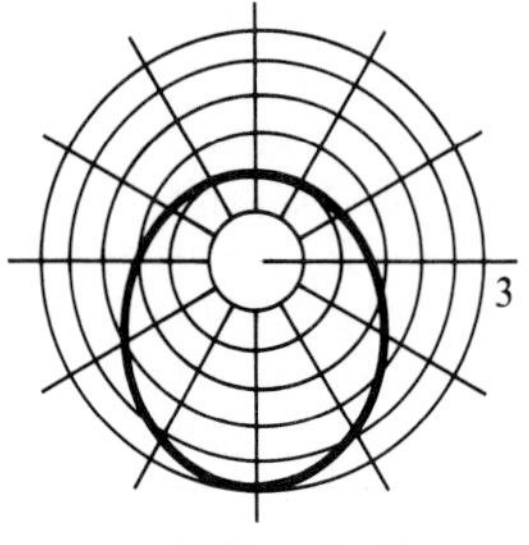

$r = 3/(2 + \sin\theta)$

27.

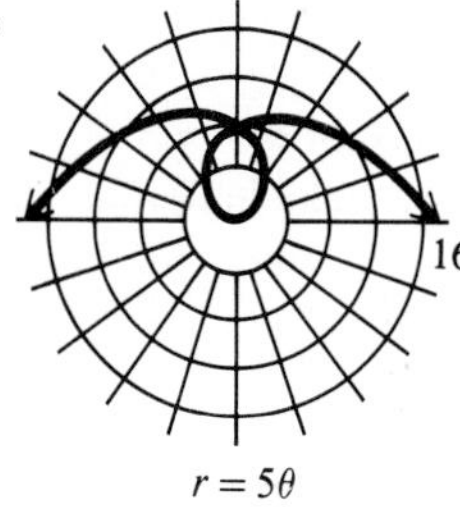

$r = 5\theta$

29.

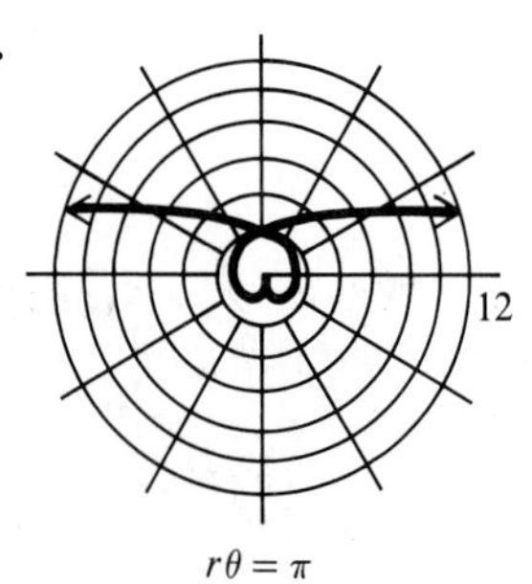

$r\theta = \pi$

31.

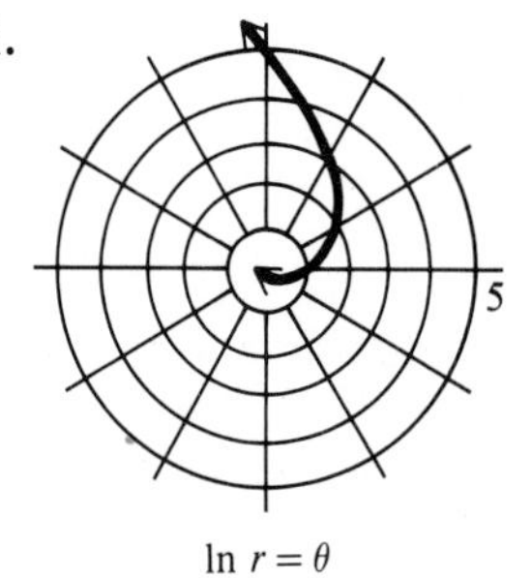

$\ln r = \theta$

33. $x^2 + (y - 1)^2 = 1$

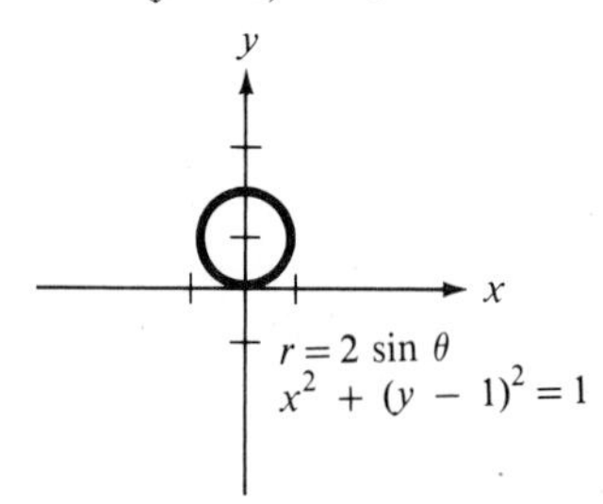

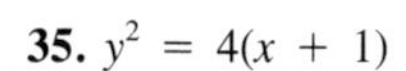

35. $y^2 = 4(x + 1)$

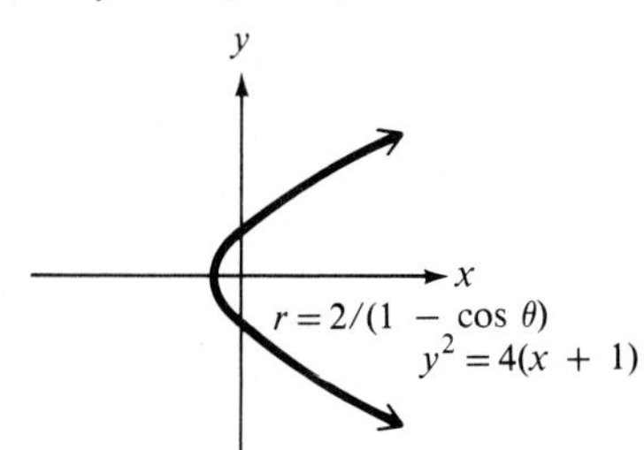

37. $x^2 + y^2 + 2x + 2y = 0$ or $(x + 1)^2 + (y + 1)^2 = 2$

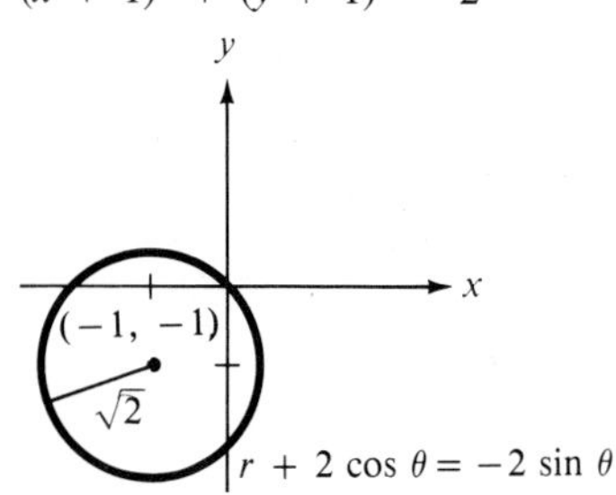

39. $x = 2$

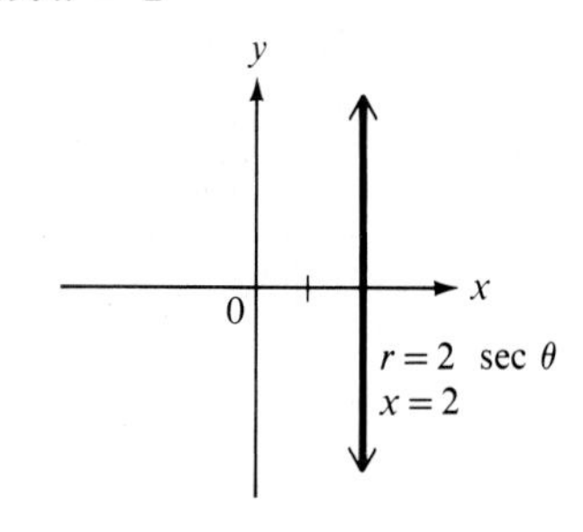

41. $x + y = 2$

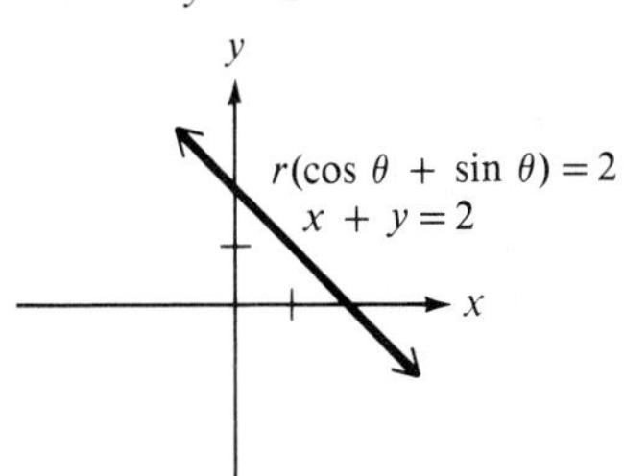

43. $y = -2$

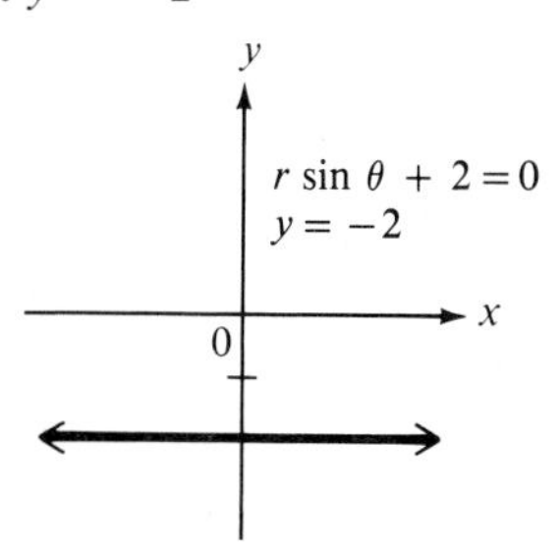

45. $r(\cos\theta + \sin\theta) = 4$ **47.** $r = 4$ **49.** $r = 2\csc\theta$ or $r\sin\theta = 2$

Chapter 8 Review Exercises (page 339)

1. $3i$ **3.** $2i\sqrt{5}/5$ **5.** $-2 - 3i$ **7.** $5 + 4i$ **9.** $29 + 37i$ **11.** $-32 + 24i$ **13.** $-2 - 2i$ **15.** $(8/5) + (6/5)i$ **17.** $(-3/26) + (11/26)i$ **19.** $-30i$ **21.** $(-1/8) + (\sqrt{3}/8)i$ **23.** $8i$ **25.** $(-1/2) - (\sqrt{3}/2)i$ **27.** 2, 3 **29.** 2, −1 **31.** $5 + 4i$

33.

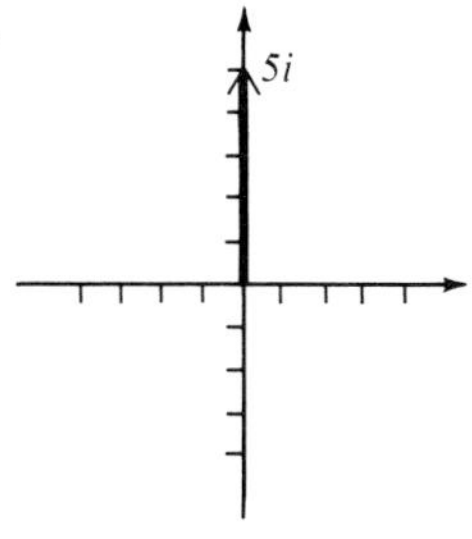

35.

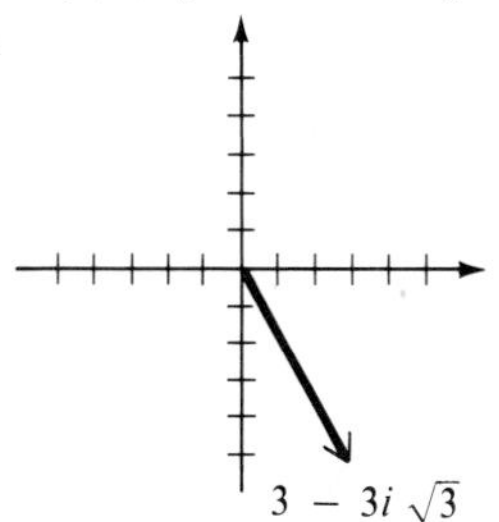

37. $2\sqrt{2}(\cos 135° + i\sin 135°)$

39. $-\sqrt{2} - i\sqrt{2}$ **41.** $\sqrt{2}(\cos 315° + i\sin 315°)$ **43.** $4(\cos 270° + i\sin 270°)$ **45.** $8^{1/10}(\cos 27° + i\sin 27°)$, $8^{1/10}(\cos 99° + i\sin 99°)$, $8^{1/10}(\cos 171° + i\sin 171°)$, $8^{1/10}(\cos 243° + i\sin 243°)$, $8^{1/10}(\cos 315° + i\sin 315°)$ **47.** $\cos 0° + i\sin 0°$; $\cos 36° + i\sin 36°$; $\cos 72° + i\sin 72°$; $\cos 108° + i\sin 108°$; $\cos 144° + i\sin 144°$; $\cos 180° + i\sin 180°$; $\cos 216° + i\sin 216°$; $\cos 252° + i\sin 252°$; $\cos 288° + i\sin 288°$; $\cos 324° + i\sin 324°$ **49.** $5(\cos 60° + i\sin 60°)$, $5(\cos 180° + i\sin 180°)$, $5(\cos 300° + i\sin 300°)$ **51.** $3(\cos 45° + i\sin 45°)$, $3(\cos 135° + i\sin 135°)$, $3(\cos 225° + i\sin 225°)$, $3(\cos 315° + i\sin 315°)$

53.

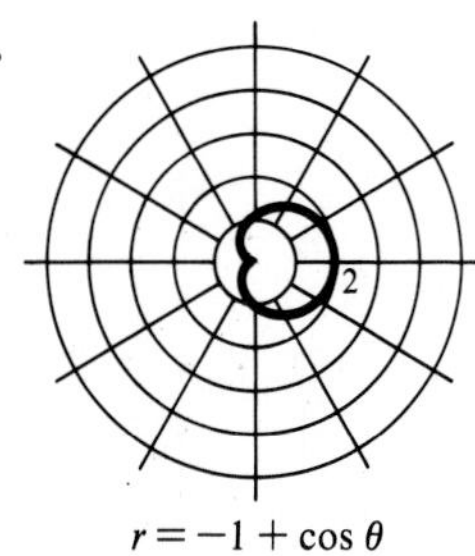

55.

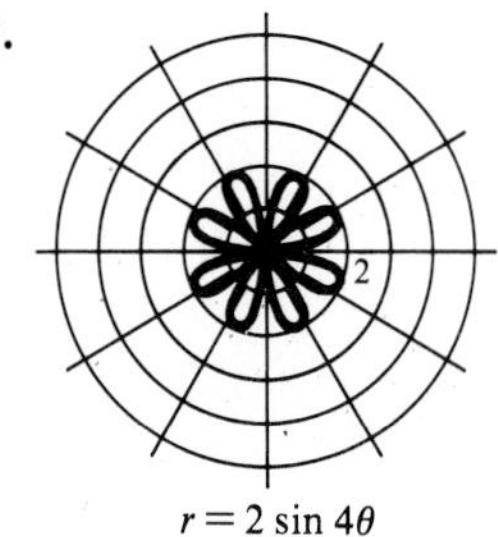

57.

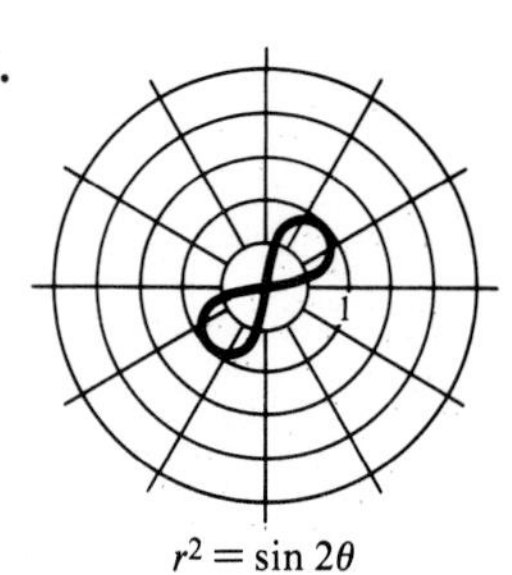

59.

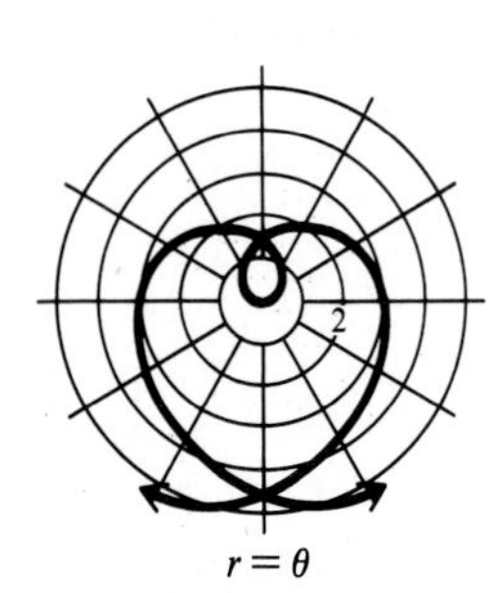

61. $y^2 + 6x - 9 = 0$ **63.** $x^2 + y^2 = x + y$
65. $r\cos\theta = -3$ **67.** $r = \tan\theta \sec\theta$

CHAPTER 9

Section 9.1 (page 347)

1. 3^{11} **3.** 7 **5.** 6^3 **7.** 8^{18} **9.** 2^8 **11.** $1/2^3$ **13.** 3 **15.** $1/9^2$ **17.** 5
19. 8 **21.**

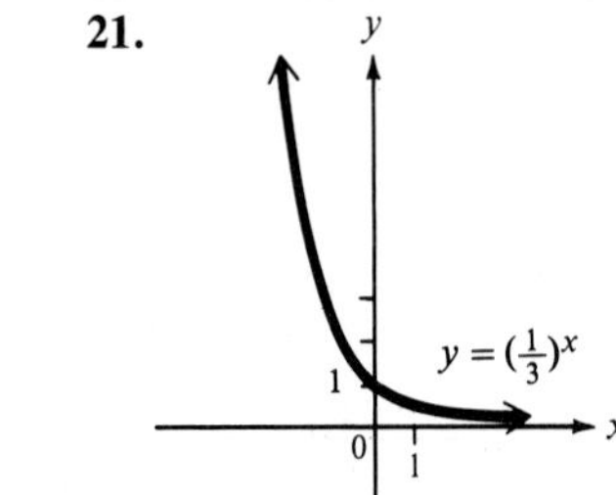

23.

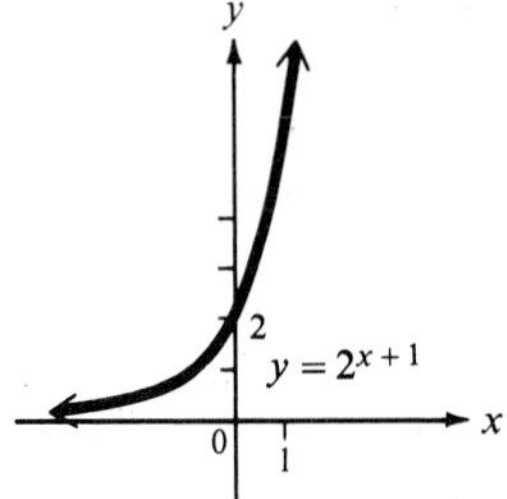

25.

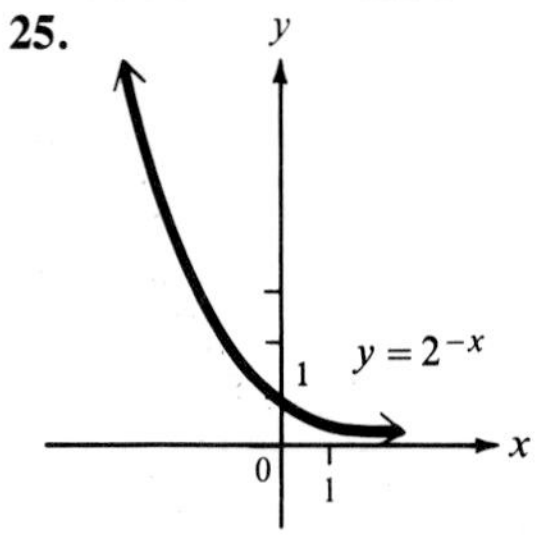

27.

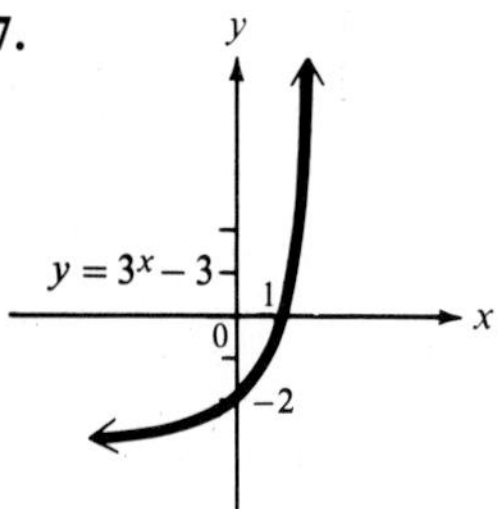

29. $\log_3 243 = 5$ **31.** $\log_{10} 10{,}000 = 4$ **33.** $\log_6 1/36 = -2$ **35.** $\log_{2/3} 9/4 = -2$
37. $4^2 = 16$ **39.** $10^3 = 1000$ **41.** $(3/4)^2 = 9/16$ **43.** $4^{-2} = 1/16$

45.

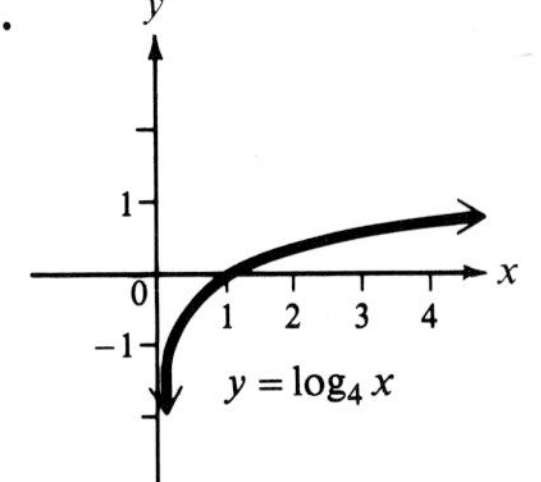

47.

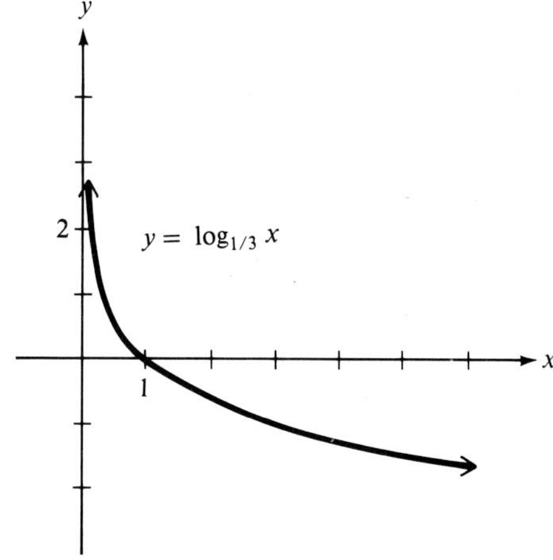

49.

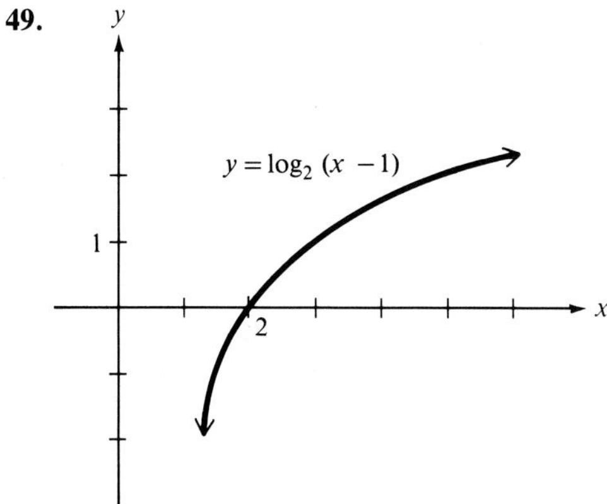

51.

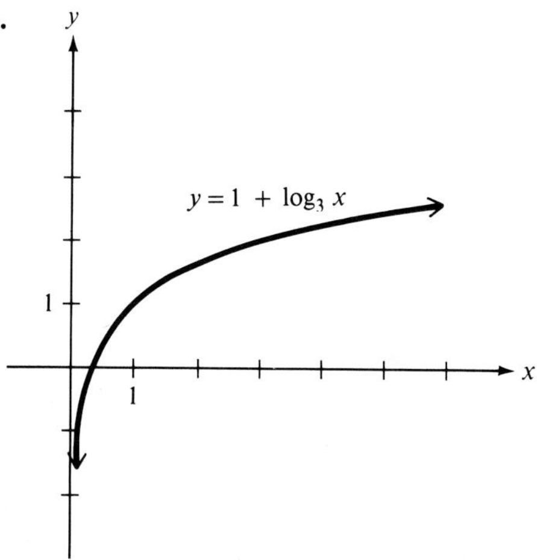

53. $\log_6 45$ **55.** $\log_5 12/7$ **57.** $\log_2 28/5$ **59.** $\log_3 16$ **61.** $\log_8 \sqrt{7}$ **63.** $\log_2 81/25$
65. 1.1461 **67.** 1.4471 **69.** 2.4080 **71.** 3.1242 **73.** .3471

Section 9.2 (page 352)
1. .4456 **3.** 4.9926 **5.** 5.8745 **7.** .7716 − 3 **9.** 722 **11.** 2330 **13.** .114
15. .00571 **17.** .6899 **19.** 4.5797 **21.** −1.5357 or .4643 − 2 **23.** 2.743
25. 2634 **27.** .005081 **29.** .0000187 **31.** 9.58 **33.** 70.8 **35.** .249
37. 3.2189 **39.** 4.4543 **41.** 17.1477 **43.** −2.5257 **45.** −.5447

Section 9.3 (page 356)
1. 3 **3.** 1/5 **5.** 3/2 **7.** 4/3 **9.** 8 **11.** 3 **13.** .477 **15.** −.824
17. 1.29 **19.** 1.26 **21.** 1.78 **23.** −1.43 **25.** 2.99 **27.** 16 **29.** 5
31. $\sqrt[3]{10}$ or 2.15 **33.** 1/100 or .01 **35.** **(a)** 2 **(b)** 2 **(c)** 2 **(d)** 3 **(e)** About 24,100
37. **(a)** About 961,000 **(b)** About 7 years **(c)** About 17.3 years **39.** **(a)** 23 days **(b)** 46 days

Chapter 9 Review Exercises (page 359)

1. 3^3 or 27 **3.** 2^{11} **5.** 8

7.

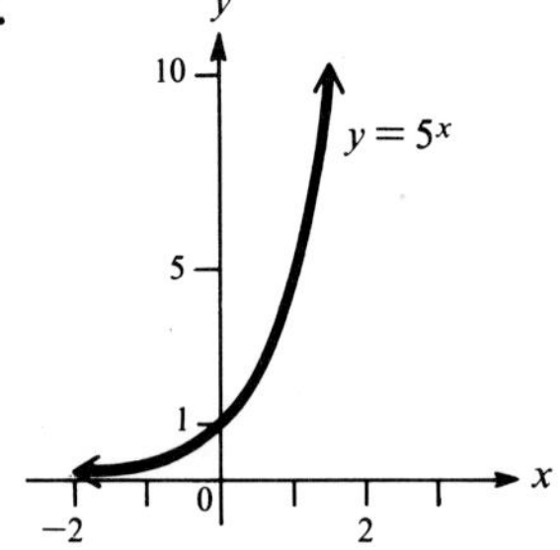

9.

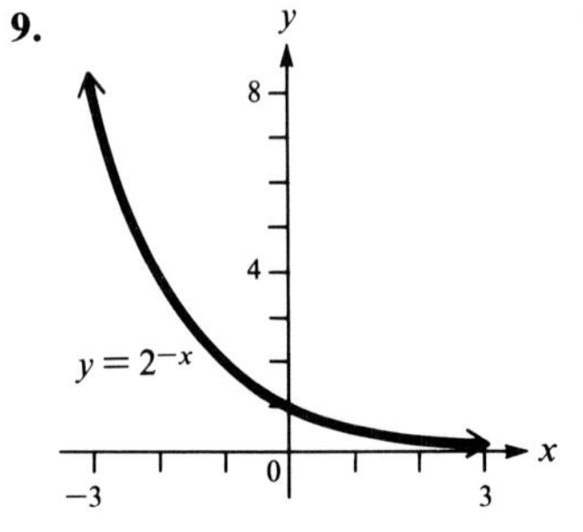

11.

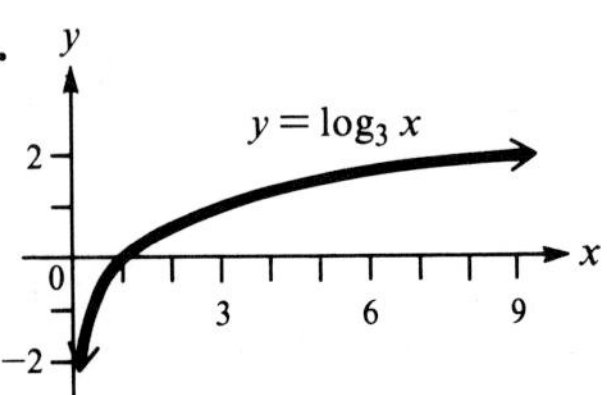

13.

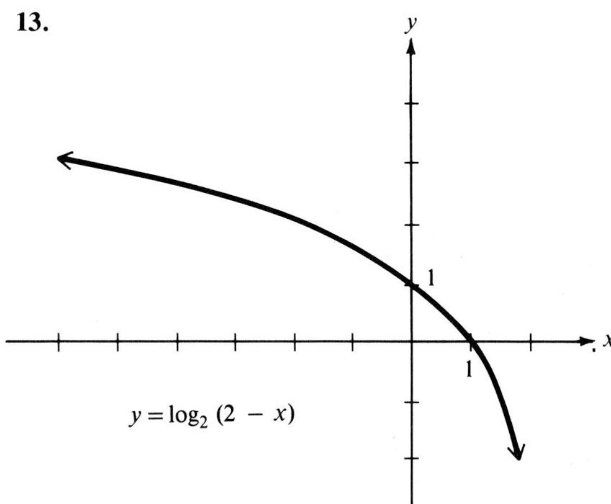

15. $\log_2 512 = 9$ **17.** $\log_{2/3}(3/2) = -1$ **19.** $3^4 = 81$ **21.** $(1/3)^{-1} = 3$ **23.** $\log_5 12$
25. $\log_2 x^5$ **27.** $\log_3 x^2y^4$ **29.** .3945 **31.** 3.0934 **33.** 6.7218 **35.** 2700
37. .0169 **39.** 1.2284 **41.** 3.55 **43.** .0104 **45.** 2.7081 **47.** -2.1203 **49.** 3
51. $-4/5$ **53.** 2.17 **55.** 11.61 **57.** 4 **59.** About 31.5 days

Index

Absolute value, 5
  of a complex number, 316
Acceleration due to gravity, 212
Acute angle, 12
Acute triangle, 20
Addition of complex numbers, 310
Addition of ordinates, 162
Adjacent side, 46
Airspeed, 300
Alternate interior angles, 25
Alternating current, 258, 313, 322
Ambiguous case of the law of sines, 277
Amplitude, 139
Amplitude modulation, 145
Angle, 10
  acute, 12
  complementary, 12
  coterminal, 16
  of depression, 78
  of elevation, 78
  of inclination, 206
  initial side, 10
  negative, 11
  obtuse, 12
  positive, 11
  quadrantal, 15
  reference, 62, 120
  right, 12
  standard position, 15
  straight, 12
  supplementary, 12
  terminal side, 11
  vertex, 11
Angular velocity, 126
Antilogarithm, 349
Arccos, 240
Archimedes, spiral of, 333
Arc length, 109
Arcsin, 240
Area of a sector, 111
Area of a triangle, 274
Argument, 149, 316
Asymptote, 152
Axis
  horizontal, 2
  imaginary, 314
  polar, 328
  real, 314
  vertical, 2

Bearing, 84

Cardioid, 330
Cartesian equation, 330
Characteristic, 349
Circular function, 118
Cis θ, 316
Cofunction, 48
Cofunction identities, 48
Common logarithm, 348
Complementary angles, 12
Complex numbers, 308
  absolute value, 316
  addition, 310
  argument, 316
  conjugate, 311
  equality, 312
  modulus, 316
  *n*th roots, 324
  polar form, 316
  product, 311
  quotient, 311
  standard form, 308
  subtraction, 310
  trigonometric form, 316
  Component, 293
  horizontal, 293
  vertical, 293
Conditional equation, 177, 249
Congruent triangles, 21
Conjugate, 311
Coordinate system, polar, 328
Cosecant function, 29
  graph, 153
  period, 152
Cosine function, 29
  amplitude, 139
  graph, 139
  period, 140

Cosines, law of, 282
Cosine wave, 139
Cotangent function, 29
  graph, 158
  period, 158
Coterminal angle identities, 58
Coterminal angles, 16

Damped oscillatory motion, 168
Decimal degrees, 14
Degree, 11
  converting to radians, 102
  decimal, 14
  of modulation, 144
De Moivre's theorem, 323
Dependent variable, 4
Depression, angle of, 78
Difference identity
  for cosine, 195
  for sine, 201
  for tangent, 201
Digits, significant, 71
Distance formula, 3
Domain, 5
Double-angle identities, 208

e, 351
Elevation, angle of, 78
Equal vectors, 292
Equation
  Cartesian, 330
  conditional, 177
  exponential, 353
  growth or decay, 355
  inverse trigonometric, 258
  logarithmic, 353
  polar, 330
  rectangular, 330
  trigonometric, 249
Equilateral triangle, 20, 52
Equilibrant, 297
Eratosthenes, 117
Exact number, 72
Exponential equation, 353
Exponential function, 342
Exponents, 341
  fractional, 342
  negative, 342
  properties of, 341

Fourier analysis, 165
Four-leaved rose, 332
Fractional exponents, 342
Frequency, 144, 169
Frequency modulation, 145
Function, 4
  circular, 118
  exponential, 342
  horizontal line test, 234
  inverse, 234
  inverse trigonometric, 232
  logarithmic, 344
  one-to-one, 233
  periodic, 136
  trigonometric, 29, 47
  vertical line test, 7
Fundamental identities, 177, 180

Graphs
  combined functions, 162
  complex numbers, 314
  cosecant function, 153
  cosine function, 139
  cotangent function, 158
  inverse functions, 236
  inverse trigonometric functions, 239–41
  polar, 330–33
  secant function, 153
  sine function, 137
  tangent function, 157
Groundspeed, 300
Growth or decay equations, 355

Half-angle identities, 213
Harmonic motion, 168
Heron's area formula, 287
Horizontal component, 293
Horizontal line test, 234
Horizontal translation, 148
Hypotenuse, 2

i, 308
Identities, 34
  cofunction, 48
  coterminal angles, 58
  double-angle, 208
  equation, 177
  fundamental, 177, 180
  half-angle, 213
  negative-angle, 180
  product, 219, 220
  Pythagorean, 39, 179
  quotient, 178
  reciprocal, 34, 178
  reduction, 223
  sum, 221
  sum and difference for cosine, 195
  sum and difference for sine, 201
  sum and difference for tangent, 201
  verifying, 187
Imaginary axis, 314
Imaginary number, 308
Impedance, 313
Inclination, angle of, 206
Independent variable, 4
Initial point, 291
Initial side, 10
Interpolation, 93, 123, 350
Inverse cosine function, 240
Inverse function, 234
  solving for, 235
  tests for, 237
Inverse sine function, 240
Inverse trigonometric equation, 258
Inverse trigonometric functions, 239–41
Isosceles triangle, 20

Latitude, 109
Law of cosines, 282

Law of sines, 270
  ambiguous case, 277
Legs of a right triangle, 2
Lemniscate, 330
Length of arc, 109
Line, 10
Linear interpolation, 93
Linear speed, 168
Linear velocity, 126, 168
Line segment, 10
Logarithm, 341
  common, 348
  definition, 344
  natural, 351
  properties of, 346
Logarithmic equation, 353
Logarithmic function, 344

Mach number, 218
Mantissa, 349
Mauna Loa, 289
Medicine Wheel, 116
Minute, 14
Modulation, 144
Modulus, 316
Moiré pattern, 18

Natural logarithm, 351
Nautical mile, 258
Negative-angle, 11
  identity, 180
Negative exponent, 342
Newton, 295
Notation, scientific, 72
*n*th roots of a complex number, 324

Oblique triangle, 269
Obtuse angle, 12
Obtuse triangle, 20
One-to-one function, 233
Opposite side, 46
Opposite vector, 293
Ordered pair, 1
Ordinate, 162
Origin, 2
Oscilloscope, 148

Pair, ordered, 1
Parallelogram rule, 293
Pendulum, 171
Period, 140
  of sine and cosine, 140
  of tangent, 156
Periodic function, 136
Phase shift, 149
Polar axis, 328
Polar coordinates, 328
Polar coordinate system, 328
Polar equation, 330
Polar form, 316
Pole, 328
Positive angle, 11
Primitive cell, 83
Product identities, 219, 220
Product of complex numbers, 311
Product theorem for complex numbers, 319
Properties of
  exponents, 341
  logarithms, 346
Protractor, 13
Pythagorean identities, 39, 179
Pythagorean theorem, 2

Quadrant, 2
Quadrantal angles, 15
  values for, 32
Quotient identities, 178
Quotient of complex numbers, 311
Quotient Theorem for complex numbers, 320

Radian, 100
  converting to degrees, 102
Range, 5
Rationalizing the denominator, 35
Ray, 10
Real axis, 314
Reciprocal, 34
  identities, 34, 178
Rectangular equation, 330
Reduction identity, 223
Reference angles, 62, 120
  chart of, 64
Reference numbers, 120
Relation, 4
  domain, 5
  range, 5
Resultant, 293
Right angle, 12
Right triangle, 20
Roots of a complex number, 324

Scalar, 291
Scalar product, 294
Scalene triangle, 20
Scientific notation, 72
Secant function, 29
  graph, 153
  period, 152
Second, 14
Sector of a circle, 111
  area of, 112
Segment, 10
Semiperimeter, 287
Side adjacent, 46
Side opposite, 46
Significant digits, 71
Signs of function values, 35
Similar triangles, 21
Simple harmonic motion, 168
Sine function, 29
  amplitude, 139
  graph, 137
  period, 140
Sines, law of, 270
Sine wave, 137
Sinusoid, 137
Slope, 206
Snell's law, 70, 267
Solving a triangle, 76, 270
Special angles, 52

Spiral of Archimedes, 333
Square wave, 165
Standard form, 308
Standard position, 15
Straight angle, 12
Subtraction
  of complex numbers, 310
  of vectors, 293
Sum identities, 221
Sum identity
  for cosine, 195
  for sine, 201
  for tangent, 201
Sum of vectors, 292
Supplementary angles, 12

Tangent function, 29
  graph, 157
  period, 156
Terminal point, 291
Terminal side, 11
Translation
  horizontal, 148
  vertical, 143
Transversal, 25
Triangle
  acute, 20
  area of, 274, 287
  congruent, 21
  equilateral, 20, 52
  isosceles, 20
  oblique, 269
  obtuse, 20
  right, 20
  scalene, 20
  semiperimeter, 287
  similar, 21
  solving, 76, 270
Trigonometric equation, 249
  inverse, 258
  solving, 252
  with multiple angles, 254
Trigonometric form of a
    complex number, 316
Trigonometric functions
  alternate definition of, 47
  definition of, 29
  graphing, 141, 154, 159
  using a calculator, 68
  using a table, 69
Trigonometric identities
  See Identities

Unit cell, 83
Unit circle, 118

Variable
  dependent, 4
  independent, 4
Vector, 291
component, 293
equal, 292
equilibrant, 297
horizontal component, 293
initial point, 291
opposite, 293
parallelogram rule, 293
quantities, 291
resultant, 293
scalar product, 294
subtraction, 293
sum, 292
terminal point, 291
vertical component, 293
zero, 293
Velocity
  angular, 126
  linear, 126, 168
Vertex, 11
Vertical asymptote, 152
Vertical component, 293
Vertical line test, 7
Vertical translation, 143
Von Frisch, Karl, 334

Wave cosine, 139
Wave, sine, 137

x-axis, 2

y-axis, 2

Zero vector, 293

45°–45° Right Triangle

In a 45°–45° right triangle, the hypotenuse has a length that is $\sqrt{2}$ times as long as the length of either leg.

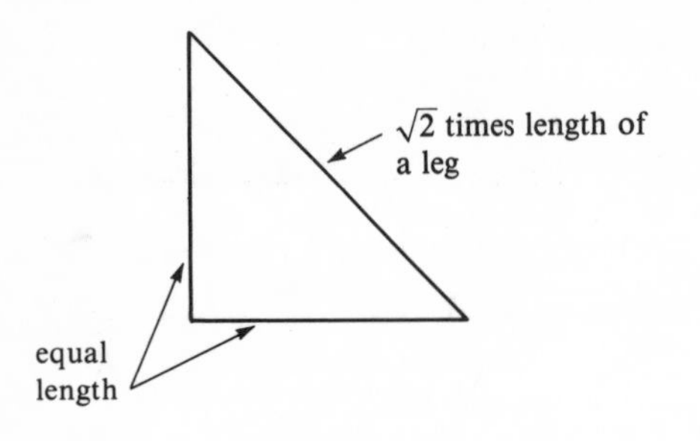

30°–60° Right Triangle

In a 30°–60° right triangle, the hypotenuse is always twice as long as the shorter leg, and the longer leg has a length that is $\sqrt{3}$ times as long as that of the shorter leg. Also, the shorter leg is opposite the 30° angle, and the longer leg is opposite the 60° angle.

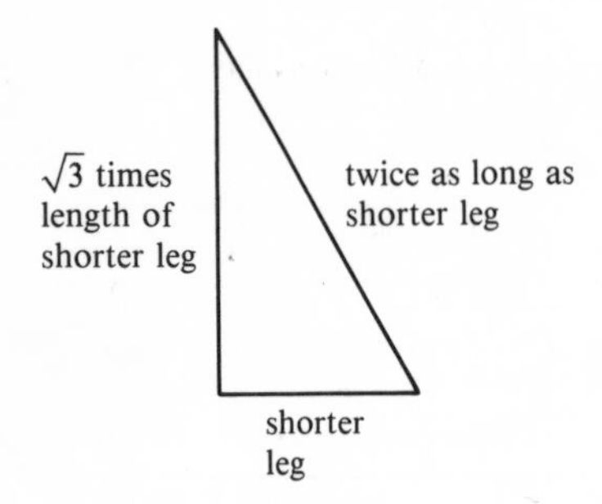

Fundamental Identities

$$\tan\theta = \frac{\sin\theta}{\cos\theta} \qquad \cot\theta = \frac{\cos\theta}{\sin\theta}$$

$$\cot\theta = \frac{1}{\tan\theta} \qquad \csc\theta = \frac{1}{\sin\theta} \qquad \sec\theta = \frac{1}{\cos\theta}$$

$$\sin^2\theta + \cos^2\theta = 1$$

$$\sin^2\theta = 1 - \cos^2\theta \qquad \cos^2\theta = 1 - \sin^2\theta$$

$$\tan^2\theta + 1 = \sec^2\theta \qquad 1 + \cot^2\theta = \csc^2\theta$$

$$\sin(-\theta) = -\sin\theta \qquad \cos(-\theta) = \cos\theta$$

$$\tan(-\theta) = -\tan\theta$$

Sum and Difference Identities

$$\cos(A + B) = \cos A\cos B - \sin A\sin B$$

$$\cos(A - B) = \cos A\cos B + \sin A\sin B$$

$$\sin(A + B) = \sin A\cos B + \cos A\sin B$$

$$\sin(A - B) = \sin A\cos B - \cos A\sin B$$

$$\tan(A + B) = \frac{\tan A + \tan B}{1 - \tan A\tan B}$$

$$\tan(A - B) = \frac{\tan A - \tan B}{1 + \tan A\tan B}$$

Cofunction Identities

$$\sin A = \cos(90° - A)$$

$$\cos A = \sin(90° - A)$$

$$\tan A = \cot(90° - A)$$

Multiple-Angle and Half-Angle Identities

$$\cos 2A = \cos^2 A - \sin^2 A$$

$$\cos 2A = 1 - 2\sin^2 A$$

$$\cos 2A = 2\cos^2 A - 1$$

$$\sin 2A = 2\sin A\cos A$$

$$\tan 2A = \frac{2\tan A}{1 - \tan^2 A}$$

$$\cos\frac{A}{2} = \pm\sqrt{\frac{1 + \cos A}{2}}$$

$$\sin\frac{A}{2} = \pm\sqrt{\frac{1 - \cos A}{2}}$$

$$\tan\frac{A}{2} = \pm\sqrt{\frac{1 - \cos A}{1 + \cos A}}$$

$$\tan\frac{A}{2} = \frac{\sin A}{1 + \cos A}$$

$$\tan\frac{A}{2} = \frac{1 - \cos A}{\sin A}$$